Green Energy and Technology

For further volumes:
http://www.springer.com/series/8059

Zhiqun Lin · Jun Wang
Editors

Low-cost Nanomaterials

Toward Greener and More Efficient
Energy Applications

 Springer

Editors
Zhiqun Lin
School of Materials Science
  and Engineering
Georgia Institute of Technology
Atlanta
USA

Jun Wang
Chemical and Materials Science Center
National Renewable Energy Laboratory
Golden, CO
USA

ISSN 1865-3529   ISSN 1865-3537 (electronic)
ISBN 978-1-4471-6999-4   ISBN 978-1-4471-6473-9 (eBook)
DOI 10.1007/978-1-4471-6473-9
Springer London Heidelberg New York Dordrecht

Printed on acid-free paper

Springer is part of Springer Science+Business Media (www.springer.com)

Contents

Introduction

Zhiqun Lin and Jun Wang

Increasing worldwide demand on energy and the limited amounts of nonrenewable fossil fuels have stimulated intense research and development efforts on renewable energy, in the areas of solar cells, energy storage, fuel cells, and water splitting, to name a few. However, widespread applications of renewable energy have not been very successful mainly due to the high cost of materials and underdeveloped processing and fabrication techniques. When compared to their bulk counterparts, nanomaterials possess well-defined nanostructures and exceptional physical and chemical properties. They exhibit great potential for developing low-cost, high performance renewable energy sources. For example, rational engineering on the nanostructure of photoanodes in dye-sensitized solar cells (DSSCs) has led to largely enhanced device performance; quantum dots that show multiple exciton generation (MEG) capability may offer great promise in the fabrication of solar cells with power conversion efficiency exceeding the Shockley–Queisser limit; development of advanced synthetic methods for pyrite nanoparticles has led to solution-based fabrication of low-cost, environmental friendly thin film solar cells; investigation on the Indium tin oxide (ITO)-free transparent conductive films will help realize the commercialization of flexible electronic devices; design and fabrication of nanostructured electrodes have provided the opportunity for developing high performance, low-cost fuel cells, efficient water splitting, high density hydrogen storage, high performance batteries, and supercapacitors; and the development of nanoscale phase change materials (PCMs) has contributed to the preparation of more efficient heat transfer fluids. Despite tremendous efforts on incorporating nanomaterials into a variety of renewable energy sources, the high

Z. Lin (✉)
School of Materials Science and Engineering, Georgia Institute of Technology,
Atlanta, GA, USA
e-mail: zhiqun.lin@mse.gatech.edu

J. Wang
Woodbury, MN, USA
e-mail: jwangnano@gmail.com

Z. Lin and J. Wang (eds.), *Low-cost Nanomaterials*, Green Energy and Technology,
DOI: 10.1007/978-1-4471-6473-9_1, © Springer-Verlag London 2014

1

fabrication cost, small synthesis scales, and underdeveloped processing methods associated with nanomaterials prevent them from progressing to commercial scale applications. How to go about affecting this transition from lab bench to industrial scales will be the primary research focus of future work in this field. Clearly, it is necessary to have a comprehensive collection of review chapters covering the current state-of-art research progresses, ongoing challenges, and possible future directions in which to apply nanomaterials for renewable energy applications.

In this book, researchers actively working on renewable energy contribute their views on how nanomaterials would be beneficial for the development of high performance yet low-cost renewable energy sources. With comprehensive coverage of fundamental knowledge, in-depth background information, status of current research and development, and an outlook for future directions, this book aims to provide general information for undergraduate and graduate students interested in nanomaterials and their applications in renewable energy, and serves as a handbook and reference for advanced readers such as materials scientists and researchers working on renewable energy or related fields.

An overview of each chapter included in this book is given in the following paragraphs under different categories, which we expect to provide readers with a brief idea about the content of this book and help them navigate to areas of interest.

1 Dye-Sensitized Solar Cells

Solar cells, also known as photovoltaic devices (PVs), directly convert incident solar photons to electricity, and are one of the most studied solar energy systems because they employ clean and abundant resources and show great potential to satisfy future global energy demands. The current PV market is dominated by single crystal silicon-based solar cells that deliver power conversion efficiencies of 15 % or higher. However, these first-generation solar cells still suffer from several inherent deficiencies, such as high fabrication cost, heavy weight, and inevitable use of toxic chemicals. DSSCs, a new generation of high performance and low-cost solar cells, have attracted tremendous attention in the past decades owing to several advantages such as low fabrication cost, low toxicity, and high power conversion efficiency. A typical DSSC consists of several key components, including an electrically conductive support (e.g., transparent conductive film), a nanostructured semiconductor film (e.g., TiO_2), a sensitizer (e.g., ruthenium dye N719), an electrolyte (e.g., iodide/triiodide couple), and a counter electrode (e.g., Pt-coated electrode). In order to develop DSSCs with sufficiently high performance for commercial scale fabrication, optimization on the above-mentioned components should be carried out to achieve higher light absorption, better charge collection and transport, minimal recombination, and long-term device stability. Among all semiconductors studied for DSSCs, TiO_2 has been regarded as the most promising material for photoanodes in DSSCs due to its appropriate electronic band structure, photostability, chemical inertness, and well-established synthesis methods. Modification of TiO_2 has been extensively investigated in recent years. The aim is to improve charge collection and charge transport in DSSCs, including the development of TiO_2 nanostructures with high surface area and unique morphologies and the preparation of doped TiO_2 nanostructures with improved optical and electronic properties.

In the chapter by Lin et al., various synthetic methods for the preparation of nanostructured TiO_2 are presented, including sol–gel, hydrothermal/solvothermal, electrochemical anodization, and electrospinning/electrospray methods. These methods produce a wide range of TiO_2 nanostructures that meet the requirements of many different applications. In the second section of this chapter, an overview of how the modification of TiO_2 influences its chemical/physical properties in different optoelectronic devices is presented.

In the chapter by Ma et al., nitrogen-doped TiO_2 nanostructures and their effect on the performance of DSSCs are described. Various doping methods, nanostructures, the resulting physiochemical properties of N-doped TiO_2, and the effect of N-doped TiO_2 photoanode on the performance of DSSCs devices are discussed in detail. In the last section of this chapter, an analysis of the electron kinetic behaviors (i.e., charge transport, electron lifetime, and charge recombination) in DSSCs based on N-doped TiO_2 photoanodes is presented.

In another chapter contributed by Ma et al., recent progress on the development of Pt-free counter electrodes for DSSCs is comprehensively reviewed. Platinum-based counter electrodes are commonly used in current DSSCs. However, its high cost has motivated the search for low-cost, high performance alternatives, including carbon materials, conductive polymers, transition metal compounds, and composite catalysts. The advantages and disadvantages of each Pt-free counter electrode alternative are subsequently reviewed.

An analog to DSSCs, quantum dot-sensitized solar cells (QDSSCs) in which quantum dots play the role of "dye" in DSSCs have been extensively explored in recent years primarily due to the exceptional optoelectronic properties of quantum dots (e.g., size-dependent optical properties and MEG characteristics) and the well-established synthetic methods for the preparation of high quality quantum dots with tuneable morphologies and compositions. In a chapter by Mora-Seró et al., recent advances in the development of QDSSCs are presented with a focus on highlighting the differences between quantum dot- and dye-sensitized solar cells (QDSSCs vs. DSSCs) in several aspects, including the preparation of sensitizers, nanostructured electrodes, hole transporting materials, counter electrodes, and recombination and surface states. Through such comparisons, further improvements on QDSSCs can be envisioned. One example is the recent breakthrough in photovoltaics with organometallic halide perovskites which came about through the intensive study on QDSSCs. This is presented in detail in the last section of this chapter.

2 Pyrite Solar Cells

Pyrite (iron disulfide (FeS_2), also known as fool's gold) has long been proposed as a green solar cell material owing to its optimum band gap (i.e., 0.95 eV), high optical absorption coefficient (i.e., greater than 10^5 cm^{-1}, two orders of magnitude greater than the absorption coefficient of silicon), and abundance on earth, making it possible to fabricate highly efficient thin film solar cells with largely decreased consumption of raw materials. However, unanswered questions about the effects of defects and techniques to grow pure crystalline material still remain. In the chapter by Ren et al., the crystal structure and fundamental properties of pyrite are first introduced to offer an overview as to why it is a promising material for photovoltaics, followed by the detailed review on various synthetic methods for the preparation of nanostructured pyrite with a focus on the most recently developed solution phase synthesis of pyrite nanoparticles. In the last section of this chapter, the applications of nanostructured pyrite for photovoltaics and photodetectors are presented.

3 Polymer-Based Solar Cells

Many organic semiconductors exhibit the electronic properties of their inorganic counterparts while carrying the advantages of plastic processing and low fabrication cost through well-established polymeric materials processing techniques, thereby yielding a variety of organic-based electronic devices such as organic photovoltaic devices (OPVs), organic light emitting diodes (OLEDs), organic thin film transistors, etc. Organic photovoltaic devices have been regarded as promising technologies for the conversion of solar energy to electricity due to their light weight, flexibility, and cost-effective processibility. Additional advantages of OPVs also include short energy payback time compared to existing photovoltaic devices, non-toxic, and abundance on earth. In a chapter contributed by Fang-Chung Chen et al., the fundamentals of OPVs are presented in detail, followed by a comprehensive review on the recent progress on conjugated polymer-based OPVs. Effects of thermal annealing, solvent annealing, and interface engineering on the performance of OPVs are also discussed. In the last section of this chapter, common optical methods used to improve light absorption in OPVs are summarized, followed by an overview of the development of low-band gap conjugated polymers for more efficient light harvesting.

Indium tin oxide is the most commonly used transparent conductive materials for a wide range of electronic devices such as OPVs. However, ITO usage is diminished due to high material cost and poor flexibility. Clearly, the development of a low-cost replacement for ITO is crucial for the commercial feasibility of OPVs and other organic-based electronic devices. In this regard, a variety of nanomaterials have been investigated that have shown great potential as an ITO replacement by providing comparable or better electronic and optical properties. In the chapter by Krebs et al., the development of ITO-free polymer solar cells is reviewed, which should help to realize commercially feasible OPV devices. In the first section of this chapter, various nanomaterials showing the possibility to substitute ITO are discussed, including metal nanogrids, metal nanowires, carbon nanotubes, and graphene; followed by a discussion on the very recent progress in the scale-up experiments on ITO-free OPVs modules.

Solution-based processing methods have been recognized as the most appropriate techniques for the fabrication of polymer-based electronic devices such as OPVs and OLEDs. However, technical challenges, including large-area processing technologies, coating quality, and long-term stability toward scalable and low-cost polymer electronics, still remain. In a chapter by Guo et al., recent advances in low-cost fabrication of OPVs and OLEDs are reviewed. Various scalable processing methods are presented by focusing on the coating quality and the resulting device performance in polymer-based electronics. This can ultimately lead to conclusions on how appropriate coating techniques can be selected based on the thickness requirement of each functional layer in polymer electronics. In the last section of this chapter, ITO-free electrodes based on polymer materials are introduced by evaluating their mechanical and optical properties, and a hybrid ITO-free transparent conductive film based on metal mesh and conjugated polymers is also introduced for the fabrication of large-area devices.

4 Hydrogen Energy and Fuel Cells

Hydrogen represents a clean and high gravimetric energy density fuel that could potentially replace fossil fuels in many applications. However, the widespread adoption of hydrogen fuels is stymied by a lack of efficient hydrogen generation and high density hydrogen storage methods. Current technology for hydrogen generation is based on the stream methane reforming and water–gas shift reaction which still relies on fossil fuels. Obviously, it is important to develop efficient, low-cost, and scalable techniques to generate hydrogen in a sustainable manner. In this context, photoelectrochemical water splitting has been considered as one of the most promising approaches as presented thoroughly in the chapter by Li et al. The most recent achievement in this area is comprehensively reviewed in this chapter, and the key factor for efficient photoelectrochemical water splitting has been identified to be the development of low-cost and efficient nanostructured photoelectrodes. In another chapter by Prieto et al. on the development of hydrogen storage materials, nanostructured magnesium and doped magnesium are described. This includes the size and shape controllable synthesis of these nanostructures, the kinetics of efficient hydrogen storage based on experimental observation and modeling, and the theoretical models that could guide experimental efforts.

Fuel cells that convert the chemical energy stored in fuels into electricity through electrochemical reactions with oxygen or other oxidizing agents have been receiving considerable attention in the past few decades. However, the widespread commercialization of fuel cells is still challenging due in part to the low catalytic performance and high cost of Pt-based electrocatalysts. In addition, efficient hydrogen storage is also critical for the commercialization of hydrogen-powered fuel cells. In the chapter by Wei Chen et al., the development of one-dimensional (1D) palladium (Pd)-based nanomaterials as efficient electrocatalysts for fuel cells is presented. Among all Pt-free catalysts, Pd has been found to be a promising substitute owing to its excellent catalytic properties and lower material cost compared to Pt. Moreover, Pd-based materials exhibit high hydrogen storage capability which is desirable for hydrogen-powered fuel cells. This chapter reviews the most recent progress in the synthesis of 1D Pd-based nanomaterials for fuel cell applications. Areas addressed include controllable synthesis of Pd-based nanostructures through various synthetic routes, their high catalytic performance for electro-oxidation of small organic molecules and oxygen reduction reaction (ORR), and the high capacities for hydrogen storage exhibited in 1D Pd-based nanomaterials.

In another chapter on fuel cells contributed by Kumar and Pillai, the development of low-cost nanomaterials for high performance polymer electrolyte fuel cells is reviewed. Proton exchange membrane fuel cells (PEMFCs) utilize a polymer electrolyte membrane to transport protons from the anode to the cathode and restrict electrons from directly going to cathode from anode. They have garnered great interest due to their easy start-up and flexible design. A typical PEMFC consists of several critical components: Pt electrocatalysts, catalyst support (e.g., carbon), gas diffusion layer or backing layer, bipolar plate, and polymer electrolyte membrane. The successful operation of PEMFCs relies on the formation of effective triple phase boundary (reactant gases, electrocatalysts, and polymer electrolyte membrane) to facilitate efficient electrochemical reactions at both anode and cathode. However, the commercialization of PEMFCs is facing obstacles due to the high materials cost associated with Pt electrocatalysts and the poor performance of existing polymer electrolyte membranes. This chapter provides the most recent progress on the development of nanomaterials for PEMFCs in both fundamental and technological aspects with special emphasis on carbon-based nanostructures such as carbon nanotubes, graphene, nanostructured Pt electrocatalysts, and bio-inspired catalysts development, followed by a sound conclusion and perspective on the future activities in developing low-cost, high performance PEMFCs.

5 Batteries

Electrochemical energy storage (EES) technologies, including flow redox batteries, super capacitors, and rechargeable batteries (Pb–acid, Ni–Cd, Na–S, and Li–ion batteries, etc.), have demonstrated significant advantages including high efficiency, low-cost, and flexibility. Li–ion batteries in particular are currently considered as one of the most promising technologies due to their long lifetime and high energy density. However, for widespread EES applications, there is an increasing concern about the costs and the limitations of natural lithium reserves. As a result, efforts have been made to explore low-cost and reliable EES technologies, among which Na–ion batteries are seen as a promising alternative due to the abundance of natural sodium resources and its similar electrochemical properties when compared with lithium. However, the realization of the Na–ion intercalation/deintercalation mechanism remains challenging as Na ions are 40 % larger in radius than Li ions. Thus, suitable host materials with high storage capacity, rapid ion uptaking rate, and long cycling life need to be developed for Na–ion batteries. In the chapter by Liu et al., the very recent progress on the development of promising electrode materials for Na–ion batteries is reviewed and discussed. The aim is to provide an overview of existing problems and future research directions in this area.

There is an ever-increasing demand for flexible portable electronic devices such as roll up displays, wearable devices, and implanted medical devices. This requires the development of flexible batteries or supercapacitors as their power sources. Typical batteries and supercapacitors are composed of several major components including electrodes (i.e., positive electrode and negative electrode), separator, and electrolyte. The two electrodes are spaced apart by a separator and all these components are soaked in solution or gel electrolyte. In order to fabricate flexible batteries and supercapacitors, electrodes with desired flexibility should be developed. In a chapter by Xue et al., the chemical routes to graphene-based flexible electrodes are discussed for applications in EES. Graphene is of interest due to its superior electronic properties and low-cost fabrication by chemical reduction. In this chapter, utilization of graphene for flexible electrodes is presented in detail by analyzing the structure–property relationships. This includes the use of graphene as dominant constituent of electrodes, the use of graphene as a conductive matrix in flexible electrodes, and the use of graphene as a functional additive to improve the performance of cellulose and carbon nanofiber papers. Evidenced by experimental examples, the development of graphene-based flexible electrodes offers new opportunities to further reduce the fabrication cost. A perspective on the future development of graphene-based flexible electrodes is presented at the end of this chapter.

6 Thermal Energy

Thermal energy storage and transfer have been two of the hottest research topics in renewable energy owing to the large abundance of thermal energy from the sun or the earth (geothermal). PCMs have garnered considerable attention for use in thermal energy storage and transfer due to their high heat storage capability during phase transitions. They offer the potential to reduce energy consumption, and in turn lower the related environmental impact. In the chapter by Yang et al., the development of nanoscale PCMs for applications in high performance heat transfer fluids is reviewed with special attention on the material synthesis and property characterizations of phase changeable fluids.

Design, Fabrication, and Modification of Cost-Effective Nanostructured TiO$_2$ for Solar Energy Applications

Meidan Ye, Miaoqiang Lv, Chang Chen, James Iocozzia, Changjian Lin and Zhiqun Lin

Abstract One of the greatest challenges for human society and civilization is the development of powerful technologies to harness renewable solar energy to satisfy the ever-growing energy demands. Semiconductor nanomaterials have important applications in the field of solar energy conversion. Among these, TiO$_2$ represents one of the most promising functional semiconductors and is extensively utilized in photoelectrochemical applications, including photocatalysis (e.g., H$_2$ generation from water splitting) and photovoltaics (e.g., dye-sensitized solar cells, DSSCs). As such, many efforts have focused on developing and exploiting cost-effective nanostructured TiO$_2$ materials for efficient solar energy applications.

1 Introduction

With worldwide economic development and population growth, global energy demands have increased dramatically. Today, such demands largely depend on nonrenewable oil and fossil fuels. The continuous rise in the price of oil and gas has motivated people to pay closer attention to issues concerning our energy supply and demand [1]. In the twentieth century alone, the world population quadrupled and energy demand went up 16 times. The exponential energy demand is exhausting fossil fuel supplies at an alarming rate. The mean global energy

M. Ye · M. Lv · C. Chen · C. Lin (✉)
State Key Laboratory of Physical Chemistry of Solid Surfaces,
Department of Chemistry, College of Chemistry and Chemical Engineering,
Xiamen University, Xiamen 361005, China
e-mail: cjlin@xmu.edu.cn

J. Iocozzia · Z. Lin (✉)
School of Materials Science and Engineering, Georgia Institute of Technology,
Atlanta, GA 30332, USA
e-mail: zhiqun.lin@mse.gatech.edu

Z. Lin and J. Wang (eds.), *Low-cost Nanomaterials*, Green Energy and Technology,
DOI: 10.1007/978-1-4471-6473-9_2, © Springer-Verlag London 2014

consumption rate was 4.1×10^{20} J, or 13 terawatts (TW) in 2000. By year 2050, the number will have doubled to 28 TW and tripled to 46 TW by the end of the century based on current estimates of population growth and energy consumption [2]. Currently, this huge energy demand is supplied by oil (35 %), coal (23 %) and natural gas (21 %), which in total yields a ratio of around 79 % from fossil fuels. While biomass provides only 8 % of the energy supply, nuclear energy only 6.5 %, and hydropower a mere 2 % share [1]. Excessive extraction is leading to gradually decreasing reserves of conventional, nonrenewable, energy resources, such as oil, coal, and natural gas. Moreover, the increasing energy production concurrently impacts the environment through the production of greenhouse gases (i.e., carbon dioxide, methane, nitrous oxide, and other gases) from fossil fuels, which contribute to climate change and even global warming concerns [3]. Therefore, in order to deal with the increasing energy demand and provide a long-term solution to the energy crisis in the future it is essential to develop environmentally and economically clean alternative energy resources in order to achieve a globally sustainable society.

Renewable energy resources, including hydroelectricity from tides and ocean currents (2.5 TW), geothermal energy (12 TW), wind power (24 TW), and solar energy striking the earth (170,000 TW) are considered highly promising options [3]. Among these, solar energy is clean, endlessly abundant, and has the largest potential to satisfy the future global demand for renewable energy sources (under ideal conditions, radiation power on a horizontal surface is $1000~\mathrm{Wm^{-2}}$) [2]. However, to date, the energy converted from sunlight remains far less than that of the total energy demand. For instance, the total average annual installed energy capacity in 2009 was about 7 gigawatts (GW), which only contributed to 0.2 % of global electricity usage [4]. Thus, the efficient, direct conversion of solar energy into electricity and fuels should, and must, be one of the most important scientific and technological pursuits of this century [5].

1.1 Applications for Solar Cells

First of all, to become a major contributor to future renewable energy, solar energy must be cost-effective and priced competitively relative to conventional energy resources, nuclear power, and other renewable energy resources [6]. Currently, the direct conversion of incident solar photons to electricity is achieved through photovoltaic (PV) devices (or solar cells) [7, 8]. Single crystal silicon-based PV devices are the first-generation solar cells, which are commercially available for installation, deliver power with a 15 % efficiency and make up about 90 % of the current PV market. However, first-generation solar cells still suffer from several inherent deficiencies as a result of the complicated and energy intensive fabrication process, inevitable use of toxic chemicals, heavy cell weight, and the high cost of manufacturing and installation [9]. A retail price of about $2 per peak watt (Wp) with a corresponding production price of about $0.5/Wp could make PV cost-competitive with electricity produced from coal. But the production cost of these first-genera-tion solar cells is presently around \$3.5/Wp, which is highly dependent on the price of the silicon material [10].

Second-generation solar cells, including amorphous (or nano-, micro-, poly-) silicon, $CuInGaSe_2$ (CIGS), and CdTe, are based on thin film technologies. These thin film solar cells possess some advantages, such as relatively simple manu-facturing processes which reduce the production cost to about \$1/Wp, multiple choices of applied materials, and the possibility of flexible substrates. However, second-generation solar cells face several shortcomings as well such as the toxicity (e.g., Cd) and low abundance (e.g., In and Se) of the component materials. Moreover, it is necessary to further increase their efficiency for practical utilization [11]. Both the first- and second-generation solar cells are based on single junction devices which must obey the Shockley-Queisser limit with a maximum thermo-dynamic efficiency of 31–33 % when the optimum band gaps fall between about 1.1 and 1.4 eV [12].

Significantly, the third-generation solar cells, including tandem cells, hot carrier cells, dye-sensitized solar cells, and organic solar cells can overcome this ther-modynamic efficiency limit. The most important thing is that these third-generation solar cells are promising to convert solar energy into electricity at a highly com-petitive price, that is, less than \$0.5/Wp [2, 10].

Among these third-generation solar cells, DSSCs offer many attractive features that facilitate market entry. They afford low production cost (i.e., inexpensive to manufacture, possibility of roll-to-roll processing and low embodied energy), low-toxicity, earth-abundant materials (except Pt and Ru), good performance in diverse light conditions (i.e., high angle of incidence, low intensity and partial shadowing), lightweight, flexible, and design feasibility (i.e., transparent, bifacial and selected colors) [2, 13]. Since the first DSSC was reported with efficiency of 7.1 % in 1991 [14], much attention has been devoted to this promising electrochemical device. After two decades of concentrated efforts, DSSCs have developed into a powerful photovoltaic technology with a recorded efficiency as high as 12.3 % [15].

Typically, a DSSC is made of five components: a conductive mechanical support (e.g., transparent conductive glass or Ti foil), a semiconductor film (e.g., TiO_2), a sensitizer (e.g., ruthenium dye N719), an electrolyte (e.g., iodide/triiodide couple), and a counter electrode (e.g., Pt-coated electrode). The operating prin-ciples of DSSCs are further described in Fig. 1a [16]. In DSSCs, electricity is created at the semiconductor film on which a monolayer of visible light absorbing dye is chemisorbed. Photo-excitation of the absorbed dye molecules generates excited electrons which are further injected into the conduction band of the semiconductor and quickly migrated to the external circuit through the conductive substrate. The original state of the dye is subsequently restored by electron donation from the electrolyte, usually an organic solvent containing a redox sys-tem, such as the iodide/triiodide (I^-/I_3^-) couple. The regeneration of the sensitizer by I^- prevents the recapture of the conduction band electron by the oxidized dye while I^- is regenerated in turn by the reduction of I_3^- at the counter electrode. The counter electrode returns charge from the external circuit back to the cycling

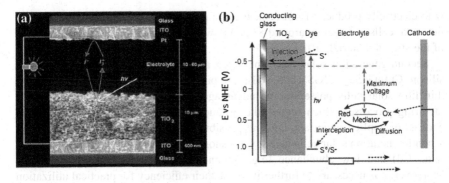

Fig. 1 **a** The physical image and **b** the simple operating principle of DSSCs based on TiO$_2$ nanomaterials. (Reprinted with permission from Ref. **a** [17], **b** [9]. Copyright © Wiley-VCH and Nature Publishing Group)

circuit in the cell [17]. The voltage generated under illumination depends on the difference between the Fermi level of the electron in the semiconductor materials and the redox potential of the electrolyte (Fig. 1b) [18, 19].

Notably, the core of the system is the nanoporous semiconductor, composed primarily of TiO$_2$ materials, which not only supplies numerous adsorption sites for dye sensitizer but also functions as an electron acceptor and electronic conductor [16]. TiO$_2$ possesses several unique chemical and physical properties which make it the most popular candidate for semiconductors in DSSCs. First, the conduction band edge of TiO$_2$ lies slightly below the excited state energy level of many sensitized dyes, which is a necessary condition for efficient electron injection. Second, TiO$_2$ also has a large dielectric constant ($\varepsilon = 80$ for anatase) for effective electrostatic shielding of the injected electrons from the oxidized dye molecules adsorbed on the TiO$_2$ surface, thus avoiding their recombination before regeneration of the dyes by the redox electrolyte. The relatively high refractive index of TiO$_2$ ($n = 2.5$ for anatase) also provides efficient diffusive light scattering inside the nanoporous film, thus significantly increasing the light harvesting potential. In addition, TiO$_2$ is stable over a wide range of environments, such as high temperature and high acidity. Lastly, TiO$_2$ is inexpensive, abundant, and nontoxic [2, 20]. In order to improve upon the aforementioned useful properties, over the past several decades extensive research interests and efforts have focused on the design, fabrication, and modification of versatile TiO$_2$ photoanodes.

1.2 Applications in Photoelectrochemical Water Splitting

In addition, another ideal way for direct conversion of solar energy into practical energy sources is through the generating of hydrogen from solar photoelectrochemical splitting of water using semiconductors as photoelectrodes [21].

Hydrogen is a well-known, potentially highly efficient and environmentally clean fuel since the chemical energy stored in the H–H bond can be easily released when the molecule reacts with oxygen to yield only water as a by-product (i.e., combustion). Moreover, hydrogen is a light gas (0.08988 g/L) with a higher energy density compared to any other fuel such as gasoline. For example, one gram of hydrogen can provide about 140 kJ of energy, which is almost four times that of methane (33 kJ/g) [22]. Therefore, large-scale and cost-effective generation of hydrogen, preferably using renewable and carbon-free resources, is highly attractive. Currently, hydrogen is produced from a variety of primary sources such as natural gas, heavy oil, methanol, biomass, wastes, coal, solar, wind, and nuclear power. Among these sources, hydrogen production from photocatalytic water splitting in the presence of semiconductor photocatalysts using solar irradiation represents one of the most promising approaches and has garnered attention because of its direct use of sunlight. In so doing, the process avoids the inefficiencies due to thermal transformation or electrolysis with the conversion of solar energy to electricity [23]. Since the pioneering work by Fujishima and Honda in 1972 on a photoelectrochemical cell (PEC) using a TiO_2 photoelectrode, water splitting under sunlight has made remarkable progress in the past 40 years [24, 25].

Water splitting is a thermodynamically uphill or endothermic process with a significantly positive change in Gibbs free energy (ΔG° = +237.2 kJ/mol, 1.23 eV per electron), and a minimum potential of 1.23 eV is needed for the reaction to proceed. Taking the recombination of excited electron-hole pairs and losses from devices such as contacts and electrode resistances into consideration, the optimal energy required for water splitting is around 2 eV [22]. In a PEC, the key components are the electrodes (i.e., cathode and anode) on which redox chemical reactions involving electron transfer take place. Typically, a PEC cell is composed of a semiconductor photoanode and a Pt counter electrode in an electrolyte solution. As shown in Fig. 2a, incident light irradiation with the photon energy matching or greater than the forbidden band gap energy (E_g) of the semiconductor generates electron-hole pairs and the photo-excited electrons are then promoted from the valence band (VB) to the unoccupied conduction band (CB), which then migrate to the cathode and react with protons to generate hydrogen ($2H^+ + 2e^- \rightarrow H_2$). Concurrently, the holes accumulate on the surface of the photoanode and split water molecules to produce oxygen ($H_2O + 2h^+ \rightarrow 2H^+ + 1/2O_2$) [26].

To effectively split water for hydrogen generation, the match of the band gap and the potential of the conduction and valence bands are important, that is, the E_g of the semiconductor should be larger than 1.23 eV ($\lambda < 1000$ nm) to realize water splitting. However, when using visible light, E_g should be less than 3.0 eV ($\lambda > 400$ nm) [27]. In addition, the semiconductor photoanode with a conduction band edge more negative than the H_2 evolution potential and a valence band edge more positive than the O_2 evolution potential is also required. Other requirements include stability under light irradiation and in aqueous solution, excellent absorption in the solar spectrum region, high-quality structure for effective charge transport, and low production cost [28]. Unfortunately, no such material has been found that can satisfy all the requirements simultaneously.

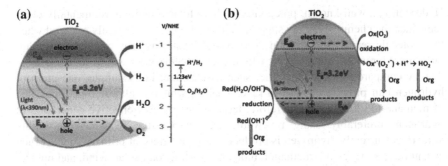

Fig. 2 a Overview principle of TiO$_2$-based photocatalytic water splitting for hydrogen generation. **b** Simplified principle of TiO$_2$-based photocatalysis of organic pollutants. (Reprinted with permission from Ref. [29]. Copyright © Simplex Academic Publishers)

Numerous semiconductor materials, for instance (TiO$_2$, Fe$_2$O$_3$, SrTiO$_3$, ZrO$_2$, WO$_3$, CdS, ZnO, NaTaO$_3$, BaTi$_4$O$_9$, BiVO$_4$, CeO$_2$) have been explored for water splitting into H$_2$ and O$_2$ under ultraviolet (UV) or visible light irradiation [23, 28, 29]. Among the various candidates for the photoanode, TiO$_2$ is considered to be one of the most promising semiconductor materials as it can fulfill the task of photocatalytic water splitting in a clean, environmentally friendly and low-cost way, owing to its favorable band gap energy (3.2 eV in anatase), high photochemical stability, non-toxic property, and relatively inexpensive cost [25]. There are however several drawbacks to TiO$_2$. The conversion efficiency of this technology is still low and thus it is currently only in the research stage. Also, the band gap of anatase TiO$_2$ is around 3.2 eV, implying that only ultraviolet (UV) light can be used for water splitting. Lastly, the PEC cell concurrently undergoes a rapid recombination of photogenerated electron-hole pairs [30]. Consequently, many research efforts are made to reduce these drawbacks in the TiO$_2$-based PEC cell.

1.3 Applications in Photocatalytic Degradation of Pollutants

Lastly, solar light may have the potential to solve the environmental contamination problem. Wastewater pollutants have become a worldwide environmental threat as a result of burgeoning industry and population [31]. To date, many techniques have been developed for the decontamination of many artificial or anthropogenic organic pollutants, especially those with high toxicity but very low concentration. A better alternative should be able to chemically transform the organic pollutants into environmentally benign compounds in an efficient manner [31].

Recently, numerous studies have been performed on the application of photocatalysis in the degradation of organic pollutants from wastewaters due to its ability to completely mineralize the toxic organic chemicals into nontoxic inorganic minerals [29]. Photocatalysis is triggered by semiconductor catalysts which function as sensitizers for light-induced redox-reactions and have the unique electronic structure of a filled valence band, and an empty conduction band. On irradiation (Fig. 2b), electrons in the valence band of semiconductors are excited to their conduction band, thus leaving holes behind. The resulting electron-hole pairs can recombine or react separately with other molecules. The holes may migrate to the surface and react either with electron donors in the solution or with hydroxide ions to produce powerful oxidizing species like hydroxyl or superoxide radicals. Meanwhile, the conduction band electrons can reduce an electron acceptor [32]. Consequently, semiconductor materials can act as either an electron donor or as an electron acceptor for molecules in the surrounding medium, depending on the charge transfer to the adsorbed species [32].

Semiconductor nanomaterials are promising options for inexpensive and environmentally friendly decontamination systems in which the correlated chemical reagents, energy source, and catalysts are abundant, inexpensive, non-toxic, and produce no secondary pollution byproducts [33]. Compared to other semiconductors (e.g., ZnO, Fe_2O_3, CdS, and ZnS), TiO_2 is the most widely used semiconductor catalyst in photocatalysis due to its chemical and biological inertness, photostability, relative ease of manufacture and utilization, reaction catalysis efficiency, low cost, and nontoxicity. It does however have the disadvantage of solely ultraviolet (UV) activation and not visible [32].

2 Fabrication of Cost-Effective Nanostructured TiO_2 Materials

As mentioned above, TiO_2 is one of the most promising photovoltaic materials because of its appropriate electronic band structure, photostability, chemical inertness, and commercial availability [34]. TiO_2 exists in nature in three different polymorphs, namely, rutile, anatase, and brookite. In addition, other synthetic phases, for example, $TiO_2(B)$, $TiO_2(H)$, and $TiO_2(R)$ as well as several high-pressure polymorphs have also been reported. Each phase shows different physical and chemical properties for different functionalities [34].

Of these phases, rutile and anatase are the most practically important crystal structures for energy applications (Table 1). In general, both anatase and rutile-type TiO_2 have a tetragonal crystal structure. The difference is that anatase TiO_2 follows a bipyramidal habit, while rutile TiO_2 obeys a prismatic habit [33]. These two tetragonal structures can be constructed by the chains of TiO_6 octahedra, where each Ti^{4+} ion is close to six O^{2-} ions. Anatase and rutile crystal structures are different in the distortion of each octahedron along with the assembly pattern of the octahedron chains. Specifically, the octahedron in rutile exhibits a slight orthorhombic distortion, while the octahedron in anatase is largely distorted, leading to lower symmetry compared to that of orthorhombic. Also, anatase has larger Ti–Ti distances but shorter Ti–O distances than those in rutile. Finally, the

Table 1 Physical and structural properties of anatase and rutile TiO_2. (Reprinted with permission from Ref. [33]. Copyright © Elsevier)

Property	Anatase	Rutile
Molecular weight (g/mol)	79.88	79.88
Melting point (°C)	1,825	1,825
Boiling point (°C)	2,500–3,000	2,500–3,000
Light absorption (nm)	<387.5	<413.3
Band-gap energy (eV)	~3.2	~3.0
Mohr's hardness	5.5	6.5–7.0
Dielectric constant	31	114
Density (g/cm^3)	3.79	4.13
Crystal structure	Tetragonal	Tetragonal
Space group	I4$_1$/amd	P4$_2$/mnm
Lattice constant (Å)	a = 3.78	a = 4.59
	c = 9.52	c = 2.96
Ti-0 bond length (Å)	1.94 (4)	1.95 (4)
	1.97 (2)	1.98 (2)

rutile structure is assembled by joining each octahedron with ten neighboring octahedrons in the form of two sharing edge oxygen pairs and eight sharing corner oxygen atoms. While for the anatase structure, each octahedron couples to eight neighbors with four sharing an edge and four sharing a corner. Thus, these differences in lattice structures result in different electronic properties between rutile and anatase TiO_2 [35]. Generally, anatase is considered to be more photoactive than rutile, largely due to the differences in the extent and nature of the surface hydroxyl groups present in the low temperature structure and the Fermi levels, where anatase ($E_g = 3.2$ eV) is about 0.2 eV higher than that of rutile ($E_g = 3.02$ eV). However, rutile TiO_2 still has some advantages over anatase such as higher chemical stability, higher refractive index, and cheaper production costs. Furthermore, rutile films have recently been used in DSSCs and have shown comparable performance to anatase DSSCs [36].

It is well accepted that the performance of devices is highly dependent on the chemical and physical properties of their as-prepared material components. Thus, today the demand for efficient, high performance devices has stimulated increasing attention in advanced functional materials [3]. Conventional micrometer-sized bulk materials suffer from inherent limitations in performance and fail to satisfy the increasing requirements of new devices. Consequently, nanostructured materials are becoming ever more important in many fields and hence have attracted great interest in recent years [34].

In brief, nanostructured materials exhibit quantum confinement effects when the electronic particles of these materials are confined by potential barriers to very small regions of space. The confinement can be in one dimension (i.e., quantum wells), in two dimensions (i.e., quantum wires, tubes or rods), or in three dimensions (i.e., quantum dots (QDs)). A variety of nanometer size-dependent properties have been found in the materials used in electrochemical energy conversion and storage devices. For example, the increasing surface-to-volume ratio and special surface area facilitate sufficient reaction or interaction between the nanostructured material and surrounding electrolyte [37]. Energy conversion and transport in nanostructured materials are also different from those in bulk materials due to the quantum size effects on energy carriers such as photons, electrons, and molecules [38]. For example, efficient light harvesting to create charge carriers in materials happens at the scale of several hundreds of nanometers, near the wavelength of light.

However, the mean free path of the excited charge carriers is shorter than the wavelength of light, requiring the small length scales afforded by nanostructures. Considering the need for both efficient photon absorption and effective collection of excited charge carriers in devices, it is necessary to design structures on a scale commensurate with both the wavelength of light and charge migration lengths simultaneously. One such option is one-dimensional (1D) nanostructures (e.g., nanotubes or nanowires), which have one dimension larger than the wavelength of light, and another dimension shorter than the mean free path of charge carriers [39]. Accordingly, current efforts in nanoscience and nanotechnology for energy applications are concentrating on utilizing these nanoscale effects to produce efficient energy technologies such as solar cells, fuel cells, and batteries. Researchers are eager to exploit cost-effective process to prepare high-performance nanostructures for a more sustainable energy economy [6, 40−44].

Many properties of nanostructured TiO_2 films, such as surface area, shape, grain size, and grain boundary density, will significantly impact the performances of energy conversion devices [45−47]. The methodology used to fabricate the nanostructured films is an essential factor to tailor the properties of TiO_2 nanostructures [48]. To date, significant progress has been achieved in the preparation of TiO_2 nanomaterials. A variety of film preparation techniques have been developed and employed for the formation of diversiform nanostructured TiO_2, including nanoparticles [49], nanorods [50], nanowires [51], nanotubes [52], nanosheets [53], and mesoporous structures [54]. In most cases, nanostructured TiO_2 materials can be prepared either by dry or wet processes. In the past decades, a number of methods, such as sol–gel [55], hydrothermal/solvothermal processes [56], electrochemical anodization [57], electrospinning [58], electrospray [59], electrodeposition [60], directional chemical oxidation [61], ultrasound and microwave irradiation [62], laser pyrolysis [63], and chemical/physical vapor deposition [64], have been developed to control the size, morphology, and uniformity of TiO_2 nanostructures simultaneously. The following sections further elaborate on some of the above-mentioned preparation methods for preparing cost-effective, high-performance TiO_2 nanostructures for energy applications, such as DSSCs, hydrogen generation, and photocatalysis.

2.1 Preparation Methods

2.1.1 Sol–Gel Method

The sol–gel method is a powerful technique in the development of nanostructured semiconductor films with diverse morphologies. Sol–gel method is typically a two-step process of synthesis and deposition, and can easily tailor the material properties during synthesis under optimized conditions [65, 66]. In a typical sol–gel process, a colloidal suspension or sol is first obtained from the hydrolysis and polymerization reactions of the precursors, which are usually inorganic metal salts or metal organic compounds such as metal alkoxides. The phase conversion from the liquid sol into the solid gel phase is realized by complete polymerization and evaporation of the solvent. Finally, thin films of the resulting solid gel can be achieved via a variety of deposition techniques such as spin-coating, dip-coating, spray pyrolysis, doctor blade, electrophoretic, and template assisted deposition onto a conducting glass substrate (e.g., fluorine-doped tin oxide, FTO or indium-tin oxide, ITO glasses) or other preferred surface [67–69]. The film is then annealed at 200–450 °C in air for 1–2 h, which is essential to reduce grain boundaries and enhance crystallinity, thus improving electrical conductivity.

Nanostructured TiO_2 materials are usually prepared from the hydrolysis of a titanium precursor, which is normally performed via an acid-catalyzed hydrolysis step of titanium (IV) alkoxide (e.g., titanium isopropoxide, TTIP) or halide precursor (e.g., $TiCl_4$) followed by condensation. Properties of the resulting TiO_2 nanomaterials, including the crystallinity, morphology, structure size, surface area, porosity, degree of dispersion, and crystal phase, heavily depend on the reaction conditions, including the temperature, reaction time, solvent, solution pH, type of precursor, drying conditions, and post-treatment [20, 35, 70]. Among various methods, the sol–gel method is one of the most widely used methods for TiO_2 nanostructure preparation, due to its relatively low cost, flexibility of substrate, and diversity of nanostructures such as nanoparticles, nanorods, nanoporous films, and nanowires [1].

Template-assisted sol–gel processes are a well-established strategy for the design of various functional TiO_2 nanostructures. This method utilizes the morphological properties of known and characterized templates in order to assemble materials with a particular morphology by reactive deposition or dissolution methods. By controlling the morphology of the template material, it is possible to prepare numerous new materials with a regular and controllable morphology (e.g., mesoporous structure) on the nano and micro scale [71, 72]. The template-assisted method can be generally divided into three sequential steps as shown in Fig. 3a: (1) assembly of a template (e.g., composed of polymer latex spheres), (2) infiltration and deposition of TiO_2 nanoparticles or titanium precursor, and (3) selective removal of the used template to yield an inverse porous structure [73]. The used templates can be soft templates (surfactants or block polymers) or hard templates (porous silica, polystyrene spheres, or porous carbon) [74].

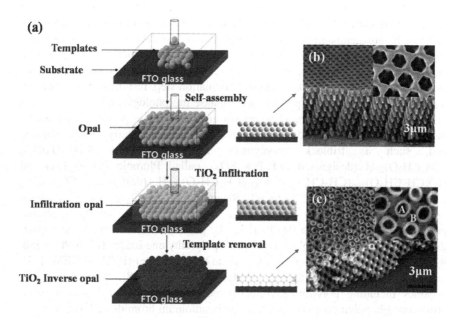

Fig. 3 a Schematics of the sol-gel fabrication procedure of a TiO_2 inverse opal via a template-assisted method. SEM images of SU8 templates (**b**) and TiO_2 inverse opal (**c**). (Reprinted with permission from Ref. **a** [92], **b**, **c** [72]. Copyright © Wiley-VCH and American Chemical Society)

In the past decade, mesoporous TiO_2 materials have attracted considerable interest for many applications due to their continuous particle framework, which contain nanoparticles distributed throughout structures of adjustable pore size and high specific surface area. Mesoporous TiO_2 materials with tailored pore size, high specific surface area, and well-defined crystalline structure in particular have potential applications in solar cells, photocatalysis, and water splitting [75, 76]. In photocatalytic applications, a tunable pore size can facilitate the diffusion rate of reactants toward adsorption sites, while a high surface area can maximize the interface between the reactant and the catalyst surfaces [77]. For DSSCs, meso-porous TiO_2 can enhance light harvesting within the electrodes without sacrificing the accessible surface for dye loading [78].

Generally, the sol–gel process using organic surfactants as assisting templates represents the most widely used route for the synthesis of mesoporous TiO_2 and involves a complicated mechanism called evaporation-induced self-assembly (EISA) [79]. The EISA process produces an ideal grid-like morphology consisting of a continuous, ordered network of anatase TiO_2 with a high surface area by condensation of a titanium precursor around self-organized organic templates in a gel phase, followed by removal of the templates via heat treatment. This simple process uses a wide range of surfactants as structure directing agents to prepare ordered mesoporous TiO_2. The structure directing agent in this method is an amphiphilic macromolecule (block copolymer) which microphase-separates into ordered domains on the 5–50 nm length scale. This is driven by the incompatibility of its covalently linked macro-molecular blocks. The inorganic material is typically selectively incorporated into one of the polymer domains in the form of a nanoparticle sol. The structure-direction controlled by macro-molecular self-assembly undergoes a high-temperature calcination step, resulting in an inorganic material that resembles the polymer microphase morphology [80].

Commonly the used organic templates are amphiphilic poly(alkylene oxide) block copolymers, composed of two different polymers covalently connected at one end, such as triblock copolymers, $HO(CH_2CH_2O)_{20}(CH_2CH(CH_3)O)_{70}$ $(CH_2CH_2O)_{20}H$ (designated $EO_{20}PO_{70}EO_{20}$, called Pluronic P-123) [81] and $HO(CH_2CH_2O)_{106}(CH_2CH–(CH_3)O)_{70}(CH_2CH_2O)_{106}H$ (designated $EO_{106}PO_{70}$ EO_{106}, called Pluronic F-127) [79, 82]. Recently diblock polymers have also been used as structure-directing agents. Some examples include polystyrene-block-poly(4-vinylpyridine) (PS-b-P4VP) [83, 84], polystyrene-block-poly(2-vinylpyridine) (PS-b-P2VP) [85], polystyrene-block-poly(ethylene oxide) (PS-b-PEO) [86, 87], poly(vinyl chloride)-g-poly(oxyethylene methacrylate) (PVC-g-POEM) [67], and poly(isoprene-block-ethylene oxide) (PI-b-PEO) [88]. Moreover, other organics including poly(dimethylglutarimide) (PMGI), hydroxyl styrene-based cross-linkable polymers [89], cetyltrimethylammonium bromide (CTAB) micelles [90], cellulose [65], sodium alginate [91] and polyethylene glycol (PEG) [40] are also employed to direct the formation of mesoporous TiO_2 for solar energy applications. Using such sol-gel process, electrodes can be prepared directly on Si or FTO substrates. Furthermore, the process is highly scalable because it can be performed at low temperatures without any expensive or complicated equipment.

Among the various new microstructured electrodes designed recently, three-dimensional (3D) photonic crystal electrodes (Fig. 3c) made from colloidal crystal templates or inverse opals (Fig. 3b) have been well investigated due to their unique advantages [92, 93]. Photonic crystal (PC) materials exhibit periodicities in their refractive index on the order of the wavelength of light, and thus provide many interesting possibilities for "photon management" [94, 95]. Bragg diffraction in a periodic lattice, localization of heavy photons near the edges of a photonic band gap, multiple scattering at disordered regions in the photonic crystal, and the formation of multiple resonant modes are some of the phenomena that are exhibited by photonic crystals and can greatly enhance the effective light path within the active layer [68, 96]. 3D photonic colloidal crystals are of particular interest for enhancing light harvesting in DSSCs because these TiO_2 crystals can be both porous and significantly enhance light-matter interactions on the long wavelength side of the stop band. More importantly, the 3D ordered porous electrode with relatively large porosity is beneficial in applications that include polymeric electrolytes with high viscosities and relatively large molecular weight. Furthermore, 3D connected TiO_2 networks can provide an organized electron path, which may facilitate charge transport and thus enhance the collection efficiency of back-contact electrodes [72, 92]. In this regard, several hard templates, such as polystyrene (PS) [97, 98], poly(methyl methacrylate) (PMMA) [73], and SU8 photoresist [72] can be used to build the 3D backbone.

Interestingly, some designs are created using two kinds of templates. For example, a double layer couples a high surface area mesoporous anatase underlayer prepared using PI-b-PEO block polymer as template, with an optically and electrically active 3D periodic TiO_2 PC overlayer reversed from a PS template. Such a design is expected to allow effective dye sensitization, electrolyte infiltration, and charge collection from both the mesoporous and the PC layers, which acts as the photoanode in DSSCs [68]. Additionally, mesoscale colloidal PS particles and lithographically patterned SU8 macropores can be used as the dual templates, with the colloidal particles assembled within the macropores. The as-prepared template produces hierarchical TiO_2 electrodes for DSSCs with sufficient surface area from mesoscale pores and effective light scattering from micropores [99].

Alongside the effective organic templates, several inorganic materials including porous anodic alumina (PAA) [100], ZnO [101], $Cd(OH)_2$ [102] and SiO_2 [103] have been successfully employed as templates to construct TiO_2 architectures, such as ordered nanowires/nanotubes, mesoporous hollow spheres, and hierarchical nanoplates. One of the most commonly used templates, a PAA membrane with high density and high aspect ratio pores, is prepared by anodic oxidation of an aluminum sheet in a solution of sulfuric, oxalic, or phosphoric acids. It is frequently used as a template in nanowire or nanotube array synthesis through the sol–gel process described in Sect. 2.2.2 [104]. The PAA allows nanowires or nanotubes to grow on transparent conducting indium tin oxide (ITO) substrates without bending or breaking. This technique is capable of achieving a surface roughness factor equal to a typical nanoparticle layer [100]. In particular, a number of hybrid templated processes have been developed to construct attractive TiO_2 nanostructures. For example, 3D interconnected nanoporous TiO_2 nanotube arrays on fluorine-doped tin oxide (FTO) glass were prepared using a sol-gel process assisted by PVC-g-P4VP block copolymer and a ZnO nanorod template [105]. An organic–inorganic hybrid template (Fig. 4), Al_2O_3-POEM, is specially designed to fabricate a crack-free, organized mesoporous TiO_2 photoanode with high surface area, good interconnectivity, and uniform pores, yielding a high energy conversion efficiency of 7.3 % in DSSCs [78]. Other synthesis methods have also been assisted by templates, including hydrothermal [106, 107], atomic layer deposition (ALD) [108–110], microwave, and sonochemical methods [77].

2.1.2 Hydrothermal/Solvothermal Method

The hydrothermal/solvothermal methods have also been used to prepare a variety of TiO_2 nanomaterials [111–113]. In such processes (Fig. 5), the synthesis reaction generally occurs in aqueous/organic solutions containing required material precursors within temperature controlled steel pressure vessels called autoclaves. The reaction temperature can be raised above the boiling point of the water/organic solvent, reaching the pressure of vapor saturation. The temperature and the amount of solution added to the autoclave largely affect the internal pressure produced. The solvothermal method is almost identical to hydrothermal method except that

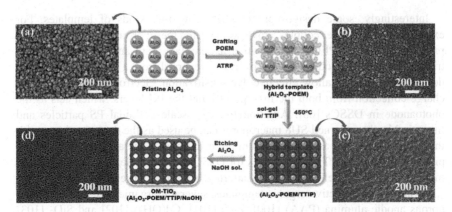

Fig. 4 Schematic of the preparation of the TiO₂ electrodes using hybrid templates. (Reprinted with permission from Ref. [78]). Copyright © Wiley-VCH

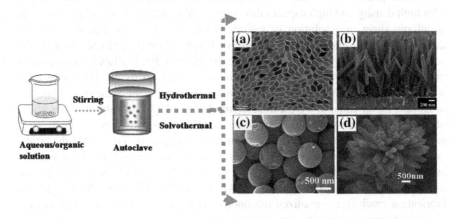

Fig. 5 Simplified schematic of hydrothermal and solvothermal synthesis and some typical structures prepared by both process: **a** nanosheets, **b** nanorods, **c** hollow microspheres and **d** flower-like microspheres. (Reprinted with permission from Ref. **a** [114], **b** [122], **c** [8] **d** [125]. Copyright © American Chemical Society and Wiley-VCH)

the solvents used for solvothermal synthesis are nonaqueous, while water is the primary hydrothermal solvent [114, 115]. Accordingly, the temperature in solvothermal process can be increased much higher than that in hydrothermal method because a variety of organic solvents with high boiling points can be adopted. Normally, the solvothermal method possesses better control over the size, structure, size distribution, and degree of crystallinity of TiO₂ nanomaterials compared to hydrothermal method [35].

In comparison with other methods, hydrothermal/solvothermal synthesis is a facile route to prepare a highly crystalline oxide under moderate reaction conditions, including low temperature (in general less than 250 °C) and relatively short reaction time. It provides a convenient and effective reaction environment for the formation of nanocrystalline TiO_2 with high purity, good dispersion, and well-controlled crystallinity [116]. By employing this method, the calcination process, which is essential to the TiO_2 transformation from amorphous phase to crystalline phase, can be eliminated. Similarly, by tuning the hydrothermal conditions (such as temperature, pH, reactant concentration, molar ratio, and additives), crystalline products with different compositions, sizes, and morphologies can be achieved [117]. Many high-performance TiO_2 nanostructures, including novel nanocrystals [114], nanosheets (Fig. 5a) [22], 1D morphologies (e.g., nanotubes, nanowires, nanorods (Fig. 5b)) [118–122], mesoporous structures (Fig. 5c) [116, 123], and especially 3D hierarchical architectures (Fig. 5d) [29, 115, 124, 125] have been extensively developed via cost-effective hydrothermal/solvothermal methods and widely applied in DSSCs, H_2 generation and photocatalysis.

Recently, anatase TiO_2 single-crystalline nanosheets (NSs) with a high percentage of reactive (001) facets have attracted much attention as a dominant source of active sites for various applications such as photovoltaic cells, photodegradation of organic molecules, and photocatalytic water splitting [112, 126, 127]. Both theoretical and experimental studies indicate that the anatase TiO_2 (001) facets in particular are more reactive than (101) facets, which could be mainly attributed to different coordination numbers of Ti in (001) and (101) facets. In the (001) facets, each Ti atom coordinates with five oxygen atoms (i.e., with 100 % five-coordinate Ti (Ti_{5c}) atoms). While in the (101) facet, each Ti atom is coordinated with either five or six oxygen atoms with a 50 % probability for either case. Thus, the (001) facets display an enhanced number of oxygen vacancies in comparison with the (101) facets owing to the low coordination number of Ti with oxygen [128, 129]. However, anatase TiO_2 crystals are usually dominated by (101) facets, which are thermodynamically stable due to their lower surface energy than that of (001) facets (i.e., surface energy: $\gamma(001) = 0.90$ $J/m^2 > \gamma(100) = 0.53$ $J/m^2 > \gamma(101) = 0.44$ J/m^2). Consequently (101) facets account for more than 94 % of the total exposed surface according to the Wolff construction [4].

An important breakthrough in the preparation of anatase TiO_2 sheets with exposed (001) facets was achieved by Yang et al. in 2008. The synthesis of anatase TiO_2 single crystals with 47 % exposed (001) facets was realized by using hydrofluoric acid (HF) as a shape controlling agent to stabilize the (001) facets (Fig. 6a–c) [130]. Thereafter, numerous studies have been conducted toward the preparation and application of different shaped anatase TiO_2 micro- or nanocrystals with exposed (001) facets [131–133]. For example, Zheng et al. demonstrated a simple solvothermal synthesis, where tetrabutyltitanate, $Ti(OBu)_4$, was mixed with absolute ethanol and 40 % hydrofluoric acid solution, to produce TiO_2 microspheres assembled of anatase TiO_2 nanosheets with 83 % dominant (001) facets (Fig. 6d). These TiO_2 microspheres exhibited excellent photocatalytic activity for the degradation of methyl orange (MO) as shown in Fig. 6e [134]. Yang et al. demonstrated controllable hydrothermal synthesis of ultra-thin TiO_2 NSs with a thickness of only 1.6–2.7 nm and up to 82 % (001) facet coverage. It is found that the concentration of HF used as a capping agent significantly affected the thickness and side length of the

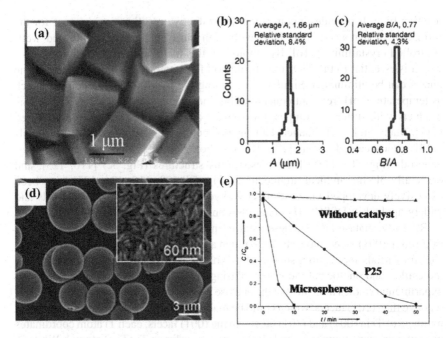

Fig. 6 SEM images of anatase single crystals (**a**) and TiO₂ microspheres composed of nanosheets with (001) facets (**d**), *inset* is the magnified surface of a microsphere. **b** The size distribution of anatase single crystals in (**a**). **c** The degree of truncation (B/A, A and B denote lengths of the side of the bipyramid and the side of the square {001} 'truncation' facets, respectively) of anatase single crystals in (**a**). **e** The corresponding photodegradation of methyl orange (MO) of (**d**). (Reprinted with permission from Ref. **a–c** [130], **d**, **e** [134]. Copyright © Nature Publishing Group and Wiley-VCH)

resulting TiO₂ nanosheets. Due to the high percentage of reactive (001) facets, the synthesized NSs loaded with 1 wt% Pt from photochemical reduction of water in the presence of methanol as a scavenger showed a H₂ evolution rate of 7381 μmol h⁻¹ g⁻¹ under UV-vis light [22]. Xiang et al. reported that the flower-like TiO₂ microsphere films with about 30 % exposed (001) facets were directly synthesized on the Ti foil in a dilute aqueous HF solution by a simple one-pot hydrothermal process and exhibited tunable photocatalytic selectivity toward decomposition of azo dyes by modifying the surface of TiO₂ microspheres as well as by varying the degree of etching of (001) facets [133]. Zhang et al. synthesized anatase TiO₂ microspheres with exposed mirror-like plane (001) crystalline facets via a facile low-temperature hydrothermal method. The improved light harvesting efficiency, attributed to the superior light scattering ability of the TiO₂ microsphere film, gave such microsphere photoanodes better overall light conversion efficiency of 7.91 % in DSSCs, which is 1.2 times the overall efficiency (6.73 %) obtained from the P25 photoanode [135].

Most of the hydrothermal syntheses reported so far have been carried out in the acidic environment with the addition of concentrated hydrofluoric acid, which is

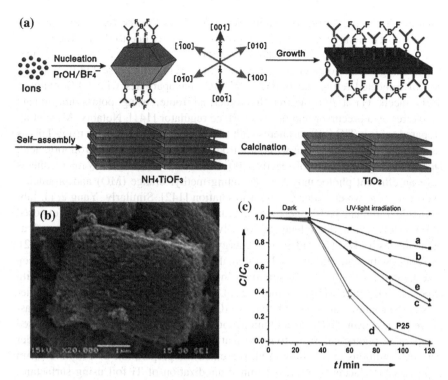

Fig. 7 a Schematic illustration of the formation mechanism for layered TiO₂. **b** SEM images of layered TiO₂ produced via the (**a**) process. **c** Photodegradation curves of RhB (20 mgL⁻¹) under UV-light irradiation over the TiO₂ samples prepared in different concentrations of 2-propanol (PrOH): a: 0, b: 1.5, c: 2.8, d: 4.0, and e: 8.0 M. (Reprinted with permission from Ref. [140]. Copyright © Wiley-VCH)

highly toxic and corrosive in both liquid and vapor forms [136]. In response to the safety and health concerns, a number of studies have investigated the preparation of anatase TiO₂ single crystals with clean and exposed (001) facets via hydrofluoric acid-free processes [137–139]. Recently, Yu et al. reported the synthesis of a layered TiO₂ structure composed of several nanosheets with exposed (001) facets, by a simple hydrothermal method in the presence of (NH₄)₂TiF₆, H₃BO₃, and 2-propanol, followed by calcination treatment (Fig. 7a). The resulting layered TiO₂ with (001) facet nanosheets (Fig. 7b) exhibited excellent photocatalytic activity for the degradation of rhodamine B (RhB) (Fig. 7c) [140]. Fang et al. synthesized submicrometer-sized yolk@shell hierarchical spheres composed of a permeable shell self-assembled by ultrathin anatase TiO₂ NSs with nearly 90 % exposed (001) facets and a mesoporous inner sphere with a high specific surface area. This process used the solvothermal method in the presence of diethylenetriamine (DETA), isopropyl alcohol (IPA) and titanium (IV) isopropoxide (TTIP), also followed by calcination treatment. Compared to the conventional (001) faceted TiO₂ NSs (4.01 %) and standard Degussa P25 (4.46 %), such anatase TiO₂ yolk@shell hierarchical sphere photoanodes (6.01 %) showed higher overall light conversion efficiency (η) for DSSC applications mainly due to their large surface area (up to 245.1 m^2 g^{-1}) and their promotion of light scattering in the visible region using submicrometer scale features [113]. Han et al. presented the fabrication of truncated tetragonal bipyramidal TiO_2 nanoparticles with 60 % exposed (001) facets via a fluorine-free hydrothermal route, using potassium titanate nanowires as a precursor and urea as surface regulator [141]. Notably, Miao et al. prepared anatase TiO_2 microspheres with >90 % exposed (001) facets using TiF_4 as capping and stabilizing agent without introducing any additional hydrofluoric acid. Due to the high exposure of reactive (001) facets, the anatase TiO_2 microspheres gave an efficient photocatalyst in degrading methyl orange (MO) and producing hydrogen from water under UV light irradiation [142]. Similarly, Yang et al. fabricated hierarchical TiO_2 spheres made of thin nanosheets with over 90 % exposed (001) facets via a diethylene glycol-solvothermal route, which were used as photoanodes for DSSCs, and gave an energy conversion efficiency of 7.51 % [112].

Along with hydrothermal/solvothermal methods, other strategies have been investigated for the green preparation of highly reactive TiO_2 nanomaterials with exposed (001) facets [143]. Amona et al. found that faceted decahedral single-crystalline anatase particles with reactive (001) facets could be fabricated by a gas-phase process using $TiCl_4$ as a titanium source and that such a structure possessed comparable photocatalytic activity to that of P25 for reactions performed under various conditions [144]. Jung et al. presented the preparation of TiO_2 NTs with exposed (001) facets by electrochemical anodization of Ti foil using surfactant-assisted processes with poly(vinyl pyrrolidone) (PVP) and acetic acid. The DSSC assembled using a TiO_2 nanotube photoanode with 77 % exposed (001) facets demonstrated an overall conversion efficiency of 3.28 % [145]. Xie et al. developed a solid-state precursor strategy for preparing a hollow anatase TiO_2 box-like structure enclosed by six single-crystalline TiO_2 plates with highly exposed reactive (001) facets by sintering a cubic $TiOF_2$ solid precursor at 500–600 °C. The formation of these particular nanostructures is attributed to the hard self-template restriction and the adsorption of F^- ions from the $TiOF_2$. Due to the high percentage of reactive (001) facets, such novel TiO_2 boxes exhibited good performance in photocatalytic H_2 evolution (7.55 mmol g^{-1} h^{-1}) [146].

Furthermore, in addition to exposed (001) facets, TiO_2 nanocrystals with other important active facets have also been studied recently. For example, single crystalline anatase TiO_2 rods with dominant reactive (010) facets (also with 100 % five-coordinate Ti (Ti5c) atoms) were directly synthesized by hydrothermally treating $Cs_{0.68}Ti_{1.83}O_4$/$H_{0.68}Ti_{1.83}O_4$ particles. The nano-sized rods showed a 7.73 % conversion efficiency in DSSCs. This was comparable to the 7.67 % benchmark for P25 TiO_2 nanocrystals [147]. The preparation of tetragonal faceted-nanorods of single-crystalline anatase TiO_2 with predominately exposed higher energy (100) facets was performed by hydrothermal transformation of sodium titanate in alkaline solution [138]. Significantly, anatase TiO_2 nanocuboids wholly exposed with high-energy (001) and (100) facets were successfully synthesized by an environmentally benign synthetic strategy using low-cost acid-delaminated vermiculite (DVMT)

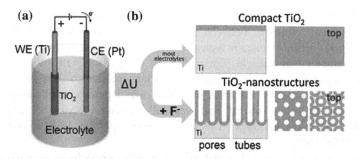

Fig. 8 **a** Schematic diagram of anodization experiments. **b** Anodization leads to the formation of a compact oxide on Ti (in most neutral and acidic electrolytes). If dilute fluoride electrolytes are used, highly ordered nanoporous or nanotubular anodic oxide layers can be formed. In the case of nanoporous oxide, the layer consists of vertically aligned nanosized channels in an oxide matrix while in the case of nanotubular morphology the layer is composed of an ordered array of oxide nanotubes. (Reprinted with permission from Ref. [150]. Copyright © Wiley-VCH)

and tetramethylammonium hydroxide (Me$_4$NOH) as synergistic morphology-controlling reagents [148]. In contrast to anatase TiO$_2$ with low-index facets like (101), (001), and (100), recently Li et al. reported the fabrication of an anatase TiO$_2$ nanosheet array assembled by (116) facet-oriented nanocrystallites. The (116) facets were made parallel to the surface of a FTO substrate via a two-step process, which exhibited 50 % higher photocatalytic activity than (001) facet-oriented nanosheet arrays in the degradation of methyl blue (MB) under UV light irradiation [149].

2.1.3 Electrochemical Anodization Method

Anodic oxidation represents a facile and well-established method to form nano-structures in a self-organiting way [104]. The controlled oxidation of titanium metal under electrochemical anodization provides another method of TiO$_2$ nanomaterial production [52]. In general, the anodization process is conducted in a two-electrode electrochemical cell at a constant potential in aqueous or organic electrolyte with Ti foil as an anode and platinum foil as a cathode (Fig. 8a) [150]. Under optimized anodized conditions, highly ordered nanoporous or nanotubular architectures with high aspect ratios could be successfully achieved [151]. Anodization is particularly useful in the synthesis of TiO$_2$ nanotubes from titanium foil. Anodization has been extensively studied after pioneering work in 2001 by Gong et al. in which they reported the formation of nanotubes up to 0.5 mm in length by electrochemical anodization of titanium foil in HF aqueous electrolyte [152].

Among the various forms of semiconductor nanostructures, one-dimensional (1D) highly ordered architectures, such as nanowires, nanorods, nanofibers, nanotubes with high surface area-to-volume ratios possess useful and unique properties compared to their bulk counterparts [39]. The highly ordered nature of these 1D nanostructures gives them excellent electron percolation pathways for directional charge transfer between interfaces. For example, the mobility of electrons in 1D nanostructures is typically several orders of magnitude higher than in semiconductor nanoparticle films [153]. In comparison with other 1D morphologies, nanotubes provide a larger interfacial area due to their external and internal surfaces, which is beneficial for surface area-dependent applications [154]. Studies on TiO_2 have shown that vertically oriented nanotube arrays are remarkably efficient when applied in sensors, water splitting, DSSCs, and photocatalysis [155–160].

In the past decade, self-organized oxide nanotube arrays have attracted extensive scientific and technological interest. Thus, TiO_2 nanotube arrays have been synthesized by a variety of methods, including template deposition (e.g., AAO templates, ZnO nanorod templates, or organic templates) [161], electrochemical anodization [162], and hydrothermal techniques [163]. Among these, an inexpensive and straightforward approach that leads to well-behaved nanotubes is the anodization method, which enables precise control over the resulting tube diameter, tube length, and overall morphology by adjusting various parameters such as the pH, concentration and composition of electrolyte, applied potential, growth time, and temperature of the anodization process [164–167]. Interestingly, the addition of fluoride ions tends to control the overall development of nanotube architecture (Fig. 8b). There are several excellent reviews detailing the growth mechanism in anodic oxidation [168–171].

Normally, the anodization process can be divided roughly into three stages: (1) electrochemical oxidization of the titanium surface which results in the formation of an initial TiO_2 barrier layer, corresponding to the first current drop; (2) chemical etching of TiO_2 by F^- to form TiF_6^{2-}, resulting in nanotube formation that leads to a current increase; and (3) the growth of nanotubes, which results in a slow current decrease [172]. Briefly, the nanotube growth is determined by the equilibrium between anodic oxidation and chemical dissolution. The anodic oxidation rate is mainly controlled by the anodic potential, while the chemical dissolution rate is controlled by the electrolyte acidity and F^- concentration [173]. As previously mentioned, since the discovery by Gong and co-workers in 2001, TiO_2 nanotube fabrication has been intensively investigated over the past decade. Consequently, a variety of nanotubular architectures have also been explored. In particular, by varying the voltage during the growth, new self-organized TiO_2 morphologies could be obtained: bamboo-type nanotubes [174], branched nanotubes [175], periodic nanotubes [176], ridged nanotubes [177], double-walled nanotubes [178], and multilayer nanotubes [179]. In the following sections, some representative fabrication processes developed recently for high-performance photovoltaic applications are discussed in greater detail.

TiO_2 nanotubes prepared via anodization of Ti foil are attached to the Ti substrate with closed bottom. In most cases, the use of Ti foil leads to TiO_2 nanotube arrays supported on Ti substrate. However, in many applications, detached TiO_2 nanotube layers are required. TiO_2 nanotube arrays grown in situ on opaque titanium foil are difficult to apply to high-efficiency DSSCs because the

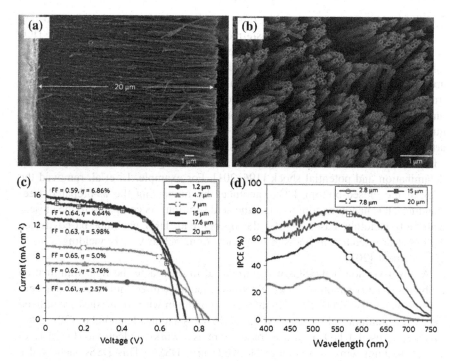

Fig. 9 SEM images of a 20-μm-long TiO₂ nanotube array from anodization of the Ti film sputter-deposited on FTO-coated glass: **a** cross-sectional view and **b** top view. The current-voltage characteristics (**c**) and the incident photon-to-current conversion efficiency (IPCE) spectra of DSSCs fabricated using transparent nanotube array films of various lengths. (Reprinted with permission from Ref. [188]. Copyright © American Chemical Society)

incoming light using back illumination mode is partially reflected by the counter electrode and partially absorbed by the counter electrode and iodine in the electrolyte before striking TiO₂ nanotubes, leading to a loss of ∼25 % of the incident solar energy [180]. Furthermore, sintering of the TiO₂ for transformation from amorphous phase to crystalline phase (i.e. anatase or rutile) can introduce the formation of a barrier layer between the nanotubes and the underlying Ti substrate. This enables recombination of electrons and holes when the nanotube layers are used for photoelectrochemical water splitting [181]. Fortunately, several strategies have been explored to solve this deadlock. One straightforward alternative solution is to deposit titanium as a thin film on a transparent substrate (e.g., FTO glass) before anodizing. The deposition process is usually performed by physical methods, for example, radio-frequency (RF) or direct-current (DC) magnetron sputtering [182–187]. Through anodization of sputtered titanium into nanotube layers on transparent substrate (Fig. 9a, b), a power conversion efficiency of 6.9 % for the resulting DSSCs was obtained (Fig. 9c, d) [188]. The length of the nanotubes is limited by the difficulty of growing a high quality, thick Ti film on the conductive glass via sputter deposition [189]. Therefore, a large-area free-standing

TiO$_2$ nanotube film with tunable tube lengths was developed by a two-step anodization process and then transferred onto FTO glass via a layer of TiO$_2$ nanoparticle paste [190–195]. This process, however, results in closed-bottom tube-ends. This is problematic because the interface between the TiO$_2$ nanotube arrays and the TiO$_2$ nanoparticle layer might cause near-UV light absorption and front surface light reflection and block the diffusion of the redox reagents into the underlying TiO$_2$ nanoparticles coating on the collecting FTO substrate. Thus, a number of methods have been developed to fabricate free-standing TiO$_2$ nanotube arrays, including critical point drying, dissolution of the Ti substrate in water-free CH$_3$OH/Br$_2$ solution, solvent evaporation, ultrasonic agitation, chemically assisted delamination and potential shock [196–201]. For example, Li et al. removed the caps of the closed-bottom TiO$_2$ nanotubes by immersing the as-prepared free-standing TiO$_2$ nanotube film in an oxalic acid solution. As compared to the closed-end TiO$_2$ nanotube-based DSSC, the opened-end TiO$_2$ nanotube-based device exhibited an increase in one-sun efficiency from 5.3 to 9.1 %, yielding a 70 % enhancement [202].

A novel version of DSSCs was introduced to overcome the light illumination problem of TiO$_2$ nanotube-based electrodes, namely, three-dimensional dye-sensitized solar cells (3D DSSCs). In this system titanium wires or meshes are utilized instead of titanium foils or sheets to fabricate TiO$_2$ anodized nanotubes [203–208]. Misra et al. used a TiO$_2$ nanotube-based wire as a working electrode (Fig. 10a–c) and a platinum wire as a counter electrode in a DSSC. This DSSC achieved a conversion efficiency of 2.78 % under AM 1.5 simulated solar light (Fig. 10d). The prototype device is capable of achieving long distance transport of photo-generated electrons and multi-directional light harvesting from surrounding to generate electricity [205]. Wang et al. developed a new type of 3D DSSCs with double deck cylindrical Ti meshes as the substrates. Here, one of the Ti meshes was anodized to in situ synthesize the self-organized TiO$_2$ nanotube layer to serve as the photoanode. Another Ti mesh was platinized through electrodeposition as the counter electrode. This all-Ti 3D DSSC exhibited the highest conversion efficiency of 5.5 % under standard AM 1.5 sunlight [204].

In addition, vertically ordered TiO$_2$ nanotube arrays also face the serious problem of insufficient surface area due to the large diameter of nanotubes and considerable free space between nanotubes. This leads to poor dye adsorption capacity when applied in DSSCs [209]. In light of this limitation, many strategies have been explored [46, 103, 210–212]. For example, a common combination of TiO$_2$ nanotubes and nanoparticles was realized by treating the as-anodized nanotubes with a TiCl$_4$ solution which hydrolyzed to yield nanoparticles. This can increase the nanotube surface area and bridge any cracks resulting from annealing, and thus improve the conversion efficiency of nanotube-based DSSCs [213, 214]. Recently, novel hierarchical-structured TiO$_2$ nanotube arrays have been prepared by combining the two-step electrochemical anodization with a hydrothermal process. The resulting DSSCs exhibited good performance and applicability [212].

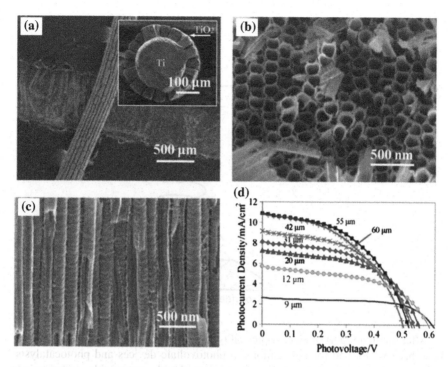

Fig. 10 SEM images of TiO$_2$ nanotube arrays grown around a Ti wire: Top view (**a**), *inset* is cross-sectional view, magnified top view (**b**) and magnified cross-sectional view (**c**). The current–voltage characteristics (**d**) of DSSCs using TiO$_2$ nanotube arrays with different lengths. (Reprinted with permission from Ref. [205]. Copyright © American Chemical Society)

2.1.4 Electrospinning and Electrospray Methods

Electrospinning is a simple and versatile nanofabrication technique for preparing several continuous 1D nanofibers, including polymers, ceramics, composites and metals, with controllable diameters ranging from a few nanometers to several micrometers [215, 216]. Electrospinning works using the principle of asymmetric bending of a charged liquid jet accelerated by a longitudinal electric field (Fig. 11) [217]. A diversity of soluble and fusible polymers (e.g., polyvinylpyrrolidone (PVP), polyacrylonitrile (PAN), polyurethane (PU) and polyvinyl acetate (PVAc)) can be electrospun to form nanofibers from their precursor solutions [218, 219]. If the polymeric solution contains the inorganic precursors (e.g., TiO$_2$, SnO$_2$ and ZnO), organic/inorganic composite nanofibers are obtained. These are subsequently calcinated at high temperature to thermally decompose organic components, and produce inorganic nanofibers with minimal morphological change [135, 220]. The diameter, alignment, and morphology of these nanofibers can be tailored by controlling the liquid injection rate, intensity of the electric field, and shape of the collector surface, respectively. The diameter of nanofibers also depends on the intrinsic properties of the polymeric solution such as the viscosity and surface charge.

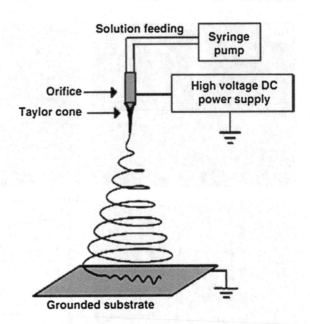

After the first report of electrospun TiO$_2$ nanofibers in 2003 [221], many studies have been performed for applications in photovoltaic devices and photocatalysis since the electrospun nanofibers can be fabricated with controllable size, density, and orientation [222, 223]. More importantly, electrospinning is simple, controllable, inexpensive, and scalable for industrial production [224]. For instance, Yang et al. prepared a bi-layer TiO$_2$ nanofiber photoanode via combining both small and large-diameter TiO$_2$ nanofibers in the nozzleless electrospinning by changing the applied voltage, electrode separation distance, and electrospinning time. It is expected that the smaller-diameter nanofiber layer with a high surface-to-volume ratio can load enough dye molecules and directly transport electrons within the 1D channels. While the larger-diameter nanofiber layer can serve as a light scattering layer, further adsorbing more dye molecules, and even providing higher porosity to facilitate electrolyte diffusion. With the bi-layer photoanode, the efficiency (η) of DSSCs can be improved from 7.14 % for the single-layer to 8.40 % for the bi-layer TiO$_2$ nanofiber photoanode [225]. As shown in Fig. 12a, multi-scale porous TiO$_2$ nanofibers were fabricated from HF etching of TiO$_2$/SiO$_2$ composite nanofibers, which were electrospun from a hybrid solution of dissolved SiO$_2$ colloidal solution, TTIP and PVP, followed by calcination. Such fiber-based DSSCs showed an η of 8.5% (Fig. 12b), which is greater than those of conventional photoelectrodes made of TiO$_2$ nanoparticles (6.0 %) [226].

Other useful structures, including nanorods [227], rice-like shapes [228], and hollow fibers [229], have been derived from electrospinning methods. TiO$_2$ nanorod-based photoelectrodes (Fig. 12c) were prepared by sintering TiO$_2$ composite nanofibers that were electrospun from a solution mixture of TiO$_2$ sol-gel and

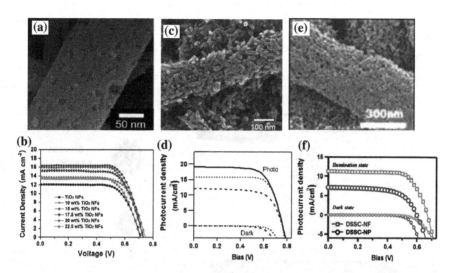

Fig. 12 SEM images of TiO$_2$ multi-scale porous nanofibers (**a**), electrospun nanorods (**c**) and hierarchically structured nanofibers (**e**). The current–voltage characteristics (**b, d, f**) of DSSCs based on (**a, c, e**), respectively. (Reprinted with permission from Ref. **a, b** [226], **c, d** [230], **e, f** [233]. Copyright © American Chemical Society)

PVAc in dimethyl formamide (DMF). The efficiency of such nanorod-based DSSCs was optimized to achieve 9.52 % under masked illumination of simulated solar light, AM 1.5 Global (Fig. 12d) [230]. Rice grain-shaped TiO$_2$ films were produced by 500 °C calcination of the nanofibers prepared from electrospinning [231]. The photovoltaic and photocatalytic performance of the rice grain-shaped structures are superior to commercially available P25 [232]. Furthermore, it has also been demonstrated that the rice grain-like structures are also superior to the nanofibers in scattering light with a 15.7 % enhancement in efficiency compared to 9.63 % for nanofibers [228]. Attractive hierarchical structures can also be obtained from a combination of electrospinning and other powerful methods, such as hydrothermal processes [214]. Hierarchically structured TiO$_2$ nanofibers prepared from electrospinning followed by a stepwise calcination treatment exhibited a unique morphology in which microscale core fibers were interconnected and numerous nanorods were deposited onto the fibers (Fig. 12e). This nanorod-in-nanofiber morphology possessed high porosity at the mesopore and macropore levels, facilitating the infiltration of plastic crystal electrolytes in DSSCs and yielding an optimized η up to 7.93 % (Fig. 12f) [233]. In addition, a hierarchically heterostructured TiO$_2$ nanocomposite composed of rutile nanosheets standing perpendicular on anatase nanofibers, can be successfully prepared through a combination of electrospinning and solvothermal processes [26].

Recently, the electrostatic spray (e-spray) technique has attracted a lot of attention because it is a simple, versatile, and cost-effective technology that can be applied in a variety of fields for the fabrication of semiconductive ceramics,

(a)

(b)

Fig. 13 **a** Schematic diagram of electrostatic spray and **b** formation of hierarchically structured TiO$_2$ nanospheres by electrostatic spray. (Reprinted with permission from Ref. **a, b** [59]. Copyright © American Chemical Society)

polymer coatings, protein films, and micro-patterns [234, 235]. In the electrospray process, a jet of liquid is sprayed toward a collector under a strong electric field (Fig. 13a). During e-spray deposition, known as induction or conduction charging, the droplets can be charged and atomized by mechanical forces in the presence of electric field between the solution and the depositing substrates. The electric field provides an electric charge on the liquid surface and the charge is carried out by the droplets detaching from the jet. The deposition efficiency of the charged droplets is usually higher than that of the uncharged droplets, which can improve the adhesion between the materials and substrates (Fig. 13b) [59]. In brief, three parameters typically determine the structures created by the electrospray method: (1) solution conditions such as the solvents and precursors, (2) processing conditions such as applied voltages, distance between the tip of the needle and collector and flow rates, and (3) ambient conditions such as temperature and humidity [235, 236]. As a deposition technique, e-spray has many advantages, such as large treatment areas, compatibility with various substrate geometries, and high deposition rate. Moreover, electrospray devices can perform at room temperature under atmospheric pressure, making it an energy efficient and low-cost technology [237].

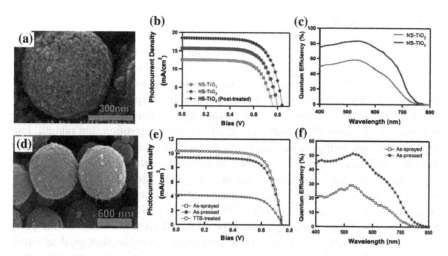

Fig. 14 SEM images of hierarchically structured TiO₂ microspheres (**a, d**). The current–voltage characteristics and IPCE spectra of DSSCs (**b, c**), and flexible DSSCs (**e, f**) based on the TiO₂ microspheres. (Reprinted with permission from Ref. **a–c** [59], **d–f** [240]. Copyright © American Chemical Society)

Accordingly, e-spray could be a promising method for the preparation of photoanodes for the industrial-scale manufacture of DSSCs. Based on reported works, several significant TiO₂ nanostructures including hierarchically structured mesoporous spheres, hollow hemispherical aggregates, and nanoflakes have been developed using electrospray methods [211, 236, 238]. Of these structures, mesoporous spheres are the most advantageous on account of their large surface areas for abundant dye loading, high porosity for efficient electrolyte diffusion, and microscale framework for effective light scattering [59]. In comparison with other methods (e.g., hydrothermal/solvothermal process) for the preparation of mesoporous TiO₂ spheres, e-spray is advantageous because this technique can directly realize the formation of aggregated spheres by purely physical force, regardless of chemical composition and surface chemistry (Fig. 13b) [235]. Moreover, it has inherent flexibility in controlling the diameter and diameter distribution of the spheres by simply tuning the electrospray parameters, such as the feeding rate and the applied voltage [238, 239]. In addition, e-spray facilitates the adjustment of stacking density and porosity of the spheres on substrates through the proper selection of polymer additives [239]. For example, Hwang et al. have performed an extensive study on the fabrication of hierarchically structured mesoporous TiO₂ spheres by e-spray method (Fig. 14a, d), and further applied such photoanodes in solar cells, achieving efficiency as high as 10.57 % for DSSCs (Fig. 14b, c) [59], 5.57 % for flexible DSSCs (Fig. 14e, f) [240], and 7.59 % for solid-state DSSCs [241], respectively.

2.1.5 Other Methods

Vapor deposition describes any process in which materials in a vapor state are condensed to form a solid phase material, usually conducted within a vacuum chamber. If no chemical reaction occurs, this process is called physical vapor deposition (PVD); otherwise it is called chemical vapor deposition (CVD) [20, 35]. In PVD, materials are evaporated followed by condensation to form a solid material. The primary PVD methods include thermal deposition, ion plating, ion implantation, sputtering, laser vaporization, and laser surface alloying [20, 35]. PVD is considered an effective method to fabricate uniform, high-quality semi-conductors, especially composite and doped semiconductors. Moreover, PVD enables the formation of uniform nanostructured thin films, in which the size and shape can be precisely controlled [242]. Two versatile PVD methods, oblique angle deposition (OAD) and glancing angle deposition (GLAD), are based on the geo-metric shadowing effect, and are widely applied to prepare well-aligned nanorod arrays [243]. In OAD, the incident vapor flux is directed onto a substrate at a nonzero angle θ with respect to the substrate normal. When the vapor incident angle θ is large (i.e., $\theta > 70$), a well-defined nanorod array tilted toward the direction of the vapor flux can be obtained [244]. The GLAD method is similar to OAD in that it uses a large incident vapor angle, but the substrate is rotated azimuthally at a constant speed during the deposition. The result is vertically aligned nanorod arrays [245]. By changing the deposition parameters in these PVD methods, one can easily fabricate specific nanostructured porous array films such as cylinders, helices, spheres, and zigzags all with controllable surface areas [242, 243, 246, 247].

In addition, femtosecond and nanosecond pulsed laser deposition (PLD) at different wavelengths is also widely used to construct nanoparticle-assembled TiO_2 films. This technique is useful for controlling the dimensions and the crystalline phase of nanoparticles by varying the laser parameters and the deposition condi-tions. It is also suitable for depositing TiO_2 films at a high deposition rate and low cost [248–250]. Recently, a novel forest-like architecture consisting of hierarchical assemblies of tree-like nanocrystalline particles of anatase TiO_2, were grown on FTO substrates via pulsed laser deposition (PLD) at room temperature by ablation of a Ti target in a background O_2 atmosphere. The resulting architecture was proposed to be beneficial to reduce the electron recombination and also control mass transport in the mesopores, and thus achieved a 4.9 % conversion efficiency in a DSSC [251].

However, in CVD processes, thermal energy heats the gases in the coating chamber to induce the deposition reaction. Typical CVD approaches include electrostatic spray hydrolysis, diffusion flame pyrolysis, thermal plasma pyrolysis, ultrasonic spray pyrolysis, laser-induced pyrolysis, and atmospheric pressure and ultronsic-assisted hydrolysis [20, 35, 252, 253]. Several TiO_2 nanostructures prepared though CVD have been reported. For example, TiO_2 nanoparticles with sizes below 10 nm were prepared by pyrolysis of TTIP via CVD in a mixed helium/oxygen atmosphere [254]. TiO_2 nanorods were grown on a Si substrate using TTIP as the precursor by metal organic CVD (MOCVD) [64].

2.2 Modifications of TiO$_2$

As a promising semiconductor material, TiO$_2$ has attracted significant attention over the past several decades. However, the practical photovoltaic and photocatalysis applications of TiO$_2$ are limited by its wide band gap and the serious recombination of photogenerated electrons and holes [33, 34]. Because of its large band gap (3.2 eV for anatase, 3.0 eV for rutile), pure TiO$_2$ only absorbs ultraviolet light shorter than 387.5 nm for the anatase and 413.3 nm for the rutile [33]. Unfortunately, UV light constitutes less than 5 % of the solar energy that reaches the surface of the earth. This reduces the effective use of sunlight since visible light ($\lambda = 400$–700 nm) accounts for about 50 % of solar energy. Thus, it is necessary to develop titania-based photocatalysts which are active under visible light (i.e., broad spectrum). Furthermore, recombination of photogenerated charge carriers is another major limitation in TiO$_2$ semiconductor materials since it reduces the overall quantum efficiency of devices. In photocatalysis applications, recombination occurs when the excited electron reverts to the valence band without reacting with adsorbed species and the energy, non-radioactively or radioactively, dissipates as either light or heat [34]. However, in DSSC applications, photoexcited electron recombination in the electron transport process, including electron injection from the excited dye to the TiO$_2$ conduction band and electron transport from the conduction band to the conductive substrate, is regarded as one of the major obstacles to achieving high solar-to-electricity conversion efficiencies [255]. Recombination may occur either on the surface or in the bulk and is generally made worse by the presence of impurities, defects, and all other factors which introduce bulk or surface imperfections into the crystal.

To solve these problems, extensive efforts have been devoted to creating TiO$_2$-based visible-light-active photocatalytic materials and modifying nanostructured TiO$_2$ photoanodes to alleviate electron recombination [33, 256]. Currently, research interests focus mainly on modifying TiO$_2$ materials via (1) doping with cations (e.g., Fe [257–259], Cr [260], Eu [261], La [262], V [263], Mg [255], In [264]) or anions (e.g., S [265], C [266], F [267], B [268] and N [269]); (2) sensitization with organic dyes (e.g. N3, N719) [270], conducting polymers (e.g., poly(3-hexylthiophene) (P3HT) [271], nafion (perfluorinated polymer with sulfonate groups) [272, 273], polyaniline (PANI) [274], and carbon nitride polymer [275]), organic–inorganic hybrid dyes (e.g. copper(II) phthalocyanine) [276], or other semiconductors that absorb visible light (e.g. CdS [277–279], Cu$_2$O [280], Ag$_2$O [281], CdSe [282], PbZr$_{0.52}$Ti$_{0.48}$O$_3$(PZT) [283], Bi$_2$O$_3$ [284], BiOI [285], Bi$_2$WO$_6$ [286], CdTe [287], PbS [288], CuInS$_2$ [289], SnS [290], SnS$_2$ [291]); (3) decoration with noble metals (e.g. Au [292–294], Ag [65, 295–297], Pd [298–300], Pt [301–303]); (4) combination with other semiconductors (e.g., SiO$_2$ [304, 305], Al$_2$O$_3$ [100, 306], MgO [307], Fe$_2$O$_3$ [308], SrTiO$_3$ [309], Nb$_2$O$_5$ [310], SnO$_2$ [311], WO$_3$ [312], ZnO [313], and ZrO$_2$ [314]); and (5) synthesis of reduced TiO$_2$ (TiO$_{2-x}$, containing Ti^{3+} or O vacancies) [235, 315–318].

Among these methods, ion doping has been widely adopted to adjust the position of either the conduction band (CB) or valence band (VB) of TiO_2. This could have the desired effect of making the electrons excitable under visible light irradiation to produce photoelectron-hole pairs. The mechanism of the visible absorption is produced from interaction between the $2p$ orbitals of the dopant and the $2p$ orbitals of the oxygen in the newly formed valence band or the creation of dopant isolated states above the valence band maximum [319–321]. Nonmetal ion doping is usually considered more promising than metal ion doping because metal ion doping can introduce undesirable defects which can serve as the recombination centers for photoelectron–hole pairs, and thus reduce the pollutant degradation efficiency [322]. Nitrogen is the most promising dopant and can be easily introduced into the TiO_2 structures, due to its comparable atomic size with oxygen, small ionization energy and high stability [323]. Since Sato et al. found that the addition of NH_4OH to TiO_2 sol, followed by calcination of the precipitated powder, could generate a material which showed a visible light response, many strategies have been developed to produce N-doped TiO_2 materials [324, 325]. In the past decades, doping of nitrogen into TiO_2 structures has been realized via both wet and dry preparation methods. Physical techniques (i.e., sputtering and ion implantation) based on the direct treatment of TiO_2 with energetic nitrogen ions have also been developed [321, 326]. Meanwhile, gas phase reaction methods (i.e., atomic layer deposition and pulsed laser deposition) have been successfully used to prepare N-doped TiO_2 materials [269, 327]. The sol–gel method has proven to be the most versatile technique for the synthesis of N-TiO_2 nanoparticles because of its low cost, relatively simple equipment, and easy control of the resulting nanostructures [328]. In brief, the simultaneous growth of TiO_2 and N doping can be realized by hydrolysis of titanium precursors (i.e., titanium tetrachloride, titanium tetra-isopropoxide, tetrabutylorthotitanate) in the presence of nitrogen sources (i.e., aliphatic amines, nitrates, ammonium salts, ammonia, and urea) [328–330].

Recent studies have also revealed that doping TiO_2 with other elements, such as S, F, C, and B shifts the optical absorption edge to longer wavelengths [267, 331–333]. For example, F-doped flower-like TiO_2 nanostructures (Fig. 15a) have been synthesized in the presence of HF by a mild hydrothermal process and exhibited high photoelectrochemical activity for water-splitting and the photodegradation of organic pollutants (Fig. 15b) [334]. Mesoporous C-doped TiO_2 materials were prepared by a hydrothermal synthetic approach using sucrose as a carbon-doping source, followed by a post-thermal treatment. The resulting C-doped TiO_2 photocatalyst showed reduced recombination of electron–hole pairs due to the reduction of surface defects and promoted visible-light photocatalytic activity (Fig. 15c, d) [335]. In order to further improve the photocatalytic activity, co-doping TiO_2 with double non-metal elements (i.e., N–S [336], N–B [329, 337], F–N [338], C–N [339]) has attracted more attention. For example, F–N co-doped TiO_2 nanoparticles with dominant (001) facets were prepared by calcination a $TiOF_2$ precursor in NH_3 gas flow. The resulting nanoparticles showed drastically enhanced absorption and excellent water oxidation performance under visible light irradiation [338].

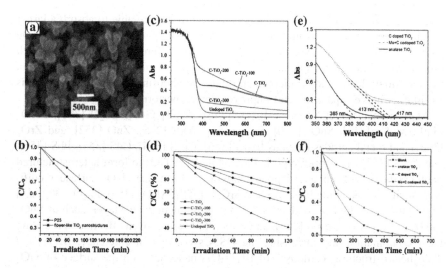

Fig. 15 SEM image of F-doped TiO_2 flowers (**a**). UV-visible absorption spectra of C-doped (**c**) and Mo, C-codoped TiO_2 (**e**), respectively. **b** The variation of 4-nitrophenol concentration by photoelectrocatalytic reaction with flower-like TiO_2 (**a**) and P-25. The visible light photocatalytic activities (**d**) of C-doped TiO_2 (**b**) treated at different temperatures in the degradation of gaseous toluene. Photodegradation of MB (**f**) by photochemical reaction with Mo, C–codoped TiO_2 (**e**). (Reprinted with permission from Ref. **a, b** [334], **c, d** [335], **e, f** [343]. Copyright © American Chemical Society)

It has been shown that nonmetal-doped TiO_2 shows a redshift of the onset and a higher absorption in the visible light spectrum. Metal doped TiO_2 would possess lower energy levels so that electrons and holes can be excited by low energy photons, which also increases the absorption of visible light [340]. For example, $Mo–TiO_2$ core-shell nanoparticles were prepared by the arc-discharge method and showed enhanced photocatalytic activity under visible light, due to the Mo-doping in (001) TiO_2 from diffusion at the shell–core interface [341], while many studies have revealed that single doping will increase recombination sites inside the TiO_2, which will therefore increase the charge recombination [260]. Conversely, it has also been demonstrated that co-doping TiO_2 with both nonmetal anions and metal cations can reduce the recombination sites because of the neutralization of positive and negative charges inside TiO_2. This can effectively improve the charge transport efficiency and thus enhance the photocatalytic activity [342]. For example, the Mo–C co-doped TiO_2 powders prepared by thermal oxidation of a mixture of TiC and MoO_3 in the air have the potential of visible light harvesting (Fig. 15e) and effective photoexcited charge separation, and can thus exhibit higher photocatalytic activity when compared with anatase TiO_2 (Fig. 15f) [343].

The fabrication of semiconductor heterostructures is one of many effective methods developed in recent years to photoexcite and separate the electro-hole pairs. Compared to a single semiconductor, heterogeneous semiconductors are ideal for light-harvesting devices such as photovoltaic and photoelectrochemical cells, because heterogeneous semiconductor electrodes are able to absorb a larger fraction of the solar spectrum and thus generate more photoinduced electron-hole pairs [344, 345]. Moreover, the coupling of two different semiconductors with proper conduction band potentials facilitates the transfer of electrons from an excited small band gap semiconductor into neighboring semiconductors. This facilitates charge separation and improves device performance.

Many metal oxides (such as Bi_2O_3 [346], Al_2O_3 [347], Cu_2O [348], Fe_2O_3 [349], MoO_3 [312], SnO_2 [350], SiO_2 [351], WO_3 [245], ZnO [352], and ZrO_2 [353]) and metal sulfides (such as Bi_2S_3 [346], Cu_2S [354], CdS [355], PbS [356], and Ag_2S [357]) have been reported to couple with TiO_2 to form heterostructured photocatalysts with enhanced photocatalytic performance. TiO_2 surface modification with an insulating layer, such as $SrCO_3$ [358], Al_2O_3 [359], La_2O_3 [302] and MgO [361], or a higher conduction band edge semiconductor layer, such as SnO_2 [362], In_2O_3 [314], Nb_2O_5 [226], and ZnO [363], was proven effective in reducing the recombination and increasing the DSSCs conversion efficiency. $Bi_4Ti_3O_{12}/$ TiO_2 heterostructures composed of $Bi_4Ti_3O_{12}$ nanosheets on the surface of TiO_2 submicron fibers were prepared via a facile two-step synthesis route combining an electrospinning method and hydrothermal process. These heterostructures showed a higher degradation rate of rhodamine B (RhB) than the pure TiO_2 submicron fibers under visible light. This is largely due to the extended absorption in the visible light spectrum resulting from the $Bi_4Ti_3O_{12}$ nanosheets, and the effective separation of photoexcited charges driven by the photoinduced potential difference generated at the $Bi_4Ti_3O_{12}/TiO_2$ interface [364]. TiO_2 nanotube arrays sensitized with $ZnFe_2O_4$ nanocrystals (Fig. 16a) were successfully fabricated by a two-step process of anodization and vacuum-assisted impregnation followed by annealing. It has been shown that the $ZnFe_2O_4$ sensitization enhanced the photoinduced charge separation (Fig. 16b, c) and extended the range of the photoresponse of TiO_2 nanotube arrays from the UV to the visible region [365]. TiO_2-multiwalled carbon nanotube (MWCNT) nanocomposites (Fig. 16d) synthesized by hydrothermal processes possess a 50 % enhancement in the conversion efficiency (4.9–7.37 %) of DSSCs compared to hydrothermally synthesized TiO_2 without MWCNTs and Degussa P25. Efficient charge transfer in the nanocomposites is a possible reason for the enhancement (Fig. 16e, f) [366].

As the most recently discovered carbonaceous material, graphene has attracted extensive attention as a useful material for solar energy applications. With a unique sp^2 hybrid carbon network, a large theoretical specific surface area ($2,630 \ m^2g^{-1}$), a high thermal conductivity ($5,000 \ Wm^{-1}K^{-1}$), a large intrinsic electron mobility ($200, \ 000 \ cm^2V^{-1}s^{-1}$), and good mechanical stability, it has applications in sensors, catalysts, and energy conversion [223]. Graphene can serve as a strong electron collector and carrier in a TiO_2/graphene composite system because their energy levels and physical properties are compatible [367]. Graphene-TiO_2 composites are also highly desirable for their promising energy and environmental remediation applications.

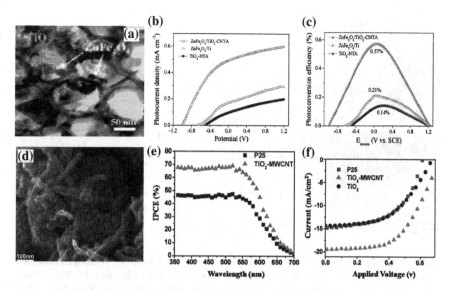

Fig. 16 SEM images of $ZnFe_2O_4/TiO_2$ nanotubes (**a**) and TiO_2 MWCNT nanocomposite (**d**). **b** Variation of the photocurrent density versus bias potential (versus SCE) and **c** Photoconversion efficiency as a function of the applied potential (versus SCE) based on (**a**). IPCE curves (**e**) and current–voltage characteristics (**f**) of DSSCs based on TiO_2 MWCNT nanocomposite (**c**). (Reprinted with permission from Ref. **a–c** [365], **d–f** [366]. Copyright © American Chemical Society)

Several synthetic strategies have been designed to fabricate graphene-TiO_2 photocatalysts. In the first method, well-defined TiO_2 structures are deposited on the surface of graphene oxide (GO) under vigorous stirring or ultrasonic agitation [368–370]. The site-specific oxygenated groups on GO favor a uniform distribution of TiO_2 across the surface. Graphene-TiO_2 photocatalysts are obtained after the reduction of GO in the composite [371, 372]. Yang et al. packed TiO_2 and graphene nanosheets into a 2D unit (TiO_2/graphene) that is structurally similar to a thylakoid in the chloroplast of photosynthetic plants. In this 2D unit, TiO_2 performs as a photo-electric conversion center to absorb light and excite the electrons, while graphene is like the cytochrome b6f complex capturing electrons and transporting them out of the circuit. Such a novel structure was formed by stacking TiO_2 nanosheets and GO nanosheets using a layer-by-layer (LBL) assembly technique in the presence of charged poly(diallyldimethylammonium chloride) (PDDA) which supplied the counter-ions. GO was reduced to graphene by hydrazine and annealed under argon flow at 400 °C and the PDDA was then removed by calcining at 450 °C in air. The graphene-TiO_2 stacking film can produce an anodic current 20 times larger than pure TiO_2 stacking films. Interestingly, the current further increased with thicker films [373]. Another significant example, graphene-wrapped anatase TiO_2 nanoparticles (Fig. 17a) with a significant reduction in the band gap (2.80 eV, Fig. 17b) were prepared by wrapping

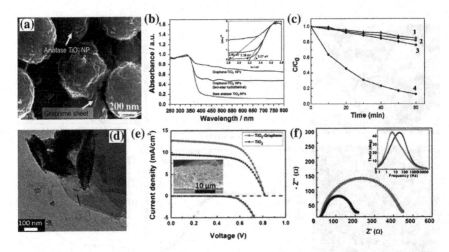

Fig. 17 **a** SEM image, **b** UV-visible spectra (the *inset* is the band gap (E_g) of samples estimated from the absorption edge), and **c** photodegradation of MB under visible light (1: P25, 2: bare anatase TiO_2 NPs, 3: graphene-TiO_2 NPs (two-step hydrothermal), 4: graphene-TiO_2 NPs). **d** TEM image of TiO_2-graphene composite, **e** current–voltage characteristic, and **f** Nyquist plots of DSSCs using TiO_2-graphene composite (the *inset* is the corresponding Bode plots in electrochemical impedance spectra (EIS) test). (Reprinted with permission from Ref. **a–c** [374], **d–f** [377]. Copyright © American Chemical Society and Wiley-VCH)

amorphous TiO_2 NPs with GO, followed by continuous GO reduction and TiO_2 crystallization via hydrothermal treatment. The graphene-TiO_2 nanoparticles possess excellent photocatalytic properties under visible light for the degradation of MB (Fig. 17c) [374].

In the second method, graphene-TiO_2 photocatalysts are fabricated by an in situ growth of TiO_2 on graphene sheets [375, 376]. By introducing cetyltrimethylammonium bromide (CTAB)-functionalized DMF-soluble graphene into the polymeric solution for electrospinning, graphene was successfully integrated into the TiO_2 rice-shaped nanostructures (Fig. 17d). The obtained composites displayed enhanced photovoltaic and photocatalytic properties compared to pure TiO_2 nanorices when used in DSSCs (Fig. 17e, f) and in the photocatalytic degradation of methyl orange (MO) [377]. In addition, a series of graphene-TiO_2 composites with different graphene contents can be controllably synthesized by a sol–gel method [378]. Graphene-TiO_2 composites demonstrated a higher photocatalytic activity compared to P25 with respect to hydrogen generation from water splitting. The highest photocatalytic activity was observed for the sample with 5 % graphene, suggesting that an excess of graphene will decrease the activity by introducing electron–hole recombination centers into the composite [378].

In the third method, TiO_2 structures are grown in situ onto GO followed by the reduction of GO in a subsequent reaction step using UV light or microwave irradiation [379, 380]. For example, novel hollow spheres consisting of $Ti_{0.91}O_2$ nanosheets and graphene nanosheets (Fig. 18a) were successfully fabricated by a

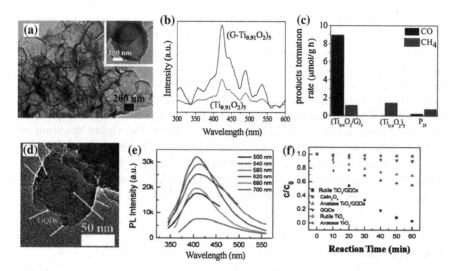

Fig. 18 TEM images of (G-Ti$_{0.91}$O$_2$)$_5$ hollow spheres (**a**) and rutile TiO$_2$/graphene quantum dot composites (**d**). **b** PL emission spectra of (Ti$_{0.91}$O$_2$)$_5$ hollow spheres and (G-Ti$_{0.91}$O$_2$)$_5$ hollow spheres (**a**). **c** Photocatalytic CH$_4$ and CO evolution rates for (Ti$_{0.91}$O$_2$)$_5$ hollow spheres (G-Ti$_{0.91}$O$_2$)$_5$ hollow spheres (**a**) and P25. **e** Upconverted PL spectra of the graphene quantum dots (**b**) at different excitation wavelengths and **f** photocatalytic degradation of MB under visible light irradiation using different catalysts. (Reprinted with permission from Ref. **a–c** [380], **d–f** [388]. Copyright © Wiley-VCH and American Chemical Society)

layer-by-layer assembly technique with polymer PMMA beads as sacrificial templates. Subsequently, microwave irradiation was used to simultaneously remove the template and reduce graphene oxide into graphene. The sufficiently compact stacking of ultrathin Ti$_{0.91}$O$_2$ nanosheets with graphene nanosheets facilitated the photogenerated electron to quickly transfer from the Ti$_{0.91}$O$_2$ nanosheets to graphene and enhance the lifetime of the charge carriers and improve photocatalytic activity (Fig. 18b, c) [380].

In the fourth method, the in situ growth of TiO$_2$ structures and reduction of GO are simultaneously accomplished through a simple one-pot growth method [381, 382]. This strategy is a particularly convenient procedure for the fabrication of graphene-TiO$_2$ photocatalysts. The reducing environment can be achieved by using either a reducing solvent or by directly adding a reducing agent [383, 384]. For example, exposed (001) facet TiO$_2$-graphene composite photocatalysts were successfully produced via a direct one-step hydrothermal method in an ethanol–water solvent. The resulting composite exhibited an extended visible light absorption range due to the formation of a chemical Ti–O–C bond and enhanced charge separation by virtue of the formation of nano-sized Schottky interfaces at the contacts between TiO$_2$ and graphene. This leads to significant improvement in photodegredation of MB (or MO) dye when compared to P25 films under both UV and visible light irradiation [367].

Recently, graphene quantum dots (GQDs) have become an active area of research. Compared to graphene with 2D nanosheets and 1D nanoribbon structures, 0D GQDs possess strong quantum confinement and edge effects when their sizes approach 10 nm or less. At this scale, new physical properties become apparent [385–387]. Zhuo et al. presented the synthesis of GQDs with excellent PL properties using a direct and simple ultrasonic reaction process. The composite photocatalysts (rutile TiO_2/GQD and anatase TiO_2/GQD systems) were prepared by mixing GQDs with TiO_2 nanoparticles (Fig. 18d) in order to harvest the visible spectrum of sunlight using the upconversion photoluminescence (PL) properties of GQDs (Fig. 18e). It was observed that the photocatalytic activity of the rutile TiO_2/GQD composite was superior to that of the anatase TiO_2/GQD composite under visible light ($\lambda > 420$ nm) irradiation in the degradation of MB (Fig. 18f) [388].

3 Conclusion

The challenge of effectively converting solar energy into useful energy will require more advanced materials. As one of the most promising functional materials for efficient, high performance, and cost-effective solar energy applications, TiO_2 nanostructures of various morphologies will continue to be an area of active research. It has been shown that every structure possesses its own advantages, such as large surface area for nanoparticle systems, effective charge separation, and transport for vertically ordered nanotubes or nanorods, a high percentage of reactive (001) facets for nanosheets, and efficient light scattering for 3D hierarchical structures, to name a few. Correspondingly, the different morphologies are derived from different characteristic synthesis methods. The sol–gel method, for example, is a simple and low-cost process to prepare nanoparticles. The hydrothermal method can be used to fabricate several structures (e.g., nanosheet, nanorod and hierarchical architecture), whereas the facile electrochemical anodization can lead to oriented nanotube arrays. Clearly, numerous strategies have been exploited to modify pure TiO_2, which suffers from the shortcomings of a narrow light absorption range (restricted to UV) and rapid charge recombination. With more efforts geared toward the fabrication of high-quality TiO_2-based nanomaterials and economically feasible synthesis procedures, there is little doubt that the future of solar energy technology and its practical applications will remain bright.

References

1. Liu J, Cao GZ, Yang ZG, Wang DH, Dubois D, Zhou XD, Graff GL, Pederson LR, Zhang JG (2008) ChemSusChem 1:676–697
2. Hagfeldt A, Boschloo G, Sun LC, Kloo L, Pettersson H (2010) Chem Rev 110:6595–6663
3. Kamat PV (2007) J Phys Chem C 111:2834–2860

4. Nozik A, Miller J (2010) Chem Rev 110:6443–6445
5. Liu C, Burghaus U, Besenbacher F, Wang ZL (2010) ACS Nano 4:5517–5526
6. Serrano E, Rus G, García-Martínez J (2009) Renew Sustain Energy Rev 13:2373–2384
7. Beard MC, Ellingson RJ (2008) Laser Photonics Rev 2:377–399
8. Bisquert J, Cahen D, Hodes G, Ruhle S, Zaban A (2004) J Phys Chem B 108:8106–8118
9. Gratzel M (2001) Nature 414:338–344
10. Green MA (2001) Prog Photovoltaics Res Appl 9:123–135
11. El Chaar L, El Zein N (2011) Renew Sustain Energy Rev 15:2165–2175
12. Lewis NS (2007) Science 315:798–801
13. Gratzel M (2006) Prog Photovoltaics Res Appl 14:429–442
14. Oregan B, Gratzel M (1991) Nature 353:737–740
15. Yella A, Lee H-W, Tsao HN, Yi C, Chandiran AK, Nazeeruddin MK, Diau EW-G, Yeh C-Y, Zakeeruddin SM, Grätzel M (2011) Science 334:629–634
16. Grätzel M (2003) J Photochem Photobiol C 4:145–153
17. Toivola M, Halme J, Miettunen K, Aitola K, Lund PD (2009) Int J Energy Res 33:1145–1160
18. Halme J, Vahermaa P, Miettunen K, Lund P (2010) Adv Mater 22:E210–E234
19. O'Regan BC, Durrant JR (2009) Acc Chem Res 42:1799–1808
20. Chen XB (2009) Chin J Catal 30:839–851
21. Galinska A, Walendziewski J (2005) Energy Fuels 19:1143–1147
22. Yang XH, Li Z, Liu G, Xing J, Sun C, Yang HG, Li C (2010) CrystEngComm 13:1378–1383
23. Chen XB, Shen SH, Guo LJ, Mao SS (2010) Chem Rev 110:6503–6570
24. Fujishima A, Honda K (1972) Nature 238:37
25. Ni M, Leung MKH, Leung DYC, Sumathy K (2007) Renew Sustain Energy Rev 11:401–425
26. Zhang QJ, Sun CH, Yan J, Hu XJ, Zhou SY, Chen P (2010) Solid State Sci 12:1274–1277
27. Lee JS (2005) Catal Surv Asia 9:217–227
28. Walter MG, Warren EL, McKone JR, Boettcher SW, Mi Q, Santori EA, Lewis NS (2010) Chem Rev 110:6446
29. Ye M, Vennerberga D, Lin C, Lin Z (2012) J Nanosci Lett 2:1
30. Youngblood WJ, Lee SHA, Maeda K, Mallouk TE (2009) Acc Chem Res 42:1966–1973
31. Kaur A, Gupta U (2009) J Mater Chem 19:8279–8289
32. Akpan UG, Hameed BH (2009) J Hazard Mater 170:520–529
33. Pelaez M, Nolan NT, Pillai SC, Seery MK, Falaras P, Kontos AG, Dunlop PSM, Hamilton JWJ, Byrne JA, O'shea K (2012) Appl Catal B 125:331–349
34. Hu X, Li G, Yu JC (2009) Langmuir 26:3031–3039
35. Chen X, Mao SS (2007) Chem Rev 107:2891–2959
36. Lv M, Zheng D, Ye M, Sun L, Xiao J, Guo W, Lin C (2012) Nanoscale 4:5872–5879
37. Chou TP, Zhang QF, Russo B, Fryxell GE, Cao GZ (2007) J Phys Chem C 111:6296–6302
38. Mor GK, Varghese OK, Paulose M, Shankar K, Grimes CA (2006) Sol Energy Mater Sol Cells 90:2011–2075
39. Shankar K, Basham JI, Allam NK, Varghese OK, Mor GK, Feng XJ, Paulose M, Seabold JA, Choi KS, Grimes CA (2009) J Phys Chem C 113:6327–6359
40. Fei H, Yang Y, Rogow DL, Fan X, Oliver SRJ (2010) ACS Appl Mater Interfaces 2:974–979
41. Nozik AJ (2010) Nano Lett 10:2735–2741
42. Arico AS, Bruce P, Scrosati B, Tarascon JM, Van Schalkwijk W (2005) Nat Mater 4:366–377
43. Guo YG, Hu JS, Wan LJ (2008) Adv Mater 20:2878–2887
44. Xin X, He M, Han W, Jung J, Lin Z (2011) Angew Chem Int Ed 50:11739–11742
45. Choi SK, Kim S, Lim SK, Park H (2010) J Phys Chem C 114:16475–16480
46. Xin X, Wang J, Han W, Ye M, Lin Z (2012) Nanoscale 4:964–969
47. Xin X, Scheiner M, Ye M, Lin Z (2011) Langmuir 27:14594–14598
48. Hartmann P, Lee DK, Smarsly BM, Janek J (2010) ACS Nano 4:3147–3154

49. Alivov Y, Fan ZY (2009) J Phys Chem C 113:12954–12957
50. Li Y, Fang XS, Koshizaki N, Sasaki T, Li L, Gao SY, Shimizu Y, Bando Y, Golberg D (2009) Adv Funct Mater 19:2467–2473
51. Kumar A, Madaria AR, Zhou CW (2010) J Phys Chem C 114:7787–7792
52. Albu SP, Roy P, Virtanen S, Schmuki P (2010) Isr J Chem 50:453–467
53. Yu JG, Fan JJ, Lv KL (2010) Nanoscale 2:2144–2149
54. Bleta R, Alphonse P, Lorenzato L (2010) J Phys Chem C 114:2039–2048
55. Isley SL, Penn RL (2008) J Phys Chem C 112:4469–4474
56. Liu JJ, Qin W, Zuo SL, Yu YC, Hao ZP (2009) J Hazard Mater 163:273–278
57. Wang J, Lin ZQ (2009) J Phys Chem C 113:4026–4030
58. Zhang W, Zhu R, Ke L, Liu XZ, Liu B, Ramakrishna S (2010) Small 6:2176–2182
59. Hwang D, Lee H, Jang SY, Jo SM, Kim D, Seo Y, Kim DY (2011) ACS Appl Mater Interfaces 3:2719–2725
60. Wu MS, Tsai CH, Wei TC (2011) Chem Commun 47:2871–2873
61. Bala H, Jiang L, Fu WY, Yuan GY, Wang XD, Liu ZR (2010) Appl Phys Lett 97:153108
62. Suprabha T, Roy HG, Thomas J, Kumar KP, Mathew S (2009) Nanoscale Res Lett 4:144–152
63. Melhem H, Simon P, Beouch L, Goubard F, Boucharef M, Di Bin C, Leconte Y, Ratier B, Herlin-Boime N, Bouclon J (2011) Adv Energy Mater 1:908–916
64. Pradhan SK, Reucroft PJ (2003) J Cryst Growth 250:588–594
65. Yu DH, Yu X, Wang C, Liu XC, Xing Y (2012) ACS Appl Mater Interfaces 4:2781–2787
66. Koo B, Park J, Kim Y, Choi SH, Sung YE, Hyeon T (2006) J Phys Chem B 110:24318–24323
67. Ahn SH, Chi WS, Park JT, Koh JK, Roh DK, Kim JH (2012) Adv Mater 24:519–522
68. Guldin S, Huttner S, Kolle M, Welland ME, Muller-Buschbaum P, Friend RH, Steiner U, Tétreault N (2010) Nano Lett 10:2303–2309
69. Halaoui LI, Abrams NM, Mallouk TE (2005) J Phys Chem B 109:6334–6342
70. Ismagilov ZR, Tsikoza LT, Shikina NV, Zarytova VF, Zinoviev VV, Zagrebelnyi SN (2009) Russ Chem Rev 78:873–885
71. Han S, Choi SH, Kim SS, Cho M, Jang B, Kim DY, Yoon J, Hyeon T (2005) Small 1:812–816
72. Jin WM, Shin JH, Cho CY, Kang JH, Park JH, Moon JH (2010) ACS Appl Mater Interfaces 2:2970–2973
73. Hatton B, Mishchenko L, Davis S, Sandhage KH, Aizenberg J (2010) PNAS 107:10354–10359
74. Shopsowitz KE, Stahl A, Hamad WY, MacLachlan MJ (2012) Angew Chem Int Ed 51:6886–6890
75. Ahn SH, Park JT, Koh JK, Roh DK, Kim JH (2011) Chem Commun 47:5882–5884
76. Agarwala S, Kevin M, Wong A, Peh C, Thavasi V, Ho G (2010) ACS Appl Mater Interfaces 2:1844–1850
77. Ismail AA, Bahnemann DW (2011) J Mater Chem 21:11686–11707
78. Park JT, Chi WS, Roh DK, Ahn SH, Kim JH (2012) Adv Funct Mater 23:26–33
79. Kim YJ, Lee YH, Lee MH, Kim HJ, Pan JH, Lim GI, Choi YS, Kim K, Park NG, Lee C (2008) Langmuir 24:13225–13230
80. Yang SC, Yang DJ, Kim J, Hong JM, Kim HG, Kim ID, Lee H (2008) Adv Mater 20:1059–1064
81. Mandlmeier B, Szeifert JM, Fattakhova-Rohlfing D, Amenitsch H, Bein T (2011) J Am Chem Soc 133:17274–17282
82. Xiong Z, Dou H, Pan J, Ma J, Xu C, Zhao X (2010) CrystEngComm 12:3455–3457
83. Sjöström T, McNamara LE, Yang L, Dalby M, Su B (2012) ACS Appl Mater Interfaces 4:6354–6361
84. Jang YH, Xin X, Byun M, Jang YJ, Lin Z, Kim DH (2011) Nano Lett 12:479–485
85. Hayward RC, Chmelka BF, Kramer EJ (2005) Adv Mater 17:2591–2595

86. Zhao D, Feng D, Luo W, Zhang J, Xu M, Zhang R, Wu H, Lv Y, Asiri AM, Rahman M (2013) J Mater Chem A 1:1591–1599
87. Cha MA, Shin C, Kannaiyan D, Jang YH, Kochuveedu ST, Kim DH (2009) J Mater Chem 19:7245–7250
88. Guldin S, Docampo P, Stefik M, Kamita G, Wiesner U, Snaith HJ, Steiner U (2012) Small 3:432–440
89. Chen Y, Kim HC, McVittie J, Ting C, Nishi Y (2010) Nanotechnology 21:185303
90. Ahmed S, Du Pasquier A, Birnie DP III, Asefa T (2011) ACS Appl Mater Interfaces 3:3002–3010
91. Dutta S, Patra AK, De S, Bhaumik A, Saha B (2012) ACS Appl Mater Interfaces 4:1560–1564
92. Kwak ES, Lee W, Park NG, Kim J, Lee H (2009) Adv Funct Mater 19:1093–1099
93. Mihi A, Zhang C, Braun PV (2011) Angew Chem Int Ed 123:5830–5833
94. Campbell M, Sharp D, Harrison M, Denning R, Turberfield A (2000) Nature 404:53–56
95. Nishimura S, Abrams N, Lewis BA, Halaoui LI, Mallouk TE, Benkstein KD, van de Lagemaat J, Frank AJ (2003) J Am Chem Soc 125:6306–6310
96. Lee SHA, Abrams NM, Hoertz PG, Barber GD, Halaoui LI, Mallouk TE (2008) J Phys Chem B 112:14415–14421
97. Shin JH, Moon JH (2011) Langmuir 27:6311–6315
98. Chang SY, Chen SF, Huang YC (2011) J Phys Chem C 115:1600–1607
99. Cho CY, Moon JH (2011) Adv Mater 23:2971–2975
100. Berrigan JD, McLachlan TM, Deneault JR, Cai Y, Kang TS, Durstock MF, Sandhage K (2013) J Mater Chem A 1:128–134
101. Xu C, Shin PH, Cao L, Wu J, Gao D (2009) Chem Mater 22:143–148
102. Shao W, Gu F, Gai L, Li C (2011) Chem Commun 47:5046–5048
103. Joo JB, Zhang Q, Lee I, Dahl M, Zaera F, Yin Y (2012) Adv Funct Mater 22:166–174
104. Thompson GE (1997) Thin Solid Films 297:192–201
105. Koh JH, Koh JK, Seo JA, Shin JS, Kim JH (2011) Nanotechnology 22:365401
106. Wang H (2012) CrystEngComm 14:6215–6220
107. Bian Z, Zhu J, Cao F, Huo Y, Lu Y, Li H (2010) Chem Commun 46:8451–8453
108. Tétreault N, Arsenault E, Heiniger LP, Soheilnia N, Brillet J, Moehl T, Zakeeruddin S, Ozin GA, Grätzel M (2011) Nano Lett 11:4579–4584
109. Liu L, Karuturi SK, Su LT, Tok AIY (2010) Energy Environ Sci 4:209–215
110. Tan LK, Liu X, Gao H (2011) J Mater Chem 21:11084–11087
111. Li L, Liu C (2009) J Phys Chem C 114:1444–1450
112. Yang W, Li J, Wang Y, Zhu F, Shi W, Wan F, Xu D (2011) Chem Commun 47:1809–1811
113. Yang H, Fang W, Yang X, Zhu H, Li Z, Zhao H, Yao X (2012) J Mater Chem 22:22082–22089
114. Wu B, Guo C, Zheng N, Xie Z, Stucky GD (2008) J Am Chem Soc 130:17563–17567
115. Yu J, Xiang Q, Ran J, Mann S (2010) CrystEngComm 12:872–879
116. Wang J, Bian Z, Zhu J, Li H (2013) J Mater Chem A 1:1296–1302
117. Jun Y, Casula MF, Sim JH, Kim SY, Cheon J, Alivisatos AP (2003) J Am Chem Soc 125:15981–15985
118. Wu D, Gao Z, Xu F, Chang J, Jiang K (2012) CrystEngComm 15:516–523
119. Nian JN, Teng HS (2006) J Phys Chem B 110:4193–4198
120. Horvath E, Kukovecz A, Konya Z, Kiricsi I (2007) Chem Mater 19:927–931
121. Guo WX, Xu C, Wang X, Wang SH, Pan CF, Lin CJ, Wang ZL (2012) J Am Chem Soc 134:4437–4441
122. Liu B, Aydil ES (2009) J Am Chem Soc 131:3985–3990
123. Chen D, Huang F, Cheng YB, Caruso RA (2009) Adv Mater 21:2206–2210
124. Liu M, Piao L, Lu W, Ju S, Zhao L, Zhou C, Li H, Wang W (2010) Nanoscale 2:1115–1117
125. Ye M, Liu HY, Lin C, Lin Z (2012) Small 9:312–321
126. Zhang D, Li G, Yang X, Jimmy CY (2009) Chem Commun 4381–4383

127. Zheng Z, Huang B, Qin X, Zhang X, Dai Y, Jiang M, Wang P, Whangbo MH (2009) Chem Eur J 15:12576–12579
128. Diebold U (2003) Surf Sci Rep 48:53–229
129. Gong XQ, Selloni A (2005) J Phys Chem B 109:19560–19562
130. Yang HG, Sun CH, Qiao SZ, Zou J, Liu G, Smith SC, Cheng HM, Lu GQ (2008) Nature 453:638–641
131. Gu L, Wang J, Cheng H, Du Y, Han X (2012) Chem Commun 48:6978–6980
132. He Z, Cai Q, Hong F, Jiang Z, Chen J, Song S (2012) Ind Eng Chem Res 51:5662–5668
133. Xiang Q, Yu J, Jaroniec M (2011) Chem Commun 47:4532–4534
134. Zheng Z, Huang B, Qin X, Zhang X, Dai Y, Jiang M, Wang P, Whangbo MH (2009) Chem Eur J 15:12576–12579
135. Zhang H, Han Y, Liu X, Liu P, Yu H, Zhang S, Yao X, Zhao H (2010) Chem Commun 46:8395–8397
136. Liu M, Piao L, Zhao L, Ju S, Yan Z, He T, Zhou C, Wang W (2010) Chem Commun 46:1664–1666
137. Li J, Cao K, Li Q, Xu D (2012) CrystEngComm 14:83–85
138. Li J, Xu D (2010) Chem Commun 46:2301–2303
139. Ma XY, Chen ZG, Hartono SB, Jiang HB, Zou J, Qiao SZ, Yang HG (2010) Chem Commun 46:6608–6610
140. Yu H, Tian B, Zhang J (2011) Chem Eur J 17:5499–5502
141. Kumar EN, Jose R, Archana P, Vijila C, Yusoff M, Ramakrishna S (2012) Energy Environ Sci 5:5401–5407
142. Liu B, Miao J (2012) RSC Adv 3:1222–1226
143. Dai Y, Cobley CM, Zeng J, Sun Y, Xia Y (2009) Nano Lett 9:2455–2459
144. Amano F, Prieto-Mahaney OO, Terada Y, Yasumoto T, Shibayama T, Ohtani B (2009) Chem Mater 21:2601–2603
145. Jung MH, Chu MJ, Kang MG (2012) Chem Commun 48:5016–5018
146. Xie S, Han X, Kuang Q, Fu J, Zhang L, Xie Z, Zheng L (2011) Chem Commun 47:6722–6724
147. Pan J, Wu X, Wang L, Liu G, Lu GQM, Cheng HM (2011) Chem Commun 47:8361–8363
148. Wang L, Zang L, Zhao J, Wang C (2012) Chem Commun 48:11736–11738
149. Li F, Xu J, Chen L, Ni B, Li X, Fu Z, Lu Y (2013) J Mater Chem A 1:225–228
150. Berger S, Hahn R, Roy P, Schmuki P (2010) Phys Status Solidi B 247:2424–2435
151. Peng XS, Wang JP, Thomas DF, Chen AC (2005) Nanotechnology 16:2389–2395
152. Gong D, Grimes CA, Varghese OK, Hu WC, Singh RS, Chen Z, Dickey EC (2001) J Mater Res 16:3331–3334
153. Wu XJ, Zhu F, Mu C, Liang YQ, Xu LF, Chen QW, Chen RZ, Xu DS (2010) Coord Chem Rev 254:1135–1150
154. Peter LM, Jennings JR, Ghicov A, Schmuki P, Walker AB (2008) J Am Chem Soc 130:13364–13372
155. Xiao P, Zhang YH, Garcia BB, Sepehri S, Liu DW, Cao GZ (2009) J Nanosci Nanotechnol 9:2426–2436
156. Allam NK, El-Sayed MA (2010) J Phys Chem C 114:12024–12029
157. Lai Y, Sun L, Chen Y, Zhuang H, Lin C (2006) J Electrochem Soc 153:D123–D127
158. Gong JJ, Lai YK, Lin CJ (2010) Electrochim Acta 55:4776–4782
159. Gong JJ, Lin CJ, Ye MD, Lai YK (2011) Chem Commun 47:2598–2600
160. Guo WX, Xue XY, Wang SH, Lin CJ, Wang ZL (2012) Nano Lett 12:2520–2523
161. Rattanavoravipa T, Sagawa T, Yoshikawa S (2008) Sol Energy Mater Sol Cells 92:1445–1449
162. Wender H, Feil AF, Diaz LB, Ribeiro CS, Machado GJ, Migowski P, Weibel DE, Dupont J, Teixeira SR (2011) ACS Appl Mater Interfaces 3:1359–1365
163. Tan YF, Yang L, Chen JZ, Qiu Z (2010) Langmuir 26:10111–10114
164. Alivov Y, Pandikunta M, Nikishin S, Fan ZY (2009) Nanotechnology 20:225602
165. Bao NZ, Yoriya S, Grimes CA (2011) J Mater Chem 21:13909–13912

166. Wang J, Zhao L, Lin VSY, Lin ZQ (2009) J Mater Chem 19:3682–3687
167. Wang J, Lin Z (2012) Chem Asian J 7:2754–2762
168. Su ZX, Zhou WZ (2011) J Mater Chem 21:8955–8970
169. Roy P, Kim D, Lee K, Spiecker E, Schmuki P (2010) Nanoscale 2:45–59
170. Roy P, Berger S, Schmuki P (2011) Angew Chem Int Ed 50:2904–2939
171. Nah YC, Paramasivam I, Schmuki P (2010) ChemPhysChem 11:2698–2713
172. Schmuki P, Macak JM, Tsuchiya H, Taveira L, Aldabergerova S (2005) Angew Chem Int Ed 44:7463–7465
173. Grimes CA, Allam NK, Shankar K (2008) J Mater Chem 18:2341–2348
174. Schmuki P, Kim D, Ghicov A, Albu SP (2008) J Am Chem Soc 130:16454
175. Fei GT, Jin Z, Hu XY, Xu SH, De Zhang L (2009) Chem Lett 38:288–289
176. Lin J, Liu K, Chen XF (2011) Small 7:1784–1789
177. Xu XJ, Tang CC, Zeng HB, Zhai TY, Zhang SQ, Zhao HJ, Bando Y, Golberg D (2011) ACS Appl Mater Interfaces 3:1352–1358
178. Schmuki P, Albu SP, Ghicov A, Aldabergenova S, Drechsel P, LeClere D, Thompson GE, Macak JM (2008) Adv Mater 20:4135
179. Li SQ, Zhang GM, Guo DZ, Yu LG, Zhang W (2009) J Phys Chem C 113:12759–12765
180. Stergiopoulos T, Ghicov A, Likodimos V, Tsoukleris DS, Kunze J, Schmuki P, Falaras P (2008) Nanotechnology 19:235602
181. Sun Y, Yan KP, Wang GX, Guo W, Ma TL (2011) J Phys Chem C 115:12844–12849
182. Biswas S, Shahjahan M, Hossain MF, Takahashi T (2010) Electrochem Commun 12:668–671
183. Chen CH, Chen KC, He JL (2010) Curr Appl Phys 10:S176–S179
184. Stergiopoulos T, Valota A, Likodimos V, Speliotis T, Niarchos D, Skeldon P, Thompson GE, Falaras P (2009) Nanotechnology 20:365601
185. Tang YX, Tao J, Tao HJ, Wu T, Wang L, Zhang YY, Li ZL, Tian XL (2008) Acta Phys Chim Sin 24:1120–1126
186. Leenheer AJ, Miedaner A, Curtis CJ, van Hest M, Ginley DS (2007) J Mater Res 22:681–687
187. Mor GK, Varghese OK, Paulose M, Grimes CA (2005) Adv Funct Mater 15:1291–1296
188. Varghese OK, Paulose M, Grimes CA (2009) Nat Nanotechnol 4:592–597
189. Sadek AZ, Zheng HD, Latham K, Wlodarski W, Kalantar-Zadeh K (2009) Langmuir 25:509–514
190. Wang DA, Yu B, Wang CW, Zhou F, Liu WM (2009) Adv Mater 21:1964–1967
191. Zhang G, Huang H, Zhang Y, Chan HLW, Zhou L (2007) Electrochem Commun 9:2854–2858
192. S. H. Kang, H. S. Kim, J. Y. Kim and Y. E. Sung, Nanotechnology, 2009, 20
193. Lei BX, Liao JY, Zhang R, Wang J, Su CY, Kuang DB (2010) J Phys Chem C 114:15228–15233
194. Lin J, Chen JF, Chen XF (2010) Electrochem Commun 12:1062–1065
195. Wang J, Lin ZQ (2008) Chem Mater 20:1257–1261
196. Ali G, Yoo SH, Kum JM, Kim YN, Cho SO (2011) Nanotechnology 22:245602
197. Wang J, Lin ZQ (2010) Chem Mater 22:579–584
198. Wang DA, Liu LF (2010) Chem Mater 22:6656–6664
199. Pang Q, Leng LM, Zhao LJ, Zhou LY, Liang CJ, Lan YW (2011) Mater Chem Phys 125:612–616
200. Wang DA, Liu LF, Zhang FX, Tao K, Pippel E, Domen K (2011) Nano Lett 11:3649–3655
201. Wang J, Lin ZQ (2008) Chem Mater 20:1257–1261
202. Lin CJ, Yu WY, Chien SH (2010) J Mater Chem 20:1073–1077
203. Wang YH, Yang HX, Liu Y, Wang H, Shen H, Yan J, Xu HM (2010) Prog Photovoltaics 18:285–290
204. Wang YH, Yang HX, Lu L (2010) J Appl Phys 108:064510
205. Liu ZY, Misra M (2010) ACS Nano 4:2196–2200
206. Zou DC, Wang D, Chu ZZ, Lv ZB, Fan X (2010) Coord Chem Rev 254:1169–1178

207. Liu Y, Wang H, Li M, Hong RJ, Ye QH, Zheng JM, Shen H (2009) Appl Phys Lett 95:233505
208. Liu Y, Li M, Wang H, Zheng JM, Xu HM, Ye QH, Shen H (2010) J Phys D Appl Phys 43:205103
209. Mor GK, Shankar K, Paulose M, Varghese OK, Grimes CA (2006) Nano Lett 6:215–218
210. Hu A, Li H, Jia Z, Xia Z (2011) J Solid State Chem 184:2936–2940
211. Hu A, Xiao L, Dai G, Xia Z (2012) J Solid State Chem 190:130–134
212. Ye MD, Xin XK, Lin CJ, Lin ZQ (2011) Nano Lett 11:3214–3220
213. Chen CC, Chung HW, Chen CH, Lu HP, Lan CM, Chen SF, Luo L, Hung CS, Diau EWG (2008) J Phys Chem C 112:19151–19157
214. Shang M, Wang W, Yin W, Ren J, Sun S, Zhang L (2010) Chem Eur J 16:11412–11419
215. Ding B, Kim H, Kim C, Khil M, Park S (2003) Nanotechnology 14:532
216. Li D, Wang Y, Xia Y (2003) Nano Lett 3:1167–1171
217. Reneker DH, Chun I (1999) Nanotechnology 7:216
218. Song MY, Ihn KJ, Jo SM, Kim DY (1861) Nanotechnology 2004:15
219. Huang ZM, Zhang YZ, Kotaki M, Ramakrishna S (2003) Compos Sci Technol 63:2223–2253
220. Choi SW, Park JY, Kim SS (2009) Nanotechnology 20:465603
221. Li D, Xia Y (2003) Nano Lett 3:555–560
222. Nair AS, Zhu P, Jagadeesh Babu V, Yang S, Krishnamoorthy T, Murugan R, Peng S, Ramakrishna S (2012) Langmuir 28:6202–6206
223. Wu MC, Sápi A, Avila A, Szabó M, Hiltunen J, Huuhtanen M, Tóth G, Kukovecz Á, Kónya Z, Keiski R (2011) Nano Research 4:360–369
224. Kumar A, Jose R, Fujihara K, Wang J, Ramakrishna S (2007) Chem Mater 19:6536–6542
225. Yang L, Leung WWF (2011) Adv Mater 23:4559–4562
226. Hwang SH, Kim C, Song H, Son S, Jang J (2012) ACS Appl Mater Interfaces 4:5287–5292
227. Jose R, Kumar A, Thavasi V, Ramakrishna S (2008) Nanotechnology 19:424004
228. Zhu P, Nair AS, Yang S, Peng S, Ramakrishna S (2011) J Mater Chem 21:12210–12212
229. Zhan S, Chen D, Jiao X, Tao C (2006) J Phys Chem B 110:11199–11204
230. Lee BH, Song MY, Jang SY, Jo SM, Kwak SY, Kim DY (2009) J Phys Chem C 113:21453–21457
231. Nair AS, Shengyuan Y, Peining Z, Ramakrishna S (2010) Chem Commun 46:7421–7423
232. Shengyuan Y, Peining Z, Nair AS, Ramakrishna S (2011) J Mater Chem 21:6541–6548
233. Hwang D, Jo SM, Kim DY, Armel V, MacFarlane DR, Jang SY (2011) ACS Appl Mater Interfaces 3:1521–1527
234. Ishida M, Park SW, Hwang D, Koo YB, Sessler JL, Kim DY, Kim D (2011) J Phys Chem C 115:19343–19354
235. Chen X, Liu L, Peter YY, Mao SS (2011) Science 331:746–750
236. An HL, Ahn HJ (2012) Mater Lett 81:41–44
237. Fujihara K, Kumar A, Jose R, Ramakrishna S, Uchida S (2007) Nanotechnology 18:365709
238. Liu B, Nakata K, Sakai M, Saito H, Ochiai T, Murakami T, Takagi K, Fujishima A (2012) Catal Sci Technol 2:1933–1939
239. Hwang D, Lee H, Seo Y, Kim D, Jo SM, Kim DY (2013) J Mater Chem A 1:1359–1367
240. Lee H, Hwang D, Jo SM, Kim D, Seo Y, Kim DY (2012) ACS Appl Mater Interfaces 4:3308–3315
241. Jang SY, Hwang D, Kim DY, Kim D (2013) J Mater Chem A 1:1228–1238
242. Yang HY, Lee MF, Huang CH, Lo YS, Chen YJ, Wong MS (2009) Thin Solid Films 518:1590–1594
243. Wolcott A, Smith WA, Kuykendall TR, Zhao Y, Zhang JZ (2008) Small 5:104–111
244. Larsen GK, Fitzmorris R, Zhang JZ, Zhao Y (2011) J Phys Chem C 115:16892–16903
245. Smith W, Wolcott A, Fitzmorris RC, Zhang JZ, Zhao Y (2011) J Mater Chem 21:10792–10800
246. Wang S, Xia G, He H, Yi K, Shao J, Fan Z (2007) J Alloy Compd 431:287–291

247. Pihosh Y, Turkevych I, Ye J, Goto M, Kasahara A, Kondo M, Tosa M (2009) J Electrochem Soc 156:K160–K165
248. Gamez F, Plaza-Reyes A, Hurtado P, Guillen F, Anta JA, Martinez-Haya B, Perez S, Sanz M, Castillejo M, Izquierdo JG, Banares L (2010) J Phys Chem C 114:17409–17415
249. Sanz M, Walczak M, de Nalda R, Oujja M, Marco JF, Rodriguez J, Izquierdo JG, Banares L, Castillejo M (2009) Appl Surf Sci 255:5206–5210
250. Sanz M, Walczak M, Oujja M, Cuesta A, Castillejo M (2009) Thin Solid Films 517:6546–6552
251. Yang XF, Zhuang JL, Li XY, Chen DH, Ouyang GF, Mao ZQ, Han YX, He ZH, Liang CL, Wu MM, Yu JC (2009) ACS Nano 3:1212–1218
252. Quinonez C, Vallejo W, Gordillo G (2010) Appl Surf Sci 256:4065–4071
253. Shan AY, Ghazi TIM, Rashid SA (2010) Appl Catal A 389:1–8
254. Seifried S, Winterer M, Hahn H (2000) Chem Vap Deposition 6:239–244
255. Zhang C, Chen S, Mo L, Huang Y, Tian H, Hu L, Huo Z, Dai S, Kong F, Pan X (2011) J Phys Chem C 115:16418–16424
256. Xu L, Steinmiller EMP, Skrabalak SE (2011) J Phys Chem C 116:871–877
257. Yu H, Irie H, Shimodaira Y, Hosogi Y, Kuroda Y, Miyauchi M, Hashimoto K (2010) J Phys Chem C 114:16481–16487
258. Zhu J, Ren J, Huo Y, Bian Z, Li H (2007) J Phys Chem C 111:18965–18969
259. Wu Q, Ouyang JJ, Xiea KP, Sun L, Wang MY, Lin CJ (2012) J Hazard Mater 199:410–417
260. Di Paola A, Marci G, Palmisano L, Schiavello M, Uosaki K, Ikeda S, Ohtani B (2002) J Phys Chem B 106:637–645
261. Huang JH, Hung PY, Hu SF, Liu RS (2010) J Mater Chem 20:6505–6511
262. Liqiang J, Xiaojun S, Baifu X, Baiqi W, Weimin C, Honggang F (2004) J Solid State Chem 177:3375–3382
263. Liu Z, Li Y, Liu C, Ya J (2011) L. E., W. Zhao, D. Zhao and L. An. ACS Appl Mater Interfaces 3:1721–1725
264. Wang E, Yang W, Cao Y (2009) J Phys Chem C 113:20912–20917
265. Li H, Zhang X, Huo Y, Zhu J (2007) Environ Sci Technol 41:4410–4414
266. Wu G, Nishikawa T, Ohtani B, Chen A (2007) Chem Mater 19:4530–4537
267. Song J, Yang HB, Wang X, Khoo SY, Wong C, Liu XW, Li CM (2012) ACS Appl Mater Interfaces 4:3712–3717
268. Liu G, Yin LC, Wang J, Niu P, Zhen C, Xie Y, Cheng HM (2012) Energy Environ Sci 5:9603–9610
269. Zhao Y, Qiu X, Burda C (2008) Chem Mater 20:2629–2636
270. Peng T, Dai K, Yi H, Ke D, Cai P, Zan L (2008) Chem Phys Lett 460:216–219
271. Liao G, Chen S, Quan X, Chen H, Zhang Y (2010) Environ Sci Technol 44:3481–3485
272. Park J, Yi J, Tachikawa T, Majima T, Choi W (2010) J Phys Chem Lett 1:1351–1355
273. Park H, Choi W (2005) J Phys Chem B 109:11667–11674
274. Zhang H, Zong R, Zhao J, Zhu Y (2008) Environ Sci Technol 42:3803–3807
275. Zhou X, Peng F, Wang H, Yu H, Fang Y (2011) Chem Commun 47:10323–10325
276. Zhang M, Shao C, Guo Z, Zhang Z, Mu J, Cao T, Liu Y (2011) ACS Appl Mater Interfaces 3:369–377
277. Wang H, Bai Y, Zhang H, Zhang Z, Li J, Guo L (2010) J Phys Chem C 114:16451–16455
278. Zhu G, Pan L, Xu T, Sun Z (2011) ACS Appl Mater Interfaces 3:1472–1478
279. Wang CL, Sun L, Yun H, Li J, Lai YK, Lin CJ (2009) Nanotechnology 20:295601
280. Hou Y, Li X, Zou X, Quan X, Chen G (2008) Environ Sci Technol 43:858–863
281. Zhou W, Liu H, Wang J, Liu D, Du G, Cui J (2010) ACS Appl Mater Interfaces 2:2385–2392
282. Kim JY, Choi SB, Noh JH, Yoon SH, Lee S, Noh TH, Frank AJ, Hong KS (2009) Langmuir 25:5348–5351
283. Huang H, Li D, Lin Q, Shao Y, Chen W, Hu Y, Chen Y, Fu X (2009) J Phys Chem C 113:14264–14269
284. Murakami N, Kurihara Y, Tsubota T, Ohno T (2009) J Phys Chem C 113:3062–3069

285. Zhang X, Zhang L, Xie T, Wang D (2009) J Phys Chem C 113:7371–7378
286. Colón G, López SM, Hidalgo M, Navío J (2010) Chem Commun 46:4809–4811
287. Gao XF, Li HB, Sun WT, Chen Q, Tang FQ, Peng LM (2009) J Phys Chem C 113:7531–7535
288. Lee HJ, Leventis HC, Moon SJ, Chen P, Ito S, Haque SA, Torres T, Nüesch F, Geiger T, Zakeeruddin SM (2009) Adv Funct Mater 19:2735–2742
289. O'Hayre R, Nanu M, Schoonman J, Goossens A, Wang Q, Grätzel M (2006) Adv Funct Mater 16:1566–1576
290. Wang Y, Gong H, Fan B, Hu G (2010) J Phys Chem C 114:3256–3259
291. Zhang Z, Shao C, Li X, Sun Y, Zhang M, Mu J, Zhang P, Guo Z, Liu Y (2012) Nanoscale 5:606–618
292. Pandikumar A, Murugesan S, Ramaraj R (2010) ACS Appl Mater Interfaces 2:1912–1917
293. Z. Bian, J. Zhu, F. Cao, Y. Lu and H. Li, Chem. Commun., 2009, 3789-3791
294. Seh ZW, Liu S, Low M, Zhang SY, Liu Z, Mlayah A, Han MY (2012) Adv Mater 24:2310–2314
295. Lee SS, Oh K (2012) ACS Appl Mater Interfaces 4:5727–5731
296. Xie KP, Wu Q, Wang YY, Guo WX, Wang MY, Sun L, Lin CJ (2011) Electrochem Commun 13:1469–1472
297. Xie KP, Sun L, Wang CL, Lai YK, Wang MY, Chen HB, Lin CJ (2010) Electrochim Acta 55:7211–7218
298. Mohapatra SK, Kondamudi N, Banerjee S, Misra M (2008) Langmuir 24:11276–11281
299. Zhang N, Liu S, Fu X, Xu YJ (2011) J Phys Chem C 115:9136–9145
300. Ye M, Gong J, Lai Y, Lin C, Lin Z (2012) J Am Chem Soc 134:15720–15723
301. Wang C, Yin L, Zhang L, Liu N, Lun N, Qi Y (2010) ACS Appl Mater Interfaces 2:3373–3377
302. Chen YC, Pu YC, Hsu YJ (2012) J Phys Chem C 116:2967–2975
303. Lai YK, Gong JJ, Lin CJ (2012) Int J Hydrogen Energy 37:6438–6446
304. Yu Y, Zhang MZ, Chen J, Zhao YD (2012) Dalton Trans 42:885–889
305. Bai H, Liu Z, Sun DD (2012) J Am Ceram Soc 96:942–949
306. Smitha VS, Baiju KV, Perumal P, Ghosh S, Warrier KG (2012) Eur J Inorg Chem 2012:226–233
307. K. M. Shrestha, C. M. Sorensen and K. J. Klabundea, J. Mater. Res., 1, 1-9
308. Lü X, Huang F, Wu J, Ding S, Xu F (2011) ACS Appl Mater Interfaces 3:566–572
309. Diamant Y, Chen S, Melamed O, Zaban A (2003) J Phys Chem B 107:1977–1981
310. Furukawa S, Shishido T, Teramura K, Tanaka T (2011) ACS Catalysis 2:175–179
311. Pan J, Hühne SM, Shen H, Xiao L, Born P, Mader W, Mathur S (2011) J Phys Chem C 115:17265–17269
312. Song KY, Park MK, Kwon YT, Lee HW, Chung WJ, Lee WI (2001) Chem Mater 13:2349–2355
313. Sun L, Bu JF, Guo WX, Wang YY, Wang MY, Lin CJ (2012) Electrochem Solid-State Lett 15:E1–E3
314. Katoh R, Furube A, Yoshihara T, Hara K, Fujihashi G, Takano S, Murata S, Arakawa H, Tachiya M (2004) J Phys Chem B 108:4818–4822
315. Jiang X, Zhang Y, Jiang J, Rong Y, Wang Y, Wu Y, Pan CX (2012) J Phys Chem C 116:22619–22624
316. Naldoni A, Allieta M, Santangelo S, Marelli M, Fabbri F, Cappelli S, Bianchi CL, Psaro R, Dal V (2012) Santo. J Am Chem Soc 134:7600–7603
317. Tominaka S, Tsujimoto Y, Matsushita Y, Yamaura K (2011) Angew Chem Int Ed 50:7418–7421
318. Zuo F, Bozhilov K, Dillon RJ, Wang L, Smith P, Zhao X, Bardeen C, Feng P (2012) Angew Chem Int Ed 124:6327–6330
319. Gu D, Lu Y, Yang B (2008) Chem Commun 2453–2455
320. Sayed FN, Jayakumar O, Sasikala R, Kadam R, Bharadwaj SR, Kienle L, Schürmann U, Kaps S, Adelung R, Mittal J (2012) J Phys Chem C 116:12462–12467

321. Kitano M, Funatsu K, Matsuoka M, Ueshima M, Anpo M (2006) J Phys Chem B 110:25266–25272
322. Gu DE, Yang BC, Hu YD (2008) Catal Commun 9:1472–1476
323. Wang CL, Wang MY, Xie KP, Wu Q, Sun L, Lin ZQ, Lin CJ (2011) Nanotechnology 22:305607
324. Sato S (1986) Chem Phys Lett 123:126–128
325. Sato S, Nakamura R, Abe S (2005) Appl Catal A 284:131–137
326. Lai YK, Huang JY, Zhang HF, Subramaniam VP, Tang YX, Gong DG, Sundar L, Sun L, Chen Z, Lin CJ (2010) J Hazard Mater 184:855–863
327. Cao J, Zhang Y, Tong H, Li P, Kako T, Ye J (2012) Chem Commun 48:8649–8651
328. Bacsa R, Kiwi J, Ohno T, Albers P, Nadtochenko V (2005) J Phys Chem B 109:5994–6003
329. Li Y, Ma G, Peng S, Lu G, Li S (2008) Appl Surf Sci 254:6831–6836
330. Chen D, Jiang Z, Geng J, Wang Q, Yang D (2007) Ind Eng Chem Res 46:2741–2746
331. Chen X, Burda C (2008) J Am Chem Soc 130:5018–5019
332. Dong F, Wang H, Wu Z (2009) J Phys Chem C 113:16717–16723
333. Sun H, Liu H, Ma J, Wang X, Wang B, Han L (2008) J Hazard Mater 156:552–559
334. Wu G, Wang J, Thomas DF, Chen A (2008) Langmuir 24:3503–3509
335. Dong F, Guo S, Wang H, Li X, Wu Z (2011) J Phys Chem C 115:13285–13292
336. Wei F, Ni L, Cui P (2008) J Hazard Mater 156:135–140
337. In S, Orlov A, Berg R, García F, Pedrosa-Jimenez S, Tikhov MS, Wright DS, Lambert RM (2007) J Am Chem Soc 129:13790–13791
338. Zong X, Xing Z, Yu H, Chen Z, Tang F, Zou J, Lu GQ, Wang L (2011) Chem Commun 47:11742–11744
339. Li L, Shi J, Li G, Yuan Y, Li Y, Zhao W (2013) New J Chem 37:451–457
340. Santos RS, Faria GA, Giles C, Leite CAP, Barbosa HS, Arruda MAZ, Longo C (2012) ACS Appl Mater Interfaces 4:5555–5561
341. Liu X, Geng D, Wang X, Ma S, Wang H, Li D, Li B, Liu W, Zhang Z (2010) Chem Commun 46:6956–6958
342. Cao G, Li Y, Zhang Q, Wang H (2010) J Am Ceram Soc 93:1252–1255
343. Zhang J, Pan C, Fang P, Wei J, Xiong R (2010) ACS Appl Mater Interfaces 2:1173–1176
344. Dai G, Yu J, Liu G (2011) J Phys Chem C 115:7339–7346
345. Wang Y, Zhang Y, Zhao G, Tian H, Shi H, Zhou T (2012) ACS Appl Mater Interfaces 4:3965–3972
346. Vogel R, Hoyer P, Weller H (1994) J Phys Chem 98:3183–3188
347. Kim W, Tachikawa T, Majima T, Choi W (2009) J Phys Chem C 113:10603–10609
348. D. Zhang, G. Li, X. Yang and C. Y. Jimmy, *Chem. Commun.*, 2009, **0**, 4381-4383
349. Peng L, Xie T, Lu Y, Fan H, Wang D (2010) Phys Chem Chem Phys 12:8033–8041
350. Wang C, Shao C, Zhang X, Liu Y (2009) Inorg Chem 48:7261–7268
351. Anderson C, Bard AJ (1997) J Phys Chem B 101:2611–2616
352. William L IV, Kostedt I, Ismail AA, Mazyck DW (2008) Ind Eng Chem Res 47:1483–1487
353. Fu X, Clark LA, Yang Q, Anderson MA (1996) Environ Sci Technol 30:647–653
354. Ding S, Yin X, Lü X, Wang Y, Huang F, Wan D (2011) ACS Appl Mater Interfaces 4:306–311
355. Shao Z, Zhu W, Li Z, Yang Q, Wang G (2012) J Phys Chem C 116:2438–2442
356. Kang Q, Liu S, Yang L, Cai Q, Grimes CA (2011) ACS Appl Mater Interfaces 3:746–749
357. Liu B, Wang D, Zhang Y, Fan H, Lin Y, Jiang T, Xie T (2012) Dalton Trans 42:2232–2237
358. Wang S, Zhang X, Zhou G, Wang ZS (2012) Phys Chem Chem Phys 14:816–822
359. Kim JY, Kang SH, Kim HS, Sung YE (2009) Langmuir 26:2864–2870
360. Yu H, Xue B, Liu P, Qiu J, Wen W, Zhang S, Zhao H (2012) ACS Appl Mater Interfaces 4:1289–1294
361. Jung HS, Lee JK, Nastasi M, Lee SW, Kim JY, Park JS, Hong KS, Shin H (2005) Langmuir 21:10332–10335
362. Shinde DV, Mane RS, Oh IH, Lee JK, Han SH (2012) Dalton Trans 41:10161–10163

363. Pang S, Xie T, Zhang Y, Wei X, Yang M, Wang D, Du Z (2007) J Phys Chem C 111:18417–18422
364. Cao T, Li Y, Wang C, Zhang Z, Zhang M, Shao C, Liu Y (2011) J Mater Chem 21:6922–6927
365. Li X, Hou Y, Zhao Q, Chen G (2011) Langmuir 27:3113–3120
366. Muduli S, Lee W, Dhas V, Mujawar S, Dubey M, Vijayamohanan K, Han SH, Ogale S (2009) ACS Appl Mater Interfaces 1:2030–2035
367. Liu B, Huang Y, Wen Y, Du L, Zeng W, Shi Y, Zhang F, Zhu G, Xu X, Wang Y (2012) J Mater Chem 22:7484–7491
368. Kim H, Moon G, Monllor-Satoca D, Park Y, Choi W (2011) J Phys Chem C 116:1535–1543
369. Zhang H, Lv X, Li Y, Wang Y, Li J (2009) ACS Nano 4:380–386
370. Liu J, Bai H, Wang Y, Liu Z, Zhang X, Sun DD (2010) Adv Funct Mater 20:4175–4181
371. Hou C, Zhang Q, Li Y, Wang H (2012) J Hazard Mater 205:229–235
372. Wojtoniszak M, Zielinska B, Chen X, Kalenczuk RJ, Borowiak-Palen E (2012) J Mater Sci 47:3185–3190
373. Yang N, Zhang Y, Halpert JE, Zhai J, Wang D, Jiang L (2012) Small 11:1762–1770
374. Lee JS, You KH, Park CB (2012) Adv Mater 24:1084–1088
375. Cottineau T, Albrecht A, Janowska I, Macher N, Bégin D, Ledoux MJ, Pronkin S, Savinova E, Keller N, Keller V (2012) Chem Commun 48:1224–1226
376. Kim IY, Lee JM, Kim TW, Kim HN, Kim H, Choi W, Hwang SJ (2012) Small 7:1038–1048
377. Peining Z, Nair AS, Shengjie P, Shengyuan Y, Ramakrishna S (2012) ACS Appl Mater Interfaces 4:581–585
378. Zhang XY, Li HP, Cui XL, Lin Y (2010) J Mater Chem 20:2801–2806
379. Liu S, Liu C, Wang W, Cheng B, Yu J (2012) Nanoscale 4:3193–3200
380. Tu W, Zhou Y, Liu Q, Tian Z, Gao J, Chen X, Zhang H, Liu J, Zou Z (2012) Adv Funct Mater 22:1215–1221
381. Sun L, Zhao Z, Zhou Y, Liu L (2012) Nanoscale 4:613–620
382. Sher Shah MSA, Park AR, Zhang K, Park JH, Yoo PJ (2012) ACS Appl Mater Interfaces 4:3893–3901
383. Zhang X, Sun Y, Cui X, Jiang Z (2012) Int J Hydrogen Energy 37:811–815
384. Jiang B, Tian C, Pan Q, Jiang Z, Wang JQ, Yan W, Fu H (2011) J Phys Chem C 115:23718–23725
385. Libisch F, Stampfer C, Burgdörfer J (2009) Phys Rev B 79:115423
386. Ritter KA, Lyding JW (2009) Nat Mater 8:235–242
387. Ponomarenko L, Schedin F, Katsnelson M, Yang R, Hill E, Novoselov K, Geim A (2008) Science 320:356–358
388. Zhuo S, Shao M, Lee ST (2012) ACS Nano 6:1059–1064

Nanostructured Nitrogen Doping TiO$_2$ Nanomaterials for Photoanodes of Dye-Sensitized Solar Cells

Wei Guo and Tingli Ma

Abstract This paper presents a review of nanostructured nitrogen doping (N-doped) TiO$_2$ nanomaterials and their application into dye-sensitized solar cells (DSCs). Such N-doped TiO$_2$ nanomaterials aim at enhancing the performance of TiO$_2$ photoanodes for DSCs. Herein, we summarize the different synthesis methods, nanostructures, and physiochemical properties of N-doped TiO$_2$. Also, the differences in electron transport behavior in DSCs based on N-doped and pure TiO$_2$ photoanodes were involved. Further understanding of the nanostructured N-doped TiO$_2$ photoanodes will promote the development of energy conversion and other related areas.

Keywords Dye-sensitized solar cell · Nitrogen-doped titania · Photoanode · Charge transport

1 Introduction

Dye-sensitized solar cells (DSCs) have been extensively studied for decades as a low-cost alternative to conventional silicon solar cells since they were reported by Grätzel and co-workers [1, 2]. Encouragingly, many improvements have been achieved by introducing new dyes, electrolytes, and different morphologies to the semiconductor materials [3]. The highest energy conversion efficiency of DSCs has reached to 12.3 % [3]. However, further improving the energy conversion efficiency of DSCs is important for successful commercialization. Photoanodes made of metal oxide semiconductor are known to be one of the key components that significantly affect the overall energy conversion efficiency of DSCs.

W. Guo · T. Ma (✉)
State Key Laboratory of Fine Chemicals, School of Chemical Engineering,
Dalian University of Technology, Dalian 116024, People's Republic of China
e-mail: tinglima@dlut.edu.cn

Z. Lin and J. Wang (eds.), *Low-cost Nanomaterials*, Green Energy and Technology,
DOI: 10.1007/978-1-4471-6473-9_3, © Springer-Verlag London 2014

Generally, nanocrystalline mesoporous metal oxide semiconductor (typically TiO_2) films, which adsorb dye molecules and transport photogenerated electrons to the outer circuit, serve as electron conductors and dictate the efficiency of electron transport and collection [4]. Therefore, the oxide semiconductor of photoanodes plays a key role in the performance of DSCs. Specifically, an excellent photoanode should include: (1) a large surface area and an appropriate isoelectric point (IEP) which can guarantee a high amount and quality of dye uptake; (2) a perfect lattice and low electron trap distribution to reduce the photogenerated electron losses; and (3) a good neck-connection between nanoparticles, which facilitate the electron transport during collection to the conductive substrate.

In recent years, nanocrystalline metal semiconductor materials are extensively studied for the mesoporous photoanodes of DSCs. Nanocrystalline mesoporous photoanodes made of TiO_2 materials, as one of the most widely used semiconductors, show an excellent performance in the DSCs. Some researchers also study other types of semiconductors, such as ZnO [5], Zn_2SnO_4 [6], WO_3 [7], $SrTiO_3$ [8], Nb_2O_5 [9], SnO_2 [10], CeO_2 [11], FeS [12], and NiO [13]. However, the photovoltaic performance of DSCs based on these semiconductor materials remains low because some of these materials are not stable in dye solution, such as ZnO and so on. On the other hand, some materials have a low isoelectric point (IEP), such as SnO_2, which is not suitable for dye molecular linking. However, these non-TiO_2 materials need to be further studied and developed to get a good photovoltaic performance for DSCs. Up till now, TiO_2 is still the best choice for photoanodes in DSCs. Aiming at the further improvement for DSCs, on the one hand, we can develop more efficient and diverse structures of TiO_2 materials. On the other hand, we can modify TiO_2 to enhance its performance, such as chemical doping. As oxygen deficiencies exist in pure TiO_2 crystal structures [14–16] these oxygen deficiencies can induce TiO_2 to a visible light absorption response producing electron-hole pairs. The photoexcited TiO_2 will lead to the oxidation of iodide or dye by photogenerated holes. Such deficiencies are possible causes for the shortened lifetime of DSCs. Element doping is an effective way to improve the performance of TiO_2. We can choose metal or nonmetal to proceed with TiO_2 doping. Recently, some studies reported the modifying of pure TiO_2 with metal doping, i.e., Zn-, La-, Ta-, and Nb-doped TiO_2 [17–20]. The performance of DSCs can be improved by adjusting doping metals. However, metal doping can affect the position of conduction band (CB) of TiO_2 contributing to the change in photovoltage. Besides, metal doping also introduces more recombination sites for electron [21]. Nonmetal doping of TiO_2 materials is another good choice for finetuning of TiO_2. Nonmetal elements, such as N [22, 23], C [24], B [25], I [26] etc., are used to dope TiO_2. Especially for DSCs, nitrogen seems to be the most effective element to enhance the photovoltaic performance of DSCs.

This review summarizes the recent works on the N-doped TiO_2 materials and their application into photoanodes of DSCs. Herein, the synthesis methods and optical properties of N-doped TiO_2 are introduced briefly. Then the effect of N-doped TiO_2 photoanodes on the performance of DSCs is described in detail. Finally, we discuss the charge transport in DSCs based on N-doped TiO_2 photoanodes.

2 Synthesis and Characterizations of Nanocrystalline N-Doped TiO₂

Various methods have been reported for the synthesis of N-doped TiO_2 since the study by Asahi et al. in 2001 [22]. The methods are generally classified as: (1) sintering TiO_2 at high temperatures under an N-containing atmosphere (NH_3 gas or mixeds), which we called dry methods [23, 27, 28]; (2) chemical wet methods, which involve sol–gel and solvothermal methods [29–32], some chemical nitrogen sources are added into water or alcohol during the hydrolysis of titanium alkoxide; and (3) sputtering and implantation deposition techniques, [33, 34] that were mainly used to prepare single crystalline or polycrystalline N-doped TiO_2 thin films. Herein, we emphasize the development of the former two methods that were used to fabricate N-doped TiO_2 photoanodes of DSCs.

2.1 Dry Methods

Dry method involves a high temperature sintering and doping process. This method can be easily controlled by adjusting the N-containing atmospheres and starting materials.

In 2005, our group reported the synthesis of N-doped TiO_2 employing the dry method. The starting pristine TiO_2 was commercial anatase powders (ST-01, Ishihara Sangyo Kaisha, Ltd.), which were treated at 550 °C for 3 h under a dry N_2 and NH_3 flow [23]. Interestingly, we obtained needle-like N-doped TiO_2 crystals with excellent thermostability. Afterwards, we also used P25 (Degussa) as the starting materials to obtain N-doped P25. These N-doped ST-01 and P25 materials show good performance over pristine TiO_2 photoanodes. Additionally, we also found that the starting materials apparently affect the N-doping effect from N-doping amount and optical properties. As reported earlier, the phase transition of anatase into rutile can occur at a high sintering temperature [35]. However, our nitridation process did not affect the crystal structure of pristine TiO_2, as shown in Fig. 1. Moreover, the obtained N-doped ST-01 showed excellent thermal stability. We can see in Fig. 2 the UV-Vis absorption spectra of N-doped ST-01 powders, treated under different conditions, which suggest that after being sintered separately in air, N_2, Ar, or at high temperature up to 700 °C, the N-doped ST-01 still shows visible light absorption, which is a signal for successful N-doping.

In 2009, Yang et al. developed a set of reaction devices for the process of thermal doping treatment [36]. The TiO_2 samples were treated with NH_3 under middle pressures and controlled conditions. This synthesis route is an effective approach to adjust the nitrogen concentration and band gap of N-doped TiO_2. They obtained a series of N-doped TiO_2 materials with different nitrogen doping amount by adjusting the temperature, pressure, and time. Moreover, the anatase type N-doped TiO_2 can be obtained at the sintering temperature of 400–500 °C. When the temperature increased to 600 °C, the rutile phase can be observed.

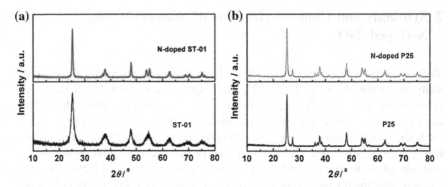

Fig. 1 X-ray diffraction patterns. **a** N-doped ST-01 and pristine ST-01 powders; **b** N-doped P25 and pristine P25 powders

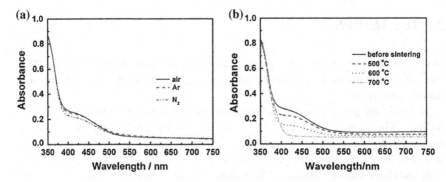

Fig. 2 UV-Vis absorption spectra of of N-doped ST-01 powders treated under different conditions: **a** in different sintering atmosphere; **b** at different temperature for 30 min [23]

The dry method is an aftertreatment process for N-doping which requires a high sintering temperature. Besides, this method can be controlled by adjusting the N-containing atmosphere and sintering temperature to get different N-doped TiO_2 materials. However, the high temperature sintering of TiO_2 may also lead to a degree of nanoparticles aggregation.

2.2 Chemical Wet Method

The wet method is widely used to synthesize TiO_2 nanomaterials. Until now, the wet method has been one the most successful methods for doping because of its convenient control of titanium sources, the nitrogen doping amount, and hydrolysis condition. In addition, simple variations in experimental conditions can lead to the required particle size and crystal structures, such as the hydrolysis rate of titanium alkoxide, pH of solvent solution, and solvent systems [37]. Besides, the nitrogen sources can be also chosen to adjust the nitrogen doping process. Inorganic and organic nitrogen dopants (such as aqueous ammonia, urea, ammonium chloride, triethylamine, and diethylamine) are widely used in the synthesis of N-doped TiO$_2$.

Our group focused on the sol–gel wet method to investigate the types of nitrogen dopants and amount on the performance of the N-doped TiO$_2$, and thereby their photovoltaic performance of the DSCs. The sol–gel method usually involves two steps: (1) hydrolysis of titanium alkoxide in solvent (water or ethanol) containing nitrogen sources; and (2) sintering the obtained precipitate under 400–500 °C for a certain amount of time. However, we found that the nitridation process and doping amount varies with each N-doping method and type of nitrogen sources. We used ammonia, triethylamine, and urea nitrogen dopants to synthesize N-doped TiO$_2$ nanocrystals, which were denoted as N-A, N-U, and N-T, respectively [38–41]. By varying initial molar ratios of N/Ti, a series of N-doped TiO$_2$ with different N dopant amounts can be also synthesized according to the N-A method [42]. As the previous literature reported, we also found that the nitrogen doping process differs for the N-A, N-U, and N-T powders. During the preparation of N-A, nitrogen doping proceeded simultaneously with the hydrolysis of the titanium alkoxide. The hydrolysis of the titanium isopropoxide (TTIP) consisted of two steps [43]: hydrolysis and concentration. Titanium hydroxide was formed in the hydrolysis and was called titanic acid, which exhibits acidity. The titanic acid then reacted with NH$_4$OH to form ammonium titanate, which when heated, dehydrated and desorbed to NH$_3$ and allowed N-doping to occur. Ammonia in the doped samples becomes oxidized by the lattice oxygen, and this oxidation allows for the uptake of nitrogen. During the preparation of N-U, simultaneous N-doping with phase formation occurred by heating a mixture of titanium hydroxide and urea. When the mixture was heated, the urea was decomposed into NH$_3$ and CO$_2$, and the generated NH$_3$ reacted with the oxygen of the TiO$_2$ to form the N-doped TiO$_2$. The N-T sample was formed by direct nitridation of the anatase TiO$_2$ nanostructures with alkylammonium salt. In this case, triethylamine was used as the alkylammonium salt, and the N-T nanocrystals were obtained by controlling hydrolysis rate of the TTIP and the pH value of the solution. As a side note, some amine groups can coordinate to the central Ti ion early during the N-doping process, and these amine linkages can be hydrolyzed by the addition of a dilute solution of acid or base, but this addition in turn adjusts the pH of the reaction mixture. Therefore, high pH values are required. By using high pH values, the Ti-bound amine groups can be easily substituted by OH$^-$ during the hydrolysis process, which results in the formation of N-doped TiO$_2$ nanoparticles. [40] Furthermore, the obtained different phases and crystallite sizes of N-doped TiO$_2$ can be ascribed to the different types of nitrogen dopants and to the hydrolysis of titanium alkoxide under controlled conditions. In the sol–gel wet method, the type of nitrogen sources not only influences the nitridation process but also the particle size and nitrogen concentration. The N dopant amounts were calculated using the XPS results and were found to be 2.77, 0.29, and 0.47 % for N-A, N-U, and N-T, respectively.

The solvothermal method is also an effective wet method to synthesis N-doped TiO_2. Dai et al. used urea as nitrogen source in the hydrolysis of TTIP [44]. The precipitation solution was treated in an autoclave at 200 °C for 10 h. They observed that the (101) peak positions of N-doped TiO_2 showed a shift compared with the undoped ones. This is also reported in Jagadale's work on N-doped TiO_2 by the sol–gel method [45].

Wet methods are the first choice to be employed to determine the suitable nitrogen dopants, and it is also a simple N-doping method. Therefore, it is necessary to seek an appropriate wet method, nitrogen sources, and N dopant amount for the large-scale production of N-doped TiO_2 nanomaterials.

2.3 Other Techniques

There are some other approaches for preparation of N-doping TiO_2 materials such as combustion, ion-implantation, and sputtering techniques.

Recently, Ogale and Gopinath et al. reported a disordered mesoporous framework of N-doped TiO_2 consisting of nanoparticles. They used a simple combustion synthesis method to prepare N-doped TiO_2 using $Ti(NO_3)_4$ as Ti precursor and urea as fuel. They found that urea/$Ti(NO_3)_4$ molar ratio of ≤ 7 leads to a biphasic (anatase and rutile) titania. A high ratio of urea/$Ti(NO_3)_4$ (≥ 9) leads to exclusive anatase phase TiO_2. The pseudo-3D nature of mesoporous N-doped TiO_2 consisting of mesoporosity and electrically interconnected nanosized crystalline particles lead to a higher efficiency in DSCs [46].

Kang et al. reported an ion-implantation technique combination with electrostatic spray to prepare hierarchical nanostructured TiO_2 clumps doped by nitrogen-ion [47]. The ion-implantation could be a straightforward tool to implant foreign atoms into the lattice. This ion doping intrinsically modifies the lattice structure and consequently the properties of host counterparts [48].

Magnetron sputtering deposition method is also a widely used technique to prepare N-doped TiO_2 thin films. We can obtain the films by depositing Ti in plasma of argon, oxygen, and nitrogen. By varying the nitrogen contents in the flow, we can get a different nitrogen concentration within TiO_2 lattice from 2.0 to 16.5 % [49]. Early in 2003, Lindquist et al. used DC magnetron sputtering to prepare nanocrystalline porous N-doped TiO_2 thin films [50]. These films displayed a porous and rough surface. The crystal structure of N-doped TiO_2 thin films varying from rutile to anatase varied with the nitrogen content. However, the thickness of films only reached to several hundred nanometers.

Therefore, many methods can be used to synthesize nanocrystalline N-doped TiO_2. However, the crystal structure, surface property, and optical property of N-doped TiO_2 are all related to the synthesis methods.

2.4 Physical and Chemical Characterization of N-doped TiO$_2$

To evaluate the optical properties of N-doped TiO$_2$, UV-Vis spectrometry is the most commonly used technique to examine the doping effects on the host metal oxide matrix [51]. Generally, after N-doping treatment, the N-doped TiO$_2$ nanomaterials show a good visible light response between 400 and 500 nm. This trend is observed by many works. In our previous work, compared with pure TiO$_2$ and P25 electrodes, the N-doped TiO$_2$ samples (N-doped ST-01, N-doped P25) exhibited new absorption peaks in the visible light region between 400 and 550 nm (Fig. 3). However, intensity of the absorption response peaks show much dependence on the preparing conditions, such as N-doping amount as well as other related factors.

The reasons for the visible light response origin of N-doped TiO$_2$ are still open questions. Some work reported that the enhanced visible light absorption derived from band gap narrowing [52] (Fig. 4): (1) the localized dopant levels near the VB and the CB; (2) broadening of the VB; (3) localized dopant levels and electronic transitions to the CB. Then it was found that the Ti^{3+} defect or oxygen vacancies can also induce the redshift absorption. Giamello et al. [30] reported that N-doped TiO$_2$ electrodes contained N$_b$ centers that were responsible for visible light absorption. Nevertheless, Serpone et al. [53] analyzed the DRS spectra (diffuse reflectance spectra) of anion- and cation-doped TiO$_2$ electrodes. They concluded that the absorption features in the visible light region originated from color centers developed during the doping process or post-treatments rather than by narrowing the intrinsic band gap for the TiO$_2$ electrode as originally proposed by Asahi and co-workers [22]. Burda et al. recently studied the electronic origins of the visible light absorption properties of C-, N-, and S-doped TiO$_2$ nanomaterials. They revealed that additional electronic states above the valence band edge existed, which could explain the redshift absorption of these materials [54]. On the basis of the above discussion, a conclusion doping mechanism is still needed to further understand the origin of visible light response.

X-ray photoelectron spectroscopy (XPS) is a powerful tool to get information about the electronic structure and chemical environment of the elements on the surface. So far, the XPS analysis is a surface characterization technique that could be affected by testing environment. The XPS result can be considered as a reference. Especially for N-doped TiO$_2$, XPS is the most reported technique to analyze the nitrogen concentration and chemical environment. What we concern most is three areas: the N 1s region, the Ti 2p region, and the O 1s region (Fig. 5).

For the N 1s region, the binding energy peaks ranged from 396 to 408 eV. However, the N 1s binding energy is highly dependent on the method of preparation. The peaks at 396 eV were not always observed. According to an earlier XPS study on the oxidation of pure TiN, the N 1s peak at 396 eV was assigned as the β-N in TiN; the 397.5 eV peak was due to the α-N$_2$, and the 400 eV and 405 eV peaks were assigned to the γ-N$_2$. In our previous work, we suggested to

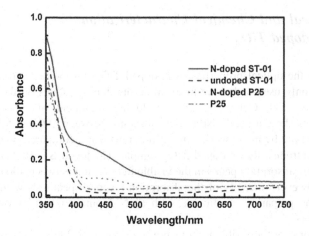

Fig. 3 UV-Vis absorption spectra of N-doped and undoped TiO_2 powders

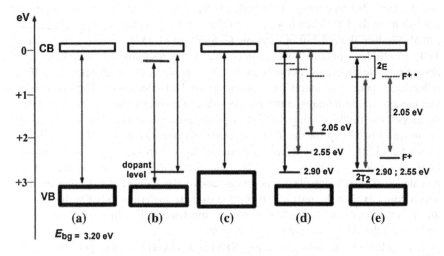

Fig. 4 Various schemes illustrating the possible changes that might occur to the band gap electronic structure of anatase TiO_2 on doping with various non-metals: **a** band gap of pristine TiO_2; **b** doped-TiO_2 with localized dopant levels near the VB and the CB; **c** band gap narrowing resulting from broadening of the VB; **d** localized dopant levels and electronic transitions to the CB; and **e** electronic transitions from localized levels near the VB to their corresponding excited states for Ti^{3+} and F^+ centers [52]

assign the peak around 398 eV to the O–Ti–N linkages in TiO_2 lattice [23]. Therefore, we concluded that nitrogen was doped into the TiO_2 lattices by substitution at the sites of the oxygen atoms. For the Ti 2p region, the Ti 2p3/2 and Ti 2p1/2 core levels appeared at 459 and 464 eV. For the O 1s region, the binding energies were around 530 eV. The Ti 2p and O 1s binding energies were similar to that in the pure TiO_2.

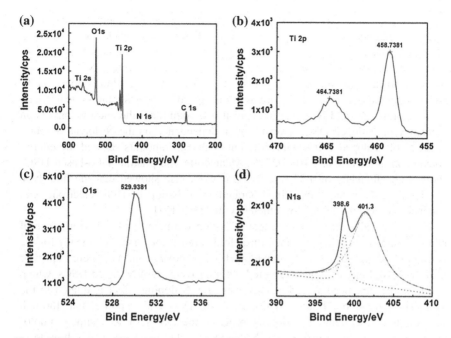

Fig. 5 XPS spectra of N-A-15: **a** survey; **b** Ti 2p; **c** O1s; **d** N1s [42]

3 Applications of N-doped TiO₂ into the DSCs

The porous semiconductor photoanode is an important part that influences the performance of DSCs. From the view of DSCs' photoanodes, the efficiency losses origin from photogenerated electron losses such as electron recombination during their transport to the substrate. The traps mainly come from the surface and lattice defect of TiO₂. Therefore, TiO₂ after nitrogen modifying is expected to decrease the electron losses. Our group and other researchers introduce N-doped TiO₂ to improve the performance of DSCs.

3.1 Effect of N-doping on the Overall Energy Conversion Efficiency

In our previous works, we reported highly efficient DSCs based on N-doped TiO₂ electrode (N-doped DSCs). The N-doped DSCs achieved a significant improvement in the energy conversion efficiency compared with the DSCs using P25 electrodes. Results show that N-doped TiO₂ electrodes could enhance the incident photo-to-current conversion efficiency (IPCE) and the overall conversion efficiency of the DSCs. Afterwards; we optimized the N-doped DSCs system, yielding a high efficiency of 10.1 % [55]. With careful evaluation of DSCs based on

N-doped and undoped TiO_2 prepared under the same dry method conditions, a 12.3 % enhancement of energy conversion efficiency was reached by N-doping. Then, we found that the nitrogen dopant type and amount influence the performance of N-doped TiO_2 photoanodes (Fig. 6) [56]. The different N dopants and wet methods affected the N-doping amount, the surface area of the N-doped TiO_2, and thereby the photovoltaic performance of the DSCs. By using the same nitrogen source (ammonia) and N-A wet method, it was found that the energy conversion efficiency of N-doped DSCs showed much dependence on the N dopant amount. A series of N-doped DSCs with different N dopant amounts showed the energy conversion efficiency of 5.01–7.27 %. Meanwhile, the pristine TiO_2-based DSCs showed an efficiency of 4.32 % only. Our work also showed that the fiber-type multiwall carbon nanotubes incorporated into N-doped TiO_2 electrode can enhance the electron collection efficiency of DSCs [57].

Yang et al. also reported the effect of N-doped amount on the performance of DSCs. Interestingly; they obtained three folds higher conversion efficiency for the optimum N-doped DSCs than the undoped ones, both J_{SC} and V_{OC} were improved [36]. In 2010, Sung et al. also reported the improvement of N-doped DSCs, which is due to the enhanced J_{SC} [58]. However, this N-doping effect is related to the synthesis method of N-doped TiO_2. In 2010, Dai et al. reported that through solvothermal treatment of N-doping process, the DSCs showed similar photovoltaic performance. However, the N-doped TiO_2 led to a more stable long-term stability and retarded electron recombination [44].

Overall, the N-doping modifying TiO_2 photoanodes contribute to the enhancement performance of DSCs either in J_{SC} or V_{OC}. We give a detailed discussion below.

3.2 Effect of N-Doping TiO$_2$ on the Short-Circuit Current (J$_{SC}$)

A significant enhancement of J_{SC} was achieved for N-doped DSCs [23, 55, 56, 58]. Our work suggested that the significantly enhanced photocurrent of the devices was found to be related to the N dopant amount and the change in surface property, which affects dye uptake amount in N-doped TiO_2 electrodes. We investigated the amount of dye adsorbed on the electrodes. The J_{SC} of the N-A and N-U solar cells were higher than that of pure TiO_2 solar cells, although the N-A electrodes possessed almost the same amount of dye as pure TiO_2 electrodes did, while the N-U electrodes obtained a lower dye uptake than that of pure TiO_2 electrodes. On the other hand, the isoelectric points of TiO_2 have an effect on the dye-loading. Surfaces with higher isoelectric points are preferable for the attachment of dye with acidic carboxyl groups [59]. During the wet method synthesis of N-doped TiO_2 precursor, hydrolysis of TTIP was conducted in solvent containing nitrogen sources with weak alkaline. Therefore, we can speculate that the pH-dependent zeta potential and the isoelectric points of N-doped TiO_2 were changed.

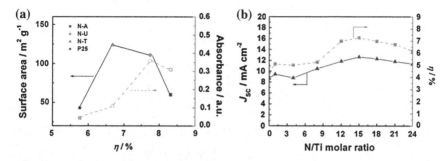

Fig. 6 **a** Effect of surface area and absorbance intensity (520 nm) on η [56]; **b** Effect of N/Ti molar ratio on J_{SC} and η [42]

Sung et al. also reported an enhanced photocurrent and efficiency in N-doped DSCs [58]; they attributed the enhancement to the increase of N-doped TiO₂ in the near-vis absorbance by nitrogen doping and partially to the morphological properties of the N-doped TiO₂ film. However, the visible light response of N-doped TiO₂ can only contribute photocurrent in tens of microamperes, which are far from enough to fill the gap caused by dye-sensitized films [55]. The efficient electron transport and retarded electron recombination can also lead to an increase in J_{SC} which will be discussed in the following section.

3.3 Effect of N-Doping TiO₂ on the Open-Circuit Voltage (V_{OC})

The increased V_{OC} of N-doped DSCs also enhanced the overall energy conversion efficiency. In theory, the V_{OC} of DSCs is determined by the difference between the Fermi level (E_F) of semiconductor and potential of redox couples [60]. It is helpful to get information about whether N-doping would cause a shift of E_F and thus the V_{OC}.

Dai et al. measured the V_{fb} of N-doped and undoped TiO₂ films [44]. They found that V_{fb} of N-doped TiO₂ shifts to the negative by 0.06 and 0.1 eV compared with that in the pure TiO₂ electrode (Fig. 7).

In the previous literature, Hashimoto et al. reported that the flatband potentials of N-doped TiO₂ tend to shift to a positive direction [61]. Kisch et al. observed that the quasi-Fermi level of electrons is anodically shifted by 0.07–0.16 eV [62]. Higashimoto et al. reported that the flatband potential of N-doped TiO₂ is not influenced by small amounts of nitrogen species doped into TiO₂ [63]. Therefore, there is still no conclusion about the change in E_F of TiO₂ after N-doping.

Our group used surface photovoltage spectroscopy (SPS) to measure the energy levels of bare N-doped TiO₂ films and dye-sensitized N-doped TiO₂ electrodes [55]. In Fig. 8a, we can see that an impurity level exists from where photoexcited electrons are injected into the conduction band, indicating that nitrogen is doped

Fig. 7 Absorbance measured
at 780 nm as a function of
applied potential for Undoped
and N-doped TiO_2 electrode
[44]

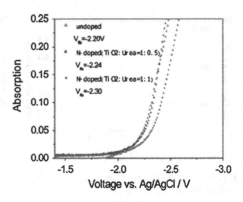

into the TiO_2 lattice, where it forms nitrogen-induced states. In Fig. 8b, the observed signal starting at around 740 nm (scanning from long to short wavelength) for N-doped and undoped TiO_2 is due to electron injection from the N719 dye into the conduction band of TiO_2. We observed that the signal for dye-sensitized N-doped ST-01 with respect to the undoped ST-01 is blueshifted at about 20–40 nm. This result may indicate the shift in the electron quasi-Fermi level in N-doped TiO_2.

We further investigated the relationship between voltage and charge using a charge extraction technique [64]. The similar slope of voltage-charge plots of the N-doped and undoped TiO_2 solar cells (N-A and TiO_2 DSCs) indicate a similar trap distribution (Fig. 9). However, if sustaining a certain voltage, more charge needs to be present in the TiO_2-based DSCs. The relationship between charge and voltage revealed that less charge is needed to get a high V_{OC} in N-doped DSCs.

We also noticed that the increase in V_{OC} cannot always be observed, which is also dependent on the synthesis method of N-doped TiO_2. The synthesis method may influence the type of N-doping, such as lattice perfection, interstitial doping, or just physical adsorption. Therefore, much effort would be made to further understand the doping mechanism in energy level of N-doped TiO_2.

3.4 Long-Term Stability

In DSCs system, photoanodes made of TiO_2 nanomaterial are not photoexcited instead of dyes. However, N-doping would cause TiO_2 visible light absorption response, which is widely applied to photocatalysis area. Whether N-doped TiO_2 can possibly accelerate the deterioration of the dye or DSC system is always a concern.

The stability test for N-doped DSCs was conducted in our previous work (Fig. 10), N-doped DSCs were examined during irradiation for 2000 h under white light illumination (100 mW/cm^2) at 25 °C [23]. The N-doped DSCs possessed good stability with efficiency and photovoltage maintaining its initial values above 90 %. Dai group also reported great stability of N-doped DSCs, the efficiency of

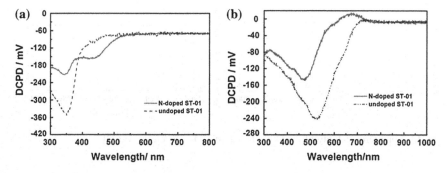

Fig. 8 Surface photovoltage spectra: **a** bare N-doped and undoped ST-01 electrodes and **b** after N719-sensitization [55]

Fig. 9 V_{OC} versus Q_{SC}
(N-A: N-doped DSCs) [56]

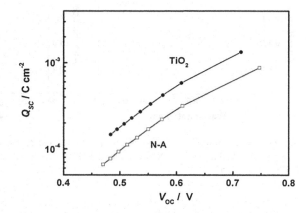

which remained at nearly at 80 % compared with the 72 % of the undoped DSC at 70 °C over 1,000 h [44].

Further investigation on the effect of N-doped TiO₂ on the deterioration of the dye was also performed [55]. The dye stability tests consisted of examining the dyes adsorbed on the N-doped TiO₂ films and the photoanodes of the solar cells. In Fig. 11a, compared with UV-Vis absorption spectra of fresh N719, the dyes desorbed from N-doped and undoped TiO₂ film were destroyed but to a similar extent. In Fig. 11b, the absorption of fresh dye is almost the same with those in N-doped and undoped TiO₂ photoanodes. These results indicate that N-doped TiO₂ did not accelerate the dye deteriorations.

3.5 Application of N-doped TiO₂ Photoanodes into QDSCs

Except for DSCs, N-doped TiO₂ nanomaterials were also applied into quantum-dot sensitized solar cells (QDSCs).

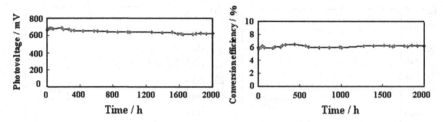

Fig. 10 Long-term stability of the DSCs based on $TiO_{2-x}N_x$ electrode [23]

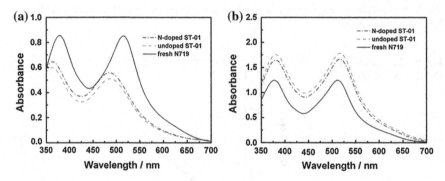

Fig. 11 UV-Vis absorption spectra of **a** fresh N19 dye and the dyes desorbed from N-doped and undoped ST-01 films; **b** fresh N19 dye and the N719 dye desorbed from N-doped and undoped DSCs devices [55]

Zhang et al. combined N-doped TiO_2 and CdSe QDs to assemble QDSCs. Compared with QDSCs using pristine TiO_2, the photocurrent of CdSe-sensitized N-doped TiO_2 QDSCs was significantly enhanced. They ascribed this photocurrent enhancement to the extra pathways for charge transfer introduced by nitrogen doping (Fig. 12) [65]. Kang et al. reported the hierarchical N-doped TiO_2 nano-clumps as photoanodes of CdSe-sensitized QDSCs. They found that the performance of QDSCs was improved by 145 % when using N-ion doping of TiO_2 photoanodes. One of the two explanations about the significant improvement of QDSCs was the increased recombination resistance at TiO_2/QDs/electrolyte interface, which was caused by the decreased surface states and oxygen vacancies after nitrogen doping in TiO_2 [47].

Fig. 12 Schematic electronic band structure of 3.5 nm CdSe with an effective band gap of 2.17 eV and nanocrystalline TiO$_2$/N with a 3.2 eV band gap [65]

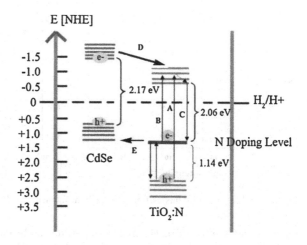

4 Electron Kinetic Behaviors in DSCs Based on N-doped TiO$_2$ Photoanodes

4.1 Charge Transport and Electron Lifetime

Understanding electron transport in the N-doped DSCs is helpful to further improve the performance of semiconductor photoanodes. We investigated the charge transport by intensity-modulated photocurrent and photovoltage spectroscopy (IMPS/IMVS) to study the effect of N-doping treatment [55]. Usually, IMPS/IMVS measurements were affected by particle size in correlation with surface area and morphology, resulting in differences in the number of particles and the quantity of the charge associated with that electrode. When comparing the doped samples with pure samples, we needed to assure they possessed similar particle sizes and surface area to obtain comparability results. Therefore, we studied the N-doping effect by using the N-doped DSCs (N-A) solar cells and pure TiO$_2$ solar cells, which had similar particle sizes (22.74 and 23.13 nm, respectively).

The electron transport time (τ_{tr}) and electron lifetime (τ_e) were deduced from IMPS and IMVS results (Fig. 13) [55]. All of the time constants followed a general trend; they decreased as the light intensity increased. Fast electron transport and short electron lifetime existed for the N-A. Commonly, fast electron transport can improve the charge-collection efficiency and thus increase the J_{SC}. In general, N-doping would cause some defects in TiO$_2$ lattice, which would cause more traps in the doped films. However, we observed similar slopes, suggesting similar trap distributions.

V_{OC} decay measurements were performed to further clarify the electron lifetime. As shown in Fig. 14a, the V_{OC} decay is slightly different in the N-A DSCs, which follows the sequence of N-A > pure TiO$_2$. Additionally, the decay is faster in the N-A solar cell, which is also in agreement with the shorter lifetime found in

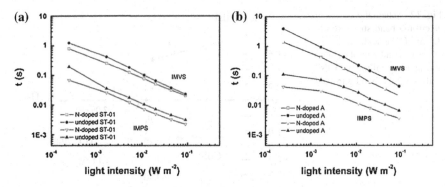

Fig. 13 IMPS time constants as functions of light intensity for N-doped and TiO$_2$ DSCs [55]

the N-A DSCs (Fig. 14b). The electron lifetime can be calculated from the voltage transients using Eq. (1) [56]:

$$\tau_e = -\frac{kT}{e}\left(\frac{dV\text{oc}}{dt}\right)^{-1} \tag{1}$$

where k is the Boltzmann constant, T the absolute temperature, and e the positive elementary charge. The calculated electron lifetimes are shown in Fig. 14b as functions of the open-circuit potential. Specifically, the lifetimes increased exponentially when the voltage decreased. However, the lifetime of the N-A solar cell was shorter than in the pure TiO$_2$ solar cells at U < 0.6 V, but it was longer when U > 0.6 V.

The electron lifetime can also be deduced from EIS spectra. The electron lifetime can be estimated by the maximum frequency in the EIS spectra as described: $\tau_e = (2\pi f_{max})^{-1}$ [47]. We conducted a linear fit of electron lifetime, which tended to decrease as the N/Ti molar ratio increased (Fig. 14c).

Therefore, the N-doping treatment can improve the electron transport but decrease the electron lifetime in DSCs.

4.2 Electron Recombination

The difference between N-doped DSCs and TiO$_2$ DSCs with respect to their charge transfer properties was also studied by EIS analysis [47–49]. The Nyquist diagram typically features three semicircles in order of increasing frequency. These three semicircles correspond to the following: the Nernst diffusion within the electrolyte, electron transport at the oxide/electrolyte interface, and redox reaction at the platinum counter electrode.

The main concerns in N-doped DSCs are the following: (1) the impedance due to electron transfer from the conduction band of the mesoscopic film to triiodide

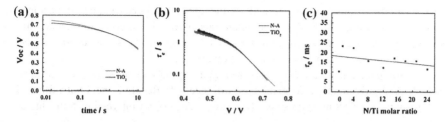

Fig. 14 **a** Open-circuit voltage decay transients of the dye-sensitized N-doped and pure TiO$_2$ solar cells; **b** calculated electron lifetime (Eq 1) versus open-circuit voltage [56]; **c** τ_e versus N/Ti molar ratio [42]

Fig. 15 EIS spectra of N-doped DSCs [44]

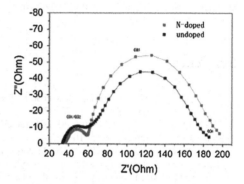

ions in the electrolyte and (2) the back reaction at the TiO$_2$/electrolyte interface, presented by the semicircle in intermediate-frequency regime. Dai et al. reported a retarded electron recombination in the N-doped DSCs. This retarded electron recombination may be due to the change in surface properties, e.g., the lattice perfection (Fig. 15). The charge transfer resistances demonstrated dependency on the N dopant amount. The electron lifetime of N-doped DSCs tended to decrease as the N dopant increased.

Overall, fast electron transport, short electron lifetime, and retarded electron recombination were found in N-doped DSCs. Moreover, a synergistic effect of high dye uptake and efficient electron transport contributed to the improvement of N-doped DSCs.

5 Summary and Outlook

In conclusion, the recent development of N-doped TiO$_2$ nanomaterials and its application into DSCs is summarized. The different synthesis approaches, nitrogen dopant types, and amount can affect the physical and chemical properties of N-doped TiO$_2$, thereby their performance in the photoanodes of DSCs. Moreover, nitrogen doping can help charge separation and transport in QDSCs. Based on these results, the synergistic effect of higher dye uptake, N dopant amount, and faster electron transport contributed to the enhanced performance of N-doped DSCs. Therefore, N-doped TiO_2 nanomaterials are good semiconductor candidates for highly efficient photoanodes of DSCs. The application of N-doped TiO_2 is also widely extended into photocatalysis and other areas.

The fundamental research on how nitrogen doping enhances the charge transport and photovoltaic performance is still needed. Besides, the controlling of nanostructure and doping process of N-doped TiO_2 is an interesting topic in the future.

Acknowledgements This work was supported by NSFC (Grant No. 50773008) and State Key Laboratory of New Ceramic and Fine Processing (Tsinghua University). This work was also supported by the National High Technology Research and Development Program for Advanced Materials of China (Grant No. 2009AA03Z220).

References

1. O'Regan B, Grätzel M (1991) A low-cost, high-efficiency solar cell based on dye-sensitized colloidal TiO_2 films. Nature 353:737–740
2. Wang Q, Ito S, Grätzel M et al (2006) Characteristics of high efficiency dye-sensitized solar cells. J Phys Chem B 110:25210–25221
3. Yella H-W, Lee HN, Tsao C et al (2011) Porphyrin-sensitized solar cells with cobalt (II/III)-based redox electrolyte exceed 12 percent efficiency. Science 334:629–634
4. Grätzel M (2001) Photoelectrochemical cells. Nature 414(6861):338–344
5. Zhang Q, Chou TP, Russo B et al (2008) Aggregation of ZnO nanocrystallites for high conversion efficiency in dye-sensitized solar cells. Angew Chem Int Ed 47:2402–2406
6. Lana-Villarreal T, Boschloo G, Hagfeldt A (2007) Nanostructured zinc stannate as semiconductor working electrodes for dye-sensitized solar cells. J Phys Chem C 111:5549–5556
7. Zheng H, Tachibana Y, Kalantar-zadeh K (2010) Dye-sensitized solar cells based on WO_3. Langmuir 26:19148–19152
8. Yang S, Kou H, Wang J et al (2010) Tunability of the band energetics of nanostructured $SrTiO_3$ electrodes for dye-sensitized solar cells. J Phys Chem C 114:4245–4252
9. Sayama K, Sugihara H, Arakawa H (1998) Photoelectrochemical properties of a porous Nb_2O_5 electrode sensitized by a ruthenium dye. Chem Mater 10:3825–3832
10. Qian JF, Liu P, Xiao Y et al (2009) TiO_2-coated multilayered SnO_2 hollow microspheres for dye-sensitized solar cells. Adv Mater 21:3663–3667
11. Turković, Z. Crnjak Orel Z (1997) Dye-sensitized solar cell with CeO_2 and mixed CeO_2/SnO_2 photoanodes. Solar Energy Mater Solar Cells 45:275–281
12. Hu Y, Zheng Z, Jia H et al (2008) Selective Synthesis of FeS and FeS_2 nanosheet films on iron substrates as novel photocathodes for tandem dye-sensitized solar cells. J Phys Chem C 112:13037–13042
13. Qin P, Linder M, Brinck T et al (2009) High incident photon-to-current conversion efficiency of P-type dye-sensitized solar cells based on NiO and organic chromophores. Adv Mater 21:2993–2996
14. Inakamura N, Negishi S Kutsuna et al (2000) Role of oxygen vacancy in the plasma-treated TiO_2 photocatalyst with visible light activity for NO removal. J Mol Catal A: Chem 161:205–212

15. Ihara T, Miyoshi M, Iriyama Y et al (2003) Visible-light-active titanium oxide photocatalyst realized by an oxygen-deficient structure and by nitrogen doping. Appl Catal B-Environ 42:403–409
16. Irie H, Watanabe Y, Hashimoto K (2003) Nitrogen-concentration dependence on photocatalytic activity of TiO$_{2-x}$N$_x$ powders. J Phys Chem B 107:5483–5486
17. Wang K-P, Teng H (2009) Zinc-doping in TiO$_2$ films to enhance electron transport in dye-sensitized solar cells under low-intensity illumination. Phys Chem Chem Phys 11:9489–9496
18. Zhang Z, Zhao X, Wang T et al (2010) Increasing the oxygen vacancy density on the TiO$_2$ surface by La-doping for dye-sensitized solar cells. J Phys Chem C 114:18396–18400
19. Liu H, Yang W, Tan X et al (2010) Photovoltaic performance improvement of dye-sensitized solar cells based on tantalum-doped TiO$_2$ thin films. Electrochim Acta 56:396–400
20. Lu X, Mou X, Wu J (2010) Improved-performance dye-sensitized solar cells using Nb-doped TiO$_2$ electrodes: efficient electron injection and transfer. Adv Func Mater 20:509–515
21. Zhang X, Wang S-T, Wang Z-S (2011) Effect of metal-doping in TiO$_2$ on fill factor of dye-sensitized solar cells. Appl Phys Lett 99:113503
22. Asahi R, Morikawa T, Ohwaki T et al (2001) Visible-light photocatalysis in nitrogen-doped titanium oxides. Science 293:269–271
23. Ma T, Akiyama M, Abe E et al (2005) High-efficiency dye-sensitized solar cell based on a nitrogen-doped nanostructured titania electrode. Nano Lett 5:2543–2547
24. Hou Q, Zheng Y, Chen J-F et al (2011) Visible-light-response iodine-doped titanium dioxide nanocrystals for dye-sensitized solar cells. J Mater Chem 21:3877–3883
25. Tian H, Hu L, Zhang C et al (2011) Enhanced photovoltaic performance of dye-sensitized solar cells using a highly crystallized mesoporous TiO$_2$ electrode modified by boron doping. J Mater Chem 21:863–868
26. Chu D, Yuan X, Qin G et al (2008) Efficient carbon-doped nanostructured TiO$_2$ (anatase) film for photoelectrochemical solar cells. J Nanopart Res 10:357–363
27. Irie H, Washizuka S, Yoshinob N et al (2003) Visible-light induced hydrophilicity on nitrogen-substituted titanium dioxide films. Chem Commun 1298–1299
28. Fu H, Zhang L, Zhang S et al (2006) Electron spin resonance spin-trapping detection of radical intermediates in N-doped TiO$_2$-assisted photodegradation of 4-chlorophenol. J Phys Chem B 110:3061–3065
29. Burda C, Lou Y, Chen X et al (2003) Enhanced nitrogen doping in TiO$_2$ nanoparticles. Nano Lett 3:1049–1051
30. Livraghi S, Paganini MC, Giamello E et al (2006) Origin of photoactivity of nitrogen-doped titanium dioxide under visible light. J Am Chem Soc 128:15666–15671
31. Mrowetz Balcerski W, Colussi AJ et al (2004) Oxidative power of nitrogen-doped TiO$_2$ photocatalysts under visible illumination. J Phys Chem 108:17269–17273
32. Etacheri V, Seery MK, Hinder SJ et al (2010) Highly visible light active TiO$_{2-x}$N$_x$ heterojunction photocatalysts. Chem Mater 22:3843–3853
33. Torres GR, Lindgren T, Lu J et al (2004) Photoelectrochemical study of nitrogen-doped titanium dioxide for water oxidation. J Phys Chem B 108:5995–6003
34. Kitano K, Funatsu M et al (2006) Preparation of nitrogen-substituted TiO$_2$ thin film photocatalysts by the radio frequency magnetron sputtering deposition method and their photocatalyticreactivity under visible light irradiation. J Phys Chem B 110:25266–25272
35. Zhang J, Li M, Feng Z et al (2006) UV Raman spectroscopic study on TiO$_2$. I. Phase transformation at the surface and in the bulk. J Phys Chem B 110:927–935
36. Wang X, Yang Y, Jiang Z et al (2009) Preparation of TiNxO2−x photoelectrodes with NH$_3$ under controllable middle pressures for dye-sensitized solar cells. Eur J Inorg Chem 2009:3481–3487
37. Zhang J, Wu Y, Xing M (2010) Development of modified N doped TiO$_2$ photocatalyst with metals, nonmetals and metal oxides. Energy Environ Sci 3:715–726
38. Nakamura R, Tanaka T, Nakato Y (2004) Mechanism for visible light responses in anodic photocurrents at N-doped TiO$_2$ film electrodes. Phys Chem B 108:10617–10620

39. Kobayakawa K, Murakami Y, Sato Y (2005) Visible-light active N-doped TiO_2 prepared by heating of titanium hydroxide and urea. J Photochem Photobiol A: Chem 170:177–179

40. Chen X, Lou Y, Samia ACS et al (2005) Formation of oxynitride as the photocatalytic enhancing site in nitrogen-doped titania nanocatalysts: comparison to a commercial nanopowder. Adv Mater 15:41–49

41. Guo W, Miao Q, Xin G et al (2011) Dye-sensitized solar cells based on nitrogen-doped titania. Key Eng Mater 451:21–27

42. Guo W, Shen Y, Wu L et al (2011) Effect of N dopant amount on the performance of N-doped TiO_2 electrodes for dye-sensitized solar cells. J Phys Chem C 115:21494–21499

43. Sato S, Nakamura R, Abe S (2005) Visible-light sensitization of TiO_2 photocatalysts by wet-method N doping. Appl Catal A Gen 284:131–137

44. Tian H, Hu L, Zhang C et al (2010) Retarded charge recombination in dye-sensitized nitrogen-doped TiO_2 solar cells. J Phys Chem C 114:1627–1632

45. Jagadale TC, Takale SP, Sonawane RS et al (2008) N-doped TiO_2 nanoparticle based visible light photocatalyst by modified peroxide sol–gel method. J Phys Chem C 112:14595–14602

46. Sivaranjani K, Agarkar S, Ogale SB et al (2012) Toward a quantitative correlation between microstructure and DSSC efficiency: a case study of $TiO_{2-x}N_x$ nanoparticles in a disordered mesoporous framework. J Phys Chem C 116(3):2581–2587

47. Sudhagar P, Asokan K, Ito E et al (2012) N-ion-implanted TiO_2 photoanodes in quantum dot-sensitized solar cells. Nanoscale 4:2416–2422

48. Stroud PT (1972) Ion bombardment and implantation and their application to thin films. Thin Solid Films 11:1–26

49. Kitano M, Funatsu K, Matsuoka M et al (2006) Preparation of nitrogen-substituted TiO_2 thin film photocatalysts by the radio frequency magnetron sputtering deposition method and their photocatalytic reactivity under visible light irradiation. J Phys Chem B 110:25266–25272

50. Lindgren T, Mwabora JM, Avendaño E et al (2003) Photoelectrochemical and optical properties of nitrogen doped titanium dioxide films prepared by reactive DC magnetron sputtering. J Phys Chem B 107:5709–5716

51. Qiu X, Burda C (2007) N-doped TiO_2: theory and experiment. Chem Phys 339:44–56

52. Serpone N (2006) Is the band gap of pristine TiO_2 narrowed by anion- and cation-doping of titanium dioxide in second-generation photocatalysts? J Phys Chem B 110:24287–24297

53. Kuznetsov VN, Serpone N (2006) Visible light absorption by various titanium dioxide specimens. J Phys Chem B 110:25203–25209

54. Chen X, Burda C (2008) The electronic origin of the visible-light absorption properties of C-, N- and S-doped TiO_2 nanomaterials. J Am Chem Soc 130:5018–5019

55. Guo W, Wu L, Chen Z et al (2011) Highly efficient dye-sensitized solar cells based on nitrogen-doped titania with excellent stability. J Photochem Photobiol A Chem 219:180–187

56. Guo W, Shen Y, Boschloo G et al (2011) Influence of nitrogen dopants on N-doped TiO_2 electrodes and their applications in dye-sensitized solar cells. Electrochim Acta 56:4611–4617

57. Guo W, Shen Y, Wu L et al (2011) Performance of dye-sensitized solar cells based on MWCNT/TiO2-xNx nanocomposite electrodes. Eur J Inorg Chem 2011:1776–1783

58. Kang SH, Kim HS, Kim J-Y et al (2010) Enhanced photocurrent of nitrogen-doped TiO_2 film for dye-sensitized solar cells. Mater Chem Phys 124:422–426

59. Kay A, Grätzel M (2002) Dye-sensitized core—shell nanocrystals: improved efficiency of mesoporous tin oxide electrodes coated with a thin layer of an insulating oxide. Chem Mater 14:2930–2935

60. Cahen D, Hodes G, Grätzel M et al (2000) Nature of photovoltaic action in dye-sensitized solar cells. J Phys Chem B 104:2053–2059

61. Irie H, Washizuka S, Watanabe Y et al (2005) Photoinduced hydrophilic and electrochemical properties of nitrogen-doped TiO_2 films. J Electrochem Soc 152:351–356

62. Kisch H, Sakthivel S, Janczarek M et al (2007) A low-band gap, nitrogen modified titania visible-light photocatalyst. J Phys Chem C 111:11445–11449

63. Higashimoto S, Azuma M (2009) Photo-induced charging effect and electron transfer to the redox species on nitrogen-doped TiO$_2$ under visible light irradiation. Appl Catal B Environ 89:557–562
64. Duffy W, Peter LM, Rajapakse RMG et al (2000) A novel charge extraction method for the study of electron transport and interfacial transfer in dye sensitised nanocrystalline solar cells. Electrochem Commun 2:658–662
65. López-Luke T, Wolcott A, Xu L et al (2008) Nitrogen-doped and CdSe quantum-dot-sensitized nanocrystalline TiO$_2$ films for solar energy conversion applications. J Phys Chem C 112:1282–1292

Low-Cost Pt-Free Counter Electrode Catalysts in Dye-Sensitized Solar Cells

Mingxing Wu and Tingli Ma

Abstract Dye-sensitized solar cells (DSCs) are potential candidates to silicon solar cells due to their merits of simple fabrication procedure, low-cost, and good plasticity. Till date, great advances have been achieved in the design of dyes, redox couples, and counter electrode (CE) catalysts for DSCs and the highest energy conversion efficiency is up to 12.3 %. In this part, our attention focuses on CE catalysts. Besides Pt, carbon materials, conductive polymers, transition metal compounds (carbides, nitrides, oxides, sulfides, phosphides, selenides), and composite catalysts, denoted as Pt-free catalysts, have been introduced into DSCs as CE catalysts. In the following sections, we give a summary of Pt-free CE catalysts and highlight the advantages and disadvantages of each variety of Pt-free catalysts.

1 Carbon materials

Carbon materials possess the merits of low-cost, high catalytic activity and electric conductivity, good thermal stability, and corrosion resistance. These merits make carbon materials ideal substitutes to the expensive Pt catalyst in many fields. In the DSCs system, they have been widely used as counter electrode (CE) catalysts. Takahashi et al. used activated carbon (Ca), a type of amorphous carbon with a

M. Wu
Department of Chemistry and Material Science, Hebei Nomal University, Shijiazhuang, China

T. Ma (✉)
State Key Laboratory of Fine Chemicals, Dalian University of Technology, Dalian, China
e-mail: tinglima@dlut.edu.cn

T. Ma
Kyushu Institute of Technology, 2-4 Hibikino, Wakamatsu-ku, Kitakyushu, 808-0196, Japan

Z. Lin and J. Wang (eds.), *Low-cost Nanomaterials*, Green Energy and Technology, DOI: 10.1007/978-1-4471-6473-9_4, © Springer-Verlag London 2014

diamond structure, as a CE catalyst in DSCs [1]. The photovoltaic performance is positively correlated with roughness factor of the Ca CE. In contrast, the roughness factor is negatively correlated with the charge-transfer resistance (R_{ct}) in the CE/ electrolyte interface. The DSC using Ca CE showed power conversion efficiency (PCE) of 3.89 %, slightly lower than that of the DSC using Pt CE (Pt-DSC) which produced a PCE of 4.30 %. Grätzel and Kay introduced graphite and carbon black (Cb) into a monolithic DSC as the CE catalyst, and the DSC produced a PCE of 6.7 % [2]. In this kind of carbon CE, graphite improved the lateral conductivity of the CE and Cb granted the CE a large surface area, resulting in high catalytic activity. They also found that the thickness of Cb film affected the fill factor (FF) and PCE significantly, while open-circuit voltage (V_{oc}) and short-circuit current density (J_{sc}) varied very little as the Cb film thickness increased [3]. The DSC using Cb CE with 14.47 μm thickness produced the highest PCE of 9.1 %.

Zou et al. fabricated fiber-shaped DSCs using carbon fiber (Cf) as CE and the DSC gave a PCE of 2.7 % [4]. Wang et al. introduced mesoporous carbon (Cm) into DSCs as CE, yielding a PCE of 6.18 % [5]. Ramasamy et al. prepared fer-rocene-derivatized large pore size mesocellular carbon foam (Fe–MCF–C) used as CE catalyst in a DSC which gave a PCE of 7.89 % [6]. Carbon nanotubes (CNTs) can be divided into single-wall CNTs (SNTs), double-wall CNTs (DNTs), and multi-wall CNTs (MNTs) and they can be designed as a semiconductor or metallic material according to the varied chiralities [7]. Using CNTs to replace Pt can endow with CE the following advantages: nanoscaled transfer channels, large specific surface area, low-cost, high catalytic activity, and light weight. The DSC using SNTs as CE yielded a PCE of 4.5 % [8]. Lee et al. applied MNTs as CE in the DSCs. The DSCs using MNTs and Pt CEs showed PCE of 7.67 % (MNTs) and 7.83 % (Pt) [9]. The high density of the defect-rich edge planes of MNTs guar-antees its high catalytic activity. The technique adopted to prepare CNTs CEs is crucial for obtaining high catalytic activity. Kim et al. prepared CNTs CEs with screen printing (SP) technique and chemical vapor deposition (CVD) technique. The CNTs (SP) were randomly oriented and woven into each other, whereas the CNTs (CVD) were grown directly on the substrate. The DSC using the CNTs (SP) CE gave a CPE of 8.03 %, while the well aligned CNTs CE made the PCE value reach to 10.04 % [10]. The advantages of CNTs (CVD) CE can be attributed to high conductivity because of the well-aligned arrangement. Besides SP and CVD techniques, spraying technique can be used to fabricate CNTs CE [11]. The CNTs film thickness was regulated by spraying time and the impact of CNTs film thickness (or spray time) on the performance of DSCs was also investigated. As the time increased from 5 to 30 s, the J_{sc} and FF increased rapidly and the PCE value improved from 0.68 to 3.39 %. The highest PCE of 7.59 % was obtained at spraying time of 200 s. In their work, the PCE increased continuously with spraying time (0–200 s). According to our experiences, there should exist an optimal spraying time (i.e., thickness).

Zhu et al. attempted to find the differences between SNTs, DNTs, and MNTs as CEs under the same conditions [12]. The DSC using SNTs CE performed best (1.46 %), compared with the other two DSCs based on DNTs (0.45 %) and MNTs

(0.62 %) CEs. The highest catalytic activity of the SNTs for regeneration of the I_3^-/I^- redox couple was caused by the one-dimensional nano-feature which provides better electron transport. Moreover, the impurities (iron, amorphous carbon) in SNTs can influence the catalytic performance dramatically. For example, the DSC using purified SNTs CE yielded a PCE of 1.46 %, while the DSC using raw SNTs CE gave a poor PCE of 0.57 %. This difference caused by catalyst poisoning, because the catalytic sites was occupied by impurities. The purification process in turn introduced many oxygen-function groups that formed new catalytic sites.

Graphene is a single layer of two-dimensional graphite with advantages of high conductivity, transparency, hardness, and corrosion resistance, and it has become a hot research topic in various fields. Grätzel et al. used graphene to fabricate optically transparent CE for DSCs [13]. It was observed that graphene CE was more suitable for ionic liquid solvent than the traditional organic solvent. The R_{ct} value of the ionic solvent is much smaller than the traditional solvent, which indicates that the mechanism for regeneration of I_3^-/I^- in the graphene surface is determined by solution events rather than viscosity. The catalytic activity of graphene is proportional to the content of active sites (edge defects and oxide groups). Finally, they suggested that graphene might be a promising substitute for Pt and also for the expensive FTO conductive layer. Aksay et al. found the catalytic activity was correlated with the concentration of the oxide group and the C/O ratio strongly influenced the catalytic activity [14]. When the C/O ratio was up to 13, graphene CE showed the highest catalytic activity, and the DSC gave a PCE of 5.0 %, close to that of the DSC using Pt CE (5.5 %). Jeon et al. synthesized graphene by reducing graphite oxide. It was found that the catalytic activity increased as the number of oxygen functional groups decreased [15]. Combining with previous results [14], we thought that there existed an optimum number of oxygen functional groups for graphene to achieve optimal catalytic activity. The PCE values for the DSCs in the above mentioned works were all less than 6 %, leaving much room for improvement by modifying the concentration of the oxygen functional groups or lattice defect. Nevertheless, to fully evaluate these statements requires more research.

Ma group compared the properties of nine kinds of carbon materials containing Ca, Cb, conductive carbon (Cc), carbon dye (Cd), Cf, CNTs, Ordered mesoporous carbon (Com), discard toner (Cp), and C_{60} at the same conditions [16]. Com is made of rows of well-ordered carbon walls with a width of approximate 10 nm. This carbon wall configuration forms many channels in the CE body which can increase the contact area of the electrolyte and the CE surface and this configuration can promote electrolyte diffusion. The DSC using Com CE yielded a high PCE of 7.5 %, the same as the DSC using Pt CE. The traditional carbon materials (Ca, Cb, Cc, CNTs, and Cf) showed decent catalytic activity and the DSCs yielded PCE values from 6.3 % to 7.0 %. Cd showed catalytic activity as high as Com, and the DSC gave a PCE of 7.5 %. Even the DSC using Cp CE showed a decent PCE of 4.3 %. When impure C_{60} CE was used in DSC, a low PCE of 2.8 % was achieved. As predicted, C_{60} should perform as effectively as the other carbon materials due to its superior electrical conductivity, stability, and other attributes. The low activity may be caused by impurities, which may result in catalyst poisoning. They indicated that the main disadvantage of carbon CEs is the poor bonding strength between the carbon films and the substrate. Adding TiO_2 can improve bonding strength between carbon films and the substrate, but it was found that adding too much TiO_2 can lower the catalytic activity, because of the bad conductivity of TiO_2. Meanwhile, the impact of carbon film thickness on the catalytic activity has been investigated. An excessively thin carbon film can lead to insufficient catalytic activity while an excessively thick carbon film can crack and detach from the substrate. To achieve high catalytic activity, carbon film with 25 μm thickness is appropriate. The thickness of the carbon films affects FF significantly.

Lee et al. studied the impact of carbon particle size on the catalytic activity [17]. The DSC using the nanocarbon (surface area, 100 $m^2 g^{-1}$) CE gave a PCE of 6.73 %, much higher than the DSC using microcarbon (surface area, 0.4 $m^2 g^{-1}$) CE. The different behavior between the two kinds of carbon materials can be attributed to the surface area and conductivity. Meng et al. prepared flexible carbon CE using Ca as the catalyst and graphite sheet as the substrate [18]. Low (series resistance) R_s and R_{ct} were obtained for this carbon CE due to high electrical conductivity of graphite and high catalytic activity of Ca. The DSC using this CE showed a PCE of 6.46 % which can match the performance of the DSC using Pt CE (6.37 %).

Carbon is indeed a qualified CEs catalyst in DSCs. However, the main disadvantage of carbon CE is still the poor bonding strength between carbon film and the substrate. This may be a potential unstable factor for long-term use. Deposited carbon film on the substrate by in situ technique could resolve this problem. Furthermore, fabricating incorporated carbon CE (carbon catalysts and carbon substrate) may be another solution. Opacity is another disadvantage of carbon CEs. We believe developing transparent carbon CEs for DSCs will become a promising research topic in the future.

2 Conductive Polymers

Transparency and flexibility are two merits for DSCs which require transparent flexible photoanode, transparent electrolyte, and transparent flexible CE. Generally, transparent flexible substrate (ITO–PET, ITO–PEN) deposited with Pt is widely used as CE. Besides Pt, conductive polymers like poly (3, 4-ethylenedioxythiophene) (PEDOT) or its derivatives can be used as CE catalysts.

PEDOT-Polystyrenesulfonate (PSS) was used as CE catalyst in quasi-solid DSCs in which this PEDOT performed better than Pt for ionic liquid electrolyte (ILE) [19]. In the EIS test, the R_{ct} value for the PEDOT–PSS/ILE was much lower than Pt/ILE. However, in the organic liquid electrolyte (OLE), the result was opposite. Yanagida et al. gave a detailed explanation for the aforementioned phenomenon [20]. ILE needs high I_2 concentration due to high viscosity and low conductivity. Therefore, the porous PEDOT CE performed better than Pt in ILE. Gîrţu et al. indicated that there existed charge transfer between PEDOT and iodide redox couple by X-ray photoelectron spectroscopy (XPS) [21]. This phenomenon was also observed by Biallozor et al. through CV test [22]. Apart from PEDOT, Poly (3,3-diethyl-3,4-dihydro-2H-thieno-[3,4-b] [1, 4] dioxepine) (PProDOT-Et$_2$) also can be used as CE catalyst for DSCs [23]. PProDOT-Et$_2$ film was directly deposited on FTO substrate by electropolymerization with various charge capacities (10, 20, 40, 80, 120, 160, 200 mC cm^{-2}). When the deposited charge capacity reached 40 mC cm^{-2}, the PProDOT-Et$_2$ film produced the largest active surface area, contributing to the high catalytic activity similar to Pt. The DSC using this polymer CE yielded a PCE of 5.20 %.

Wu et al. prepared polypyrrole nanoparticles (PPy) with particle size ranging from 40 to 60 nm [24]. The DSC using PPy CE showed a PCE of 7.66 %, higher than the photovoltaic performance of the DSC using Pt CE (6.90 %). Xia et al. fabricated PPy CEs on FTO glass with vapor phase polymerization (VPP) and electropolymerization (EP) techniques [25]. The PPy particle (VPP) size is 100–150 nm and the PPy (EP) has a larger particle size of 200–300 nm. The DSCs using these PPy CEs yielded PCE of 3.4 % (VPP) and 3.2 % (EP), slightly lower than Pt-DSC (4.4 %). In addition, PPy sphere with a uniform size of 85 nm was used as CE catalyst for DSCs and decent PCE was obtained [26]. Zhao et al. prepared transparent polyaniline (PANI) CE on FTO glass for DSCs CE [27]. At front illumination, the DSC showed a PCE of 6.54 %; at rear illumination, a PCE of 4.26 % was achieved. Moreover, Wu et al. fabricated mesoporous PANI thorough an oxidative polymerization of aniline monomer as CE for DSCs and a PCE of 7.15 % was obtained [28].

Organic polymer CEs own the advantages of transparency, high catalytic activity, easy availability, and low cost. However, the stability (chemical stability, thermal stability, and photo stability) may be an adverse property for practical application. Reports on stability tests for polymer CEs are still rare.

3 Transition Metal Compounds

Early transition metal carbides, nitrides, oxides, sulfides, and phosphides exhibit Pt–like catalytic behavior. These materials have been used in the fields of CO_2 methanation, ammonia synthesis, dehydrogenation, hydrogenation, methanol oxidation, photocatalysts, gas sensors, electrochromic devices, field emitters, among others, as a replacement for the noble metal. Very recently, these compounds have been proposed to use as CE catalysts in DSCs. Next, we will give a detailed introduction.

3.1 Carbides

Ma group introduced tungsten carbide (WC) and molybdenum carbide (Mo$_2$C) into DSCs as CE catalysts. The DSCs using WC and Mo$_2$C as CEs gave PCE of 5.35 % (WC) and 5.70 % (Mo$_2$C), lower than the PCE of the Pt-DSC (7.89 %).

This relative lower performance stems from the large size of the two carbides Mo_2C, 300 nm; WC, 190 nm [29]. Then they synthesized nano-scaled WC and W_2C using metal-urea route. The DSCs using the nanoscaled carbides as CEs achieve high PCE of 6.68 % (W_2C) and 6.23 % (WC). Compared with the large WC particle, the nano-scaled WC showed improved catalytic activity [30]. Lee et al. prepared polymer-derived WC (WC–PD) and microwave-assisted WC (WC–MW), which have been introduced in DSCs as CE catalysts. The DSCs yielded PCE of 6.61 % (WC–PD) and 7.01 % (WC–MW), lower than that of the DSC using Pt CE (8.23 %) [31]. More-over, Ma group did a systematic research on other transition metal carbides, such as TiC, VC, ZrC, NbC, Cr_3C_2, and so on [32, 33]. All of the carbides give high catalytic activity except ZrC and NbC.

3.2 Nitrides

Gao et al. prepared TiN nanotubes used as CE catalyst in DSC which produced a PCE of 7.73 % [34]. Similar to TiN, MoN, WN, and Fe_2N were synthesized by nitridation of the oxide (MoO_2, WO_3, Fe_2O_3) precursors in ammonia atmosphere, after which the nitrides were used as CE catalysts into DSCs [35]. The DSCs gave PCE values of 5.57 % (MoN), 3.67 % (WN), 2.65 % (Fe_2N), and 6.56 % (Pt). Meanwhile, Ma et al. prepared Mo_2N and W_2N films on flexible Ti sheet as CEs for DSCs which produced PCE of 6.38 % (Mo_2N) and 5.81 % (W_2N). In addition, they also introduced ZrN, VN, NbN, CrN into DSCs, and decent results were obtained [32]. Gao et al. used surface-nitrided Ni foil as the CE in DSCs, resulting in a PCE of 5.68 %, much lower than that of the Pt-DSC (8.41 %) [36], while NiN with a mesoporous structure showed high catalytic activity and the DSC yielded a PCE of 8.31 %.

3.3 Oxides

Ma et al. observed that WO_2 and WO_3 can be used as a catalyst for the reduction of I_3^- to I^- [37]. The DSC based on WO_3 CE showed a PCE of 4.67 %. WO_2 nanorod showed excellent catalytic activity and the DSC gave a high PCE of 7.25 %, close to that of the Pt-DSC (7.57 %). Recently, they synthesized H–Nb_2O_5 (hexagonal), O–Nb_2O_5 (orthorhombic), M–Nb_2O_5 (monoclinic), and T–NbO_2 (tetragonal) and then used the four niobium oxides as CE catalysts in DSCs [38]. The DSCs showed PCE values of 5.68 % (H–Nb_2O_5), 4.55 % (O–Nb_2O_5), 5.82 % (M–Nb_2O_5), and 7.88 % (T–NbO_2). Obviously, NbO_2 performs best among the four niobium oxides and the crystal forms significantly affecting the catalytic activity. Xia et al. used V_2O_5 as CE in solid DSCs. The PCE of the DSC using a 10 nm thick V_2O_5 CE reached 2.0 % [39].

As we know, WO_3, Nb_2O_5, SnO_2, and other oxides are widely used as photoanode semiconductor. Now, it has been proved that some oxides can be used as CE catalysts. If the oxide is used in the photoanode, the direct contact of oxide and I_3^-/I^- redox couples may cause the I_3^- to be reduced by the electrons injected in the conductive band of the oxide due to autocatalytic activity. This is to say, a number of the electrons injected in the conductive band cannot be collected by the substrate and, thus, flow into the external circuit. The autocatalytic activity of oxide can result in a large dark current density. This may be a key reason for the poor performance of the DSCs using TiO_2—free oxides as photoanodes [40]. In addition, Ma et al. found that the catalytic activity can be enhanced significantly by sintering the oxide CEs in N_2 atmosphere [41]. The fundamental reason for the high catalytic activity of some oxides is still unclear and requires further study.

3.4 Sulfides Selenides and Phosphides

The aforementioned carbides, nitrides, and oxides all show catalytic activity for the regeneration of the redox couples in DSCs. Similarly, transition metal sulfides are expected to behave in a similar fashion. Grätzel et al. prepared CoS CE on flexible substrate and the DSC yielded a PCE of 6.5 % [42]. Co and Ni both belong to Group VIII A metals. Meng et al. prepared NiS CEs by periodic potential reversal (PR) and potentiostatic (PS) techniques. The DSCs showed PCE of 6.83 % (PR–NiS) and 3.22 % (PS–NiS) [43]. Lin et al. introduced copper zinc tin sulfide (CZTS) to DSCs system as CE. CZTS performed well for the regeneration of iodide from triiodide. The DSC using the CZTSSe CE gave a PCE of 7.37 % higher than the corresponding photovoltaic of the Pt-DSC [44]. Ma et al. introduced MoS_2 and WS_2 into DSCs as CEs [45]. MoS_2 and WS_2 showed high catalytic activity compared to Mo (or W) carbides, nitrides, and oxides and the DSCs gave high PCE of 7.59 % (MoS_2) and 7.73 % (WS_2) which can match the performance of the Pt-DSC (7.64 %). Besides sulfides, Ma et al. used Ni_5P_4 and MoP as CEs in DSCs, which showed PCE of 5.71 % (Ni_5P_4) and 4.92 % (MoP). Meanwhile, Gao et al. introduced $Ni_{12}P_5$ as CE in DSCs, a PCE of 3.94 % was achieved [46]. Recently, selenides of $Co_{0.85}Se$ and $Ni_{0.85}Se$ were also proposed as CEs in DSCs, and both of them showed high catalytic activity [47].

4 Composites

As the name indicates, composite CE catalysts commonly comprise two or more components such as TiN/CNTs, Pt/Carbon, Carbon/PEDOT/PSS, Carbon/TiO_2, CoS/PEDOT/PSS, and so forth. The advantage of this CE is, naturally, the combination of the best qualities of all components into one composite.

To achieve highly effective catalysts, Ma et al. synthesized MoC and WC imbedded in ordered mesoporous carbon (MoC–OMC, WC–OMC) by the in situ method [29]. The PCE values of the DSCs reached 8.18 % (WC–OMC) and 8.34 % (MoC–OMC), much higher than those of the DSCs using WC (5.35 %) and Mo_2C (5.70 %) CEs. Further, they prepared WO_2 and Ni_5P_4 imbedded in mesoporous carbon and both the composite CEs catalysts showed high catalytic activity [48, 49]. Gao et al. deposited TiN nanoparticles on CNTs (TiN/CNTs) and then used the composite as a CE in DSC which produced a PCE of 5.41 %, higher than both the DSCs using TiN CE (2.12 %) and pure CNTs CE (3.53 %) [50]. The high catalytic activity can be attributed to the combination of the high catalytic activity and the high electrical conductivity into one composite. This strategy can be used to design effective catalysts in the future research. Park et al. prepared W_2C/WC composite CE in DSC, resulting in a PCE of 4.2 %, still lower than that of Pt (5.22 %) [51]. Wu et al. introduced the Pt/Cb composite catalyst into DSCs. Loading 1.5 wt% Pt on Cb was enough to achieve high catalytic activity and the DSC gave a PCE of 6.72 % [52]. Ouyang et al. prepared PEDOT–PSS/CNTs and PSS/CNTs composite CEs [53]. The DSC using PEDOT–PSS/CNTs CE yielded a PCE of 6.5 %, much higher than that of the DSC using PSS/CNTs CE (3.6 %). Kang et al. used CoS/PEDOT/PSS as a composite CE catalyst, with the corresponding DSC giving a PCE of 5.4 %, comparable to that of the DSC using Pt CE (6.1 %). The high catalytic activity of this composite CE can be attributed to the synergistic catalytic effect [54]. Lin et al. synthesized MoS_2/Graphene composite as CE in a DSC, which gave a PCE of 6.04 %, slightly lower than that of the DSC using Pt CE [55].

On current evidence, prepared composite has become an effective path to achieve CE catalysts for DSCs. However, the mechanism of high catalytic activity and the roles of the catalyst and the supporter are still in dispute.

5 Summary

In sum, a series of Pt-free CE catalysts have been introduced into DSCs, each with unique advantages and disadvantages. We think that carbon material is the most potential substitute to Pt in DSCs. Organic polymers offer a promising alternative as flexible and transparent CEs. In addition, the introduction of transition metal carbides, nitrides, oxides, sulfides, selenides, and phosphides into DSCs widens the selective scope of CE catalysts remarkably. Combining two or more proper materials into one composite is an effective path to design high-efficiency catalysts for DSCs. Now, to develop low-cost Pt-free CE catalysts has become a hot research topic to reduce the cost of DSCs, which makes DSCs more competitive among various photovoltaic devices, contributing to realize the industrialization of DSCs.

References

1. Imoto K, Suzuki N, Tkahashi Y et al (2003) Activated carbon counter electrode for dye–sensitized solar cell. Electrochemistry 71:944–946
2. Kay A, Grätzel M (1996) Low cost photovoltaic modules based on dye sensitized nanocrystalline titanium dioxide and carbon powder. Sol Energy Mater Sol Cells 44:99–117
3. Imoto K, Takahashi K, Yamaguchi T et al (2003) High–performance carbon counter electrode for dye–sensitized solar cells. Sol Energy Mater Sol Cells 79:459–469
4. Hou S, Cai X, Fu Y et al (2011) Transparent conductive oxide-less, flexible, and highly efficient dye–sensitized solar cells with commercialized carbon fiber as the counter electrode. J Mater Chem 21:13776–13779
5. Wang G, Xing W, Zhuo S (2009) Application of mesoporous carbon to counter electrode for dye-sensitized solar cells. J Power Sources 194:568–573
6. Ramasamy E, Lee J (2010) Ferrocene-derivatized ordered mesoporous carbon as high performance counter electrodes for dye-sensitized solar cells. Carbon 48:3715–3720
7. Zhu H, Wei J, Wang K et al (2009) Applications of carbon materials in photovoltaic solar cells. Sol Energy Mater Sol Cells 93:1461–1470
8. Suzuki K, Yamaguchi M, Kumagai M et al (2003) Application of carbon nanotubes to counter electrodes of dye-sensitized solar cells. Chem Lett 32:28–29
9. Lee W, Ramasamy E, Lee D et al (2009) Efficient Dye-sensitized solar cells with catalytic multiwall carbon nanotube counter electrodes. ACS Appl Mater Interfaces 1:1145–1149
10. Nam J, Park Y, Kim B et al (2010) Enhancement of the efficiency of dye-sensitized solar cell by utilizing carbon nanotube counter electrode. Scripta Mater 62:148–150
11. Ramasamy E, Lee W, Lee D et al (2008) Spray coated multi-wall carbon nanotube counter electrode for tri-iodide (I_3^-) reduction in dye-sensitized solar cells. Electrochem Commun 10:1087–1089
12. Zhu H, Zeng H, Subramanian V et al (2008) Anthocyanin–sensitized solar cells using carbon nanotube films as counter electrodes. Nanotechnology 19:65204
13. Kavan L, Yum J, Grätzel M (2011) Optically transparent cathode for dye-sensitized solar cells based on graphene nanoplatelets. ACS Nano 5:165–172
14. Roy-Mayhew J, Bozym D, Punckt C et al (2010) Functionalized graphene as a catalytic counter electrode in dye-sensitized solar cells. ACS Nano 4:6203–6211
15. Choi H, Kim H, Hwang S et al (2011) Graphene counter electrodes for dye-sensitized solar cells prepared by electrophoretic deposition. J Mater Chem 21:7548–7551
16. Wu M, Lin X, Wang T et al (2011) Low–cost dye–sensitized solar cell based on nine kinds of carbon counter electrodes. Energy Environ Sci 4:2308–2315
17. Ramasamy E, Lee W, Lee D et al (2007) Nanocarbon counter electrode for dye sensitized solar cells. Appl Phys Lett 90:173103
18. Chen J, Li K, Luo Y et al (2009) A flexible carbon counter electrode for dye-sensitized solar cells. Carbon 47:2704–2708
19. Shibata Y, Kato T, Kado T et al (2003) Quasi-solid dye sensitised solar cells filled with ionic liquid—increase in efficiencies by specific interaction between conductive polymers and gelators. Chem Commun 21:2730–2731
20. Saito Y, Kubo W, Kitamura T et al (2004) I^-/I_3^- redox reaction behavior on poly(3,4–ethylenedioxythiophene) counter electrode in dye-sensitized solar cells. J Photochem Photobiol A 164:153–157
21. Kanciurzewska A, Dobruchowska E, Baranzahi A et al (2007) Study on Poly(3,4-ethylene dioxythiophene)-Poly(styrenesulfonate) as a plastic counter electrode in dye sensitized solar cells. J Optoelectron Adv Mater 9:1052–1059
22. Biallozor S, Kupniewska A (2000) Study on poly (3,4-ethylenedioxythiophene) behaviour in the I^-/I_2 solution. Electrochem Commun 2:480–486
23. Lee K–M, Hsu C–Y, Chen P–Y et al (2009) Highly porous PProDOT–Et$_2$ film as counter electrode for plastic dye-sensitized solar cells. Phys Chem Chem Phys 11:3375–3379

24. Wu J, Li Q, Fan L et al (2008) High–performance polypyrrole nanoparticles counter electrode for dye-sensitized solar cells. J Power Sources 181:172–176
25. Xia J, Chen L, Yanagida S (2011) Application of polypyrrole as a counter electrode for a dye-sensitized solar cell. J Mater Chem 21:4644–4649
26. Jeon S, Kim C, Ko J et al (2011) Spherical polypyrrole nanoparticles as a highly efficient counter electrode for dye-sensitized solar cells. J Mater Chem 21:8146–8151
27. Tai Q, Chen B, Guo F et al (2011) In situ prepared transparent polyaniline electrode and its application in bifacial dye-sensitized solar cells. ACS Nano 5:3795–3799
28. Li Q, Wu J, Tang Q et al (2008) Application of microporous polyaniline counter electrode for dye-sensitized solar cells. Electrochem Commun 10:1299–1302
29. Wu M, Lin X, Hagfeldt A et al (2011) Low-cost molybdenum carbide and tungsten carbide counter electrodes for dye-sensitized solar cells. Angew Chem Int Ed 50:3520–3524
30. Wu M, Ma T (2012) Platinum–free catalysts as counter electrodes in dye-sensitized solar cells. ChemSusChem 5:1343–1357
31. Jang J, Ham D, Ramasamy E et al (2010) Platinum–free tungsten carbides as an efficient counter electrode for dye sensitized solar cells. Chem Commun 46:8600–8602
32. Wu M, Lin X, Wang Y et al (2012) Economical Pt-free catalysts for counter electrodes of dye-sensitized solar cells. J Am Chem Soc 134:3419–3428
33. Yun S, Wu M, Wang Y et al (2013) Pt-like behavior of high-performance counter electrodes prepared from binary tantalum compounds showing high electrocatalytic activity for dye-sensitized solar cells. ChemSusChem 6:411–416. doi:10.1002/cssc.201200845
34. Jiang Q, Li G, Gao X (2009) Highly ordered TiN nanotube arrays as counter electrodes for dye-sensitized solar cells. Chem Commum 44:6720–6722
35. Li G, Song J, Pan G et al (2011) Highly Pt-like electrocatalytic activity of transition metal nitrides for dye-sensitized solar cells. Energy Environ Sci 4:1680–1683
36. Jiang Q, Li G, Liu S et al (2010) Surface-nitrided nickel with bifunctional structure as low-cost counter electrode for dye-sensitized solar cells. J Phys Chem C 114:13397–13401
37. Wu M, Lin X, Hagfeldt A et al (2011) A novel catalyst of WO$_2$ nanorod for the counter electrode of dye-sensitized solar cells. Chem Commun 47:4535–4537
38. Lin X, Wu M, Wang Y et al (2011) Novel counter electrode catalysts of niobium oxides supersede Pt for dye-sensitized solar cells. Chem Commun 47:11489–11491
39. Xia J, Yuan C, Yanagida S (2010) Novel counter electrode V2O5/Al for solid dye-sensitized solar cells. ACS Appl Mater Interface 2:2136–2139
40. Wu M, Wang Y, Lin X et al (2012) An autocatalytic factor in the loss of efficiency in dye-sensitized solar cells. ChemCatChem 4:1255–1258
41. Wu M, Lin X, Guo W et al (2013) Great improvement of catalytic activity of oxide counter electrodes fabricated in N$_2$ atmosphere for dye-sensitized solar cells. Chem Commun 49:1058–1060
42. Wang M, Anghel A, Marsan B et al (2009) CoS supersedes Pt as efficient electrocatalyst for triiodide reduction in dye-sensitized solar cells. J Am Chem Soc 131:15976–15977
43. Sun H, Qin D, Huang S et al (2011) Dye–sensitized solar cells with NiS counter electrodes electrodeposited by a potential reversal technique. Energy Environ Sci 4:2630–2637
44. Xin X, He M, Han W et al (2011) Low-cost copper zinc tin sulfide counter electrodes for high-efficiency dye-sensitized solar cells. Angew Chem Int Ed 50:11739–11742
45. Wu M, Wang Y, Lin X et al (2011) Economical and effective sulfide catalysts for dye-sensitized solar cells as counter electrodes. Phys Chem Chem Phys 13:19298–19301
46. Dou Y, Li G, Song J et al (2012) Nickel phosphide-embedded graphene as counter electrode for dye-sensitized solar cells. Phys Chem Chem Phys 14:1339–1342
47. Gong F, Wang H, Wang Z-S (2012) In situ growth of Co(0.85)Se and Ni(0.85)Se on conductive substrates as high-performance counter electrodes for dye-sensitized solar cells. J Am Chem Soc 134:10953–10958
48. Wu M, Lin X. Wang L et al (2011) In situ synthesized economical tungsten dioxide imbedded in mesoporous carbon for dye-sensitized solar cells as counter electrode catalyst. J Phys Chem C 115:22598–22602

49. Wu M, Bai J, Wang Y et al (2012) High-performance phosphide/carbon counter electrode for both iodide and organic redox couples in dye-sensitized solar cells. J Mater Chem 22:11121–11127
50. Li G, Wang F, Jiang Q et al (2010) Carbon nanotubes with titanium nitride as a low-cost counter-electrode material for dye-sensitized solar cells. Angew Chem Int Ed 49:3653–3656
51. Ko A-R, Oh J-K, Lee Y-W et al (2011) Characterizations of tungsten carbide as a non-Pt counter electrode in dye–sensitized solar cells. Mater Lett 65:2220–2223
52. Li P, Wu J, Lin J et al (2009) High–performance and low platinum loading Pt/carbon black counter electrode for dye-sensitized solar cells. Solar Energy 83:845–849
53. Fan B, Mei X, Sun K et al (2008) Conducting polymer/carbon nanotube composite as counter electrode of dye-sensitized solar cells. Appl Phys Lett 93:143103
54. Sudhagar, P Nagarajan S, Lee Y-G et al (2011) Synergistic catalytic effect of a composite (CoS/PEDOT:PSS) counter electrode on triiodide reduction in dye-sensitized solar cells. ACS Appl Mater Interfaces 3:1838–1843
55. Liu C, Tai S, Chou S et al (2012) Facile synthesis of MoS_2/graphene nonocomposite with high catalytic activity toward triiodide reduction in dye-sensitized solar cells. J Mater Chem 39:21057–21064

Quantum Dot-Sensitized Solar Cells

P. Sudhagar, Emilio J. Juárez-Pérez, Yong Soo Kang
and Iván Mora-Seró

Abstract Quantum dot sensitized solar cells, but in general semiconductor sensitized photovoltaic devices, have erupted in recent years as a new class of systems, differentiated for several reasons of the most common dye-sensitized solar cells. In this chapter, we review the enormous potentialities that have impelled the research in this field. We highlight the differences between quantum dot and dye-sensitized solar cells that we divide in five aspects: (i) Preparation of the sensitizer; (ii) Nanostructured electrode; (iii) Hole Transporting Material; (iv) Counter electrode, and (v) Recombination and surface states. Some of the optimization works performed in each one of these lines is revised, observing that further improvement can be expected. In fact, the recent breakthrough in photovoltaics with organometallic halide perovskites, originated by the intensive study on quantum dot-sensitized solar cells, is also revised, stressing the potentiality of these systems for the development of low cost photovoltaic devices.

1 Introduction

Since the apparition of the Homo Sapiens in Africa, 200,000 years ago, the evolution of the society has been closely linked to an increment of the energy consumption. From the early days of the hunters and harvesters society to the current industrial societies the energy consumption per person and day and the world population have both experienced an exponential growth [1]. The demand for

P. Sudhagar · Y. S. Kang
World Class University Program Department of Energy Engineering and Center for Next Generation Dye-Sensitized Solar Cells, Hanyang University, Seoul 133-791, South Korea

E. J. Juárez-Pérez · I. Mora-Seró (✉)
Photovoltaic and Optoelectronic Devices Group, Departament de Física, Universitat Jaume I, 12071 Castelló, Spain
e-mail: sero@uji.es

Z. Lin and J. Wang (eds.), *Low-cost Nanomaterials*, Green Energy and Technology,
DOI: 10.1007/978-1-4471-6473-9_5, © Springer-Verlag London 2014

power during the last period of industrialization is compensated with the use of nonrenewable sources, fossil fuels such as coal and oil. Their use not only represents an irreplaceable depletion of stocks, avoiding its use on other beneficial issues especially in the case of oil, but it has caused serious side effects as the greenhouse effect. Gases from combustion have a potentially catastrophic impact in the global warming and the climate change as it has been recognized since the end of the twentieth century by the United Nations [2]. In this sense, energy has become one of the most striking problems of the humankind. It is easy to recognize that the energy problem is not only important in itself but for its implications in many of the great challenges of humanity in the twenty-first century (wars, environment, food, water...). This situation strongly demands a change in the weighting of the different energy sources in the energy cocktail that powers the world. The weight of nonrenewable energy sources necessarily has to decrease, even if this decrease in nonintentional due to the stock reduction. It is debatable whether the size of the nonrenewable energy source reserves is large or small, but the irrefutable fact is that these reserves are finite. On the other hand, to maintain the economic development the energy consumption cannot be significantly decreased, and the reduction of energy from nonrenewable sources has to be compensated with power coming from renewable sources.

Undoubtedly photovoltaic energy is the renewable power source with higher potentiality, as the Earth receives from the Sun in just 1 hour the same amount of energy that it is expended by all the humankind in 1 year [3]. But the devices that we currently have to harvest energy from this huge pool, they are expensive compared to other energy sources. In this sense, a reduction of cost in photovoltaic devices is mandatory to take full advantage of the tremendous potentialities of these systems. There are two ways to attain this scope: (i) reducing the fabrication cost of the device; (ii) increase the efficiency of the photovoltaic devices. To attain both solutions nanoscience can help us.

The new technologies capable of constructing material structures with dimensions from 0.1 to 50 nm have opened numerous possibilities to investigate new devices in a domain heretofore inaccessible to the investigators and technologists. A considerable activity exists in nanoscience and technology in university and industrial laboratories around the world. The requirements of improved efficiency and versatility, and of reduced cost of the photovoltaic devices, have led to search for extending the field of traditional study in inorganic semiconductors, incorporating new materials, and structures capable of satisfying these requirements. In the field of the photovoltaic solar power, the challenge of the future for the new concepts on nanoscale is to reduce the present cost of the devices by a factor 10–100. Candidates to lead this revolution are the new nanocomposite solar cells, formed by coatings deposited from liquid solutions. These procedures avoid the need of treatment in vacuum, and improve substantially the time and cost of production. Also the highly energetic stages of the process of production are avoided, improving the time of recovery of the spent energy. With the advanced optimization on nanoscale, and the application of solution low cost, it is possible to achieve the efficient use of minuscule quantities of matter.

Nanoscale considerations have been also important implications in the development of low cost photovoltaic devices, from the point of view of the fundamental principles. Conventional Si solar cells rely on high quality materials since the carriers generated in the device after photon absorption remain in the same material until they are extracted at the selective contacts [4]. This involves the use of sophisticated technologies with high production cost in order to avoid carrier recombination before their extraction. Conversely, nanoscale absorbers can quickly separate the photogenerated carriers into two different media which allows for a less-demanding materials quality and therefore cost reduction, as transport of electrons and holes is occurring in different materials and recombination is reduced. These are the fundamentals of the new paradigm introduced to the photovoltaic conversion by the sensitized solar cells that have received a major attention since the seminal paper of O'Regan and Grätzel at the beginning of the 1990s [5]. Most of the work carried out in sensitized solar cells has been made using organic and/or metalorganic molecular dyes acting as light absorbing materials. The use of inorganic semiconductors as nanoscale light harvesters has been minority. But the undoubted photovoltaic properties of these materials have pushed the research in this field with a continuous growing interest in the last few years. This interest has resulted in a constant growth of the reported efficiencies. This fact with the easy preparation and its low cost make of the inorganic semiconductor sensitized solar cells one of the fields of photovoltaics that is experienced higher growth. This chapter is devoted to the study of these systems focusing mainly in two aspects the low cost preparation and the role of nanoscale design and the advanced structures for solar cell preparation

This chapter is structured in nine sections. The first one is this brief introduction. In the second, we analyze the fundamentals of sensitized solar cells, explaining the specificities of dye, semiconductor, and quantum dot solar cells, and reviewing the state of the art. In Sect. 3, the experimental techniques used for semiconductor light absorbing material deposition are revised, stressing the fact of their low cost and easy industrial implementation. In the fourth part, new promising substrate structures in the case of inorganic semiconductor absorber are revised, in order to highlight the role of nanoscale and the possibilities of design. In Sect. 5, the hole transporting material is revised for both liquid electrolytes of all solid devices. In Sect. 6, the counter electrode material is determined by the hole transporting material and proper choice has to be done as we analyze in this section. In the seventh, the effect of carrier recombination before being collected at the output terminals, the role of surface stats in the light absorbing materials has overviewed. In the penultimate part, a new class of light absorbing materials "halide perovskites" has arisen and made revolution in this field. Despite there are several proves indicating that this kind of devices are not fully behaving as a sensitized device, it has been the intense work in sensitized solar cells that have allowed to achieve this breakthrough. Therefore, we consider that it is important to include these systems in this chapter. This new hot topic in photovoltaics is also taking benefit of the knowledge and characterization techniques developed in the last decades for sensitized devices. Finally, very briefly conclusions are highlighted.

2 Sensitized Solar Cells

As commented in the introduction the fundamentals of sensitized solar cells is to decouple light absorption and charge transports making that these processes occur in different media. When light is absorbed by molecular dyes the devices are called Dye-Sensitized Solar Cells (DSSCs), and are the most extensively studied class of sensitized devices [6]. But here we are going to focus our attention in the case where inorganic semiconductor are used as light absorbing materials instead of dyes giving place to the Semiconductor-Sensitized Solar Cells (SSSCs) [7]. When the size of the semiconductor material is small enough to observe effects due to quantum confinement (particle radius lower than the Bohr radius) the devices is called Quantum Dot-Sensitized Solar Cells (QDSSCs) [8–10]. The line separating SSSCs from QDSCs is fuzzy and many times the denomination QDSSCs is preferred even in cases where quantum confinement is not observed.

The working principle of QDSSCs is fundamentally similar to DSSCs and they are represented in Fig. 1. A wide bandgap semiconductor material is sensitized with an semiconductor with a bandgap in the visible or near IR region. The most extensively used wide bandgap semiconductor has been TiO_2, but several examples of the utilization of other ones as ZnO [11] or SnO_2 [12–14] have been reported. Light irradiation photo excites electron-hole pairs from the QD Valence Band (VB) to the QD Conduction Band (CB). Photoexcited carries are injected into two different transporting media. Electrons are injected into the CB of the wide bandgap semiconductor, while hole are injected into a hole transporting material (HTM). Then both carriers diffuse to their respective contacts. In order to optimize cell performance, recombination of diffusing carriers should be avoided. As the light absorbing layer is extremely thin effective surface area is significantly enhanced with the use of a nanostructured electrode, increasing consequently the light harvesting [5]. This description is perfectly valid also for DSSCs. In this sense, in a first analysis it could be thought that semiconductor QDs are only one more of the thousand of dyes that have been checked in DSSCs. But we want to highlight here that it is not the case. The different nature of QDs in comparison with molecular dyes makes that the complete design of the solar cell device have to be rethought.

Main differences between DSSCs and QDSSCs could be divided into five aspects:

(i)   Preparation of the sensitizer.
(ii) Nanostructured electrode.
(iii) Hole Transporting Material (HTM).
(iv) Counter electrode.
(v)   Recombination and surface states.

In the next five sections, we will develop each of these points reviewing the current state of the art of QDSSCs.

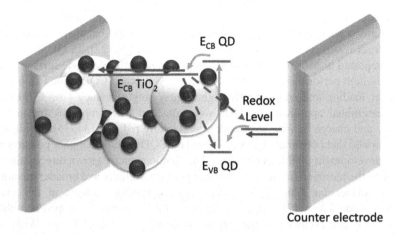

Counter electrode

Fig. 1 Cartoon representing the fundamental principles of sensitized solar cells. Light photoexcites electron-hole pairs at the QD (vertical *green arrow*). Photoexcited electron is injected into the CB of the wide bandgap semiconductor, while hole is regenerated from the redox level of the liquid electrolyte (or the VB of a solid hole transporting material). Electron in wide bandgap semiconductor is transported to the electron contact (*orange arrow*). Redox sytem is regenerated at the counter electrode and electrons diffuse to the working electrode (*purple arrow*). During the transport process electron can recombine with holes in the QDs or in the HTM (*blue dashed arrows*)

3 Sensitization with Inorganic Semiconductor Quantum Dots

Considering the sensitized synthesis, an inorganic semiconductor can be prepared by several ways and the growth mechanism determines dramatically the performance of the photovoltaic device [15, 16]. In this chapter, we want to highlight the potentiality of QDSSCs for the development of low cost photovoltaic devices. In particular, we will review some of the techniques, most commonly, employed for the sensitization with inorganic semiconductors. Interestingly, these techniques do not require any special experimental conditions like vacuum or high temperatures; this fact is especially attractive for industrial development as it is an effective way for cost reduction.

Two major strategies have been practically used to sensitize QDs onto the electrodes (a) in situ preparation of QDs on the electrode surface and (b) attachment of presynthesized Colloidal Quantum Dots (CQDs) on the electrode surface. The former method includes chemical bath deposition, [17] successive ionic layer adsorption/reaction, [18] electrochemical deposition, [19] and photochemical deposition [20]. The latter exploits, but is not limited to, linker molecules, [8, 21–23] direct absorption, [16, 22, 24] spray pyrolysis deposition, [25, 26], and electrophoretic deposition [27–30]. There are also some examples of the combination of these two approximations [31, 32].

The use of CQDs allows the production of material with a high degree of crystallinity a high control of QD properties (as size, shape, bandgap...) but in the first reports relatively low QD loading was obtained [16, 22]. In QDSSCs, it is widely reported historically that directly grown QDs on the electrode surface show relatively high performance than that of indirectly attached CQDs, due to a higher sensitizer loading and consequently higher harvesting efficiency [16, 33]. Using direct assembled approach, the QDs were attached to mesoporous framework robustly and charge transfer resistance was possibly reduced [34, 35]. But it was demonstrated that potentially QDSSCs based on CQDs exhibit a higher potentiality for the development of efficient devices [16]. Semiconductor grown directly on the surface of the mesoporous electrode present poor crystallinity and broader quantum dot size dispersion than CQDs. In addition, deleterious effects in the grain boundaries could arise [7]. Recently, the group of Zhong has developed a method based in a ligand exchange of the capping molecules in CQDs [36–38]. With this method higher QD loading is attained and consequently higher efficiency in fact the record efficiency for QDSSCs has been reported by this group with $CdSe_xTe_{1-x}$ CQDs [38].

Taking into account these considerations here we review different sensitization methods for both CQDs and semiconductors grown directly on the electrode surface. The following coating methods were identified as low cost methods compared to physical and vacuum techniques:

3.1 Sensitization with Colloidal Quantum Dots

For the sensitization with CQDs two consecutive steps are required. In first step, presynthesized CQDs are grown with an accurate control of the crystalline quality, size, and shape (and consequently bandgap) distribution or capping [39]. Second, CQDs should be attached to the nanostructured electrode. However, there is no single procedure to attach CQDs to the photoanode and it is possible to differentiate between assisted and direct sensitization. In the assisted sensitization process, bifunctional linker molecules are used to anchor CQDs to the nanostructured photoanode [8, 23]. These linker molecules generally have a functional carboxylic group which first attach to one side of the TiO_2 and the other side of the linker (generally a thiol group) was connected to the CQD [8, 23]. In addition to carboxyl linkers, other functional groups can also be applied as bifunction linkers for fixing CQD to the nanoporous electrode (as amine group in cysteine), and is realized that this linker molecules play an important role in PEC performance [21]. On the other hand, CQDs can be directly attached to the electrode without the use of any specific linker. For example, if toluene solvent for CdSe CQDs is substituted by dichloromethane, CQDs can be directly absorbed on TiO_2 surface simply by dipping the substrate in the dichloromethane solution with CQDs [22, 24]. The use of linker molecules is benefit during the CQDs synthesis which capped with, generally, long organic molecules (oleic acid), thus offer controlling QD size and avoid agglomeration [39]. It has been shown that the ligand exchange helps anchoring QDs to mesoporous electrode, for example, TOPO molecules coating CdSe QDs were substituted by pyridine in order to enhance the QD loading of sensitized electrode [40]. This strategy has been employed with significant success by the group of Zhong, where oleic acid is substituted by mercaptopropionic acid (MPA) [41]. Then aqueous solution with MPA-capped QDs is pipetted directly on the electrode surface, where it stayed for 2 h before rinsing sequentially with water and ethanol and then drying with nitrogen. This procedure has produced QDSSCs with efficiencies higher than 5 % [36, 37] or even 6 %, [38] as it has been already commented.

The methods presented so far have the drawbacks that need long duration for sensitization in most of the cases, as the dipping process to attach the linker molecule and/or the QDs requires several hours for an optimum loading [8]. Sensitization time can be significantly reduced by employing the electrophoretic technique. Basically, in this technique an electrical field is applied between two electrodes dipped in a QD solution. Ionized QDs are attracted to the electrodes and attached in one or both electrodes [42]. CQD deposition rate on the mesoporous electrode depends on the applied voltage, high CQD loading can be obtained in few hours [28, 29] or even in few minutes [27]. Even, electrophoresis has been used to attach rod shape sensitizers (nanorods) to TiO$_2$ electrodes, [29] to prepare electrodes with QDs of different sizes [30] or to assemble CdSe QDs and fullerene for an innovative solar cell [43]. Nevertheless the efficiencies reported for QDSSCs using electrophoresis are shown below the performance reported for other attaching modes with CQDs.

3.2 Chemical Bath Deposition

The second main approximation for the sensitization with inorganic semiconductor is direct growth of light absorbing material on the surface of the wide bandgap semiconductor electrode. The remaining section will be dedicated to this last approximation with three low cost techniques that will be overviewed: (a) Chemical bath deposition (b) successive ionic layer adsorption and reaction (SILAR), and (c) electrodeposition. Chemical bath deposition (CBD) is a convenient method to assemble the QDs on a variety of substrates (conducting and nonconducting) at elevated temperatures compared with most other semiconductor QDs deposition methods [44]. The simplicity of CBD process together with the inherent low temperature operations results in superior control over the QDs particle size [17]. Several articles explain the variety of QDs assembly by CBD method, i.e., CdS, [18, 45, 46] CdSe, [34, 47, 48] Sb$_2$S$_3$ [49–52], CdTe, [53–56] PbS [27, 57–60], etc. The two main mechanisms for the CBD process are (1) ion-by-ion deposition onto a coating surface without bulk precipitation in the deposition solution and (2) bulk precipitation (or colloid formation) with diffusion of the bulk semiconductor clusters to the coating surface [44]. The complex agents were usually utilized to maintain the pH of the chemical bath, which directly control the reaction rate and hence overcome the bulk precipitation in the chemical bath [61]. For example, the reaction mechanism of in situ coating of CdSe QDs onto mesoporous TiO_2 is explained as follows: [62]

$$Cd^{2+} + HSe^- + OH^- \leftrightarrow CdSe(s) + H_2O \tag{1}$$

$$(CdSe)_m + Cd^{2+} + HSe^- + OH^- \leftrightarrow (CdSe)_m + H_2O \tag{2}$$

$$(CdSe)_m + (CdSe)_n \leftrightarrow (CdSe)_{m+n} \tag{3}$$

$$(Cd)^{2+} + 2OH^- \leftrightarrow Cd(OH)_2(s) \tag{4}$$

$$Cd(OH)_2(s) + HSe^- \rightarrow CdSe + OH^- + H_2O \tag{5}$$

Reactions (1)–(2) describe the formation of CdSe without the formation of $Cd(OH)_2$ seed layer. Subsequently, Cd^{2+} and Se^{2-} species are forming larger clusters by coalescence or aggregation with other clusters (3). Coalescence refers to a combination of two clusters to form one single crystal, while aggregation means the formation of two or more separate but contacting crystals. The latter was suggested to be the terminating step in crystal growth for CBD CdSe films. As for CdSe, the reverse direction of reaction (5) will lead to redissolution of the solid phase and will control the size of the $Cd(OH)_2$ crystallites (i.e., colloidal particles) in the solution. Thus, lower temperatures result in a higher concentration of small $Cd(OH)_2$ nucleii and results in the formation of CdSe crystal with smaller size. The effect of Se^{2-} ions at a $Cd(OH)_2$ surface will be replaced by a selenide due to much lower solubility of CdSe compared with $Cd(OH)_2$. The Cd(OH)* nucleus will eventually be converted to CdSe (reaction 1). Thus, the reaction will continue by formation of hydroxide and its subsequent conversion into CdSe (4 and 5). Therefore, the nucleation process on coating surface takes vital role in determining QDs formation mechanism, which is often controlled by precursor concentration and chemical bath temperature and is explained elsewhere by Gorer et al. [62] (Fig. 2). Different QD size, as can be deduced from the different bandgap, can be obtained by varying temperature and concentration in the CBD process, see Fig. 2.

The typical chemical bath deposition of CdSe QDs is as follows [63, 64]: a chemical bath solution was prepared by mixing 80 mM $CdSO_4$ and 80 mM sodium selenosulfite (Na_2SeSO_3) solution with 120 mM nitriloacetic acid. The mesoporous TiO_2 electrodes were immersed in the chemical bath at 10 °C for 12 hours. Finally, the films were soaked in deionized water and dried with nitrogen. Recently, antimony sulfide (Sb_2S_3) QDs sensitizers prepared by CBD method showed efficient PCE performance 4–5 % in both liquid-type and solid-state hybrid solar cells [65–68]. Deposition of Sb_2S_3 coating by CBD is cheaper than the other available techniques. Typical coating of Sb_2S_3 QDs is explained elsewhere [67, 68]. Briefly, the mixture of 1 M solution of $SbCl_3$ in acetone and 1 M $Na_2S_2O_3$ cold aqueous solution addition with cold water is adjusted to have a final

Fig. 2 Optical transmission spectra of CdSe films deposited at various temperatures (**a**) 10 °C (**b**) 40 °C, and (**c**) 80 °C with various NTA:Cd ratios (NTA: nitrilo triacetic acid applied as complex agent) (Reprinted permission with Gorer et al. [62])

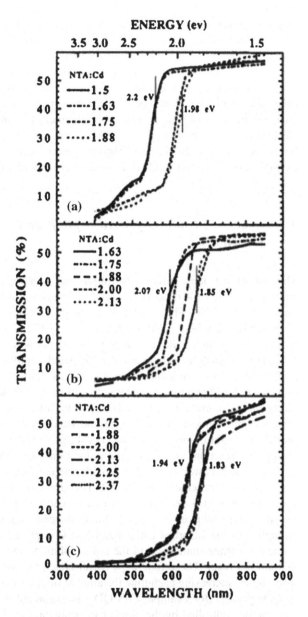

concentration of Sb^{3+} and S_2O_3 of ~0.025 and ~0.25 M, respectively. The CBD solution is quickly poured in the CBD recipient, where the electrodes were placed vertically, and the recipient was left in a refrigerator at ~7 °C for 2 h. Then, the samples are rinsed with deionised water and dried by flowing nitrogen. Efficiencies as high as 6.3 % have been reported in all-solid QDSSCs with Sb_2S_3 [67].

From the earlier reports, it is found that the deposition of QDs by CBD method densely covers the mesoporous metal oxide surface and enhances the recombination resistance [10, 63, 69, 70]. Despite the effective coverage on TiO_2 surface, CBD method generally result in nanocrystalline films pattern rather than particles coating. Therefore, it is mostly suitable for wide-pore-nanostructured electrodes (nanotube, nanowire, nanotube, and inverse opal) [71]. The nanocrystalline film coatings may block the narrow pore-channel of conventional nanopaticulate electrodes having limited pore size (5–7 nm). Also, this method takes several hours for coating (12–48 h) and high roughness surface requires seed layers, for example CdS is widely applied as seed layer for CdSe deposition [64].

3.3 Successive Ionic Layer Adsorption and Reaction (SILAR)

The successive ionic layer adsorption and reaction (SILAR) method is an emerging method for depositing variety of semiconductor quantum dots both as binary (CdS, CdSe, PbS, CdTe, CuS, Sb_2S_3, Sb_2Se_3, Bi_2S_3, etc.) and ternary compounds (CdS_xSe_{1-x}, CuInS, $CuIn_2S_3$ etc.) [72, 73]. The SILAR method is inexpensive, simple and convenient for large area deposition. There is no restriction of coating surface and can be widely applied to all kind of materials such as insulators, semiconductors, and metals. One of the advantages of SILAR process is the likelihood of achieving coatings at low temperature which avoids oxidation and corrosion of the substrate. The following parameters are indispensible in controlling the particle size and bandgap of QDs; concentration of the precursors, nature of complexing agent, pH of the precursor solutions and adsorption, reaction, and rinsing time durations etc. [74]. The reviews by Mane et al. [44] and Pawar et al. [75] provide the detailed picture of semiconducting chalcogenide films coating by SILAR. The growth mechanism involves four most important steps: (a) specific adsorption of cationic precursor (b) rinsing of the nonspecifically adhered chemicals, and (c) the chemical reaction between the most strongly specific adsorbed cations and less strongly adsorbed anions by the subsequent substrate immersion in the anion solution; and (d) rinsing of the species that did not react. These four steps constitutes a SILAR cycle, it can be repeated many times, increasing each time the amount of deposited material. Various stages of QDs growth (for example CdS QDs) is explained in Fig. 3. The average QD size can be controlled by the number of deposition cycles. This method has been used specifically to prepare metal sulfides, but recently SILAR process is expanded to prepare a variety of metal selenides and tellurides [76].

In SILAR method, the particle size of QDs is controlled by number of coating cycles. The number of coating cycles depends on the concentration of chemical bath. As explained in CBD method, increasing the bath concentration above the critical concentration (R_c) limits the number of SILAR cycles where the QDs are

Fig. 3 Schematic illustration
of SILAR process
(**a**) Adsorption of cationic
ions (Cd²⁺) (**b**) rinsing (I)
removes excess,
nonspecifically adsorbed
Cd²⁺ (**c**) reaction of anionic
(S⁻) with preadsorbed Cd²⁺
ions to form CdS and
(**d**) rinsing (II) to remove
excess and unreacted species
and form the solid solution
CdS on surface of the
substrate. The coverage of
CdS at higher cycles (**e**) 10
and (**f**) 30

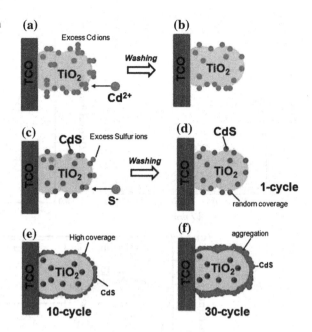

redissolved into chemical bath and QDs films might feel-off from the coating
surface. During first SILAR cycle, the seed layer of QDs was formed on the
coating surface and directing further growth for successive coating cycles. The
influence of coating cycles on growth of QDs can be studied by optical absorption
spectra. Figure 4a–c explains the influence of coating cycles on optical absorption
of CdS, CdSe [77], and Sb₂S₃ [78], respectively. From Fig. 4a, it is clearly
understood that the absorption of CdS and CdSe is found to increase by gradually
improving the coating cycles. Interestingly, nucleation and growth of CdSe on
TiO₂ can be greatly accelerated with a CdS underlayer, where CdS is rather a
promoter for the preferential growth of CdSe (Fig. 4b) as it has been also observed
for CBD.

Typically, semiconductor QDs by SILAR process has been demonstrated under
aqueous medium, but the high surface tension causes poor wetting ability on a
solid surface, which leads to poor penetration of the solution in a porous matrix.
Therefore, low surface tension solvent is recommended like alcohol solutions for
efficient QDs coating. Since it has high wettability and superior penetration ability
on the mesoscopic TiO₂ film, well-covered QDs on the surface of mesopores is
achieved easily. The high coverage of QDs results by alcohol solvent showed high
absorbance than that of aqueous solvent (Fig. 4c).

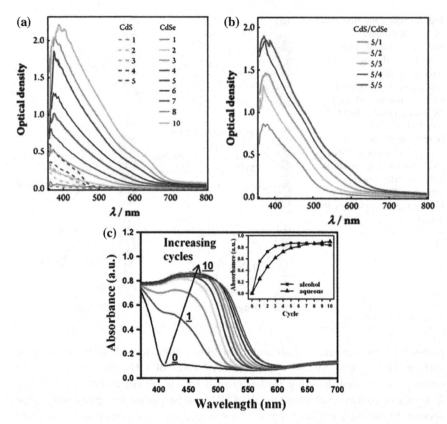

Fig. 4 UV-visible optical density (OD) spectra of as-prepared (**a**) nCdS and nCdSe, and (**b**) 5CdS/nCdSe-coated 2.4 mm TiO$_2$ electrodes, where n is the number of SILAR deposition cycles. The OD spectra represent the net light absorption by the sensitizers as the substrate absorption (mesoporous TiO$_2$) was subtracted from the absorption spectra of corresponding nCdS, nCdSe and 5CdS/nCdSe-sensitized TiO$_2$ electrodes [Hossian et al. reprinted with permission from RSC] [77] (**c**) UV-vis absorption spectra of TiO$_2$ with different coating cycles of CdS QDs in alcohol solutions. The inset shows the absorbance of excitonic peaks after various cycles of the CBD process for alcohol and aqueous systems [Chang et al. [45] reprinted permission from AIP publishers]

3.4 Electrodeposition

Electrodeposition (ED) is an emerging technique for synthesizing semiconductor thin films and nanostructures, especially chalcogenides and oxides [79–85]. One of the great advantages of ED method is that they are more suitable for solar cells application [86, 87] since it allows the possibility of easily altering both the bandgap and lattice constant by composition modulation through the control of growth parameters such as applied potential, pH, and temperature of the bath. Thus, it is at least in principle possible to easily grow large areas of tandem cells designed for the most efficient conversion of the solar spectrum. A large number of semiconductors CdS, [19, 88] PbS, [89–91], and CdSe, [85, 92, 93] etc., have been electrodeposited with varying coating parameters such as electrolyte concentration, pH, and applied potential, etc. The ED method is restricted to electrically conductive materials. In the electrochemical deposition, the substrate (mesoporous TiO_2-coated TCO) is submerged in a liquid solution (electrolyte). When an electrical potential is applied between a conducting area of the substrate and a counter electrode (usually platinum) in the liquid, a chemical redox process takes place resulting in the formation of a semiconductor QDs layer on the substrate. The schematic of hybrid electrochemical/chemical deposition of CdS QDs are presented in Fig. 5 [94].

Recently, Wang group, [13] and X-Y Yu et al., [19] demonstrated the electrodeposition of CdSe QDs on SnO_2 and ZnO nanostructures, respectively. The completely covered CdSe QDs by electrodeposition on mesoporous SnO_2 photoanodes showed 17.4 mA cm^{-2} with 3.68 % PCE, which is relatively a high performance compared to the previous reports on SnO_2 photoanodes-based QDSSCs. The growth methodology of electrodeposited CdSe QDs at different cycles is explained in Fig. 6. The size of QDs was influenced by number of coating cycles. However, more number of coating cycles may block the pores.

. In this section, we have summarized the most important sensitization methods for QDSSCs. We observe the relative simplicity of the methods, that they do not require vacuum or high temperature conditions. Therefore, chemical approach of QDs sensitization method is inherently low cost and remarkable efficiency higher than 6 % has been reported. This fact advocates that QDSSCs have tremendous potential for the future development of low cost photovoltaic devices.

4 Photoanodes in QDSCs

The photoanode of a QDSSC functions as selective contact for electrons [4]. In addition, it also works as an electron "vehicle" to transport the injected electrons from the excited QDs sensitizers to the outer circuit. In general, the photoanode materials need to satisfy several properties: first, the energy gap of the semiconductor could match with that of the QDs-sensitizer to ensure an effective injection of the photo-induced electrons from the QDs to the semiconductor, ensuring in addition electron selectivity blocking holes. Second, the semiconductor electrode must have a high surface area to accommodate more QDs, so as to harvest as much photon as possible. The photoanode material is the indispensible component of QDSSCs which plays crucial role in sensitizer loading, electron injection, transportation, and collection, and therefore exhibits significant influence on the photocurrent, photovoltage, and the power conversion efficiency. Recently, remarkable efforts have been paid to the design of the chemical composition, structure, and morphology of the semiconductor photoanode.

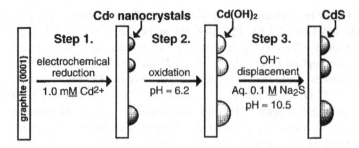

Fig. 5 Schematic diagram illustrating the three step electrochemical/chemical (E/C) synthesis of epitaxial CdS nanocrystals [Anderson et al. [94] Reprint permission from ACS]

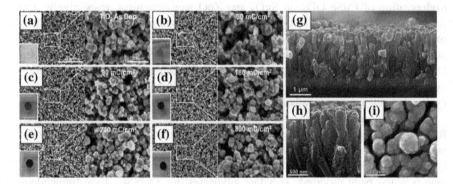

Fig. 6 SEM images of low and high magnification of CdSe-free mesoporous film of TiO$_2$ (a) and for different CdSe electrodeposition times from 30 to 300 mC cm^2 (b)–(f). [Sauvage et al. [183], reprint permission from IOP publishers]. (g)–(i) CdSe QDs electrodeposited TiO$_2$ ruitle nanowires [Kamat et al. [184], reprint permission from Wiley VCH]

In sensitizer-type solar cells, the photoanode should have high surface area to accommodate more amount of sensitizer loading to enhance light harvesting. However, the recombination process is proportional to the electrode surface area. The open-circuit potential (V_{oc}) in sensitized solar cells is significantly affected by the recombination process. A balance between recombination and light harvesting is therefore needed to maximize sensitized solar cell performance. With these considerations on mind inorganic semiconductors are extremely attractive due to their high extinction coefficient, in many cases, see Fig. 7.

In Fig. 7, a comparison of the extinction coefficients of one of the most employed dyes in DSSCs, N719, with the inorganic semiconductor Sb$_2$S$_3$. This implied that the effective surface area of QDSSCs covered with QDs may not need as much increase as in DSSCs for achieving high light harvesting. In this sense, QDSCs can take advantage of the development of new electrode structures to control the recombination process with lower effective area and consequently lower recombination. In addition, despite the high extension coefficient of

Fig. 7 Extinction coefficient of molecular dye N719, one of the most popular dyes in DSSCs, and Sb₂S₃ inorganic semiconductor [Boix et al. [185] reprint permission from ACS]

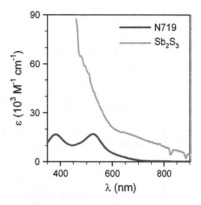

semiconductor QDs sensitizers, it is relatively larger in size than dye molecules; therefore it is difficult to penetrate deeper parts of TiO₂ electrode and thus limiting the sensitizer loadings. Therefore, large-pore network is prerequisite to afford effective QDs loading. On the other hand, such photoanodes could demonstrate high charge transport from sensitizer to a charge collector, ultimately, overwhelming the charge recombination at photoanode/electrolyte interface. Therefore, to achieve (a) high sensitizer loading (b) fast electron transport channel, and (c) good electrolyte pore-filling, establishing multifunctional photoanode frame work is the promising approach in QDs-sensitized solar cells. The following nanostructures has been identified as futuristic architectures in QDSSCs (a) highly interconnected, spatially assembled 1-D network (b) branched nanowires with highly conducting backbone which directly attached to charge collector, and (c) three-dimensionally ordered pore arrays with high scattering capability. Some of the electrode preparation methods also present a high interest for industrial application, as electrospinning.

In this context, nanofibrous membrane (NF), inverse opal (IO) and hierarchal nanowire (HN) electrodes receive great deal of attention as three-dimensional (3-D) photoanodes for next generation DSSCs or QDSSCs. In this section, we will discuss the fabrication and advantageous of 3-D nanostructured photoanodes in QDSCs.

4.1 Directly Assembled Continuous Fibrous Electrodes

Utilization of the wide pore-structured nanofibers is receiving great attention for superficial electrolyte penetration through their vertical pores yield effective interfacial contacts with TiO₂–sensitizer interfaces [95, 96]. In particular, fibrous electrodes prepared by an electrospinning technique showed remarkable performance in DSSCs since it support large-scale anode fabrication at low cost. Archana et al. [97] reported that a 1-D fibrous film results in a high diffusion coefficient

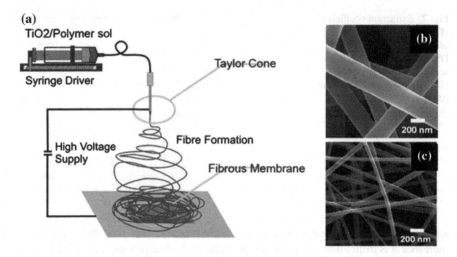

Fig. 8 a Schematic illustration of electrospinning set up. The TiO$_2$ fibrous channels were directly deposited onto collector (FTO substrates); SEM images of (**b**) as prepared TiO$_2$ fibers and (**c**) calcinated fibers (An et al. [186] Reprinted permission from Elsevier)

($D_n \approx 4.6 \times 10^{-4}$ cm^2 s^{-1}), which is nearly 3 times higher than that of the particulate electrodes ($D_n \approx 1.5 \times 10^{-4}$ cm^2 s^{-1}) under identical photoexcitation density. Furthermore, they observed less charge transport resistance and transit time in the fibrous electrode over the nanoparticle electrode. The high electron diffusive transport characteristics of fibrous electrodes are fostering as effective photoanodes for achieving feasible charge collection at QDSCs compared to conventional nanoparticle electrodes. In addition, fibers have less grain boundary density, which can overcome the trapping–detrapping loss unlike conventional nanoparticulate electrodes and beneficial advantages in attenuating the scattering loss [98]. On the other hand, the inclusion of nanofibers in conventional nanoparticle electrodes (composite) substantially improves the light harvesting through promoting the light-scattering pattern (analogous to antenna lobes) [99]. The first QDs (CdS:CdSe)-sensitized solar cells using direct assembled 3-D fibrous electrodes has been demonstrated by Sudhagar et al. [47] Followed that significant work has been developed on fibrous QDSSCs. The schematic structure of electrospinning (ES) set up is shown in Fig. 8.

The detailed working principles of ES technique have been explained elsewhere [100–102]. Briefly a continuous TiO$_2$ fibrous membrane is directly assembled on a FTO substrate using the electrospinning technique (Fig. 8). The typical solution is used for spinning: 0.53 g of polyvinyl acetate (PVAc) dissolved in 4.45 g of dimethyl formamide (DMF). After completely dissolving the PVAc in DMF, 1 g of titanium(IV) propoxide and 0.5 g of acetic acid were mixed well for 30 minutes. The resultant solution for spinning is loaded into a syringe driver and connected to a high-voltage power supply. An electrical potential of 15 kV was applied over a collection distance of 12 cm. The electrospun fibrous membranes were collected continuously on the grounded FTO substrate for 20 minutes. Subsequently, the films were sintered at 450 °C for 30 minutes in air. The diameter of the fibrous membranes can be easily modified by adjusting the Taylor cone size through (a) sol concentration (b) flow rate, and (c) applied potential.

Most fibrous electrodes have been fabricated using a two-step method. After fiber synthesis, substrates are coated with the synthesized fibers, but in this case poor particle interconnectivity is detected [96, 103]. Continuous electrospun fibers can solve this problem but peel off limitations has to be overcome. Various approaches have been demonstrated for this purpose. Song et al. utilized a hot press method to enhance the adhesion of the nanofiber to the substrate [95]. Similarly, chemical treatment was also employed to relax the nanofiber to yield an improved adhesion of the nanofiber to the substrate [104]. Onozuka et al. [105] applied a dimethyl formamide (DMF) treatment to TiO_2 fibrous electrodes, where DMF induced the swelling of the polymeric substances in the composite film and reinforced the adhesion between the substrate and fibrous membranes. This DMF treatment has improved the charge collection efficiency of DSCs by about 20 % compared to untreated electrodes. Analogous DMF-treated electrospun spheroidal electrodes were tested in QDSSCs, where the efficiencies increased from 0.85 to 1.2 % through improving the electrical contact between the fibers and the Transparent Conducting Oxide (TCO) substrate [106]. Despite the good electrical contact achieved by chemical treatment, it could affect the fibrous morphology thus lower the pore volume of the electrode, which severely affect the QDs loading and electrolyte penetration. In view of maintaining electrode pore volume as well as good electrical contact between TCO and fibrous network concurrently, it would be a better choice to choose chemical vapor treatment instead of direct chemical treatment.

Recently, we studied the above said hypothesis of surface treatment (chemical treatment and chemical vapor treatment) in fibrous electrodes and provide the detailed insights in controlling recombination rate at QDSCs [63]. Figure 9a–c compares the SEM images of different surface-treated TiO_2 fibrous electrodes. It clearly indicates that the untreated fibers (Fig. 9a), show a smooth surface with 70–100 nm diameters and several micrometers length. After surface pretreatment with DMF (Fig. 9b), the fibrous channels were intertwined and became a rather compact structure. This may be due to the fact that DMF treatment swells the polymer content of the fibers, thus resulting in a coagulated fiber structure. This coagulated fiber structure reduces the interpore distance between each fiber channel, reducing consequently the effective surface area of the electrode for electrolyte penetration. However, it could improve the electrical contact between the FTO substrate and the fibrous TiO_2 layer. In the case of tetrahydrofuran (THF) vapor treatment (Fig. 9c), the fibrous surface seems etched, which could consequently improve fibers inter-junction points. Also, the interpore distance between fibrous channels is partially retained after the THF pretreatment. As a consequence, THF treatment produces electrodes to half way between UT and DMF electrodes.

From Fig. 9d THF-QDSSCs result high open-circuit voltage $V_{oc} = 0.57$ V with slight improved short circuit photocurrent $J_{sc} = 9.74$ mA cm^{-2} compares to untreated fibrous electrode based QDSSCs, while the DMF cell presents a lower

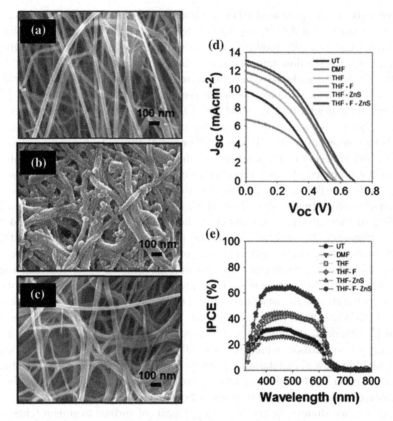

Fig. 9 SEM images of (**a**) untreated (**b**) DMF treated and (**c**) THF vapor-treated TiO_2 fibers; **d** current–voltage (*J–V*) measurements and (**f**) IPCE spectra of QDSSCs using different surface treated TiO_2 fibers electrodes. The electrode thickness is 2.4 μm. (Sudhagar et al. [63] reprint permission from RSC publishers)

photocurrent. The observed higher performance in THF vapor-treated cells may attribute to the improved physical adhesion of fibers with the TCO substrate. The recombination parameters of this system have been studied and found that chemical vapor treatment offer high charge recombination resistance to the electrodes. Interestingly, these TiO_2 fibrous membranes afford high feasibility in altering their Fermi level under post-doping process. Under the post-doping of fluorine ion (in Fig. 9d, THF+F) results upward Fermi level shift in TiO_2 conduction band. This Fermi level upward shift promotes the V_{oc} as well as fill factor (FF) of the device. We found that the combined pretreatment (THF) and post-treatment (F ion doping and ZnS QDs decoration on CdS/CdSe QDs) the 3-D TiO_2 fibrous photoanode yielded a device performance of 3.2 % with a remarkable $V_{oc} = 0.69$ V compared with most of the reports existing in the literature.

Yet another attempt on improving the electrical contact between fibrous electrode with TCO substrates has been demonstrated by Samadpour et al. [107].

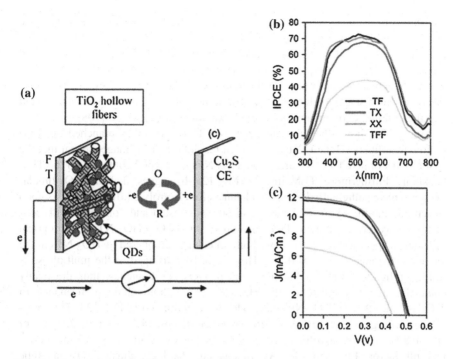

Fig. 10 a Schematic view of the hollow nanofibrous anode based QDSCs, IPCE; corresponding
(**b**) IPCE and (**c**) J–V curves of the SILAR sensitized cells. (Samadpour et al. [107] reprint
permission from RSC publishers)

They explored that the inclusion of nanoparticle glue layer in between the hollow
nanofiber promotes the electrical contact between fiber channels with TCO as well as
promote the surface area of the electrode. Figure 10 shows the schematic of hollow
nanofiber photoanode-based QDSC (Fig. 10a) along with JV and IPCE performance
(Fig. 10b, c). In this work, authors compared nine different photoanodes for instance
conventional particulate TiO_2 (NP), pure hollow nanofiber (NF), and mixed composite of NP and NF. From Fig. 10b, c, the efficiency of the pure hollow fiber (F)
structure is just 0.66 % (V_{oc} = 0.43 V, J_{sc} = 3.1 mA cm^{-2}, FF = 0.54), the efficiency of cells with XX structures (TiO_2 paste which contained TiO_2 nanoparticles
and fibers) increased to 3.24 % (Fig. 10c), which constitutes a 6-fold enhancement.
In this structure, the 20 nm nanoparticles act as a glue crosslinking the fibers and also
improving the adhesion with the FTO substrate. Mixing both fibers and nanoparticles
(X structures) has the additional beneficial effect of light scattering and facile
electron transport provided by the one-dimensional hollow fiber structure. The TiO_2
paste which contained TiO_2 nanoparticles and fibers (XX) showed high performance
than that of pure hollow fiber-based QDSSCs (data not shown in the figure).

Nevertheless the direct assembled fibers showed competitive performance with
nanoparticulate structures, under similar thickness, further enhancing the electrode
thickness (above 10 micron) is really a challenging task. Since, at high thickness

the fibrous membranes are found to peel off from the TCO substrate. Recently, we demonstrated the hierarchical hollow nanofiberous electrode with nanotube branches which balancing the trade-off between the fibrous thickness and mechanical stability [108]. Assembling NT arrays on a highly interconnected 3D fibrous backbone would reinforce the stability of the electrode. Figure 11 shows the fabrication stages of hierarchical 3-D hollow TiO_2 nanofibers (H-TiO_2-NFs) and the detailed experimental procedure explained in elsewhere [108].

Figure 12a shows the QD-sensitized 3-D TiO_2 nanotubes branched on TiO_2 hollow nanofibers (H-TiO_2 NFs). The high resolution TEM images reveal that the spatially decorated TiO_2 NT arrays on TiO_2 NFs have good contact with the TiO_2 backbone. Furthermore, TEM images (Fig. 12a, b) suggest that the TiO_2 tubular branches have sufficiently large pore channels for electrolyte filling as well as good structural stability. This 3-D photoelectrode was tested and compared with conventional vertically grown TiO_2 nanotube on TCO (TiO_2-NT). The optical reflectance spectra (Fig. 12c) shows high reflectance compared to TiO_2-NTs in the wavelength range of 380–800 nm. This might be attributed to the multiple scattering of incident light at the hierarchical TiO_2 NT branches, thus drastically enhancing the reflectance of the electrode. The photovoltaic performance of TiO_2-NT and 3-D H-TiO_2 NFs electrodes were compared in Fig. 12d. The TiO_2-NTs directly grown on a FTO electrode resulted in a photoconversion efficiency of $\eta = 0.9$ % with photovoltage, $V_{oc} = 0.62$ V, photocurrent, $J_{sc} = 2.5$ mA cm^{-2}, and fill factor, FF $= 58.3$ %. As anticipated, the hierarchical TiO_2 nanotube branches grown on a hollow NF backbone show unprecedentedly promoted to $\eta = 2.8$ % with $V_{oc} = 0.61$ V, $J_{sc} = 8.8$ mA cm^{-2} and F.F. $= 50.3$ %. It is clearly evident that the TiO_2 NTs spatially assembled on the hierarchical 3D-nanofibrous backbone promote the QDSSC performance by a factor of three compared to the TiO_2 NTs directly grown on a TCO substrate. We can relate the enhancement of photocurrent generation with the H-TiO_2 NF photoanodes to several contributions: (a) higher effective surface area and consequently higher QD loading and light harvesting; (b) highly efficient charge collection throughout the photoanode with fewer boundary layers and (c) the multiple scattering effects of the comb-like hierarchical NT arrays, in particular, red photon harvesting.

4.2 3-D Tree-Like Branched Hierarchical Nanowire

Designing vertically grown 1-D nanostructures such as nanowire [109–112] and nanotube [113–117] which directly attached to the charge collectors (TCO) provides high charge collection efficiency in DSSCs. However, the larger voids presence in between the nanowire/tube channels lowers the internal surface area of the electrode [118]. Promoting internal surface area through introducing hierarchal branches on 1-D nanostructured stems are beneficial approach for achieving 3-D photoanodes and offers simultaneously improved surface area and favorable electron transportation. The resultant 3-D complex nanoarchitectures show many

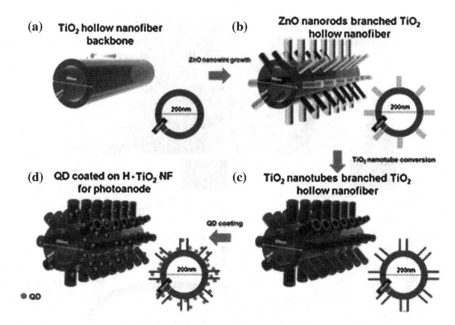

Fig. 11 Schematic illustration of H-TiO$_2$ NF photoanode fabrication stages of (**a**) TiO$_2$ hollow nanofibers (TiO$_2$-NFs), (**b**) ZnO NR templates grown on TiO$_2$-HNFs, (**c**) TiO$_2$ nanotube branches grown on TiO$_2$-NFs through ZnO NR templates, and (**d**) QD-sensitized H-TiO$_2$ NF photoanode. (Han et al. [108] reprint permission from RSC publishers)

advantages in sensitizer-type solar cells including: (i) the enhancement in the light-harvesting probability through large surface area for sensitizer loading and photon localization arising from the random multiple scattering of the light within the pine tree network (ii) the direct charge extraction pathways throughout the device thickness (fast electron transport from sensitizer to collecting terminal), and (iii) the huge 3-D porous network that allows better electrolyte filling, with possible beneficial implications for preparation of solid devices.

Among the many 3-D hierarchical anode architectures, "tree-like" morphology is a versatile candidate, since the tree-like configuration can be fabricated in large scale and each electronic interfaces being assembled at separate stages without any complication. The schematic structure of proximally coated sensitizers on tree-like framework is illustrated in Fig. 13. Under irradiation the excited electrons at sensitizer can inject to wide bandgap branches (WB-PA), and subsequently the electrons were being rapidly reached to collector substrate through conducting backbone nanowire (stem) [119].

Macroscopically, assembling tree-like crystalline framework results 3-D "nanoforest" structure, which offer ample room for more quantity of sensitizer loading compare to branch-free nanowire electrodes. This approach is simply mimicking branched plant structures, ultimately, to capture more sunlight.

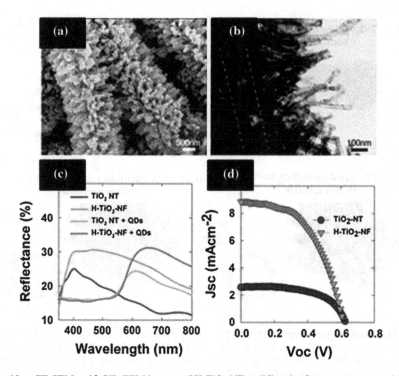

Fig. 12 **a** FE-SEM and **b** HR-TEM images of H-TiO$_2$ NF; **c** diffused reflectance spectra and **d** J-V plots of QDS-sensitized TiO$_2$ nanostructures. (Han et al. [108] reprint permission from RSC publishers)

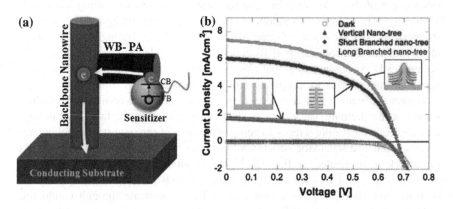

Fig. 13 **a** Schematic structure of electronic interfaces in 3-D tree-like photoanodes; **b** the J-V performance of branched nano-tree photoanodes compares with branch-free TiO$_2$ photoanodes (Herman et al. [187] reprint permission from IOP publishers)

Figure 13b clearly explains the advantage of tree-like photoanodes compared to conventional nanowire electrodes. From Fig. 13b the photocurrent of the DSSC has monotonically increased by extending the length of branches in the electrode. The similar photocurrent enhancement in hierarchical 3-D tree-like photoanodes has been widely found in the DSSCs literature [120–122]. The efficiency enhancement of 3-D tree-like electrodes is mainly due to enhanced surface area enabling higher sensitizer loading and light harvesting and also is due to reduced charge recombination by direct conduction along the stem of the nano-tree [123]. The hierarchical branches in tree-like electrodes has been grown either by seed layer-assisted method, [124–126] or self-catalyst-based mechanism [127]. The stem of the electrode (back bone) may be nanowire or nanotube. However, the overall photoconversion efficiency of the 3D tree-like photoanode strongly depends on their growth mechanism. Wu et al. compared the performance of hierarchical TiO_2 nanowire array photoanodes (hydrothermally grown without seed layer) with P25 nanoparticle under similar dye uptaking condition [127]. The J_{sc} and V_{oc} of P25-based DSSC is 12.0 mA·cm^{-2} and 794 mV, respectively, which is much lower than those of 3hr grown hierarchically TiO_2 nanowire arrays (13.9 mA·cm^{-2} and 826 mV). By increasing the branch length through increasing hydrothermal duration time to 9 h, the efficiency of the device promotes to 7.34 %. Further extending nanowire branch growth duration to ~ 12 h the conversion efficiency found to be decreasing to 6.35 %, which may be ascribed to the formation of more recombination pathways at nanowire/electrolyte interfaces which lower the conversion efficiency of the device. Interestingly, diffusion coefficient of iodine redox shuttle is ~ 2 orders higher in hierarchical nanoforest electrodes (D $(I_3^-) = 3.3 \times 10^{-8}$ cm^2/s) than NP electrode (D (I_3^-)=5.4 $\times$ 10^{-10} cm^2/s) [128]. It is understood that the enhanced mass transport of the redox shuttle anticipated to result high V_{oc} in the device and is realized in many tree-like anode-based DSSCs.

Recently, we demonstrated the QDSSCs with high open-circuit voltage as high as 0.77 V in ZnO nanowires array electrodes [70]. The performance of the cell can even be increased to a promising 3 %, using a novel photoanode architecture of "pine tree" ZnO nanorods (NRs) on Si NWs hierarchical branched structure. The different stages of fabricating hierarchical ZnO nanorods on Si backbone nanowires are schematically shown in Fig. 14.

The typical experimental procedure of pine tree-like Si/ZnO electrodes is as follows: in first stage, the backbone Si nanowires were fabricated via the vapor–liquid–solid (VLS) process on the Au nanoparticle decorated FTO glass by using a chemical vapor deposition (CVD) system. By introducing a 10 % silane (SiH$_4$) in Hydrogen (H$_2$) mixture in a tube furnace, Si nanowires were synthesized under the reactor temperature and pressure were kept at 480–520 °C and 40 torr for 3 min. This Si NW thickness is assumed as total thickness of the photoanode. The n-type doping of Si NWs was achieved by using PH$_3$ gaseous precursor as in situ doping sources. A more detailed description of the Si NWs growth conditions and their electrical characteristics can be found elsewhere [70]. At second stage, ZnO hierarchical rods were grown on Si NWs. The 25 nm thick ZnO film, as a seed layer, was deposited on FTO glass by using magnetron sputtering. The sample was

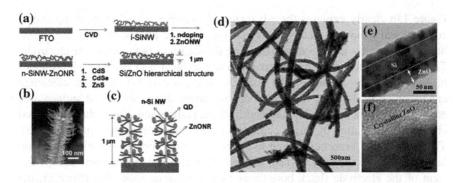

Fig. 14 **a** Fabrication scheme of Si/ZnO hierarchical structure **b** SEM image of n-Si-NW/ZnO-NR, and **c** schematic Si/ZnO hierarchical structure for photoanode **d** HRTEM images of ZnO seed-coated Si Nanowire, and **e** and **f** their corresponding high magnification TEM images. (Sudhagar et al. [70] ©Reprint permission from ACS Publishers)

transferred to a beaker containing 25 mM zinc nitrate and 25 mM methenamine in water for the growth of ZnO NWs and was kept in the oven at 85 °C. After 12 h, the electrodes were removed from the beaker and rinsed with water then dried by N_2. The ZnO NWs present an average length and diameter of 2.2–2.5 µm and 100–150 nm, respectively. Figure 14d, e shows the TEM images of bare and ZnO seed particle coated Si nanowire, respectively.

Figure 15a clearly depicted that the branched ZnO NRs directly grown on the Si-NWs forming a pine tree structure. HR-TEM image (Fig. 15b) at the top of ZnO NRs reveals that the ZnO NRs are single crystalline and were grown along the [001] direction. The porosity of the electrode has been maintained even after QDs coating (Fig. 15c), which is beneficial for effective electrolyte penetration through entire device. The optical reflectance property of ZnO-NR and 3-D n-Si-NW/ZnO-NR HBS has been studied by diffuse reflectance (Fig. 15e). Both sensitized electrodes present a decrease in the reflectance for the wavelength shorter than 620 nm due to light absorption of CdS/CdSe sensitizer. Analyzing the reflectance values at longer wavelength, before the absorption threshold, the improved scattering properties of the pine tree HBS, in comparison with ZnO NWs, are manifested in the higher reflection values at these wavelengths. The IPCE presents (Fig. 15f) a similar threshold to reflectance. It is important to highlight that no significant IPCE is observed at wavelengths longer than 700 nm, indicating a negligible contribution of the Si-NWs to QDSC performance. This fact also suggests that the light absorbed by Si-NWs is negligible and consequently the differences observed in reflectance values at longer wavelength in (Fig. 15e) are in fact due to the scattering properties. The enhanced light scattering is also responsible to the higher IPCE obtained for pine tree HBS in the wavelength range of 500–650 nm. By sensitization with CdS/CdSe QDs, the devices yielded an overall efficiency of ~3 %, which was higher than that of ZnO nanowire arrays due to increased QDs loading and enhanced light scattering by the 3D geometry. It was also noted that the doping in

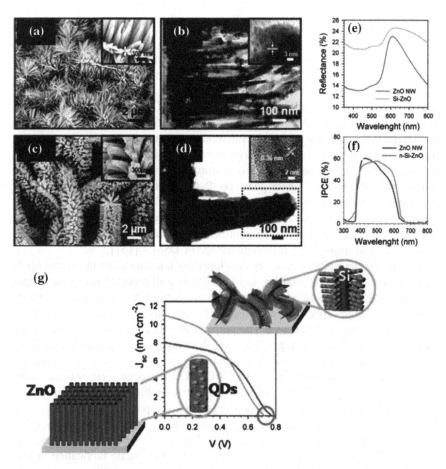

Fig. 15 a and **c** SEM images of pure and QDs-sensitized Si/ZnO hierarchical structure, respectively; **b** and **d**, corresponding TEM images; **e** diffused reflectance spectra and **f** IPCE spectra of Si/ZnO hierarchical anode based QDSCs; **g** J-V performance of conventional ZnO nanowire array and Si/ZnO hierarchical anode-based QDSCs compared with corresponding schematic anode structure (Sudhagar et al. [70] ©Reprint permission from ACS Publishers)

the Si backbone nanowire played an important role in the performance: the ZnO NR had a J_{sc} of 8.0 mA/cm^2 and efficiency of 2.7 %, whereas the 3-D pine tree QDSC resulted in J_{sc} of 11.00 mA/cm^2 and efficiency of 3.00 % (Fig. 15g). It is indicated, pine tree structures have the additional advantage of the fast electron transport through n-Si NWs. More importantly, we demonstrated that it is possible to obtain higher V_{oc} QDSCs with an appropriate treatment of the recombination process, highlighting the fact that QDSCs have to be rethought, separately of DSCs, to optimize their performance.

4.3 Inverse Opal Architecture Photoanodes

The disordered geometrical structure of conventional nanoparticle-based photoanode is often limiting the performance of the sensitized-type solar cells due to
interfacial interference for electron transport. Mainly, trap-limited diffusion process in randomly connected networks can be affected by recombination with the
oxidizing species in the electrolyte during trapping process. Therefore, designing
the anode frameworks with highly interconnected morphology is a promising
approach in achieving superior charge transport and high penetration of both
sensitizers and redox couples. In this context, a nanostructure which contains
bottom-up 3D host–passivation–guest (H–P–G) electrode has been realized as
promising candidate in DSSCs [129, 130]. Since, 3-D H-P-G electrode offer good
structural control in the electron extraction and the recombination dynamics, this
new type of H-P-G electrode has significantly promoted the photocurrent, fill
factor, and most importantly the photovoltage of DSSCs [131].

The H-P-G electrode is basically developed by micromolding in inverse opals
(IO) structures using colloidal crystals [132]. It is well reported that inverse opal
TiO_2 has large interconnected pores that lead to a better infiltration, also it exhibits
a photonic bandgap (photonic crystal), which depends on the filling fraction of
TiO_2 in the inverse opal structure. The preparation procedure of IO electrodes has
been demonstrated in two stages. First, the host layer is self-assembled on TCO
substrates, subsequently secondary coating of guest (TiO_2) layer coated on the
host. The preparation methods of 3-D TiO_2 IO electrodes were schematically
explained in the Fig. 16a. The host layer may be assembled by either chemical or
physical technique. It is widely demonstrated that the polymer microspheres are
utilized as host layer (polystyrene, co-polymers, etc.) since it is easily removable
without altering the final TiO_2 morphology. Figure 16a shows the deposition of
TiO_2 layer by chemical method, which support for large-scale fabrication at low
cost. Besides, it is difficult to control the thickness of the TiO_2 layer in this
approach. The undesired thick TiO_2 coating could clog the mesoscopic pores of
the host layer, which would inhibit sensitization and electrolyte infiltration. This
can be overcome by physical coating like "atomic layer deposition" (ALD), where
the thin guest layer can be conformably coated on the entire polymer microspheres
(Fig. 16b) without clogging the pore structure [129, 133–135]. Finally, polymer
beads were removed (Fig. 16c) by high temperature annealing or with solvents at
room temperature, we obtained a 3D host backbone that is well connected to the
underlying TCO substrate. The resultant direct electronic connection with TCO
facilitates the charge extraction throughout the interconnected 3D H–P–G electrode. The IO photonic crystal are playing crucial roles in DSSCs as a (a) dielectric
mirror for wavelengths corresponding to the stop band and (b) medium for
enhancing light absorption on the long-wavelength side of the stop band [136, 137].
The advantages of light interaction in these structures ultimately enhance the
backscattering of the device through localization of heavy photons near the edges of
a photonic gap. This scattered light increases the probability of light absorption

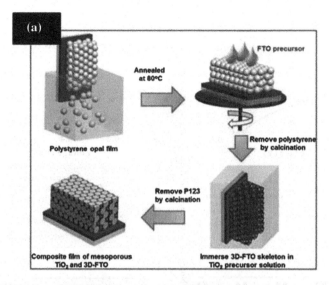

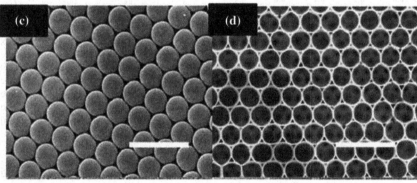

Fig. 16 a Schematic illustration of the forming 3-D TiO$_2$ Inverse Opal Structures by chemical solution method; SEM images of (**c**) polystyrene beads (host) and (**d**) TiO$_2$ inverse opal (after removing polystyrene beads); (Reprint permission from Science Direct, [142] and ACS publisher

especially in the red part of the solar spectrum. Wide range of IO-based photoanodes were exemplified in DSSCs as a (a) effective scaffold for high sensitizer loading [138] (b) scattering layer [139, 140] (c) window layer for more light photon harvesting [141], and (d) high electron-collection 3-D conductive grid [142]. Among many characteristics of IO, particularly opal size (diameter) [133, 143] and thickness of the electrode [131, 138, 144] has influencing the device performance of DSSCs.

The QDSCs with IO electrodes has been well demonstrated by Prof. Toyoda group. The influence of the QDs deposition time at IO surface was studied in detail [145]. Diaguna et al. studied the influence of opal diameter on photovoltaic performance [146]. The J-V performance of two different TiO$_2$ inverse opal

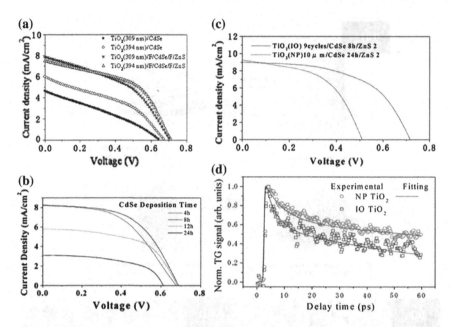

Fig. 17 a J-V characteristic of two different TiO$_2$ inverse opal electrodes made from latex template with diameters of 309 and 394 nm [146]; **b** Photocurrent density–photovoltage characteristics of CdSe QD-sensitized inverse opal TiO$_2$ solar cells with different adsorption times [145]; **c** Photocurrent density–photovoltage characteristics of CdSe QD-sensitized inverse opal TiO$_2$ and nanoparticulate TiO$_2$ solar cells with the same electrode thickness [145]; **d** Time dependence of the TG signal intensity of CdSe QD-sensitized inverse opal (IO) and nanoparticulate (NP) TiO$_2$ solar cells [145] (Reprint permission from IOP publishers, [146] and RSC Publishers [145])

electrodes made from latex template with diameters of 309 and 394 nm was tested (Fig. 17a). The wide pore size of opal 394 nm results slightly higher efficiency 2.7 % than the smaller pore size opal-based cell (2.4 %) [146]. Further, Toyoda et al. explored how the wide pore nature of IO supports the efficient QDs coating [145]. Figure 17b shows the JV performance of IO-based QDSSCs with different amount of QDs loading. From Fig. 17b it clearly understands that 8 h chemical bath deposited CdSe QDs coating performed higher than that of 4 h QDs coating. The former QDs coating result 3.1 % of PCE with $J_{sc} = 8.3$ mA cm^{-2}, $V_{oc} = 0.69$ V and FF = 0.57. In the case of later one (4 h CdSe QD coating) results 2.8 % of PCE with $J_{sc} = 8.2$ mA cm^{-2}, $V_{oc} = 0.68$ V and FF = 0.54. Further enhancing the QDs deposition to 24 h, the PCE was found to be reduced to 1.0 % ($J_{sc} = 3.1$ mA cm^{-2}, $V_{oc} = 0.61$ V and FF = 0.55) due to pore blockage, which severely affect the electrolyte penetration. Thus, result poor hole scavenging at QDs/IO interfaces, and lower the V_{oc} and J_{sc} [145]. In order to understand the feasibility of IO electrode in QDSCs, the typical PV performance of IO is compared with NP electrode in Fig. 17c. Both IO and NP electrode thickness is about 9 μm. The IO-based QDSSC showed a higher efficiency (3.5 %) than that of the nanoparticulate-based QDSSC (2.4 %) due to the higher V_{oc}, although the amount of CdSe QDs in the inverse opal case might be a half of that in the nanoparticulate case due to the difference of surface area. The higher V_{oc} in the IO QDSSCs was ascribed to the larger fraction of electron injection to the TiO_2 resulting in a higher quasi Fermi level, detected with ultrafast optical measurements, Fig. 17d. In addition to the above comparative results, Samadpour et al. also compared the performance of IO-based QDSSCs along with variety of nanostructured electrodes (nanotube, nanofiber, and nanoparticulate) [71]. They suggested that choosing the semiconductor deposition strategy (CBD or SILAR) is more important where the pore blockage takes a key role which influences the efficiency of QDSSCs.

5 Hole Transporting Material

Concerning the hole transporting material, it is a major difference with liquid DSSCs as many of the most employed semiconductors for light harvesting are not stable in solutions with the conventional I^-/I_3^- redox couple [7]. Stability problems could be solved by the use of Co redox couples, but the photocurrents obtained were relatively low [40, 60, 76]. The other approximation used to solve this issue was the coating of the semiconductor QDs with a protecting layer of amorphous TiO_2 [147]. However, obtained photocurrent was not much appreciable. In order to solve these issues, the most common redox couple employed for QDSSCs is the polysulfide (S^{2-}/S_n^{2-}), generally in a aqueous solution [148]. Polysulfide redox helps in the stability of semiconductor light absorber in the liquid devices, it also allows high photocurrents (photocurrents as high as 22 mA/cm^2 has been recently reported for PbS/CdS QDSSCs [149]), but it introduces an additional problem of a bad charge transfer with the platinized counter electrodes [21] commonly used in DSSCs. As a consequence very bad fill factors, FF, are commonly reported with platinized electrodes and polysulfide redox, see for example, refs. [21, 146]. In order to replace this inconvenient Pt catalytic electrodes, the fourth aspect commented previously have been developed for polysulfide electrolyte. We will come back with this issue later.

There are three interesting aspects that is worthy to comment related with polysulfide redox electrolytes. The first one is related with a deceptive practice that unfortunately is reported in a relatively significant number of papers. One of the requirements that an electrolyte for sensitized solar cells has to fulfil is the stability during a complete charge extraction process. Reduced species of the redox couple are oxidized during the hole regeneration of the sensitizer. Oxidized species diffuse to the counter electrode where they are reduced again in a fully regenerative process. This is the case of aqueous polysulfide electrolyte, but it is not estrange to find a paper where methanol is added as solvent. Methanol acts as a hole scavenger regenerating oxidized sensitizers, but it is not regenerated itself at the counter electrode, as the redox couples. As a consequence cells prepared with methanol in the electrolyte increases artificially the photocurrents, but this effect disappear when the methanol is exhausted. Obviously, this electrolyte cannot be used in a solar cell that demands a long device lifetime.

Related with this aspect, it is also important to work in the long-term stability of QDSSCs. There are very few report in this issue, probably as the first aspects that the authors wants to optimize is the efficiency as for most of the QDSSCs studied it lays significantly below than the efficiency reported for DSSCs. The main problem of liquid devices is the volatility of the electrolyte. It forces to use very sophisticated (and sometimes expensive) sealing processes in order to avoid electrolyte leakage. A common procedure systematically used in DSSCs is to employ ionic liquid, with significantly lower volatility that the standard solvents, as electrolytes. The reduced volatility eases the process of sealing that cheapens the cell at the same time that extend the device lifetime. The price of this improvement is a reduction in efficiency due to the increase in the diffusion of species in the electrolyte. A similar approximation has been developed for QDSSCs but using ionic liquids based on sulfide/polysulfide instead of the ionic liquids based on iodine employed on DSSCs [150]. Despite, further work is required to optimize the results sulfide/polysulfide ionic liquids could present promising results in terms of cell stability.

Nevertheless at this moment there is no optimum electrolyte for QDSSCs. Proof of this is the fact that the efficiencies obtained in liquid QDSSCs are no higher than those obtained in solid cells, while in the case of DSSCs the efficiencies of liquid cells is sensibly higher than solid ones. At least until the apparition of the cells using lead halide perovskite as light absorbing material, we will discuss this especial case in the Sect. 8. With the exception of perovskites the semiconductor providing the highest efficiency in all-solid sensitized devices is Sb_2S_3, with an efficiency of 6.3 % [67]. The study of this material in solid configuration was approached for the first time by Larramona and co-workers [151] and independently by Messina et al. [152]. In the case of Larramona CuSCN, a wide band gap p-type semiconductor, was used as HTM, while in Messina's work they contacted directly Sb_2S_3 for hole extraction. Very encouraging results were reported later by the groups of Hodes, [49] and Hodes and Grätzel, [153] in this last one using spiro OMeTAD as HTM. But the higher results have been reported by the groups of Seok and Grätzel [68, 154] using conductive polymers as HTM that boosted the efficiency to unprecedented values, at that moment, values of 5–6 %, even higher than liquid QDSSCs.

6 Counter Electrode

The last consideration concerning polysulfide electrolyte is related with the counter electrode. We have already discussed that platinized counter electrodes are one of the cause of poor FF obtained for QDSSCs with polysulfide electrolyte [21, 146]. This problematic cause was already discussed at the beginning of the 1980s by

Fig. 18 Effect of counter electrode material on the QDSSC performance. The effect is especially evident in the FF (Reprinted permission with Giménez et al. [24])

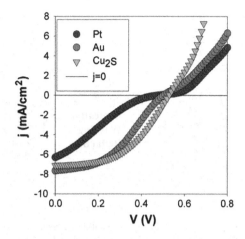

Hodes et al. [155] suggesting PbS, CoS, and Cu_2S as alternative counter electrodes for a good charge transfer with aqueous polysulfide. All three counter electrode materials have been investigated in QDSSCs. PbS counter electrodes increases significantly the cell performance of QDSSCs with polysulfide electrolyte, especially by the increase of FF and photocurrent [57]. PbS has been also studied in combination with carbon black [156]. Carbon foam was also checked [157]. Cu_2S, CoS, and combination of both materials have been also employed in the development of QDSSCs [158]. But probably Cu_2S has been the material most used in the preparation of efficient counter electrodes for polysulfide electrolyte and the record cells with polysulfide, 6.36 % efficiency with $CdSe_xTe_{1-x}$ colloidal quantum dots, [38] has been prepared using this material as the base of the counter electrode. This material presents higher performance that platinized and Au counter electrodes, see Fig. 18. Generally, it is prepared from a thin foil [24]. But in this format the sealing of active electrode and counter electrode becomes difficult. To solve this inconvenience Cu_2S pastes that can be deposited on transparent conductive oxide substrates has been developed [159]. An interesting alternative is the use of Cu_2S-reduced graphene oxide composite [160]. Briefly, graphene oxide is reduced by adding Cu ions in its structure, see Fig. 19. Cu_2S is finally formed with the help of sulfur atoms (just by using a polysulfide solution). Finally, a powder is obtained and properly treated can be deposited by doctor blade in the conductive substrate. This counter electrode has been used in the preparation of record cell in that moment with 5.4 % efficiency [161] (Fig. 20).

7 Recombination and Surface States

In this sense, Sb_2S_3 exhibits an interesting potentiality for the development of efficient QDSSCs, but in order to improve the obtained results, it is needed to reduce the recombination in photovoltaic devices using this material as light

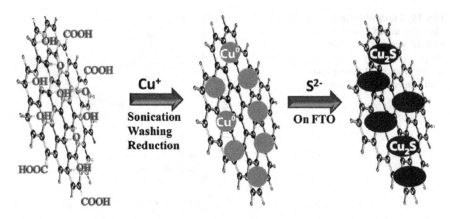

Fig. 19 Scheme of the preparation of reduced graphene oxide with Cu₂S nanocomposite for applications as counter electrode material in QDSSCs (Reprinted permission with Radich et al. [160])

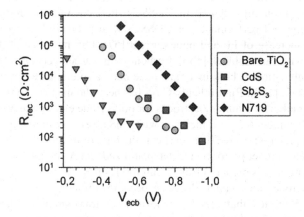

Fig. 20 Recombination resistance of bare mesoporous electrode of TiO_2 and three sensitized TiO_2 electrodes. Electrodes have been sensitized with the molecular dye N719 and with two inorganic semiconductors CdS and Sb_2S_3. Recombination resistance has been measured with impedance spectroscopy in three electrode configuration with liquid electrolyte. Recombination resistance is inversely proportional to the recombination rate and it is plotted as function of the equivalent conduction band potential, V_{ecb}, where the shift of the conduction band [188] of the different electrodes has been removed (Reprinted permission with Boix et al. [189])

absorber. This fact highlights the last of the five aspect in which QDSSCs shows sensible differences with DSSCs. In Fig. 19 recombination resistance, R_{rec}, of TiO_2 electrodes bare or sensitized with different materials is depicted. Note that the recombination is reduced, in comparison with bare TiO_2, when the electrode is sensitized with the molecular dye N719, as the molecule acts as a blocking layer for the recombination of electrons in TiO_2. In contrast, only a slight decrease in recombination (slight increase in R_{rec}) is observed when CdS is used as sensitizer. But, in the case of Sb_2S_3 recombination is even enhanced in comparison with bare electrode. This fact indicates that the inorganic sensitizer itself has an active role in recombination process, in other word the sensitizer introduces active recombination site where photoinjected electron in the TiO_2 find a pathway for recombination. Here a clear difference with DSSC is highlighted, while molecular dyes only presents the HOMO and LUMO levels for shelter electrons, in the case of QDs or inorganic semiconductors, surface states could play an important role in the QDSSCs performance [7]. In fact the main effect of the well-known ZnS treatment that enhances significantly QDSSCs in many cases, [10, 24, 34] is to passivate the surface states of the inorganic sensitizer [162]. Thus, the recombination though the sensitizer in QDSSCs has to be reduced in order to improve the photovoltaic conversion efficiency, passivation treatments of TiO_2 surface [163], or of the light absorbing material itself [164, 165] have shown a tremendous effect in the cell performance, but further analysis in this direction is needed to continue the optimization of QDSSCs. On the other hand, each kind of inorganic semiconductor could require a different passivation process in terms of stability and efficiency of the semiconductor layer.

Finally as introduction for the next section, it has to be highlighted that in addition one of the effective approaches to control the recombination rate in QD-sensitized solar cells (QDSSCs) is to densely cover the QDs on mesoporous-nanostructured electrodes [10] without naked sites, that directly block the physical contact between electrolyte and TiO_2 [166, 167] In this configuration, semiconductor sensitizer layer is acting as a barrier between electrons in the TiO_2 and accepting species in the electrolyte hindering this recombination path way, but recombination though sensitizer semiconductor is still possible as it has been previously commented. Taking into account this consideration, the procedure in with electrodes sensitized also plays an important role in the recombination of QDSSCs [71].

8 The Perovskite Revolution

Recently, a new type of photovoltaic cells with organometallic halide perovskites as light harvester materials have burst onto the scene gaining great attention in the scientific community. It is due mainly because a rapid succession of occurrences of record devices in a short-time period, which it is now reaching above of 14 % of certified efficiency, [168] and even further increase has been announced. Here, we briefly review the most striking results in the current hottest topic on photovoltaics.

8.1 Structure, Chemical Composition, and Optoelectronic Properties of Halogenated Perovskites

The halogenated hybrid perovskites have been previously studied as semiconducting materials processable in solution at low temperature for thin-film field-effect transistors, where they have shown to have a high mobility of carriers compared with organic materials [169]. Hybrid perovskites of type ABX_3 ($A = CH_3NH_3^+$, $B = Pb^{2+}$, $X = Cl^-$, Br^-, I^-) used in the newly developed photovoltaic devices are formed by inorganic layers of lead halide corner-sharing octahedrals interpenetrated by an alkylammonium cation network where the size of the organic cation plays an important role to define the final type of perovskite structure formed, Fig. 21.

The most used cations in the cells reported to date were methylammonium and lead (II) for the positions A and B, respectively. $(CH_3NH_3)PbI_3$ compound has a direct bandgap of 1.51 eV determined experimentally [170] and theoretically [171]. Also, this hybrid organometal halide perovskite has a high absorption coefficient compared to the N719 dye [172]. The work function studied by photoelectron spectroscopy of spin-coated polycrystalline films showed valence-band levels of $(CH_3NH_3)PbI_3$ and $(CH_3NH_3)PbBr_3$ at -5.44 and -5.38 eV versus the vacuum level, respectively and the conduction band levels calculated from the optical absorption edges are at -4.0 and -3.36. Therefore, these profiles mean that electron injection to the TiO_2 conduction band is favored. Figure 22 summarizes the main optoelectronic properties of $(CH_3NH_3)PbI_3$.

8.2 Construction of Solid-State Devices Using Wet Methods

Perovskites have been used as light harvester since 2009 when Miyasaka et al. [173] reported that these materials could be an alternative to binary chalcogenide based on QDSSCs, reaching an efficiency of 3.8 %, Fig. 23. An interesting aspect of the manufacture of these devices is that the active material is solution processable using a stoichiometric solution of $(CH_3NH_3)I$ and PbI_2 at room temperature and without employ vacuum techniques. However, the liquid electrolyte used as hole transport layer rapidly degrades the active material of sensitized cell. Approximately 2 years later, a work from Park and collaborators developed a cell which had twice the efficiency of Miyasaka cell due mainly to the use of more concentrated perovskite precursor solutions [172]. However, the cell was also deteriorated rapidly by the liquid electrolyte.

The breakthrough in efficiency (9.7 %) and stability (> 500 h) was done by the same group in 2012 using a solid electrolyte, the spiro-MeOTAD, instead of liquid electrolyte, Fig. 24 [170]. At the same time, devices whose active layer is also

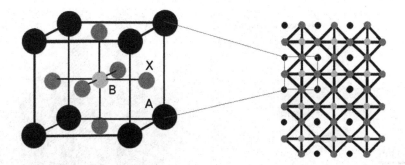

Fig. 21 3D structure for ABX$_3$ type perovskites

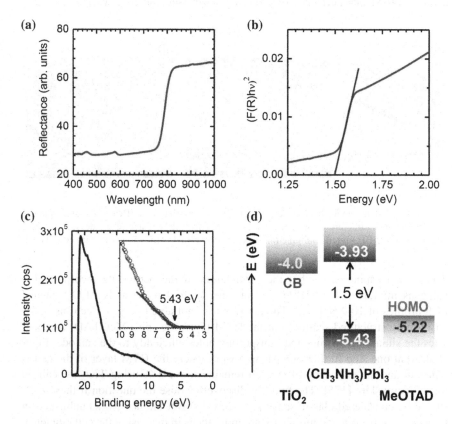

Fig. 22 Diffuse reflectance and UPS spectra for (CH$_3$NH$_3$)PbI$_3$ perovskite sensitizer. **a** Diffuse reflectance spectrum of the (CH$_3$NH$_3$)PbI$_3$-sensitized TiO$_2$ film. **b** Transformed Kubelka-Munk spectrum of the (CH$_3$NH$_3$)PbI$_3$-sensitized TiO$_2$ film. **c** UPS spectrum of the(CH$_3$NH$_3$)PbI$_3$-sensitized TiO 2 film. **d** Schematic energy level diagram of TiO$_2$ (CH$_3$NH$_3$)PbI$_3$, and spiro-MeOTAD (Reprinted permission with Kim et al. [170])

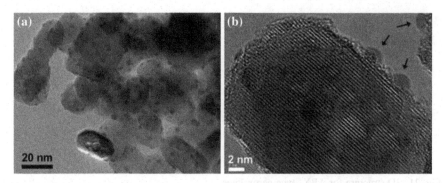

Fig. 23 TEM micrographs of (**a**) wide view of (CH₃NH₃)PbI₃ deposited TiO₂ (**b**) magnified image of (CH₃NH₃)PbI₃ deposited TiO₂. (Reprinted permission with Im et al. [172])

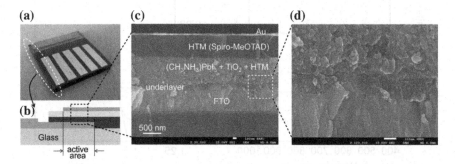

Fig. 24 Solid-state device and its cross-sectional mesostructure. **a** Real solid-state device. **b** Cross-sectional structure of the device. **c** Cross-sectional SEM image of the device. **d** Active layer-underlayer-FTO interfacial junction structure (Reprinted permission with Kim et al. [170])

deposited by spin-coating but the formulation of the perovskite incorporates a halogens mixture (CH₃NH₃)PbI₂Cl, and the scaffold material is Al₂O₃ reached PCE profiles of 10.9 % [174]. The perovskite is not able to inject electrons in the Al₂O₃-mesostructured scaffold but the HTM layer, the spiro-MeOTAD, compensates the situation injecting and transporting holes efficiently to the cathode. Etgar et al. went one step further publishing a work where the HTM layer of the cell is removed and the efficiency of the cell remains above 5 % using TiO₂ as scaffold for the perovskite [175]. The two last discoveries raise the question if these cells should be considered classic sensitized dye cells. But the last works published in such devices to early and mid-2013, are more focus in increasing the cell efficiency that in attempting to elucidate the mechanisms of storage, transportation and injection of electrons and holes. High efficiency has been also obtained by using polymeric HTM, [176] different morphologies of mesostructured layer, [177] perovskites with a mixture of halogens, [178, 179] and an impressive manufacturing technique employing sintering temperatures below 150 °C and efficiencies

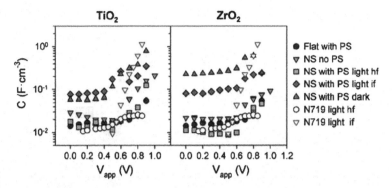

Fig. 25 *Left* graph plots capacitance of: flat sample with perovskite (PS); blank NS TiO$_2$ (0.35 μm thickness) with no PS; NS TiO$_2$ (0.55 μm thickness) with PS extracted from measurement under dark and under 1 sun illumination (light) conditions; and all-solid DSSC (2.2 μm thickness) with N719 as dye and spiro-MeOTAD as HTM. Right graph plots capacitance of flat sample with perovskite (PS); blank NS ZrO$_2$ (0.39 μm thickness) with no PS; NS TiO$_2$ (0.36 μm thickness) with PS extracted from measurement under dark and under 1 sun illumination (light) conditions; and all-solid DSSC (2.2 μm thickness) with N719 as dye and spiro-MeOTAD as HTM. Capacitance has been normalized to the electrode volume. Capacitance for both graphs has been extracted by fitting the impedance spectroscopy spectra from the intermediate frequency (*if*) region if nothing else is indicated. In some cases capacitance has been extracted from the high frequency (*hf*) region as it is indicated in the legend (Reprinted permission with Kim et al. [182])

of 12.3 % [180]. In addition impressive V_{oc} of 1.3 V has been also reported for (CH$_3$NH$_3$)PbBr$_3$ [181].

Finally, a recent study [182] focusing on the analysis of capacitances in perovskite solar cells using photoanodes with both TiO$_2$ and ZrO$_2$ mesoporous photoanodes, have shown a similar behavior despite the significant differences between both (electrons cannot be injected from (CH$_3$NH$_3$)PbI$_3$ into ZrO$_2$ as in the case of Al$_2$O$_3$). In both cases a capacitance that can be attributed to the perovskite has been detected (Blue triangles and green diamonds in Fig. 25). This capacitance indicates that charge accumulation is occurring in the perovskite itself in contrast with conventional all-solid DSSC where the charge accumulation is produced in the TiO$_2$ after photoinjection (red inverted triangles in Fig. 25). This is, the first observation of charge accumulation in the light absorbing material for nanostructured solar cells, indicating that it constitutes a new kind of photovoltaic device, halfway between sensitized and thin-film solar cell for NS TiO$_2$ and a thin-film solar cell with ZrO$_2$ scaffold for NS ZrO$_2$.

In summary, perovskite (CH$_3$NH$_3$)PbI$_3$-sensitized or mesoscopic solid-state solar cells using different HTMs (polymeric or molecular materials), scaffolds and halides are a promising type of solar cells to be converted in real-life manufactured devices in short time. However, the fast succession of record devices prepared last year, Fig. 26, is not being balanced with a full understanding of these cells work, and more studies in the theoretical plane are needed.

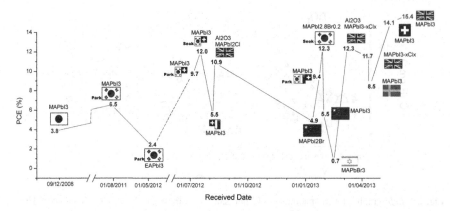

Fig. 26 Time evolution of published papers of hybrid perovskite solar cells (received date by the journal) versus PCE reported profiles of record cells; (*red*) liquid electrolyte (*black*), solid electrolyte

8.3 Conclusions

Despite the limited efficiencies obtained in the early stages of the study of QDSSCs, the potentialities of semiconductor thin layer or quantum dots has impelled the research on this devices in the last few years. We noted that the problems of such devices are different in many respects to that happening in DSSC, while both types of solar cells work according to the same principles. In QDSSC, the sensitizer material can be prepared in several ways affecting the final device performance. Photoanode nanostructure, hole transporting material, counter electrode, and recombination processes (affected by semiconductor trapping states) have to be optimized in order to enhance the reported efficiencies. As a result of the intense work on QDSSCs a real breakthrough, with organometallic halide perovskites, has occured very recently. Solar cells with efficiencies as high as 15 % are reported, and it looks that further increase will be attained soon. In addition to the impressive efficiencies one of the strongest points of these new cells is that they can be prepared from solution techniques, as the once developed for QDSSCs, and consequently the production cost of the photovoltaic device can be dramatically reduced.

References

1. Tyler Miller Jr G (2007) Living in the environment. Thomson Brooks/Cole, London
2. United Nations Framework Convention on Climate Change (Text) (1992) (UNEP/WMO, Climate Change Secretariat, Geneva, 1992)
3. Barber J (2009) Photosynthetic energy conversion: natural and artificial. Chem Soc Rev 38(1):185–196

4. Bisquert J, Cahen D, Hodes G, Rühle S, Zaban A (2004) Physical chemical principles of photovoltaic conversion with nanoparticulate, mesoporous dye-sensitized solar cells. J Phys Chem B 108:8106–8118

5. O' Regan B, Grätzel M (1991) A low-cost high-efficiency solar cell based on dye-sensitized colloidal TiO$_2$ films. Nature 353:737–740

6. Peter LM (2011) The Grätzel cell: where next? J Phys Chem Lett 2:1861–1867

7. Hodes G (2008) Comparison of dye- and semiconductor-sensitized porous nanocrystalline liquid junction solar cells. J Phys Chem C 112(46):17778–17787

8. Robel I, Subramanian V, Kuno M, Kamat PV (2006) Quantum dot solar cells. harvesting light energy with CdSe nanocrystals molecularly linked to mesoscopic TiO$_2$ films. J Am Chem Soc 128:2385–2393

9. Zaban A, Micic OI, Gregg BA, Nozik AJ (1998) Photosensitization of nanoporous TiO$_2$ electrodes with InP quantum dots. Langmuir 14(12):3153–3156

10. Mora-Seró I, Giménez S, Fabregat-Santiago F, Gómez R, Shen Q, Toyoda T, Bisquert J (2009) Recombination in quantum dot sensitized solar cells. Acc Chem Res 42(11):1848–1857

11. Leschkies SK, Divakar R, Basu J, Enache-Pommer E, Boercker JE, Carter CB, Kortshagen UR, Norris DJ, Aydil ES (2007) Photosensitization of ZnO nanowires with CdSe quantum dots for photovoltaic devices. Nano Lett 7(6):1793–1798

12. Hossain A, Yang GW, Parameswaran M, Jennings JR, Wang Q (2010) Mesoporous SnO2 spheres synthesized by electrochemical anodization and their application in CdSe-Sensitized Solar Cells. J Phys Chem C 114(49):21878–21884

13. Hossain MA, Jennings JR, Koh ZY, Wang Q (2011) Carrier generation and collection in CdS/CdSe-sensitized SnO2 solar cells exhibiting unprecedented photocurrent densities. ACS Nano 5(4):3172–3181

14. Hossain MA, Koh ZY, Wang Q (2012) PbS/CdSe-sensitized mesoscopic SnO$_2$ solar cells for enhanced infrared light harnessing. Phys Chem Chem Phys 14(20):7367–7374

15. Samadpour M, Giménez S, Boix PP, Shen Q, Calvo ME, Taghavinia N, zad AI, Toyoda T, Míguez H, Mora-Seró I (2012) Effect of nanostructured electrode architecture and semiconductor deposition strategy on the photovoltaic performance of quantum dot sensitized solar cells. Electrochimica Acta 75(0):139–147

16. Giménez S, Xu X, Lana-Villarreal T, Gómez R, Agouram S, Muñoz-Sanjosé V, Mora-Seró I (2010) Determination of limiting factors of photovoltaic efficiency in quantum dot sensitized solar cells: correlation between cell performance and structural properties. J Appl Phys 108:064310

17. Vogel R, Hoyer P, Weller H (1994) Quantum-sized Pbs, Cds, Ag2 s, Sb2s3, and Bi2s3 particles as sensitizers for various nanoporous wide-bandgap semiconductors. J Phys Chem 98(12):3183–3188

18. Lin SC, Lee YL, Chang CH, Shen YJ, Yang YM (2007) Quantum-dot-sensitized solar cells: Assembly of CdS-quantum-dots coupling techniques of self-assembled monolayer and chemical bath deposition. Appl Phys Lett 90(14)

19. Yu XY, Liao JY, Qiu KQ, Kuang DB, Su CY (2011) Dynamic study of highly efficient CdS/CdSe quantum dot-sensitized solar cells fabricated by electrodeposition. ACS Nano 5(12):9494–9500

20. Hu HW, Ding JN, Zhang S, Li Y, Bai L, Yuan NY (2013) Photodeposition of Ag2S on TiO$_2$ nanorod arrays for quantum dot-sensitized solar cells. Nanoscale Res Lett 8

21. Mora-Sero I, Gimenez S, Moehl T, Fabregat-Santiago F, Lana-Villareal T, Gomez R, Bisquert J (2008) Factors determining the photovoltaic performance of a CdSe quantum dot sensitized solar cell: the role of the linker molecule and of the counter electrode. Nanotechnology 19(42)

22. Guijarro N, Lana-Villarreal T, Mora-Sero I, Bisquert J, Gomez R (2009) CdSe quantum dot-sensitized TiO$_2$ electrodes: effect of quantum dot coverage and mode of attachment. J Phys Chem C 113(10):4208–4214

23. Watson DF (2010) Linker-assisted assembly and interfacial electron-transfer reactivity of quantum dot-substrate architectures. J Phys Chem Lett 1:2299–2309

24. Giménez S, Mora-Seró I, Macor L, Guijarro N, Lana-Villarreal T, Gómez R, Diguna LJ, Shen Q, Toyoda T, Bisquert J (2009) Improving the performance of colloidal quantum dot sensitized solar cells. Nanotechnology 20:295204

25. Im SH, Lee YH, Seok SI, Kim SW, Kim SW (2010) Quantum-dot-sensitized solar cells fabricated by the combined process of the direct attachment of colloidal CdSe quantum dots having a ZnS glue layer and spray pyrolysis deposition. Langmuir 26(23):18576–18580

26. Zhu GA, Lv TA, Pan LK, Sun Z, Sun CQ (2011) All spray pyrolysis deposited CdS sensitized ZnO films for quantum dot-sensitized solar cells. J Alloy Compd 509(2):362–365

27. Benehkohal NP, Gonzalez-Pedro V, Boix PP, Chavhan S, Tena-Zaera R, Demopoulos GP, Mora-Sero I (2012) Colloidal PbS and PbSeS quantum dot sensitized solar cells prepared by electrophoretic deposition. J Phys Chem C 116(31):16391–16397

28. Salant A, Shalom M, Hod I, Faust A, Zaban A, Banin U (2010) Quantum dot sensitized solar cells with improved efficiency prepared using electrophoretic deposition. ACS Nano 4(10):5962–5968

29. Salant A, Shalom M, Tachan Z, Buhbut S, Zaban A, Banin U (2012) Quantum rod-sensitized solar cell: nanocrystal shape effect on the photovoltaic properties. Nano Lett 12(4):2095–2100

30. Santra PK, Kamat PV (2012) Tandem-layered quantum dot solar cells: tuning the photovoltaic response with luminescent ternary cadmium Chalcogenides. J Am Chem Soc 135(2):877–885

31. Lin S-C, Lee Y-L, Chang C-H, Shen Y-J, Yang Y-M (2007) Quantum-dot-sensitized solar cells: assembly of CdS-quantum-dots coupling techniques of self-assembled monolayer and chemical bath deposition. Appl Phys Lett 90:143517

32. Lee Y-L, Huang B-M, Chien H-T (2008) Highly efficient CdSe-Sensitized TiO₂ Photoelectrode for quantum-dot-sensitized solar cell applications. Chem Mater 20(22): 6903–6905

33. Martínez-Ferrero E, Mora-Seró I, Alberoa J, Giménez S, Bisquert J, Palomares E (2010) Charge transfer kinetics in CdSe quantum dot sensitized solar cells. Phys Chem Chem Phys 12:2819–2821

34. Shen Q, Kobayashi J, Diguna LJ, Toyoda T (2003) Effect of ZnS coating on the photovoltaic properties of CdSe quantum dot-sensitized solar cells. Journal of Appl Phys 103(8)

35. Pernik DR, Tvrdy K, Radich JG, Kamat PV (2011) Tracking the adsorption and electron injection rates of CdSe quantum dots on TiO₂: linked versus direct attachment. J Phys Chem C 115(27):13511–13519

36. Pan Z, Zhang H, Cheng K, Hou Y, Hua J, Zhong X (2012) Highly efficient inverted type-I CdS/CdSe Core/Shell Structure QD-sensitized solar cells. ACS Nano 6(5):3982–3991

37. Zhang H, Cheng K, Hou YM, Fang Z, Pan ZX, Wu WJ, Hua JL, Zhong XH (2012) Efficient CdSe quantum dot-sensitized solar cells prepared by a postsynthesis assembly approach. Chem Commun 48(91):11235–11237

38. Pan Z, Zhao K, Wang J, Zhang H, Feng Y, Zhong X (2013) Near infrared absorption of CdSexTe1–x alloyed quantum dot sensitized solar cells with More than 6 % efficiency and high stability. ACS Nano 7(6):5215–5222

39. Alivisatos AP (1996) Semiconductor clusters, nanocrystals, and quantum dots. Science 271:933–937

40. Lee HJ, Yum J-H, Leventis HC, Zakeeruddin SM, Haque SA, Chen P, Seok SI, Grätzel M, Nazeeruddin MK (2008) cdse quantum dot-sensitized solar cells exceeding efficiency 1 % at full-sun intensity. J Phys Chem C 112:11600–11608

41. Liu L, Guo X, Li Y, Zhong X (2010) Bifunctional multidentate ligand modified highly stable water-soluble quantum dots. Inorg Chem 49(8):3768–3775

42. Jia S, Banerjee S, Herman IP (2008) Mechanism of the electrophoretic deposition of CdSe nanocrystal films: influence of the nanocrystal surface and charge. J Phys Chem C 112:162–171
43. Brown P, Kamat PV (2008) Quantum dot solar cells. Electrophoretic deposition of CdSe-C-60 composite films and capture of photogenerated electrons with nC(60) cluster shell. J Am Chem Soc 130(28):8890-+
44. Mane RS, Lokhande CD (2000) Chemical deposition method for metal chalcogenide thin films. Mater Chem Phys 65(1):1–31
45. Chang CH, Lee YL (2007) Chemical bath deposition of CdS quantum dots onto mesoscopic TiO$_2$ films for application in quantum-dot-sensitized solar cells. Appl Phys Lett 91(5)
46. Jung SW, Kim JH, Kim H, Choi CJ, Ahn KS (2010) CdS quantum dots grown by in situ chemical bath deposition for quantum dot-sensitized solar cells. J Appl Phys 110(4)
47. Sudhagar P, Jung JH, Park S, Lee YG, Sathyamoorthy R, Kang YS, Ahn H (2009) The performance of coupled (CdS:CdSe) quantum dot-sensitized TiO$_2$ nanofibrous solar cells. Electrochem Commun 11(11):2220–2224
48. Lee YL, Huang BM, Chien HT (2008) Highly efficient CdSe-Sensitized TiO$_2$ photoelectrode for quantum-Dot-sensitized solar cell applications. Chem Mater 20(22): 6903–6905
49. Itzhaik Y, Niitsoo O, Page M, Hodes G (2009) Sb2S3-sensitized nanoporous TiO$_2$ solar cells. J Phys Chem C 113(11):4254–4256
50. Boix PP, Larramona G, Jacob A, Delatouche B, Mora-Sero I, Bisquert J (2012) Hole transport and recombination in all-solid Sb2S3-sensitized TiO$_2$ solar cells using CuSCN as hole transporter. J Phys Chem C 116(1):1579–1587
51. Maiti N, Im SH, Lim CS, Seok SI (2012) A chemical precursor for depositing Sb2S3 onto mesoporous TiO$_2$ layers in nonaqueous media and its application to solar cells. Dalton Trans 41(38):11569–11572
52. Gui EL, Kang AM, Pramana SS, Yantara N, Mathews N, Mhaisalkar S (2012) Effect of TiO$_2$ mesoporous layer and surface treatments in determining efficiencies in antimony sulfide-(Sb2S3) sensitized solar cells. J Electrochem Soc 159(3):B247–B250
53. Lan GY, Yang ZS, Lin YW, Lin ZH, Liao HY, Chang HT (2009) A simple strategy for improving the energy conversion of multilayered CdTe quantum dot-sensitized solar cells. J Mater Chem 19(16):2349–2355
54. Samadpour M, Zad AI, Taghavinia N, Molaei M (2011) A new structure to increase the photostability of CdTe quantum dot sensitized solar cells. J Phys D Appl Phys 44(4)
55. Yang ZS, Chang HT (2010) CdHgTe and CdTe quantum dot solar cells displaying an energy conversion efficiency exceeding 2 %. Sol Energy Mater Sol Cells 94(12):2046–2051
56. Yu XY, Lei BX, Kuang DB, Su CY (2011) Highly efficient CdTe/CdS quantum dot sensitized solar cells fabricated by a one-step linker assisted chemical bath deposition. Chem Sci 2(7):1396–1400
57. Tachan Z, Shalom M, Hod I, Ruhle S, Tirosh S, Zaban A (2011) PbS as a highly catalytic counter electrode for polysulfide-based quantum dot solar cells. J Phys Chem C 115(13):6162–6166
58. Ju T, Graham RL, Zhai GM, Rodriguez YW, Breeze AJ, Yang LL, Alers GB, Carter SA (2010) High efficiency mesoporous titanium oxide PbS quantum dot solar cells at low temperature. Appl Phys Lett 97(4)
59. Lee H, Leventis HC, Moon SJ, Chen P, Ito S, Haque SA, Torres T, Nuesch F, Geiger T, Zakeeruddin SM, Gratzel M, Nazeeruddin MK (2009) PbS and US quantum dot-sensitized solid-state solar cells: old concepts, new results. Adv Funct Mater 19(17):2735–2742
60. Lee HJ, Chen P, Moon SJ, Sauvage F, Sivula K, Bessho T, Gamelin DR, Comte P, Zakeeruddin SM, Il Seok S, Gratzel M, Nazeeruddin MK (2009) Regenerative PbS and CdS quantum dot sensitized solar cells with a cobalt complex as hole mediator. Langmuir 25(13):7602–7608

61. Lu P, Shi ZW, Walker AV (2010) Selective formation of monodisperse CdSe nanoparticles on functionalized self-assembled monolayers using chemical bath deposition. Electrochim Acta 55(27):8126–8134
62. Gorer S, Hodes G (1994) Quantum-size effects in the study of chemical solution deposition mechanisms of semiconductor-films. J Phys Chem 98(20):5338–5346
63. Sudhagar P, Gonzalez-Pedro V, Mora-Sero I, Fabregat-Santiago F, Bisquert J, Kang YS (2012) Interfacial engineering of quantum dot-sensitized TiO$_2$ fibrous electrodes for futuristic photoanodes in photovoltaic applications. J Mater Chem 22(28):14228–14235
64. Niitsoo O, Sarkar SK, Pejoux C, Ruhle S, Cahen D, Hodes G (2006) Chemical bath deposited CdS/CdSe-sensitized porous TiO$_2$ solar cells. J Photochem Photobiol A-Chem 181(2–3):306–313
65. Heo JH, Im SH, Kim HJ, Boix PP, Lee SJ, Seok SI, Mora-Sero I, Bisquert J (2012) Sb2S3-sensitized photoelectrochemical cells: open circuit voltage enhancement through the introduction of poly-3-hexylthiophene interlayer. J Phys Chem C 116(39):20717–20721
66. Lim CS, Im SH, Rhee JH, Lee YH, Kim HJ, Maiti N, Kang Y, Chang JA, Nazeeruddin MK, Gratzel M, Seok SI (2012) Hole-conducting mediator for stable Sb2S3-sensitized photoelectrochemical solar cells. J Mater Chem 22(3):1107–1111
67. Chang JA, Im SH, Lee YH, Kim H-J, Lim C-S, Heo JH, Seok SI (2012) Panchromatic photon-harvesting by hole-conducting materials in inorganic-organic heterojunction sensitized-solar cell through the formation of nanostructured electron channels. Nano Lett 12(4):1863–1867
68. Chang JA, Rhee JH, Im SH, Lee YH, Kim H-J, Seok SI, Nazeeruddin MK, Grätzel M (2010) High-performance nanostructured inorganic—Organic heterojunction solar cells. Nano Lett 10(7):2609–2612
69. Rodenas P, Song T, Sudhagar P, Marzari G, Han H, Badia-Bou L, Gimenez S, Fabregat-Santiago F, Mora-Sero I, Bisquert J, Paik U, Kang YS (2013) Quantum dot based heterostructures for unassisted photoelectrochemical hydrogen generation. Adv Energy Mater 3(2):176–182
70. Sudhagar P, Song T, Lee DH, Mora-Sero I, Bisquert J, Laudenslager M, Sigmund WM, Park WI, Paik U, Kang YS (2011) High open circuit voltage quantum dot sensitized solar cells manufactured with ZnO Nanowire arrays and Si/ZnO branched hierarchical structures. J Phys Chem Lett 2(16):1984–1990
71. Samadpour M, Gimenez S, Boix PP, Shen Q, Calvo ME, Taghavinia N, Zad AI, Toyoda T, Miguez H, Mora-Sero I (2012) Effect of nanostructured electrode architecture and semiconductor deposition strategy on the photovoltaic performance of quantum dot sensitized solar cells. Electrochim Acta 75:139–147
72. Niesen TP, De Guire MR (2002) Review: deposition of ceramic thin films at low temperatures from aqueous solutions. Solid State Ionics 151(1–4):61–68
73. Pathan HM, Lokhande CD (2004) Deposition of metal chalcogenide thin films by successive ionic layer adsorption and reaction (SILAR) method. Bull Mater Sci 27(2):85–111
74. Froment M, Cachet H, Essaaidi H, Maurin G, Cortes R (1997) Metal chalcogenide semiconductors growth from aqueous solutions. Pure Appl Chem 69(1):77–82
75. Pawar SM, Pawar BS, Kim JH, Joo OS, Lokhande CD (2011) Recent status of chemical bath deposited metal chalcogenide and metal oxide thin films. Curr Appl Phys 11(2):117–161
76. Lee H, Wang MK, Chen P, Gamelin DR, Zakeeruddin SM, Gratzel M, Nazeeruddin MK (2009) Efficient CdSe quantum dot-sensitized solar cells prepared by an improved successive ionic layer adsorption and reaction process. Nano Lett 9(12):4221–4227
77. Hossain MA, Jennings JR, Shen C, Pan JH, Koh ZY, Mathews N, Wang Q (2012) CdSe-sensitized mesoscopic TiO$_2$ solar cells exhibiting > 5 % efficiency: redundancy of CdS buffer layer. J Mater Chem 22(32):16235–16242
78. O'Mahony FTF, Lutz T, Guijarro N, Gomez R, Haque SA (2012) Electron and hole transfer at metal oxide/Sb2S3/spiro-OMeTAD heterojunctions. Energy Environ Sci 5(12):9760–9764

79. Sahu SN (1989) Aqueous electrodeposition of Inp semiconductor-Films. J Mater Sci Lett 8(5):533–534
80. Allongue P, Souteyrand E (1989) Semiconductor electrodes modified by electrodeposition of discontinuous metal-films. 1. Role of the film morphology. J Electroanal Chem 269(2):361–374
81. Switzer JA, Hung CJ, Bohannan EW, Shumsky MG, Golden TD, VanAken DC (1997) Electrodeposition of quantum-confined metal semiconductor nanocomposites. Adv Mater 9(4):334–000
82. Riveros G, Gomez H, Henriquez R, Schrebler R, Marotti RE, Dalchiele EA (2001) Electrodeposition and characterization of ZnSe semiconductor thin films. Sol Energy Mater Sol Cells 70(3):255–268
83. Rajeshwar, K.; deTacconi, N. R., Electrodeposition and characterization of nanocrystalline semiconductor films. Semiconductor Nanoclusters- Physical, Chemical, and Catalytic Aspects 1997, 103, 321-351
84. Lévy-Clément C, Katty A, Bastide S, Zenia F, Mora I, Munoz-Sanjose V (2002) A new CdTe/ZnO columnar composite film for Eta solar cells. Physica E 14:229–232
85. Lévy-Clément C, Tena-Zaera R, Ryan MA, Katty A, Hodes G (2005) CdSe sensitized p-CuSCN/Nanowire n-ZnO heterojunctions. Adv Mater 17:1512–1515
86. Lincot D (2005) Electrodeposition of semiconductors. Thin Solid Films 487(1–2):40–48
87. Savadogo O (1998) Chemically and electrochemically deposited thin films for solar energy materials. Sol Energy Mater Sol Cells 52(3–4):361–388
88. Behar D, Rubinstein I, Hodes G, Cohen S, Cohen H (1999) Electrodeposition of CdS quantum dots and their optoelectronic characterization by photoelectrochemical and scanning probe spectroscopies. Superlattices Microstruct 25(4):601–613
89. Han W, Cao LY, Huang JF, Wu JP (2009) Influence of pH value on PbS thin films prepared by electrodeposition. Mater Technol 24(4):217–220
90. Fernandes VC, Salvietti E, Loglio F, Lastraioli E, Innocenti M, Mascaro LH, Foresti ML (2009) Electrodeposition of PbS multilayers on Ag(111) by ECALE. J Appl Electrochem 39(11):2191–2197
91. Takahashi M, Ohshima Y, Nagata K, Furuta S (1993) Electrodeposition of Pbs films from acidic solution. J Electroanal Chem 359(1–2):281–286
92. Fantini MCA, Moro JR, Decker F (1988) Electrodeposition of Cdse films on Sno2-F coated glass. Solar Energy Mater 17(4):247–255
93. Kutzmutz S, Lang G, Heusler KE (2001) The electrodeposition of CdSe from alkaline electrolytes. Electrochim Acta 47(6):955–965
94. Anderson MA, Gorer S, Penner RM (1997) A hybrid electrochemical/chemical synthesis of supported, luminescent cadmium sulfide nanocrystals. J Phys Chem B 101(31):5895–5899
95. Song MY, Kim DK, Ihn KJ, Jo SM, Kim DY (2004) Electrospun TiO_2 electrodes for dye-sensitized solar cells. Nanotechnology 15(12):1861–1865
96. Fujihara K, Kumar A, Jose R, Ramakrishna S, Uchida S (2007) Spray deposition of electrospun TiO_2 nanorods for dye-sensitized solar cell. Nanotechnology 18(36)
97. Archana PS, Jose R, Vijila C, Ramakrishna S (2009) Improved electron diffusion coefficient in electrospun TiO_2 nanowires. J Phys Chem C 113(52):21538–21542
98. Ghadiri E, Taghavinia N, Zakeeruddin SM, Grätzel M, Moser JE (2010) Enhanced electron collection efficiency in dye-sensitized solar cells based on nanostructured TiO_2 hollow fibers. Nano Lett 10(5):1632–1638
99. Joshi P, Zhang LF, Davoux D, Zhu ZT, Galipeau D, Fong H, Qiao QQ (2010) Composite of TiO_2 nanofibers and nanoparticles for dye-sensitized solar cells with significantly improved efficiency. Energy Environ Sci 3(10):1507–1510
100. Martins A, Reis RL, Neves NM (2008) Electrospinning: processing technique for tissue engineering scaffolding. Int Mater Rev 53(5):257–274
101. Bhardwaj N, Kundu SC (2010) Electrospinning: a fascinating fiber fabrication technique. Biotechnol Adv 28(3):325–347

102. Ramaseshan R, Sundarrajan S, Jose R, Ramakrishna S (2007) Nanostructured ceramics by electrospinning. J Appl Phys 102(11)
103. Yang SY, Nair AS, Jose R, Ramakrishna S (2010) Electrospun TiO$_2$ nanorods assembly sensitized by CdS quantum dots: a low-cost photovoltaic material. Energy Environ Sci 3(12):2010–2014
104. Song MY, Kim DK, Ihn KJ, Jo SM, Kim DY (2005) New application of electrospun TiO$_2$ solid-state dye-sensitized solar electrode to cells. Synth Met 153(1–3):77–80
105. Onozuka K, Ding B, Tsuge Y, Naka T, Yamazaki M, Sugi S, Ohno S, Yoshikawa M, Shiratori S (2006) Electrospinning processed nanofibrous TiO$_2$ membranes for photovoltaic applications. Nanotechnology 17(4):1026–1031
106. Sudhagar P, Jung JH, Park S, Sathyamoorthy R, Ahn H, Kang YS (2009) Self-assembled CdS quantum dots-sensitized TiO$_2$ nanospheroidal solar cells: Structural and charge transport analysis. Electrochim Acta 55(1):113–117
107. Samadpour M, Gimenez S, Zad AI, Taghavinia N, Mora-Sero I (2012) Easily manufactured TiO$_2$ hollow fibers for quantum dot sensitized solar cells. Phys Chem Chem Phys 14(2):522–528
108. Han H, Sudhagar P, Song T, Jeon Y, Mora-Sero I, Fabregat-Santiago F, Bisquert J, Kang YS, Paik U (2013) Three dimensional-TiO$_2$ nanotube array photoanode architectures assembled on a thin hollow nanofibrous backbone and their performance in quantum dot-sensitized solar cells. Chem Commun 49(27):2810–2812
109. Li ZD, Zhou Y, Bao CX, Xue GG, Zhang JY, Liu JG, Yu T, Zou ZG (2012) Vertically building Zn2SnO4 nanowire arrays on stainless steel mesh toward fabrication of large-area, flexible dye-sensitized solar cells. Nanoscale 4(11):3490–3494
110. Tao RH, Wu JM, Xue HX, Song XM, Pan X, Fang XQ, Fang XD, Dai SY (2010) A novel approach to titania nanowire arrays as photoanodes of back-illuminated dye-sensitized solar cells. J Power Sources 195(9):2989–2995
111. Law M, Greene LE, Johnson JC, Saykally R, Yang PD (2005) Nanowire dye-sensitized solar cells. Nat Mater 4(6):455–459
112. Xu CK, Wu JM, Desai UV, Gao D (2011) Multilayer assembly of nanowires for dye-sensitized solar cells. J Am Chem Soc 133(21):8122–8125
113. Liu XL, Lin J, Chen XF (2013) Synthesis of long TiO$_2$ nanotube arrays with a small diameter for efficient dye-sensitized solar cells. Rsc Adv 3(15):4885–4889
114. Lin J, Liu XL, Guo M, Lu W, Zhang GG, Zhou LM, Chen XF, Huang HT (2012) A facile route to fabricate an anodic TiO$_2$ nanotube-nanoparticle hybrid structure for high efficiency dye-sensitized solar cells. Nanoscale 4(16):5148–5153
115. Kang TS, Smith AP, Taylor BE, Durstock MF (2009) Fabrication of highly-ordered TiO$_2$ nanotube arrays and their use in dye-sensitized solar cells. Nano Lett 9(2):601–606
116. Xie ZB, Adams S, Blackwood DJ, Wang J (2008) The effects of anodization parameters on titania nanotube arrays and dye sensitized solar cells. Nanotechnology 19(40)
117. Shankar K, Bandara J, Paulose M, Wietasch H, Varghese OK, Mor GK, LaTempa TJ, Thelakkat M, Grimes CA (2008) Highly efficient solar cells using TiO$_2$ nanotube arrays sensitized with a donor-antenna dye. Nano Lett 8(6):1654–1659
118. Zhang QF, Cao GZ (2011) Nanostructured photoelectrodes for dye-sensitized solar cells. Nano Today 6(1):91–109
119. Xu F, Dai M, Lu YN, Sun LT (2010) Hierarchical ZnO nanowire-nanosheet architectures for high power conversion efficiency in dye-sensitized solar cells. J Phys Chem C 114(6):2776–2782
120. Ko SH, Lee D, Kang HW, Nam KH, Yeo JY, Hong SJ, Grigoropoulos CP, Sung HJ (2011) Nanoforest of hydrothermally grown hierarchical ZnO nanowires for a high efficiency dye-sensitized solar cell. Nano Lett 11(2):666–671
121. Cheng CW, Fan HJ (2012) Branched nanowires: synthesis and energy applications. Nano Today 7(4):327–343

122. Qiu JJ, Li XM, Gao XD, Gan XY, Weng BB, Li L, Yuan ZJ, Shi ZS, Hwang YH (2012) Branched double-shelled TiO$_2$ nanotube networks on transparent conducting oxide substrates for dye sensitized solar cells. J Mater Chem 22(44):23411–23417

123. Chen HY, Kuang DB, Su CY (2012) Hierarchically micro/nanostructured photoanode materials for dye-sensitized solar cells. J Mater Chem 22(31):15475–15489

124. Dai H, Zhou Y, Liu Q, Li ZD, Bao CX, Yu T, Zhou ZG (2012) Controllable growth of dendritic ZnO nanowire arrays on a stainless steel mesh towards the fabrication of large area, flexible dye-sensitized solar cells. Nanoscale 4(17):5454–5460

125. Bierman MJ, Jin S (2009) Potential applications of hierarchical branching nanowires in solar energy conversion. Energy Environ Sci 2(10):1050–1059

126. McCune M, Zhang W, Deng YL (2012) High efficiency dye-sensitized solar cells based on three-Dimensional multilayered ZnO nanowire arrays with "Caterpillarlike" structure. Nano Lett 12(7):3656–3662

127. Wu WQ, Lei BX, Rao HS, Xu YF, Wang YF, Su CY, Kuang DB (2013) Hydrothermal fabrication of hierarchically anatase TiO$_2$ nanowire arrays on FTO glass for dye-sensitized solar cells. Sci Rep 3

128. Sauvage F, Di Fonzo F, Bassi AL, Casari CS, Russo V, Divitini G, Ducati C, Bottani CE, Comte P, Graetzel M (2010) Hierarchical TiO$_2$ photoanode for dye-sensitized solar cells. Nano Lett 10(7):2562–2567

129. Tétreault N, Heiniger L-P, Stefik M, Labouchère PL, Arsenault É, Nazeeruddin NK, Ozin GA, Grätzel M (2011) (Invited) Atomic layer deposition for novel dye-sensitized solar cells. ECS Trans 41(2):303–314

130. Xu CK, Wu JM, Desai UV, Gao D (2012) High-efficiency solid-state dye-sensitized solar cells based on TiO$_2$-coated ZnO nanowire arrays. Nano Lett 12(5):2420–2424

131. Halaoui LI, Abrams NM, Mallouk TE (2005) Increasing the conversion efficiency of dye-sensitized TiO$_2$ photoelectrochemical cells by coupling to photonic crystals. J Phys Chem B 109(13):6334–6342

132. Guldin S, Hüttner S, Kolle M, Welland ME, Müller-Buschbaum P, Friend RH, Steiner U, Tétreault N (2010) Dye-sensitized solar cell based on a three-dimensional photonic crystal. Nano Lett 10(7):2303–2309

133. Liu LJ, Karuturi SK, Su LT, Tok AIY (2011) TiO$_2$ inverse-opal electrode fabricated by atomic layer deposition for dye-sensitized solar cell applications. Energy Environ Sci 4(1):209–215

134. Choi JH, Kwon SH, Jeong YK, Kim I, Kim KH (2011) Atomic layer deposition of Ta-doped TiO$_2$ electrodes for dye-sensitized solar cells. J Electrochem Soc 158(6):B749–B753

135. King JS, Graugnard E, Summers CJ (2005) TiO$_2$ inverse opals fabricated using low-temperature atomic layer deposition. Adv Mater 17(8):1010-+

136. Chen JIL, von Freymann G, Choi SY, Kitaev V, Ozin GA (2008) Slow photons in the fast lane in chemistry. J Mater Chem 18(4):369–373

137. Yip C-H, Chiang Y-M, Wong C-C (2008) Dielectric band edge enhancement of energy conversion efficiency in photonic crystal dye-sensitized solar cell. J Phys Chem C 112(23):8735–8740

138. Shin JH, Kang JH, Jin WM, Park JH, Cho YS, Moon JH (2011) Facile synthesis of TiO$_2$ inverse opal electrodes for dye-sensitized solar cells. Langmuir 27(2):856–860

139. Han SH, Lee S, Shin H, Jung HS (2011) A quasi-inverse opal layer based on highly crystalline TiO$_2$ nanoparticles: a new light-scattering layer in dye-sensitized solar cells. Adv Energy Mater 1(4):546–550

140. Mihi A, Calvo ME, Anta JA, Miguez H (2008) Spectral response of opal-based dye-sensitized solar cells. J Phys Chem C 112(1):13–17

141. Nishimura S, Abrams N, Lewis BA, Halaoui LI, Mallouk TE, Benkstein KD, van de Lagemaat J, Frank AJ (2003) Standing wave enhancement of red absorbance and photocurrent in dye-sensitized titanium dioxide photoelectrodes coupled to photonic crystals. J Am Chem Soc 125(20):6306–6310

142. Yuan S, Huang H, Wang Z, Zhao Y, Shi L, Cai C, Li D (2013) Improved electron-collection performance of dye sensitized solar cell based on three-dimensional conductive grid. J Photochem Photobiol A 259:10–16

143. Cho CY, Moon JH (2012) Hierarchical twin-scale inverse opal TiO_2 electrodes for dye-sensitized solar cells. Langmuir 28(25):9372–9377

144. Kwak ES, Lee W, Park NG, Kim J, Lee H (2009) Compact inverse-opal electrode using non-aggregated TiO_2 nanoparticles for dye-sensitized solar cells. Adv Funct Mater 19(7):1093–1099

145. Toyoda T, Shen Q (2012) Quantum-dot-sensitized solar cells: effect of nanostructured TiO_2 morphologies on photovoltaic properties. J Phys Chem Lett 3(14):1885–1893

146. Diguna LJ, Shen Q, Kobayashi J, Toyoda T (2007) High efficiency of CdSe quantum-dot-sensitized TiO_2 inverse opal solar cells. Appl Phys Lett 91(2)

147. Shalom M, Dor S, Rühle S, Grinis L, Zaban A (2009) Core/CdS quantum dot/shell mesoporous solar cells with improved stability and efficiency using an amorphous TiO_2 coating. J Phys Chem C 113:3895–3898

148. Chakrapani V, Baker D, Kamat PV (2011) Understanding the role of the sulfide redox couple (S^{2-} /S_n^{2-}) in quantum dot-sensitized solar cells. J Am Chem Soc 133:9607–9615

149. Gonzalez-Pedro V, Sima C, Marzari G, Boix PP, Gimenez S, Shen Q, Dittrich T, Mora-Sero I (2013) High performance PbS quantum dot sensitized solar cells exceeding 4 % efficiency: the role of metal precursors in the electron injection and charge separation. Phys Chem Chem Phys

150. Jovanovski V, González-Pedro V, Giménez S, Azaceta E, Cabañero G, Grande H, Tena-Zaera R, Mora-Seró I, Bisquert J (2011) A sulfide/polysulfide-based ionic liquid electrolyte for quantum dot-sensitized solar cells. J Am Chem Soc 133:20156–20159

151. Choné C, Larramona G (2007) French Patent 2899385, 05

152. Messina S, Nair MTS, Nair PK (2007) Antimony sulfide thin films in chemically deposited thin film photovoltaic cells. Thin Solid Films 515:5777–5782

153. Moon S-J, Itzhaik Y, Yum J-H, Zakeeruddin SM, Hodes G, Grätzel M (2010) Sb_2s3_based mesoscopic solar cell using an organic hole conductor. J Phys Chem Lett 1:1524–1527

154. Im SH, Lim C-S, Chang JA, Lee YH, Maiti N, Kim H-J, Nazeeruddin MK, Grätzel M, Seok SI (2011) Toward interaction of sensitizer and functional moieties in hole-transporting materials for efficient semiconductor-sensitized solar cells. Nano Lett 11:4789–4793

155. Hodes G, Manassen J, Cahen D (1980) Electrocatalytic electrodes for the polysulfide redox system. J Electrochem Soc 127:544–549

156. Yang YY, Zhu LF, Sun HC, Huang XM, Luo YH, Li DM, Meng QB (2012) Composite counter electrode based on nanoparticulate PbS and carbon black: towards quantum dot-sensitized solar cells with both high efficiency and stability. Acs Appl Mater Interfaces 4(11):6162–6168

157. Sudhagar P, Ramasamy E, Cho W-H, Lee J, Kang YS (2011) Robust mesocellular carbon foam counter electrode for quantum-dot sensitized solar cells. Electrochem Commun 13(1):34–37

158. Yang, Z.; Chen, C.-Y.; Liu, C.-W.; Li, C.-L.; Chang, H.-T., Quantum Dot–Sensitized Solar Cells Featuring CuS/CoS Electrodes Provide 4.1 % Efficiency. Advanced Energy Materials 2011, 1, 259-264

159. Deng M, Huang S, Zhang Q, Li D, Luo Y, Shen Q, Toyoda T, Meng Q (2010) Screen-printed $Cu2_s$-based counter electrode for quantum-dot-sensitized solar cell. Chem Lett 39:1168–1170

160. Radich JG, Dwyer R, Kamat PV (2011) $Cu2_s$ reduced graphene oxide composite for high-efficiency quantum dot solar cells. overcoming the redox limitations of $S2'_sSn_2^-$ at the counter electrode. J Phys Chem Lett 2:2453–2460

161. Santra PK, Kamat PV (2012) Mn-doped quantum dot sensitized solar cells: a strategy to boost efficiency over 5 %. J Am Chem Soc 134(5):2508–2511

162. Santra PK, Kamat PV (2012) Mn-doped quantum dot sensitized solar cells: a strategy to boost efficiency over 5 %. J Am Chem Soc 134(5):2508–2511

163. Samadpour M, Boix PP, Giménez S, Iraji Zad A, Taghavinia N, Mora-Seró I, Bisquert J (2011) Fluorine treatment of TiO$_2$ for enhancing quantum dot sensitized solar cell performance. J Phys Chem C 115:14400–14407
164. Barea EM, Shalom M, Giménez S, Hod I, Mora-Seró I, Zaban A, Bisquert J (2010) Design of injection and recombination in quantum dot sensitized solar cells. J Am Chem Soc 132:6834–6839
165. de la Fuente MS, Sánchez RS, González-Pedro V, Boix PP, Mhaisalkar SG, Rincón ME, Bisquert J, Mora-Seró I (2013) Effect of organic and inorganic passivation in quantum-dot-sensitized solar cells. J Phys Chem Lett 4(9):1519–1525
166. Mora-Sero I, Bisquert J (2010) Breakthroughs in the development of semiconductor-sensitized solar cells. J Phys Chem Lett 1(20):3046–3052
167. Kamat PV, Tvrdy K, Baker DR, Radich JG (2010) Beyond photovoltaics: semiconductor nanoarchitectures for liquid-junction solar cells. Chem Rev 110(11):6664–6688
168. Burschka J, Pellet N, Moon S-J, Humphry-Baker R, Gao P, Nazeeruddin MK, Gratzel M (2013) Sequential deposition as a route to high-performance perovskite-sensitized solar cells. Nature (advance online publication)
169. Kagan CR, Mitzi DB, Dimitrakopoulos CD (1999) Organic-inorganic hybrid materials as semiconducting channels in thin-film field-effect transistors. Science 286:945–947
170. Kim H-S, Lee C-R, Im J-H, Lee K-B, Moehl T, Marchioro A, Moon S-J, Humphry-Baker R, Yum J-H, Moser JE, Gratzel M, Park N-G (2012) Lead iodide perovskite sensitized all-solid-state submicron thin film mesoscopic solar cell with efficiency exceeding 9 %. Natl Sci Rep 2(591):591
171. Baikie T, Fang Y, Kadro JM, Schreyer M, Wei F, Mhaisalkar SG, Graetzeld M, Whitec TJ (2013) Synthesis and crystal chemistry of the hybrid perovskite (CH3NH3)PbI3 for solid-state sensitised solar cell applications. J Mater Chem A 1:5628–5641
172. Im J-H, Lee C-R, Lee J-W, Park S-W, Park N-G (2011) 6.5 % efficient perovskite quantum-dot-sensitized solar cell†. Nanoscale 3:4088–4093
173. Kojima A, Teshima K, Shirai Y, Miyasaka T (2009) Organometal halide perovskites as visible-light sensitizers for photovoltaic cells. J Am Chem Soc 131(2):6050–6051
174. Lee MM, Teuscher J, Miyasaka T, Murakami TN, Snaith HJ (2012) Efficient hybrid solar cells based on meso-superstructured organometal halide perovskites. Science 338:643–647
175. Etgar L, Gao P, Xue Z, Peng Q, Chandiran AK, Liu B, Nazeeruddin MK, Gratzel M (2012) Mesoscopic CH3NH3PbI3/TiO$_2$ heterojunction solar cells. J Am Chem Soc 134:17396–17399
176. Heo JH, Im SH, Noh JH, Mandal TN, Lim C-S, Chang JA, Lee YH, Kim H-J, Sarkar A, Nazeeruddin MK, Gratzel M, Seok SI (2013) Efficient inorganic-organic hybrid heterojunction solar cells containing perovskite compound and polymeric hole conductors. Nat Photonics 7:486–491
177. Kim H-S, Lee J-W, Yantara N, Boix PP, Kulkarni SA, Mhaisalkar S, Gratzel M, Park N-G (2013) High efficiency solid-state sensitized solar cell-based on submicrometer rutile TiO$_2$ nanorod and CH3NH3PbI3 perovskite sensitizer. Nano Lett 16:2412–2417
178. Noh JH, Im SH, Heo JH, Mandal TN, Seok SI (2013) Chemical management for colorful, efficient, and stable inorganic- organic hybrid nanostructured solar cells. Nano Lett 7:1764–1769
179. Qiu J, Qiu Y, Yan K, Zhong M, Mu C, Yan H, Yang S (2013) All-solid-state hybrid solar cells based on a new organometal halide perovskite sensitizer and one-dimensional TiO$_2$ nanowire arrays. Nanoscale 5:3245–3248
180. Ball JM, Lee MM, Hey A, Snaith H (2013) Low-temperature processed mesosuperstructured to thin-film perovskite solar cells. Energy Environ Sci 6:1739–1743
181. Edri E, Kirmayer S, Cahen D, Hodes G (2013) High open-circuit voltage solar cells based on organic-inorganic lead bromide perovskit. J Phys Chem 4:897–902
182. Kim H-S, Mora-Sero I, Gonzalez-Pedro V, Fabregat-Santiago F, Juarez-Perez EJ, Park N-G, Bisquert J (2013) Mechanism of carrier accumulation in perovskite thin absorber solar cells. Nat Commun 4:2242

183. Sauvage F, Davoisne C, Philippe L, Elias J (2012) Structural and optical characterization of electrodeposited CdSe in mesoporous anatase TiO_2 for regenerative quantum-dot-sensitized solar cells. Nanotechnology 23(39)

184. Bang JH, Kamat PV (2010) Solar cells by design: photoelectrochemistry of TiO_2 nanorod arrays decorated with CdSe. Adv Funct Mater 20(12):1970–1976

185. Boix PP, Lee YH, Fabregat-Santiago F, Im SH, Mora-Seró I, Bisquert J, Seok SI (2012) From flat to nanostructured photovoltaics: balance between thickness of the absorber and charge screening in sensitized solar cells. ACS Nano 6(1):873–880

186. An H, Ahn HJ (2013) Fabrication of wrinkled Nb-doped TiO_2 nanofibres via electrospinning. Mater Lett 93:88–91

187. Herman I, Yeo J, Hong S, Lee D, Nam KH, Choi JH, Hong WH, Lee D, Grigoropoulos CP, Ko SH (2012) Hierarchical weeping willow nano-tree growth and effect of branching on dye-sensitized solar cell efficiency. Nanotechnology 23(19)

188. Fabregat-Santiago F, Garcia-Belmonte G, Mora-Seró I, Bisquert J (2011) Characterization of nanostructured hybrid and organic solar cells by impedance spectroscopy. Phys Chem Chem Phys 13:9083–9118

189. Boix PP, Larramona G, Jacob A, Delatouche B, Mora-Seró I, Bisquert J (2012) Hole transport and recombination in all-solid Sb_2s_3 sensitized TiO_2 solar cells using CuSCN as hole transporter. J Phys Chem C 116(1):1579–1587

The Renaissance of Iron Pyrite Photovoltaics: Progress, Challenges, and Perspectives

Alec Kirkeminde, Maogang Gong and Shenqiang Ren

Abstract Pyrite has long been proposed as a green solar cell material. Even with its promising properties, studies on pyrite have lagged behind many other semiconducting materials. Unanswered questions about the affects of defects and how to grow pure crystalline material still exist. With the rise of nanochemistry and more powerful computational methods, pyrite is seeing an explosion of new studies. This chapter first presents pyrite and its green promise as a material, followed by the materials characteristics. It then moves into synthesis of pyrite, starting with old methods and then transitioning into different methods of nanocrystal creation. Finally, photo-devices created out of pyrite materials are discussed. The chapter then wraps up with a summary and what still needs to be done for pyrite to achieve its golden status.

1 Introduction

1.1 Pyrite's Green Energy Aspects

The need for new renewable resources is growing every year. Solar cells have been proposed as an alternative means to help generate energy for the world's continuously growing needs [1, 2]. Many different semiconductor systems have been studied for solar cell application (CdS, CdTe, CuS_2, $CuInSe_2$), but the most well-developed material is silicon. Crystalline silicon (C-Si) devices have been extensively optimized, but still suffer from the high costs of creating the crystallized material. C-Si also requires high material consumption due to its poor

A. Kirkeminde · M. Gong · S. Ren (✉)
Department of Chemistry, University of Kansas, 1251 Wescoe Hall Drive, Lawrence, KS 66045-7572, USA
e-mail: shenqiang@ku.edu

Z. Lin and J. Wang (eds.), *Low-cost Nanomaterials*, Green Energy and Technology, DOI: 10.1007/978-1-4471-6473-9_6, © Springer-Verlag London 2014

absorption, requiring thicker layers of photoactive material. These drawbacks have kept C-Si solar cell energy more expensive than other methods of conventional energy generation. Other semiconducting materials have been proposed to avoid the high material consumption such as CdS, CdSe CdTe, $CuInSe_2$ due to their better absorbance. While these materials exhibit better absorbing properties, they contain toxic elements that propose environmental problems both in mass production and in widespread use. A material that exhibits high absorption while retaining low material cost and toxicity will be necessary to drive energy costs down.

Iron Pyrite (FeS_2, Fool's Gold, Iron Disulfide) exhibits promising properties for use in solar cell devices. It boasts an indirect band gap of 0.95 eV and an absorption coefficient of greater than 10^5 cm^{-1}, which is unusual for an indirect band gap [3, 4]. In comparison to silicon, another indirect band gap photovoltaic (PV) material, FeS_2 shows two magnitudes greater absorption coefficient [5]. This higher absorption coefficient means that thinner films, and thus less material, can be used while creating devices.

Pyrite is a cheaper material than many other inorganic solar materials. Since silicon is the dominant material for commercial PVs, it is logical to use for comparison. Silicon is the second most abundant material in the earth's crust, while iron trails in fourth [6]. Even so, silicon production still trails in extraction costs which is $\sim$\$1.70/kg [7]. This is 57 times higher extraction costs to that of iron (\$0.03/kg). The huge difference between the price stems from thermodynamics of converting raw material into final elemental forms. It requires 24 kWh/kg to purify silicon from its feedstock of silica (SiO_2) while it only takes 2 kWh/kg to achieve iron from hematite (Fe_2O_3) [8]. This natural barrier will always exist for silicon solar cells and will limit its use in a future where PVs need to be cost-effective.

Much research has been conducted to increase the efficiency of standard silicon solar cells to combat this natural cost of production. Yet, studies show that a crystalline silicon cell with efficiency of 19 % will still produce 10–100 times less energy than the annual global consumption [7]. When compared to FeS_2, creating a solar cell with 4 % efficiency and three times less material consumption could produce the same amount of energy. This example shows that it may be best to put aside the mentality of trying to achieve the best efficiency of silicon cells and look into mass production of cheap, less efficient cells.

It is obvious from the above that being able to use less material is conducive to keeping solar cell costs down. Thin film solar cells have been extensively researched in the past and show great promise to help improve, and have put to use both organic and inorganic active layers [9–14]. Recently, with the explosion of nanotechnology research, nanocrystals have been utilized in creating films of material that are less than 300 nm that show greater than 3 % efficiencies [15–18]. Combining the promising aspects of nanomaterial and iron pyrite's notable properties could unlock the door to creating low-cost solar cells with minimal material usage, which could drive down the cost of energy and decrease our dependence on other, less green, sources of energy.

1.2 Pyrite Crystal Structure and Properties

Pyrite (FeS_2, Fool's Gold) crystal structure is one of the best examples of cubic AB_2 structures. This structure is mainly characteristic of AB_2 compounds of pnictides, such as P, As, Sb, and chalcogenides, such as S, Se, Te. Since FeS_2 is the most prominent example of this structure, pyrite has been used to name this family. In this section, we focus mainly on FeS_2, but this structure can be used for the other members of this family. The pyrite structure can be represented by imagining the NaCl cubic crystal structure and replacing the sodium with the iron atom and the chlorine atom with dumbbells of the S_2 dimer compound. The iron atom is in a distorted octahedral coordination site surrounded by six sulfur atoms, while the sulfur atoms sit in a distorted tetrahedral coordination surrounded by three other sulfur atoms and one iron atom. The distortion of the pure NaCl structure reduces the symmetry of the pyrite structure, putting in the Pa3 point group. Lattice constants of pyrite is found to be 5.418 A with Fe-S distances being 2.26 A and S-S distances being 2.14 Å [4]. It is important to note that FeS_2 can also crystallize into an orthorhombic structure named marcasite. While the pyrite structure has corner linked coordination octahedral, marcasite exhibits edge linked octahedral. Marcasite offers different properties, which are usually detrimental to pyrite's promising characteristics for solar application.

From crystal field theory, it is known that transition metal's d orbitals are nondegenerate in an octahedral environment. In pyrite, the t_{2g} set made of the d_{xy}, d_{yz}, d_{xz} orbitals controls the valence band and the e_g set made with the remaining d_z^2 and the $d_{s^2-y^2}$ orbitals control the conduction band. Since iron is in an oxidation state of 2+ in pyrite, this leaves six electrons remaining to fill up the three t_{2g} orbitals, making it a diamagnetic low-spin semiconductor. The splitting between these two orbital sets controls the band gap of the semiconductor and for pyrite it has been found that this gap is 0.95 eV. While this band gap has mostly been accepted, solar devices made with this material exhibit low open circuit voltage, which is mostly attributed to sulfur vacancies in the crystal that pin the Fermi levels. Recently, there have been computational studies that have started debates in the literature on whether pyrite is a purely stoichiometric compound [19–22]. While this is beyond the scope of this chapter, it is important to know that this is due to its affect on creating working devices.

It has been shown both in computational studies and shape control studies of nanocrystal pyrite that the equilibrium faces that can be obtained are {100} and {111} [21, 23–25]. Other facets, such as {110} and {210}, can be obtained in macroscopic crystals, but focus is given to the nanocrystal formations in this work [21]. It is necessary to first note that pyrite crystal surface energies are an enigma compared to normal crystal theory. It has been shown that the {100} face is lower in surface energy than the {111} energy, which is the opposite of normal theory [26]. This leads to many different observables in synthesis, which will be discussed later. Here, focus is given on the difference between the two facets. Figure 1 shows models of both (100) and (111) surfaces. In all cases the {111} surface is sulfur

Fig. 1 **a** Pyrite (100) surface
showing non-polar
arrangement **b** Bulk (100)
pyrite surface **c** Bulk pyrite
(111) surface. Reprinted with
permission from [21]

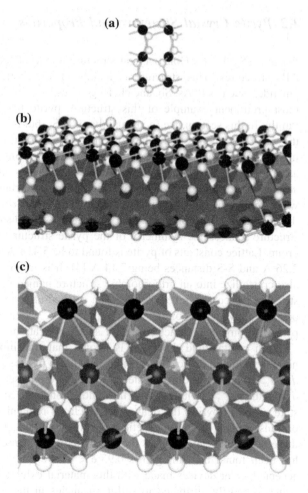

terminated and nonpolar. In the {100} facet there are three different formations that
can occur at the surface. The three different terminations are [S-Fe-S], [Fe-S-S], and
[S-S-Fe]. Both [Fe-S-S] and [S-S-Fe] are polar, and show that there is a possibility
for iron-terminated facets. [S-Fe-S] is nonpolar and iron is in a reduced coordination number of five, whose closeup view is presented in Fig. 1a. This is of interest
due to iron being in trigonal bipyramidal state, which changes its splitting in ligand
field model. At first it was believed that this could introduce gap states causing the
seen problems of open circuit voltage. With the rise of stronger computational
systems, this has been dismissed as a problem with studies showing that these
differences will not put this surface state in the gap, and is not the cause of the
V_{oc}, although it still deserves more attention [21]. It has also been shown
computationally that changing sulfur concentration during growth can control the
terminating element of the crystal facet [27]. With these two facets being quite

unique to each other, they have shown differences in activity in many different studies and are important to consider when choosing pyrite material for applications [23, 24, 28, 29].

2 Iron Pyrite Nanomaterial Synthesis

2.1 Introduction

The growth of pyrite has been an interesting topic for past geologists due to its impact in mining, as it is the main component of creating acid mine run-off [30]. It is from these papers that information can be first learned about the growth of pyrite. There are three different mechanisms that occur to produce pyrite. One involves iron (II) and polysulfide, the other uses FeS and S(-II), and the last is growth of pyrite crystal growth on pyrite seeds [31−35]. It is important to note that the first two of these reactions involve an [FeS] intermediate step. The nucleation step of this intermediate has been proposed as the limiting reaction in the rate of pyrite crystal formation. Figure 2 shows the proposed mechanisms of pyrite formation using the two different sulfur sources. The first mechanism utilizes polysulfides, while the second uses hydrogen sulfide (H_2S). For a beautifully detailed review of pyrite formation in nature, we direct the reader to the publication of Rickard and Luther [36]. While this section is about nano-synthesis of pyrite, these background studies are necessary to achieve better understanding of the system.

Tributsch and colleagues did a major portion of the initial studies of making use of pyrite for solar energy conversion that deserves attention. Studies were focused on creating thin films by first chemical vapor transportation (CVT) and then follow-up studies were done using metal organic chemical vapor deposition (MOCVD) [37−41]. It was found that they could produce pure pyrite films by adjusting pressure, temperature, and molar concentrations of the reactants. They could also control the preferred crystal growth and grain size by changing the substrate on which the pyrite was grown. A great summary on their work can be found in the review article by Ennaoui et al. [4].

More recently, the Wolden group has shown that pure phase pyrite thin films can be achieved from hematite (Fe_2O_3) using an H_2S plasma [42]. First, Fe_2O_3 nanorods are synthesized using a chemical bath deposition (CBD) method. Then the high sulfur activity created in plasma can be used to achieve sulfurization of the Fe_2O_3. By monitoring the optical band gap and Fe:S ratio's using energy-dispersive X-ray spectroscopy (EDAX) the transition from hematite to pyrite can be observed. This study has shown a new technique utilizing sulfur plasma has breathed new life into creating grown thin films.

(a) proposed polysulfide mechanism

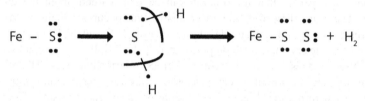

(b) proposed H₂S mechanism

Fig. 2 a Proposed mechanisms of pyrite formation through a FeS intermediate. Reprinted with permission from [36]

2.2 Hydrothermal Synthesis of Nano Pyrite

Hydrothermal methods have been used extensively in nanocrystal synthesis in the past [3, 25, 43]. This method makes use of a stainless steel digestion bomb that is typically Teflon lined. The buildup of pressure in the container allows for lower temperature to be used instead of high temperature ambient pressure synthesis; care must be used during use due to this pressure buildup, which is why these systems have earned the nickname "bomb." The use of a single precursor is usually used, and through the decomposition of the molecule produces the reactive species. Greater control of the system is achieved due to lack of variables such as injection rate and injection temperature. Having a single precursor though limits you to only a few precursors and are expensive. Another drawback of these systems is the time required to carry out the reaction ($\sim$24 h) and the lack of the ability to take timed aliquots of the sample during synthesis to study reaction progression. Even with these drawbacks, it is still used for its simplicity and control.

One of the first reports of utilizing hydrothermal synthesis methods come from Chen et al. [43]. In this chapter, an iron Diethyldithiocarbamate (Fe(S$_2$CNEt$_2$)$_3$) complex was utilized for the single precursor, and upon completing the reaction cubic FeS$_2$ crystallites with $\sim$500 nm edge lengths were obtained. It was shown that a minimum temperature of 180 °C was necessary to achieve pure phase pyrite, and that lower temperatures produced marcasite impurities. These crystallites showed an absorbance peak in the infrared (1420 nm), which is a redshift of the bulk material.

In a following study, Wadia et al. showed that the size of these pyrite cubes could be reduced in size [3]. By making use of a different precursor, iron (III) diethyl dithiophosphate ($Fe[(C_2H_5O)_2P(S)S]_3$), hexadecyltrimethylammonium bromide (CTAB), and a higher temperature nanocube with side lengths of ~ 100 nm were achieved. Purity of phase was proven by XRD and also confirmed the 0.95 eV indirect band gap with X-ray absorption and resonant X-ray emission spectroscopy. They note that in this synthesis, slightly acidic conditions and the CTAB were critical for the creation of pure phase pyrite.

The most recent study utilizing hydrothermal synthesis for pyrite nanocrystals is from Wang et al. [25]. In this report it is shown that it was possible to avoid single precursors and also that it was possible to create shapes other than cubes by the addition of polyvinylpyrrolidone (PVP) and polyvinyl alcohol (PVA) and sodium hydroxide (NaOH). For precursors, ferrous chloride tetrahydrate ($FeCl_2 \bullet 4H_2O$) and pure sulfur powder were used along with the PVA and PVP. By varying the NaOH concentration and keeping the PVA and PVP polymer concentration constant, it was shown that beautiful cubes, octahedral, and sphere-like crystallites are formed with sizes of ~ 140, ~ 220, and ~ 400 nm, respectively. Figure 3 shows scanning electron microscopy (SEM) and transmission electron microscopoy (TEM) pictures of the final particles. The reaction itself was dependent on the presence of NaOH, as when a control reaction with out any NaOH was conducted only black grease was obtained. A mechanism for formation of the FeS_2 was proposed as follows:

$$Fe^{2+} + 2OH^- \leftrightarrow Fe(OH)_2$$
$$4S + 3H_2O \rightarrow 2H_2S + H_2S_2O_3$$
$$H_2S + 2OH^- \rightarrow S^{2-} + 2H_2O$$
$$3S + 6OH^- \rightarrow 2S^{2-}SO_3^{3-} + 3H_2O$$
$$Fe^{2+} + S^{2-} \rightarrow FeS$$
$$Fe^{2+} + H_2S \rightarrow FeS + 2H^+$$
$$FeS + S \rightarrow FeS_2$$
$$FeS + H_2S \rightarrow FeS_2 + H_2$$

NaOH plays a vital role in the final formation of these particles and explains the lack of pyrite formation without it. It is also worthwhile to note that this mechanism goes through an FeS intermediate state. The reaction rate is dependent on the inclusion of NaOH, and therefore changing the concentration should affect the amount of FeS_2 seeds generated. With low concentrations of NaOH, less seeds are formed and a large amount of the polymer adheres to the seeds that result in a system where there is no preferred attachment to the different crystal facets by the polymer. In this case, the final particles then form large sphere-like composites of smaller particles most likely via a self-organization followed by intergrowing

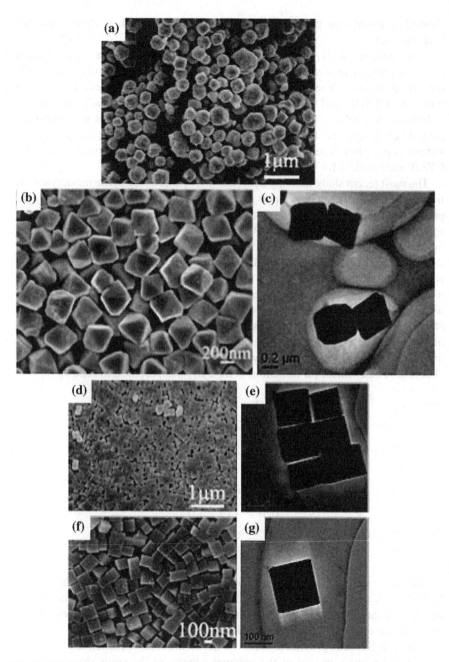

Fig. 3 SEM and TEM images of pyrite **a** spheres **b, c** octahedrals and **d–g** cubes all formed by hydrothermal methods. Reprinted with permission from [25]

between crystallites to minimize surface energy. When the NaOH concentration is increased, more FeS_2 seeds will be generated. While there are still more seeds present, the polymer will adhere to all the surfaces, though the {111} face still has a higher surface energy than the {100} even with adsorbed polymer. This facilitates faster growth along the [111] direction than the [100] direction, resulting in final formation of cubic structure. On raising the NaOH concentration, even more FeS_2 seeds are produced and the polymer to seed ratio drops. This causes the polymer starts selectively adhering to the {111} that causes [100] direction to grow faster at the expense of the [111] direction, creating octahedral structures. Dosages of PVP and PVA were also proven to be important. Holding NaOH concentration constant (corresponding to creation of octahedral structure) and changing the ratios, it was seen that the octahedral structure is still obtained, but the size distribution was much wider. This report is an excellent example of shape control utilizing ligand affinity to different surface facets and provides insight to synthetic control of pyrite nanomaterial.

2.3 Solvothermal Synthesis

Synthesis of Pyrite nanocrystals has more recently shifted from hydrothermal methods to solvothermal synthesis. This stems from the problems of hydrothermal mentioned above, such as mostly using an expensive single precursor and long reaction times. Solvothermal synthesis is much like hydrothermal, but does not utilize water as its solvent and can be underdone in both autoclaves and in standard glass reaction vessels. The solvent is usually a high boiling point organic along with ligands to help stabilize the growth of the nanoparticle. Some of the more common ligands utilized in pyrite synthesis are oleic acid and oleylamine. In this section, two different solvothermal methods are examined. The first is precipitation method (one pot method) where all the precursors are loaded into a flask and then heated up and the particles are then precipitated out after reaction. The second is the hot injection method (two pot) where the two precursors are put into separate flasks and allowed to decompose separately. One of the precursors is then quickly injected into the other, usually via syringe, to start the reaction. A key parameter is having the precursor solutions at different temperatures upon injection, as this causes a drop in temperature upon injection that causes a supersaturation of the compound. Supersaturation is relieved by precipitating out small seeds of the material, followed by growth.

2.3.1 Precipitation Method

With the goal of eliminating the use of a single precursor and the desire to create final particles quickly, researchers have started using a one-pot precipitation method to create pyrite nanocrystals. Yuan et al. showed by utilizing Hematite

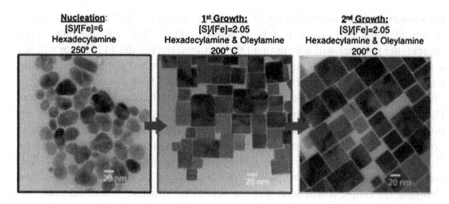

Fig. 4 TEM pictures of nucleation step to create pyrite particles and the following results of growth steps. Reprinted with permission from [23]

(Fe$_2$O$_3$) and elemental sulfur as the chalcogenide along with oleic acid (OA), oleylamine (OLA) and CTAB, cubic structures could be obtained [44]. All components are loaded into one flask and heated up to 290 °C and kept there for 1 h with stirring. Such a high temperature was necessary to completely convert the hematite into pyrite, and if the temperature was raised even higher greigite (Fe$_3$S$_4$) impurities were observed. Without the presence of the ligands, it was found that that both starting material and marcasite phases were found in the particles. Particle size length was ~100 nm, though the size distribution was quite large (±25 nm, size distribution was not reported). This report earns honorable mention due to its ability to convert iron raw material feedstock (hematite) directly to pyrite, instead of utilizing different iron precursor where the iron likely originated from hematite.

More recently, the Soldt group reported a unique synthetic method utilizing a modification of the standard one pot synthesis. In their article, it was shown that by utilizing a nucleation step FeS$_2$ that creates irregular nanocrystals followed by a growth step by adding more precursors, beautiful cubic/rectangular shapes could be obtained [23]. Figure 4 shows TEM images of the particles throughout the synthesis. From the images it can be seen that these are the cleanest edged cubic structures created to date. It is also the first report of cubic structures with side lengths {50 nm, which would allow for thinner films when creating devices.

The synthetic method starts with loading FeCl$_2$ and sulfur powder with a ratio of [S]/[Fe] equaling 6 along with hexadecylamine into a three-necked flask and then purged under argon. The material is then heated up to 250°C for 3 h. The end result of this first step was irregular-shaped pure phase nanocrystals with sizes around ~30–40 nm. They deemed this step the nucleation step, which should not be confused with the initial nucleation step that the seeds are created from which these particles grow. These irregular particles show absorbance spectra much like other spherical particles that have been made with a shoulder around ~550 nm with steady increase of absorbance to the blue. Nucleation step was followed by a growth step, where they cooled the flask to room temperature and hexadecylamine solidified. They load in on top of the solid reaction mixture oleylamine, which will act as a weak coordinating ligand and more iron and sulfur precursor with the ratio of $[S]/[Fe] = 2.05$. The flask is then repurged with argon and then heated to 200 °C for 9 h. This growth was repeated once more without the oleylamine that helped to complete the growth of the cubes (some truncated cubes were observed after just one growth). Note that this synthesis takes a total of 21 h in the reaction steps alone, but the time may be worth the price to achieve these beautiful structures. During the growth period the irregular pyrite nanocrystals were used as "seeds" to facilitate pyrite crystal growth that produced the final cube shape. The use of the weak alkylamines ligands was claimed to help make sure that small cubes were created. Hexadecylamine has strong interligand interactions that facilitate the creation of the small particles. Oleylamine is added after the nucleation step to help steer growth to cube structure, and it is found that if not added then the final product is still irregularly shaped particles, albeit bigger in size. The reduction of the ratio of sulfur to iron concentrations also played a role in the final shape. If a ratio of $[S]/[Fe] = 2$ was used then pyrrhoite(FeS) impurities were found, and if $[S]/[Fe] = 2.1$ the edges of the final cubes were more rounded. It is also important to note, even with the long growth periods associated with this synthesis, that Ostwald ripening did not occur. Interestingly, the absorbance of these final cubes mocked those of irregular shape. Most cubic pyrite nanostructures in the literature show absorbance in the NIR [24, 44]. This could show that size indeed affects the absorbance of these particles or that some other phenomenon is occurring such as plasmonics. Nonetheless, this synthetic method shows great control over particle growth and has produced some of the best crystalline particles to date.

2.3.2 Hot Injection Synthesis

Hot injection synthetic methods have been extensively used in the past to create nanoparticles of many different material systems. Peng et al. demonstrated the versatility of this method in their groundbreaking reports of creating cadmium selenide nanoparticles [45−48]. From there it has been adapted to be used in creating a whole host of metal-chalcogenide nanoparticles [49−52]. The Law group was the first to use hot injection to create spherical pyrite nanocrystals [53]. Two flasks were used in a typical synthesis, one loaded with Octadeclyamine and FeCl$_2$ while the other was filled with diphenyl ether and sulfur. The iron-containing flask was heated to 220 °C for one hour to allow for decomposition of the iron salt, while the sulfur flask was heated to 70 °C for an hour to allow for complete dissolution of the sulfur. The sulfur solution was then rapidly injected into the iron-containing flask and allowed to react for several hours. Aliquots were taken to examine the growth of the particles over time. Figure 5 shows the progression of synthesis. The seeds can be seen from the start as very small clusters and as the reaction progresses, it forms spherical-like particles. After the creation

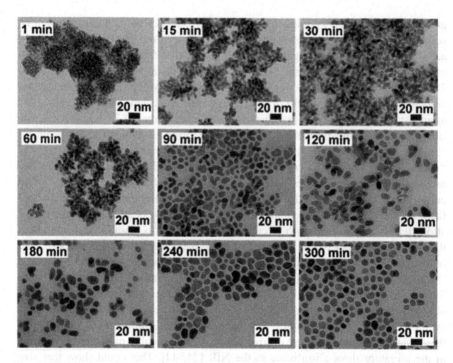

Fig. 5 TEM images of pyrite nanocrystal growth progression. Reprinted with permissions from [53]

of the separated nanocrystals the particles' overall size does not change dramatically over time, which also suggests that Ostwald ripening is not a major factor in pyrite nanosynthesis. Absorbance of these particles shows a typical absorbance shoulder around 550 nm. Thin films were created from these particles by layer-by-layer deposition, and then sintered to try to improve the properties of the material by removing sulfur deficiencies. This sintering also had the effect of increasing grain size that can be seen in Fig. 6 that left many voids in the film layer and also increased the roughness of the top surface. Although sintering may not have helped to create the best films for devices, synthesis of nanosphere particles with narrow size distribution was achieved.

In our lab, we have also taken to utilizing the hot injection method for pyrite nanocrystal synthesis [24, 54]. It has been observed that both cubes and nanospheres can be made for pyrite nanomaterial, but from the use of different synthetic routes. Being able to create both cubes and nanospheres using the same method would useful in the future testing of the material in devices. With this motivation, a model was conceived to achieve shape control with the most studied iron precursor $FeCl_2$. Figure 7 shows the proposed model. Since it has been shown that the {100} face is lower in surface energy than the {111} face, simply changing the temperature of the injection reaction could dictate the final shape. The task was then

Fig. 6 SEM images of thin films created out of pyrite nanocrystals before and after sintering. Reprinted with permission from [53]

taken to test this model by utilizing Law's synthetic method with few modifications. Like in previous works, upon injecting the sulfur into an $FeCl_2$ flask at 220 °C nanosphere particles were created, but when lowering the injection temperature to 120 °C, cubes were obtained. An intermediate state of aggregated cubes was obtained when the injection temperature of 170 °C was used. Figure 8 shows TEM images and UV-V is absorption spectrums of each of the particles. The exposed surface facets were proven to be {100} and {111} for the cubes and nanospheres, respectively, by high resolution transmission electron microscopy. We also have shown that by changing the iron precursor, the kinetics of the reaction changes due to the different decomposition rates of the iron precursor. When iron(0) pentacarbonyl ($Fe(CO)_5$) is used as the iron source, thick hexagonal sheets of pyrite are formed with the {100} surface exposed. If iron (II) acetylacetonate ($Fe(acac)_2$) is utilized for the iron source, thin random sheets with greigite (Fe_3S_4) impurities, also with the {100} face exposed, are formed. Final shape dependence can be related to the decomposition rate of the iron precursor that can be roughly estimated by hard/soft acid/base theory. This theory states that like components, either soft or hard, will bond stronger than if components are opposite. Fe^{+2} is classified as an intermediate strength acid and acetylacetonate is a very hard base [55]. This difference makes $Fe(acac)_2$ the weakest bonded of the three precursors and it will decompose readily. This means that there will be much more iron monomers available for reaction, which then explains the higher iron percent mineral greigite (Fe_3S_4) showing up as impurity. When comparing Cl^- ion to acetylacetonate the chloride ion is also a hard base but not to the same magnitude as acetylacetonate. Taking this into consideration, it does not decompose as

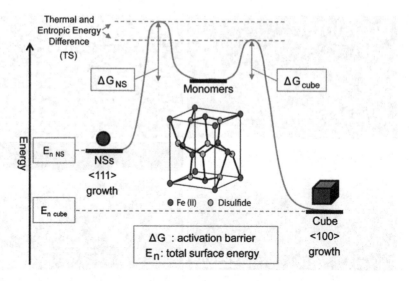

Fig. 7 Proposed growth model of pyrite nanocrystals. Reprinted with permission from [24]

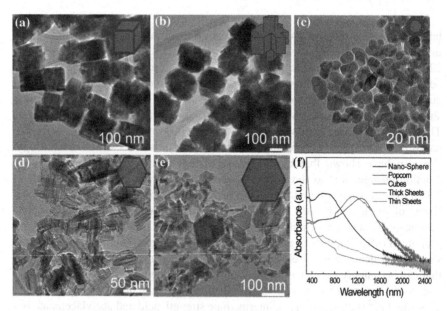

Fig. 8 TEM images of different acheieved shapes of pyrite nanocrystals and their absorbance. Reprinted with permissions from [24]

readily as Fe(acac)$_2$ allowing for pure phase pyrite nanospheres to be fabricated. Finally, both Fe(0) and carbonyl ligands are both a soft acid and soft base, respectively, meaning Fe(CO)$_5$ should decompose the slowest. Since the decomposition is slow, less seeds will be created resulting in large ligand coverage of the crystals that result in thicker/larger plates [56]. With these simple changes in synthetic procedures, final shapes of pyrite nanocrystals can be controlled.

2.4 Surface Facet Activity

Now that the different synthetic methods for creating pyrite nanocrystals have been discussed, a brief aside to discuss the different surface facet activities of pyrite will be taken. Ennaoui et. al. showed by X-ray Photoelectron Spectroscopy that adsorbates affect the materials properties and the different crystal surfaces show different activity [4]. It was shown that exposing a freshly cleaved sample to oxygen, the photovoltage increases. It is believed that the oxygen is passivating the surface and removing defect states out of the gap. When the oxygen coverage levels increase, no change or more changes occurred which suggests adsorption occurs only at defect sites. When studying the {100} surface, they see that both electron donors and electron accepters both adsorb to iron sites. H_2O shows coordination via oxygen though it was shown to de-adsorb completely at 300 K. When the surface is exposed to Br_2 at low concentrations a Br^- emission line is seen, and when the dosage is increased Br_2 emission lines are increased showing adsorption of molecular Br_2. After annealing the samples, an interesting band remains that is attributed to adsorbed bromine ions. The XPS data shows no evidence of transformation to $FeBr_3$ or $FeBr_2$, which leads to the conclusion that the Br^- ions adsorb rather strongly to the pyrite {100} surface.

We also looked at the different surface activity of pyrite surface facets by testing the nanocrystals as photocatalysts to photo degrade methyl orange dye [24]. A set amount of the particles were put in an aqueous methyl orange solution and put into a black box with a xenon light source that provided a 46 mW cm^{-2} of power. Samples were taken every 10 min to access the intensity of the methyl orange absorbance peak at ~ 475 nm. Figure 9 shows the results of the experiments. Cubes with {100} surface-exposed showed modest activity, while the {111} terminated nanospheres showed no activity at all. When the intermediate state was tested, it showed activity in between the cubes and the nanospheres, which is evidence that these particles are indeed an intermediate, and contain both surface facets. These differences in activity can be attributed to the exposed atoms on each crystal face. It has been shown that {100} has iron atoms exposed and also that this is the active site for adsorption, whereas the {111} only has sulfur which does not seem photocatalytically active.

Interesting side effects of these experiments allowed us to not only examine the photocatalytic activity, but also the photostability of the different facets. In the same figure, the peak shifts to blue as a new peak grows in next to it. It was determined that this intruding peak was caused by Fe^{+3} ions. It is known that pyrite degrades in water naturally by equation [30]:

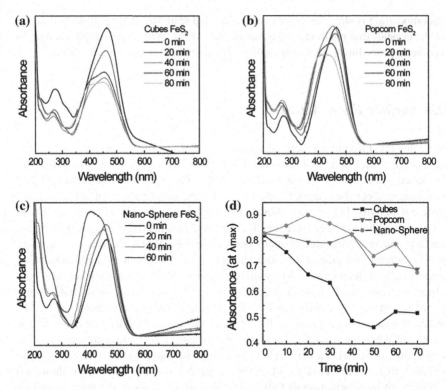

Fig. 9 a–c Methyl orange absorbance degredation measurements dependent on shape **d** change of absorbance over time for each shape. Reprinted with permissions from [24]

$$FeS_2(s) + 3.5O_2(g) + H_2O(l) \rightarrow Fe^{3+}(aq) + 2SO_4^{2-} + 2H^+(aq)$$

This oxidation reaction is a main cause of acid mine drainage which causes many environmental problems. Pyrite can also photocorrode via holes by the reaction [4]:

$$FeS_2(s) + 8H_2O(l) + 15h^+ \rightarrow Fe^{3+}(aq) + 2SO_4^{2-}(aq) + 16H^+(aq)$$

The presence of Fe^{3+} was confirmed 3-fold by a reduction of pH of the final solution, the precipitation of $Fe(OH)_3$ when an OH^- source was added, and matching of the absorbance peak to an $FeCl_3$ solution. With this knowledge, the time it took for the peak to shift and the intensity of the shifted peak can be used to examine the stability of the different surface facets of pyrite. The {100} face shows decent stability, although it eventually degrades. The {111} face shows a quicker shifting of the peak and also the intensity of the peak is greater than that of the {100} particles. Once again, the intermediate particles show an almost average of the two other surface facets. This difference in stability can be traced back to the adsorption of water onto defect sites, indicating that the {111} surface of these crystals have more defect sites to allow for the above reaction to occur, where the {100} has less. While this difference in stability exists, it is important to realize that both of the facets eventually degrade, which makes water a poor choice as a solvent when used in electrochemical/photoelectrochemial studies or devices unless these defect sites can be passivated.

3 Iron Pyrite Photovoltaic and Photodetector Technology

3.1 Introduction

Once again the first reports of using pyrite in a photovoltaic device come out of the Tributsch group. Throughout their exploration of pyrite material, they created a photoelectrochemical (PCE), Schottky barrier, and thin film-sensitized solar cells. In the PCE device, it was shown that by coupling synthetic n-type FeS_2 with an iodine electrolyte (I^-/I_3^-) that a device with 2.8 % efficiency could be obtained [4]. This is the champion device of the pyrite system at the time of writing. Figure 10 shows quantum yield measurements for the device and can be seen to exceed 90 % at points while maintaining high yield over a wide band of the energy spectrum. Figure 11 shows the J–V curve of the device, showing high short circuit photocurrent while showing the typical low V_{oc} that plagues the pyrite system of 0.2 V. It is mentioned that the electrolyte system of 4 M HI, 0.05 M I_2, and 2 M CaI_2 is vital to the performance. Previously, it was found that both I^- ions and hydrogen treatment (either molecular hydrogen or acid etching) helps to reduce dark current. Although it reduces dark current, it still limits this device's efficiency due to the dark current reducing the photovoltage.

The Schottky device was made from a pyrite/platinum interface [57]. In these devices, a 0.1 mm layer of electrochemically etched pyrite surface was covered by a thin transparent platinum film (50–120 Å) was utilized. The metal film was deposited by electron beam evaporation under vacuum, though it was thought that thin oxide layers or sulfide layers could exist, created by the evaporation, deteriorating the performance. Figure 12 shows the J–V characteristics. The device exhibited high short circuit currents of around 30 mA/cm^2 and could achieve saturation currents of 100 mA/cm^2 under higher illumination power. The quantum yield of these devices was lower than those of the PCE, only reaching $\sim$40 %. Although the system worked, it still produced low efficiency, prompting a shift to switch to thin layer-sensitized cells.

Thin layer-sensitized cells were a modification of the now well-known dye-sensitized solar cells pioneered by Gratzel [58]. The contrast between the two is the replacement of the dye in the Gratzel cell with a thin (10–15 nm) layer of pyrite to act as the absorber [59]. Figure 13 shows the schematic of such a device and the energy diagram of the overall system. The pyrite acts as a photoabsorber, where when it is excited by a photon an exiton is formed. When this exciton diffuses to the interface between the pyrite and the titanium oxide (TiO_2), it splits, injecting the electron into the TiO_2 and the hole is transported to an electrolyte

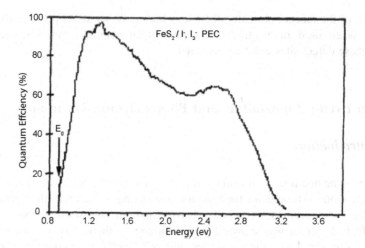

FeS$_2$/ I$^-$, I$_3^-$ PEC

Fig. 10 EQE of photoelectrochemical solar cell. Reprinted with permission from [4]

Fig. 11 J–V curve of champion pyrite photoelectrochemical cell. Reprinted with permission from [4]

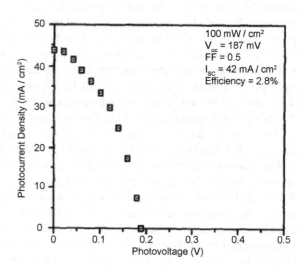

100 mW / cm²
V_{oc} = 187 mV
FF = 0.5
I_{sc} = 42 mA / cm²
Efficiency = 2.8%

solution where it is reduced. Though the cell deliverers an efficiency of 1 %, the quantum efficiency only reached 10 %. It was believed the quantum efficiency was low due to pyrite being a high absorbing material, which creates the excitons near the surface of the pyrite/electrolyte interface where they will quickly recombine. These three device setups are the main body of solar cell devices in older literature, but with the revival of interest due to nanotechnology, new photovoltaic and photodetector devices are being fabricated.

Fig. 12 J–V curve of pyrite/
platinum Schottky solar cell.
Reprinted with permission
from [57]

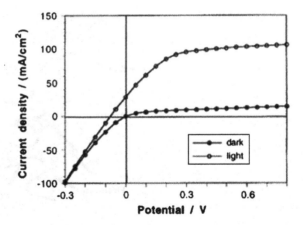

3.2 Pyrite Nanocrystal Photovoltaic Technology

The reason behind creating and controlling the shape of pyrite nanocrystals is to ultimately use the material in useful devices. Most of the recent papers on synthesis have also included attempts at creating working photovoltaic (PV) devices. No attempt has yet been made to create devices out of the beautiful cubes that the Stoldt group created. The Law group showed that after sintering the nanocrystal films were no longer viable for PV applications due to the holes left over. Steinhagen et al. has gone on to utilize pyrite nanocrystals much like the Law group's in a battery of different PV architectures with no results [60]. Figure 14 shows all the different schematics that were tested. First a pure schottky barrier device with gold and pyrite was tried. As seen, this device did not even show diode behavior. Next a CuInSe$_2$ device was tested and showed a measly response of 0.03 mA/cm^2 short circuit current and 0.041 V open circuit voltage, resulting in a negligible efficiency. A depleted heterojunction structure was also investigated by putting a layer of pyrite on top of TiO$_2$ layer to act as a hole-blocking layer that will limit recombination while simultaneously creating a depleted region. This type of cell was recently shown to produce some of the highest efficiencies for nanocrystal cells when utilizing lead sulfide [61]. When pyrite was used the cell once again showed no performance. Finally, an organic/inorganic hybrid heterojunction was created. These devices have shown promise when coupling poly(3-hexylthiophene) (P3HT), an organic polymer with other quantum dot systems [62]. When pyrite is tested, a V_{oc} of 0.041 V and J_{sc} of 0.007 mA/cm^2 was achieved, once again showing an efficiency of zero. This communication was discouraging to the hope of making use of pyrite's exceptional properties for solar devices, though recently our group has made progress in creating working devices.

Our group took this valuable information from the Korgel report and examined how we could work out the kinks in the pyrite system. The first thing that was noted was that the type of crystals that was used was {111} terminated. As stated

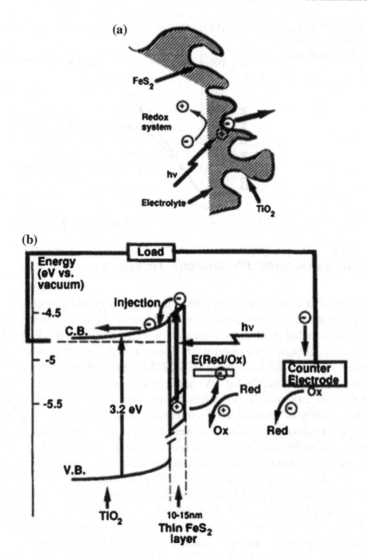

Fig. 13 a schematic diagram of TiO_2 support with a thin layer of pyrite on top **b** band diagram of thin layer sensitized solar cell. Reprinted with permission from [59]

above, it was found that the {111} face was inactive and less stable than the {100} face of pyrite. Since this is the case, it was decided to make use of the {100} terminated nano-cubes in our PV devices. Cadmium Sulfide quantum dots were chosen to create the heterojunction in the cell due to the large offset of the valence band of FeS_2 and the conduction band of CdS of 1.1 eV. It was hoped that this would help overcome the problem of low photovoltages that plague the pyrite system. The pyrite cubes did not create a nice film to create a bi-layer device, so a

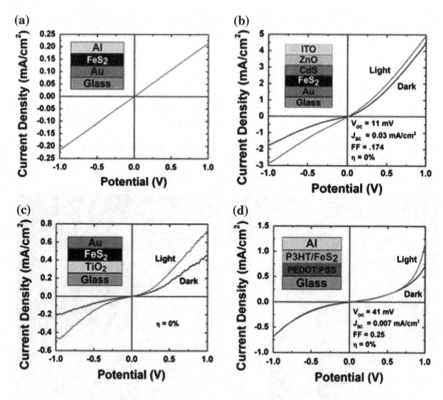

Fig. 14 J–V curves of all different pyrite solar cell schemes fabricated. Reprinted with permission from [60]

strategy from organic PV systems was adapted; the bulk heterojunction. By mixing the small CdS quantum dots with the bigger FeS_2 cubes and spincoating onto the substrate a pseudo-bulk heterojunction was created. The CdS quantum dots created a matrix around the bigger cubes that allowed for not only complete coverage of the pyrite material, but also allowed for smooth films for electrodes to be deposited onto. The band diagram, thin film absorption, and fluorescence quenching measurements are shown in Fig. 15. Absorption in the thin film showed absorption in the NIR due to the cubes absorbance (see cube synthesis above). Fluorescence quenching measurements showed that as the pyrite loading increased, better quenching occurred, indicative of better exiton separation. Figure 16 shows TEM images of the mixture of the CdS QDs and the FeS_2 cubes. The second image shows the intimate contact of the QDs with the cubes edge that allows for exceptional exiton splitting. The SEM image shows the cross-section of the active layer, where the percolation of the cubes inside of the QD matrix can be seen. J–V curves are presented in Fig. 17. Three different loading ratios of Pyrite:CdS were tested. The 1:1 (FeS_2:CdS) ratio device preformed the best with average of J_{sc} of 3.9 mA/cm^2 a V_{oc} of 0.79 V and a fill factor of 0.36 giving it an over all efficiency

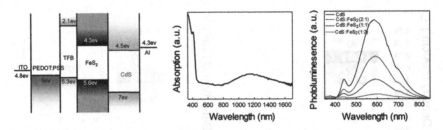

Fig. 15 From *left* to *right* Band diagram of fabricated solar cells. Thin film absorbance of active layer. Photoluminescence quenching measurements. Reprinted with permission from [61]

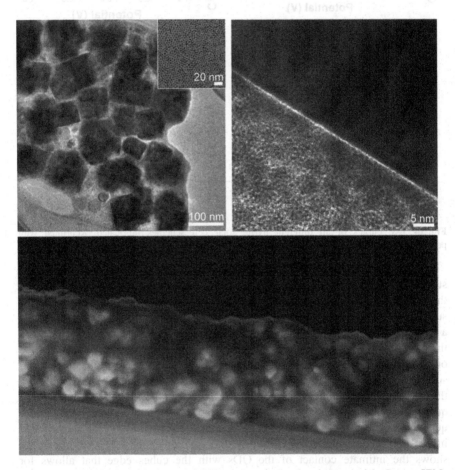

Fig. 16 *Top* TEM images of mixture of CdS quantum dots and pyrite nanocubes. *Bottom* SEM image of photoactive layer of the mixture. Reprinted with permissions from [61]

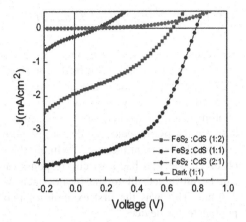

Fig. 17 J–V curves of fabricated solar cells with different FeS$_2$/CdS ratios. Reprinted with permission from [61]

Fig. 18 AFM images of layers created with **a** 2:1 and **b** 1:1 loading ratio of FeS$_2$/CdS. Reprinted with permission from [61]

of 1.1 % over 100 devices. The devices with higher loading of FeS$_2$ experienced worse performance due to roughness of the film that can be seen in the Atomic Force Microscopy (AFM) images in Fig. 18. These devices have overcome the lifetime problem of low V_{oc} by creating bulk heterojunction allowing for increased charge separation. With the devices reaching 1.1 % efficiency, the first working pyrite nanocrystal solar cell has been fabricated [63]. Focus on improving these cells and controlled synthesis of pyrite is ongoing in our lab.

3.3 Pyrite Nanocrystal Photodetector Technology

While most focus has been on creating working PV devices with pyrite, other devices can also benefit from its unique properties. Photodetectors could put to use the high absorption coefficient and broad absorption spectra of pyrite. Wang et al. reported the use of pyrite nanocrystals coupled with zinc oxide (ZnO) to create a photodetector [64]. When MoO_3 is added as electron blocking, hole transport layer between the pyrite nanocrystals and the gold electrode the dark current decreases and the photocurrent increases. Without the MoO_3 layer large leakage current was observed. Devices had a spectra response range from 450 nm to 1150 nm, which can be attributed all to the pyrite since MoO_3 and ZnO do not absorb above 450 nm. This was the first example of making use of pyrite nanomaterial in a photodetector device.

Our group has recently expanded the pyrite material system utilized in our PV work to create a photodetector [65]. Pyrite cubes, Pyrite nanospheres, and CdS material were used in the fabrication of such device, though deposition of the material was vastly different. In the photodetector work a bulk heterojunction structure was not wanted so a bi-layer structure was chosen. This was achieved by a novel micro centrifugation deposition of the pyrite nanocrystals followed by a chemical bath deposition of an over layer of CdS. It is shown that the final photodetector device showed fast photoresponse time of 10 ms and high responsivity of 174.9 A/W. Even more interestingly, these devices exhibited the ability to tune the photocurrent in the presence of a magnetic field due to the creation of a dilute magnetic semiconductor (DMS) phase that was created from the CBD method of CdS growth. It is a challenge in the field to create a photodetector with high sensitivity and quick temporal response that is visible, and every more unheard of in the NIR region. In systems utilizing pyrite nanospheres, response time was quicker than in systems using pyrite nanocubes. Pyrite nanocubes, not to be outdone, showed response in the NIR due to their photoabsorption peak around 1200 nm. Figure 19 shows current-voltage characteristics of all devices made and also EQE and absorption of thin films. Figure 20 shows the different on/off cycles to show temporal response.

Since the CBD method is done in aqueous medium, a DSM interfacial layer of CdFeS was found from the migration of Fe^{3+} ions (due to oxidation of the pyrite in water) into CdS layer. With this layer being present, it allows for the ability to turn the intensity of the current by changing the magnetic field around the device. Figure 21 shows the J–V curves of device testing when the magnet is moved a set amount away from the sample. An overall change of 72.6 % current can be achieved by simply changing the position of the magnet. This work shows the versatility of the pyrite material system, not only as a photovoltaic material, but as a potential system for photodetectors as well. Ongoing studies of creating better photodetectors, especially for NIR response, are being conducted in our lab.

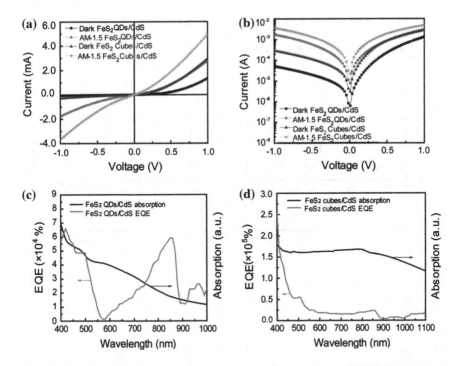

Fig. 19 *Top* J–V curves of photodetector devices. *Bottom* Absorbance and EQE measurements of both pyrite quantum dot and nanocube utilizing devices. Reprinted with permission from [63]

4 Summary

Pyrite is an interesting system full of promise, but many problems must be addressed before it can become a golden material for photodevices. It started off being investigated in the 1970s for use in thin film solar cells without much luck on final devices, although much was learned and can be learned from these well-studied films. With the explosion of nanochemistry, research into creating and controlling shapes and exposed surface facets has been bolstered. New synthetic ways, from hydrothermal to solvothermal, have been investigated with success in control of nanocrystal formation. With these materials, new working devices are now being fabricated and studied. Pyrite nanochemistry is a new subject that holds much promise for energy harvesting and photodetection. Studies of pyrite must press on and try to answer still open questions of the system such as:

- Kinetics of crystal growth and effects of ligands on creating stoichiometric pyrite without impurities.
- Shape and size control of nanocrystals and creation of quantum confined pyrite nanocrystals.

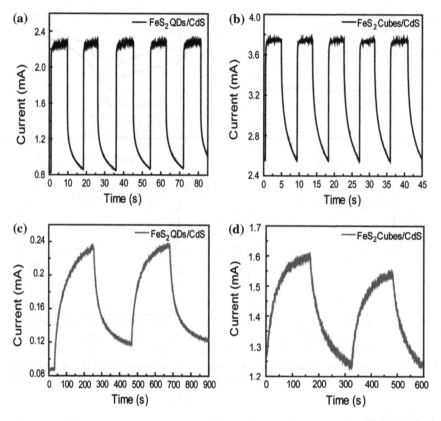

Fig. 20 On/Off cycle response in (*top*) normal AM 1.5 light and (*bottom*) IR source for both QD and cube utilizing systems. Reprinted with permissions from [63]

- Better understanding of actual effects of sulfur deficiencies and how to eliminate them.
- New synthetic techniques that allow for quicker reaction times and scalability.
- Why pyrite has such a high absorption coefficient even with an indirect band gap.
- What are the effects of doping and how will it allow tune-abliity of pyrite's band gap
- The cause for low open circuit voltage and how to eliminate this problem.

With the solving of these problems, pyrite may become a golden material that will benefit all.

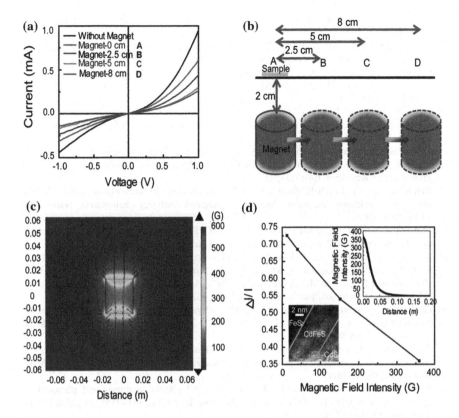

Fig. 21 a J–V curves showing dependence on magnet placement. **b** Schematic of experiment ran to obtain J–V curves. **c** Simulated image of magnetic field intensity. **d** Relative current intensity dependent on magnet distance. Inset is TEM image showing interfacial layer. Reprinted with permission from [63]

References

1. Zweibel K, Mason J, Fthenakis V (2008) By 2050 solar power could end US dependence on foreign oil and slash greenhouse gas emissions. Sci Am 298(1):64–73
2. Green M (2000) Power to the people: sunlight to electricity using solar cells. University of New South Wales Press, Sydney
3. Wadia C, Wu Y, Gul S, Volkman SK, Guo J, Alivisatos AP (2009) Surfactant-assisted hydrothermal synthesis of single phase pyrite FeS2 nanocrystals. Chem Mater 21(13):2568–2570
4. Ennaoui A, Fiechter S, Pettenkofer C, Alonsovante N, Buker K, Bronold M et al (1993) Iron disulfide for solar-energy conversion. Sol Energy Mater Sol Cells 29(4):289–370. PubMed PMID: WOS:A1993LH67000001 (English)
5. Altermatt PP, Kiesewetter T, Ellmer K, Tributsch H (2002) Specifying targets of future research in photovoltaic devices containing pyrite (FeS2) by numerical modelling. Sol Energy Mater Sol Cells 71(2):181–195

6. Emsley J (2011) Nature's building blocks: an A–Z guide to the elements. Oxford University Press, USA
7. Wadia C, Alivisatos AP, Kammen DM (2009) Materials availability expands the opportunity for large-scale photovoltaics deployment. Environ Sci Technol 43(6):2072–2077
8. Chapman P, Roberts F (1983) Metal resources and energy. Buttersworths, London
9. Huynh WU, Dittmer JJ, Alivisatos AP (2002) Hybrid nanorod-polymer solar cells. Science 295(5564):2425–2427
10. Li G, Shrotriya V, Huang J, Yao Y, Moriarty T, Emery K et al (2005) High-efficiency solution processable polymer photovoltaic cells by self-organization of polymer blends. Nat Mater 4(11):864–868
11. Contreras MA, Egaas B, Ramanathan K, Hiltner J, Swartzlander A, Hasoon F et al (1999) Progress toward 20 % efficiency in Cu(In, Ga)Se2 polycrystalline thin-film solar cells. Prog Photovoltaics Res Appl 7(4):311–316
12. Ren S, Chang L-Y, Lim S-K, Zhao J, Smith M, Zhao N et al (2011) Inorganic–organic hybrid solar cell: bridging quantum dots to conjugated polymer nanowires. Nano Lett 11(9):3998–4002
13. Luther JM, Law M, Beard MC, Song Q, Reese MO, Ellingson RJ et al (2008) Schottky solar cells based on colloidal nanocrystal films. Nano Lett 8(10):3488–3492
14. Arango AC, Oertel DC, Xu Y, Bawendi MG, Bulović V (2009) Heterojunction photovoltaics using printed colloidal quantum dots as a photosensitive layer. Nano Lett 9(2):860–863
15. Gur I, Fromer NA, Geier ML, Alivisatos AP (2005) Air-stable all-inorganic nanocrystal solar cells processed from solution. Science 310(5747):462–465
16. Gledhill SE, Scott B, Gregg BA (2005) Organic and nano-structured composite photovoltaics: an overview. J Mater Res 20(12):3167–3179. PubMed PMID: WOS:000233628600002
17. Tian B, Zheng X, Kempa TJ, Fang Y, Yu N, Yu G et al (2007) Coaxial silicon nanowires as solar cells and nanoelectronic power sources. Nature 449(7164):885–888. PubMed PMID: WOS:000250230600042
18. Wu Y, Wadia C, Ma W, Sadtler B, Alivisatos AP (2008) Synthesis and photovoltaic application of copper (I) sulfide nanocrystals. Nano Lett 8(8):2551–2555. PubMed PMID: WOS:000258440700076
19. Yu L, Lany S, Kykyneshi R, Jieratum V, Ravichandran R, Pelatt B et al (2011) Iron chalcogenide photovoltaic absorbers. Adv Energy Mater 1(5):748–753. PubMed PMID: WOS:000295140100005
20. Bronold M, Tomm Y, Jaegermann W (1994) surface-states on cubic d-band semiconductor pyrite (FES(2)). Surf Sci 314(3):L931–L936. PubMed PMID: WOS:A1994NZ92900011
21. Sun R, Chan MKY, Ceder G (2011) First-principles electronic structure and relative stability of pyrite and marcasite: implications for photovoltaic performance. Phys Rev B 83(23):235311. PubMed PMID: WOS:000291398400006
22. Sun R, Chan MKY, Kang S, Ceder G. Intrinsic stoichiometry and oxygen-induced p-type conductivity of pyrite FeS_{2}. Phys Rev B 84(3):035212
23. Macpherson HA, Stoldt CR (2012) Iron pyrite nanocubes: size and shape considerations for photovoltaic application. ACS Nano 6(10):8940–8949. PubMed PMID: WOS:000310096100053
24. Kirkeminde A, Ren S (2013) Thermodynamic control of iron pyrite nanocrystal synthesis with high photoactivity and stability. J Mater Chem A 1(1):49–54
25. Wang D, Wang Q, Wang T (2010) Shape controlled growth of pyrite FeS2 crystallites via a polymer-assisted hydrothermal route. CrystEngComm 12(11):3797–3805. PubMed PMID: WOS:000283315900078
26. Barnard AS, Russo SP (2007) Shape and thermodynamic stability of pyrite FeS2 Nanocrystals and Nanorods. J Phys Chem C 111(31):11742–11746
27. Barnard AS, Russo SP (2009) Modelling nanoscale FeS2 formation in sulfur rich conditions. J Mater Chem 19(21):3389–3394. PubMed PMID: WOS:000266269300010

28. Ennaoui A, Fiechter S, Jaegermann W, Tributsch H (1986) Photoelectrochemistry of highly quantum efficient single-crystalline n-FeS2 (Pyrite). J Electrochem Soc 133(1):97–106
29. Ennaoui A, Tributsch H (1984) Iron sulphide solar cells. Sol Cells 13(2):197–200
30. Sullivan P, Yelton J, Reddy K (1988) Iron sulfide oxidation and the chemistry of acid generation. Environ Geol 11(3):289–295
31. Rickard DT (1975) Kinetics and mechanism of pyrite formation at low-temperatures. Am J Sci 275(6):636–652. PubMed PMID: WOS:A1975AD80800002
32. Luther GW (1991) Pyrite synthesis via polysulfide compounds. Geochim Cosmochim Acta 55(10):2839–2849. PubMed PMID: WOS:A1991GK84900012
33. Rickard D, Luther GW (1997) Kinetics of pyrite formation by the H2S oxidation of iron (II) monosulfide in aqueous solutions between 25 and 125 degrees C: the mechanism. Geochim Cosmochim Acta 61(1):135–147. PubMed PMID: WOS:A1997WE86100009
34. Rickard D (1997) Kinetics of pyrite formation by the H2S oxidation of iron (II) monosulfide in aqueous solutions between 25 and 125 degrees C: the rate equation. Geochim Cosmochim Acta 61(1):115–134. PubMed PMID: WOS:A1997WE86100008
35. Harmandas NG, Fernandez EN, Koutsoukos PG (1998) Crystal growth of pyrite in aqueous solutions. Inhibition by organophosphorus compounds. Langmuir 14(5):1250–1255. PubMed PMID: WOS:000072390800040
36. Rickard D, Luther GW (2007) III. Chemistry of iron sulfides. Chem Rev 107(2):514–562. PubMed PMID: WOS:000244206600010
37. Fiechter S, Mai J, Ennaoui A, Szacki W (1986) chemical vapor transport of pyrite (FeS2) with halogen (CL, BR, I). J Cryst Growth 78(3):438–444. PubMed PMID: WOS:A1986F670100003
38. Fleming JG (1998) Growth of FeS2 (pyrite) from Te melts. J Cryst Growth 92(1–2):287–293. PubMed PMID: WOS:A1988R106400038
39. Bither TA, Bouchard RJ, Cloud WH, Donohue PC, Siemons WJ (1968) Transition metal pyrite dichalcogenides high-pressure synthesis and correlation of properties. Inorg Chem 7(11):2208–2220. PubMed PMID: WOS:A1968C016000008
40. Chatzitheodorou G, Fiechter S, Konenkamp R, Kunst M, Jaegermann W, Tributsch H (1986) Thin photoactive FeS2 (pyrite) films. Mater Res Bull 21(12):1481–1487. PubMed PMID: WOS:A1986F263200010
41. Thomas B, Ellmer K, Muller M, Hopfner C, Fiechter S, Tributsch H (1997) Structural and photoelectrical properties of FeS2 (pyrite) thin films grown by MOCVD. J Cryst Growth 170(1):808–812
42. Morrish R, Silverstein R, Wolden CA (2012) Synthesis of stoichiometric FeS2 through plasma-assisted sulfurization of Fe2O3 nanorods. J Am Chem Soc 134(43):17854–19857
43. Chen X, Wang Z, Wang X, Wan J, Liu J, Qian Y (2005) Single-source approach to cubic FeS2 crystallites and their optical and electrochemical properties. Inorg Chem 44(4):951–954
44. Yuan B, Luan W, Tu S-t (2012) One-step synthesis of cubic FeS2 and flower-like FeSe2 particles by a solvothermal reduction process. Dalton Trans 41(3):772–776. PubMed PMID: WOS:000298753800012
45. Peng XG, Manna L, Yang WD, Wickham J, Scher E, Kadavanich A et al (2000) Shape control of CdSe nanocrystals. Nature 404(6773):59–61. PubMed PMID: WOS:000085775100042
46. Peng ZA, Peng XG. Formation of high-quality CdTe, CdSe, and CdS nanocrystals using CdO as precursor. J Am Chem Soc 123(1):183–184. PubMed PMID: WOS:000166258800026
47. Peng ZA, Peng XG (2001) Mechanisms of the shape evolution of CdSe nanocrystals. J Am Chem Soc 123(7):1389–1395. PubMed PMID: WOS:000167031300016
48. Peng ZA, Peng XG (2002) Nearly monodisperse and shape-controlled CdSe nanocrystals via alternative routes: Nucleation and growth. J Am Chem Soc 124(13):3343–3353. PubMed PMID: WOS:000174793400033
49. Acharya KP, Hewa-Kasakarage NN, Alabi TR, Nemitz I, Khon E, Ullrich B et al (2010) Synthesis of PbS/TiO2 colloidal heterostructures for photovoltaic applications. J Phys Chem C 114(29):12496–12504. PubMed PMID: WOS:000280070900019

50. Baumgardner WJ, Choi JJ, Lim Y-F, Hanrath T (2010) SnSe nanocrystals: synthesis, structure, optical properties, and surface chemistry. J Am Chem Soc 132(28):9519–9521. PubMed PMID: WOS:000280086800003

51. Dai Q, Li D, Chang J, Song Y, Kan S, Chen H et al (2007) Facile synthesis of magic-sized CdSe and CdTe nanocrystals with tunable existence periods. Nanotechnology 18(40):405603. PubMed PMID: WOS:000249735400016

52. Yarema M, Pichler S, Sytnyk M, Seyrkammer R, Lechner RT, Fritz-Popovski G et al (2011) Infrared emitting and photoconducting colloidal silver chalcogenide nanocrystal quantum dots from a silylamide-promoted synthesis. ACS Nano 5(5):3758–3765. PubMed PMID: WOS:000290826800041

53. Puthussery J, Seefeld S, Berry N, Gibbs M, Law M (2012) Colloidal iron pyrite (FeS2) nanocrystal inks for thin-film photovoltaics. J Am Chem Soc 133(4):716–719

54. Kirkeminde A, Ruzicka B, Wang R, Puna S, Zhao H, Ren SQ (2012) Synthesis and optoelectronic properties of two-dimensional FeS2 nanoplates. ACS Appl Mater Interfaces. doi:10.1021/am300089f

55. Zhang JT, Tang Y, Lee K, Min OY (2010) Nonepitaxial growth of hybrid core-shell nanostructures with large lattice mismatches. Science 327(5973):1634–1638. PubMed PMID: WOS:000275970600041. English

56. Xu C, Zeng Y, Rui XH, Xiao N, Zhu JX, Zhang WY et al (2012) Controlled soft-template synthesis of ultrathin C@FeS nanosheets with high-Li-storage performance. ACS Nano 6(6):4713–4721. PubMed PMID: WOS:000305661300019 (English)

57. Buker K, Alonsovante N, Tributsch H (1992) Photovoltaic output limitation of N-FeS2 (pyrite) schottky barriers—a temperature-dependent characterization. J Appl Phys 72(12):5721–5728. PubMed PMID: WOS:A1992KC85000029

58. O'Regan B, Gratzel M (1991) A low-cost, high-efficiency solar cell based on dye-sensitized colloidal TiO2 films. Nature 353(6346):737–740

59. Ennaoui A, Fiechter S, Tributsch H, Giersig M, Vogel R, Weller H (1992) photoelectrochemical energy-conversion obtained with ultrathin organo-metallic-chemical-vapor-deposition layer of FeS2 (pyrite) on TiO2. J Electrochem Soc 139(9):2514–2518. PubMed PMID: WOS:A1992JL82500032

60. Steinhagen C, Harvey TB, Stolle CJ, Harris J, Korgel BA (2012) Pyrite nanocrystal solar cells: promising, or fool's gold? J Phys Chem Lett 3(17):2352–2356. PubMed PMID: WOS:000308342500008

61. Tang J, Kemp KW, Hoogland S, Jeong KS, Liu H, Levina L et al Colloidal-quantum-dot photovoltaics using atomic-ligand passivation. Nat Mater 10(10):765–771

62. Ren S, Chang L-Y, Lim S-K, Zhao J, Smith M, Zhao N et al Inorganic–organic hybrid solar cell: bridging quantum dots to conjugated polymer nanowires. Nano Lett 11(9):3998–4002

63. Kirkeminde A, Scott R, Ren S (2012) All inorganic iron pyrite nano-heterojunction solar cells. Nanoscale 4(24):7649–7654. PubMed PMID: MEDLINE:23041909

64. Wang D-Y, Jiang Y-T, Lin C-C, Li S-S, Wang Y-T, Chen C-C et al (2012) Solution-processable pyrite FeS2 nanocrystals for the fabrication of heterojunction photodiodes with visible to NIR photodetection. Adv Mater 24(25):3415–3420

65. Gong M, Kirkeminde A, Xie Y, Lu R, Liu J, Wu JZ et al (2012) Iron pyrite (FeS2) broad spectral and magnetically responsive photodetectors. Adv Funct Mater 78–83

High-Performance Bulk-Heterojunction Polymer Solar Cells

Fang-Chung Chen, Chun-Hsien Chou and Ming-Kai Chuang

Abstract Most of the high-efficiency organic photovoltaic devices (OPVs) reported to date have been fabricated based on the concept of a bulk heterojunction, where a conjugated polymer (donor) and a soluble fullerene (acceptor) form an interpenetrating network featuring a large donor–acceptor interfacial area. In this chapter, we first introduce the fundamentals of OPVs and then review the recent progress related to OPVs based on conjugated polymers. We then discuss the annealing approaches that have been used to optimize the morphologies of the photoactive layers, including thermal annealing and solvent annealing, and describe the engineering of the interfaces at the contacts between the polymer blends and the metal electrodes. Next, we outline the two most common optical methods for improving the light absorption efficiency of OPVs: the use of optical spacers and the triggering of surface plasmons. Finally, we summarize the development of low-band-gap polymers for the absorption of long-wavelength photons from solar irradiation and provide a brief outlook of the future use of OPVs.

1 Introduction

1.1 Overview

Many organic semiconductors exhibit the electronic properties of their common inorganic counterparts while providing the advantages of plastic processing and low cost. To date, several organic-based electronic devices, including organic

F.-C. Chen (✉)
Department of Photonics and Display Institute, National Chiao Tung University,
Hsinchu 300, Taiwan
e-mail: fcchen@mail.nctu.edu.tw

C.-H. Chou · M.-K. Chuang
Department of Photonics and Institute of Electro-optical Engineering,
National Chiao Tung University, Hsinchu 300, Taiwan

Z. Lin and J. Wang (eds.), *Low-cost Nanomaterials*, Green Energy and Technology,
DOI: 10.1007/978-1-4471-6473-9_7, © Springer-Verlag London 2014

light-emitting diodes, organic thin-film transistors, and organic photodiodes have been demonstrated for practical applications in various fields [1–4]. More recently, organic photovoltaic devices (OPVs) have been recognized as promising technologies for the utilization of solar energy because of their light weight, flexibility, and cost-effective production with simple processability [5–9]. Furthermore, the energy payback time of OPVs is amazingly short, relative to that of photovoltaic cells based on silicon or other inorganic semiconductors, because they can be fabricated at low temperatures [9]. In addition, OPV technologies employ non-toxic, earth-abundant materials in the manufacturing process. At present, OPVs based on conjugated polymers are exhibiting power conversion efficiencies (PCEs) of up to approximately 10 % [8]. Although such efficiency remains low relative to that of commercial inorganic cells, recent rapid improvements in efficiency have generated considerable interest in OPVs for practical and truly low-cost energy production.

Most of the high-efficiency OPVs reported to date have been fabricated based on the bulk heterojunction (BHJ) concept, where a conjugated polymer (donor) and a soluble fullerene (acceptor) form an interpenetrating network possessing a large donor–acceptor interfacial area. Various approaches have been proposed to achieve high efficiencies from BHJ devices. In this chapter, we provide an introduction to the development of BHJ OPVs, especially for those made of conjugated-polymer materials. We first discuss the basic fundamental principles behind OPVs and then review some recent progress in the field, focusing on methods for optimizing the morphologies of the photoactive layers through various annealing treatments. We also review the engineering of the interface and optical effects in the devices. We then outline the two most common optical methods for improving the light absorption efficiency of OPVs: the use of optical spacers and the triggering of surface plasmons. Finally, we examine the development of low-band-gap polymers for the absorption of long-wavelength photons from solar irradiation. Because our main purpose for this chapter is to introduce the basic concepts and recent development of OPVs, rather than completely reviewing all of the literature in this field, by necessity we have had to overlook many outstanding contributions.

1.2 Basic Principles

Figure 1a illustrates the typical device architecture of an OPV. The photoactive layer, consisting of donor and acceptor materials, is sandwiched between the anode and cathode. The working principle of an OPV can be divided into six processes: (i) photon absorption, (ii) exciton generation, (iii) exciton diffusion, (iv) exciton dissociation, (v) charge transport, and (vi) charge collection (Fig. 1b) [10]. Upon the absorption of photons in an organic material, excitons are generated. The exciton binding energy in an organic material is, however, usually very large relative to that of an inorganic semiconductor. Therefore, excitons in organic

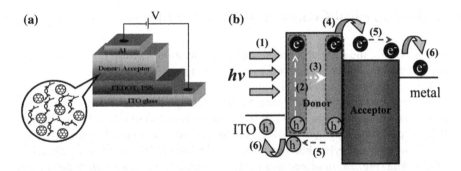

Fig. 1 **a** Device architecture of a typical BHJ OPV. **b** Schematic representation of the working principle of an OPV device: *1* light absorption *2* exciton generation *3* exciton diffusion *4* exciton dissociation *5* charge transport, and *6* charge collection [10]

materials can dissociate into free carriers only when they diffuse to the donor–acceptor junctions. In a so-called BHJ structure, the large area of the donor–acceptor interface ensures efficient exciton dissociation to produce a high density of free charge carriers. After exciton dissociation, the free charges are transported through the donor and acceptor materials, respectively, to the electrodes. The electrodes then collect the carriers, resulting in a photocurrent.

Based on the six-step mechanism, the external quantum efficiency (EQE) of an OPV can be expressed using the equation [10]

$$EQE(\lambda) = \eta_A(\lambda) \times \eta_G(\lambda) \times \eta_C(\mu) \tag{1}$$

where $\eta_A(\lambda)$, $\eta_G(\lambda)$, and $\eta_C(\mu)$ are the absorption efficiency, carrier generation efficiency, and charge collection efficiency, respectively. For an OPV prepared using the BHJ concept, the large area of the donor–acceptor interface throughout the photoactive layer suggests that the carrier generation efficiency could, in theory, be increased to close to 100 %. On the other hand, the charge collection efficiency ($\eta_C(\mu)$) strongly depends on the morphology of the photoactive layer because this morphology will affect the nature of the conducting channel. Furthermore, the quality of the interface between the metallic electrodes and the organic materials also plays an essential role in determining the performance of the device. In other words, the interfaces strongly affect the charge collection efficiency. The other key issue is sufficient harvesting of sunlight. Although the use of a thicker active layer is a straightforward approach toward obtaining high absorption efficiency ($\eta_A(\lambda)$), this method inevitably results in increased device resistance, due to the low carrier motilities of organic materials. Therefore, in real devices, it is difficult to decouple the relationship between these factors; simultaneous improvement of both $\eta_A(\lambda)$ and $\eta_C(\mu)$ is hard to achieve. In the following section, we outline strategies for improving the efficiencies of OPVs and review some recent progress in this area.

2 Morphological Control in Photoactive Layers

In BHJ OPVs, the light absorption, exciton generation, exciton dissociation, and charge transport and collection processes are strongly influenced by the morphology of the photoactive layer [11–20]. The most common way of controlling the morphology of the active polymer blend has been through the application of annealing processes. For example, in 2005 Erb et al. studied the effect of thermal annealing on thin films containing poly(3-hexylthiophene) (P3HT) and 1-(3-methoxycarbonyl)propyl-1-pheny[6,6]methanofullerene (PCBM) [13]. Their X-ray diffraction (XRD) data (Fig. 2a) indicated that the (100) peak had increased significantly after thermal annealing, suggesting the formation of crystalline P3HT domains. The authors inferred that thermal annealing led to changes in the morphology of the P3HT:PCBM thin film, as illustrated in Fig. 2b. Hence, the device performance had improved as a result of increased crystallinity.

Meanwhile, Ma et al. also investigated the effects of post-annealing treatment on device performance [14]. After thermal annealing at 150 °C for 30 min, they observed a higher degree of crystallinity of P3HT (Fig. 3a). Furthermore, the authors used atomic force microscopy (AFM) to investigate the surface morphology after removal of the Al cathode (Fig. 3b). The rougher surface indicated that adhesion between the active layer and the Al cathode was also enhanced. In other words, the quality of the polymer–Al contact improved after post-annealing, leading to decreased contact resistance. Combining these effects (higher P3HT crystallinity and lower contact resistance), the series resistance of the device decreased significantly, from 113 to 7.9 Ω cm^2, after post-annealing. Overall, the optimized post-annealing conditions resulted in an excellent PCE, approaching 5 %.

In addition to thermal annealing, Li et al. reported another important approach: "solvent annealing" [15, 16]. They controlled the growth rate of the active layer (P3HT/PCBM) from solution to the solid state, resulting in a decrease in series resistance (R_s) and an increase in optical absorption. The $J–V$ characteristics of devices prepared using various solvent evaporation times (t_{evp}) (Fig. 4a) revealed that the value of J_{sc} increased upon increasing the solvent evaporation time. Absorption spectra suggested that slower growth of the film led to increased absorption and a red shift (Fig. 4b). The pronounced vibronic shoulders for the slowly grown layer indicated a higher degree of polymer ordering. The authors concluded that self-organization originating from the slow growth of the active layer plays an important role in improving device efficiencies [15].

From the discussion above, we find that one of the keys to high device performance is improving the self-organization of the polymers. Other methods for achieving higher degrees of polymer ordering have also been proposed. For instance, Jin et al. found that using P3HT:PCBM blends dissolved in a cosolvent during device fabrication could improve the device efficiency [17]. Moulé et al. also revealed that the PCEs of devices could be improved after the addition of a high-boiling-point solvent (e.g., nitrobenzene) to the base solvent, chlorobenzene [18]. Furthermore, our group also developed a new solvent mixture system

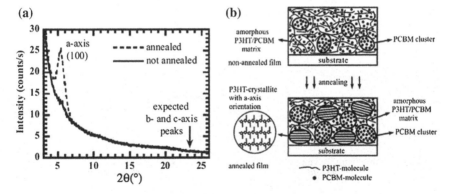

Fig. 2 **a** Grazing-incidence XRD patterns of P3HT:PCBM thin films before and after thermal annealing. **b** Graphical representation of the microscopic behavior in P3HT/PCBM blends during thermal annealing [13]

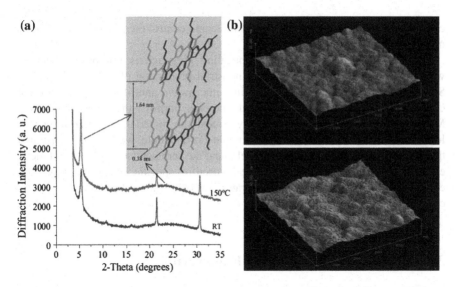

Fig. 3 **a** XRD diagram of P3HT/PCBM blends before and after thermal annealing at 150 °C for 30 min. Inset: P3HT crystalline structure. **b** AFM images of P3HT/PCBM blends annealed at 150 °C for 30 min after the removal of the Al cathode: (*top*) annealed prior to Al deposition; (*bottom*) annealed after Al deposition [14]

consisting of 1-chloronaphthalene (Cl-naph) and *o*-dichlorobenzene (DCB) [19]. Because the vapor pressure of Cl-naph is lower (0.029 mm Hg) than that of DCB (1.2 mm Hg), the addition of Cl-naph decreases the volatility of the solvent for the photoactive layer and extends the drying time of the polymer film. Therefore, the polymers have a longer time during which to undergo self-organization, thereby increasing the degree of crystallinity. Such approaches using solvent mixtures or

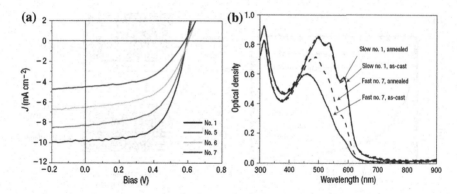

Fig. 4 Effects of the growth rates of the photoactive blends. **a** J–V characteristics under illumination at 100 mW cm^{-2} for devices prepared at solvent evaporation times (t_{evp}) of 20 min (No. 1), 3 min (No. 5), 40 s (No. 6), and 20 s (No. 7). **b** Absorption spectra for slowly (No. 1) and rapidly (No. 7) grown films [15]

cosolvents to slow down the evaporation rate and control the polymer morphology may be called "cosolvent annealing."

Another interesting, noncontact annealing approach is microwave heating, which might be suitable for efficient industrial production of OPV devices [20]. Similar to thermal and solvent annealing, microwave annealing can increase the crystallinity of the polymer layer and, thereby, improve the device efficiency. Moreover, this method can selectively anneal the material, minimizing energy loss, during the period of heating treatment. For example, we have demonstrated that microwave annealing can selectively heat the semiconducting layer and metal electrodes. Because the energy can be focused on specific layers, this rapid and energy-saving approach should be attractive for the future development of low-carbon products [20].

3 Electrode Modification and Engineering

The nature of the interfaces between the polymer active layer and the electrodes directly affects the efficiency of charge collection. Ideally, both the anode and cathode contacts should be ohmic (i.e., the contact resistances are negligible relative to the bulk resistance of the polymer materials). To achieve ohmic contacts, many functional interlayers have been inserted into the interface between the metals and the polymer films; for example, poly(3,4-ethylenedioxythiophene):polystyrenesulfonate (PEDOT:PSS) [21], metal oxides (V_2O_5 and MoO_3) [21], and graphene oxide [22] are efficient buffer layers at the anodic interface for collecting holes, while alkali metal complexes (LiF [23], CsF [24], and Cs_2CO_3 [25]), ZnO modified with self-assembled monolayers (SAMs) [26], and poly(ethylene oxide) (PEO) [27] are effective cathode interlayers for collecting

Fig. 5 **a** *J–V* characteristics of typical MDMO-PPV/PCBM solar cells incorporating LiF/Al electrodes of various LiF thicknesses, compared with the performance of the device featuring a pristine Al electrode. **b, c** Box plots of the FFs and values of Voc obtained from six separate solar cells [23]

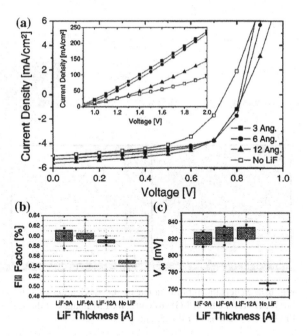

electrons efficiently. Figure 5 displays a classic example of the function of an interlayer: here, both the open-circuit voltage (V_{oc}) and fill factor (FF) of the OPV improved, yielding increased PCEs after the insertion of a layer of LiF [23]. This photoactive layer comprised PCBM and poly[2-methoxy,5-(3'/7'-dimethyloctyl-oxy)]-*p*-phenylenevinylene (MDMO-PPV); the FF increased by up to approximately 20 % relative to that of the reference cell featuring a pristine Al electrode. Together with a short-circuit current of 5.25 mA cm^{-2} and a value of V_{oc} of 0.825 V, the PCE reached 3.3 % under illumination with white light. The authors suggested that the formation of a dipole moment across the interface was the main mechanism behind the improved device performance [23].

The preparation of most of the above-mentioned interlayers required an additional step, the deposition of the thin film, thereby complicating the device fabrication process and possibly increasing the cost. More recently, our group conceptually demonstrated a novel method for interface modification toward preparing self-organized bilayer structures in OPVs [28, 29]. In this approach, only a single step is required to fabricate both the active and buffer layers, thereby simplifying the fabrication process. The buffer layer material we employed, poly(ethylene glycol) (PEG), is an insulating polymer. After chemical interaction with Al atoms, the PEG molecules formed an effective cathode contact with the photoactive layer. Figure 6a outlines the procedure used for the fabrication of an OPV featuring such a self-organized bilayer (P3HT:PCBM/PEG) structure. The PEG molecules were directly blended with P3HT and PCBM; the device was fabricated using a conventional spin-coating process. During the drying process,

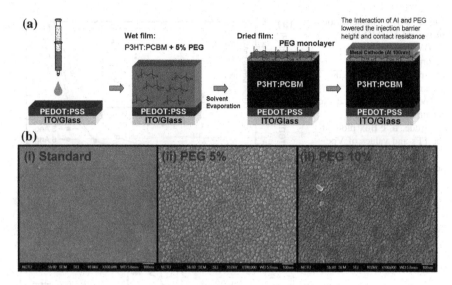

Fig. 6 **a** Schematic representation of the fabrication of solar cells featuring a self-organized bilayer (P3HT:PCBM/PEG) structure. **b** SEM images of the surfaces of active layers prepared at PEG concentrations of *i* 0 *ii* 5, and *iii* 10 % [28]

PEG spontaneously migrated to the surface of the active layer. As a result, a nanoscale functional interlayer was created. Scanning electron microscopy (SEM) revealed morphological evidence supporting the formation of vertically-phase-separated PEG molecules (Fig. 6b). The corresponding thin film containing no PEG exhibited a smooth surface morphology. After blending with 5 or 10 % PEG, we clearly observed additional "dot-like" phases on the surfaces of the films. We assigned this new phase to assemblies of PEG molecules. In terms of device performance, the addition of 5 wt% PEG into the active layer improved the PCE from 2.21 to 3.97 %. Furthermore, OPVs fabricated using this method also exhibited superior device stability under illumination, presumably because a PEG interface might have lower sensitivity toward moisture and oxygen from the atmosphere. In addition, because the PEG molecules preferred to segregate to the top of the polymer blend, interdiffusion through the contact was inhibited, forming a thermodynamically stable interface [28].

4 Light Trapping Technologies

As we stated in the introduction to this chapter, it is difficult to improve the absorption efficiency and charge collection efficiency simultaneously. Therefore, many light trapping strategies have been proposed to increase the absorption efficiency without affecting the charge collection efficiency [30–43]. For example, ray optics approaches, such as the use of folded device architectures [30, 31], microlenses [32], antireflection (AR) coatings [33], and collector mirrors [34], have been proposed to effectively improve the light harvesting efficiency. Nanostructures, including photonic crystals [35, 36] and metallic structures for triggering surface plasmons (SP) [37, 38], have also been adopted in OPVs to strength their light absorption ability. In the following section, we focus on the two most common methods—optical spacers and SPs—developed for OPVs.

4.1 Optical Spacers

Typically, the incorporation of an optical spacer will redistribute the optical electrical field ($|E|^2$) in a thin film device [39–42]. As indicated in Fig. 7a, the maximum optical electric field distribution inside a conventional device is located near the glass–indium tin oxide (ITO) interface, because of the interference effect between the incoming photons and the reflected ones by the Al electrode. As a result, the exciton generation rate, which is related directly to the optical electric field of the photoactive layer, decreases. Furthermore, excitons produced near the ITO/PEDOT:PSS surface may be quenched by the electrode [39, 41]. To overcome these problems, Kim et al. introduced a layer of optical spacer between the polymer layer and the Al electrode [39]. This optical spacer, titanium oxide (TiO_x), was deposited, using a solution-based sol–gel process, on top of the photoactive layer. After incorporating the TiO_x layer, the optical-electric field changed inside the photoactive layer, with the maximum strength of the field distribution tuned to fall into the photoactive layer (Fig. 7a). The IPCE spectra of these devices exhibited significant enhancements (ca. 40 %) over the entire spectral range (Fig. 7b), attributable to the increased number of photogenerated charge carriers that resulted from the optimized redistribution of the light intensity. Figure 7c displays the device performance before and after incorporation of the optical spacers. The conventional device exhibited a value of J_{sc} of 7.5 mA cm^{-2}, a value of V_{oc} of 0.51 V, and a FF of 0.54, resulting in a PCE of 2.3 %. The device featuring a TiO_x layer achieved a much higher PCE of 5.0 % ($J_{sc} = 11.1$ mA cm^{-2}; $V_{oc} = 0.61$ V; FF $= 0.66$), suggesting a promising future for optical spacers [39].

In addition to TiO_x, Gilot et al. investigated the insertion of a layer of ZnO between the active layer and the reflective electrode as the optical spacer [40]. When the thickness of the P3HT:PCBM layer was 40 nm, insertion of a 39-nm-thick ZnO layer shifted the position of the maximum into the photon-absorbing layer, suggesting that a higher photocurrent would be obtained when using this optical spacer. The authors found, however, that cells made with a larger film thickness (70–130 nm) exhibited contrasting results. They speculated that the active layer was already optimized; the incorporation of an optical spacer shifted the maximum electric field intensity away from the active layer. From both experimental and calculation results, Gilot et al. concluded that the absorption efficiency could be enhanced by using optical spacers only if the thickness of the

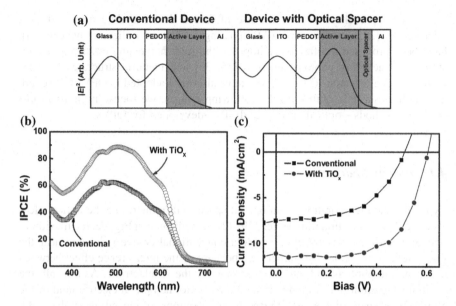

Fig. 7 a Schematic representation of the spatial distributions of the optical electric field strengths $|E|^2$ inside devices prepared with (*top*) and without (*down*) an optical spacer. **b** Incident monochromatic photon-to-current collection efficiency (IPCE) spectra for devices prepared with and without a TiO_x optical spacer layer. **c** Current density–voltage characteristics of polymer solar cells prepared with (*circles*) and without (*squares*) a TiO_x optical spacer, recorded under AM1.5 illumination from a calibrated solar simulator (90 mW cm^{-2}) [39]

P3HT:PCBM layer was less than approximately 60 nm. Similarly, Andersson et al. performed simulations, using the transfer matrix method and the finite element method for OPVs prepared with various band gaps. [42]. They found that no beneficial effect could be expected when incorporating optical spacers in devices prepared with an already-optimized active layer thickness. The effect of an optical spacer depends strongly on the thickness of the photoactive layer. In other words, optical spacers can function well only when an optically optimized thickness of the polymer layer cannot be achieved. For example, the mobilities of the charge carriers might limit the use of thick films for certain polymer systems. OPVs fabricated with these materials might result in lower charge collection efficiencies. In these cases, optical spacers can be expected to enhance the absorption efficiency.

Another beneficial feature of an optical spacer is that redistribution of the optical electrical field can decrease the level of exciton quenching near the electrodes [39, 41]. For example, our group used ITO as an optical spacer to improve the efficiency of inverted polymer solar cells [41]. The thickness of the P3HT:PCBM layer had been optimized in this case; the best thickness of the active layer was 180 nm. Figure 8a presents the electrical characteristics of inverted devices incorporating ITO spacers of various thicknesses; the inset displays the device structure of the OPV. Because ITO has high electrical conductivity and high transparency, it appears to be a suitable candidate for use as an optical spacer.

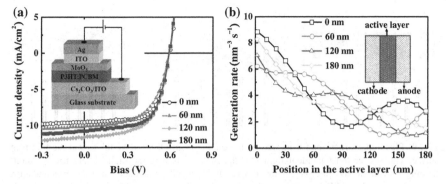

Fig. 8 a Electrical characteristics, recorded under AM 1.5G illumination (100 mW cm^{-2}), of inverted OPVs incorporating ITO optical spacers of various thicknesses; inset: device architecture of an OPV incorporating an ITO optical spacer. **b** Calculated distribution profiles of the exciton generation rate within the active layer for OPV devices incorporating optical spacers of various thicknesses; inset: schematic representation of the layer stack [41]

Unfortunately, the work function of ITO (ca. 4.7 eV) was somehow misaligned with the HOMO energy levels of the polymers, imposing an energy barrier for hole collection at the electrodes. Therefore, we incorporated a layer of MoO$_3$, which has a high work function (ca. 5.3 eV), to decrease the contact resistance. The reference device exhibited a value of V_{oc} of 0.59 V, a value of J_{sc} of 9.54 mA cm^{-2}, and a FF of 0.67, yielding a PCE of 3.76 %. The value of V_{oc} of the device remained at 0.59 V after incorporating the optical spacer. By tuning the thickness of the ITO layer, the value of J_{sc} increased to 11.49 mA cm^{-2}. Although the FF decreased slightly to 0.62, presumably due to the increased resistance arising from the presence of ITO and/or possible sputtering damage, its effect was overwhelmed by the much higher photocurrent. Overall, the PCE improved to 4.20 %. To understand the mechanism responsible for the enhanced device performance, we calculated the ideal exciton generation rate within the active layer (Fig. 8b). After integrating the area beneath the curves, we found that the optical spacers failed to increase the total number of excitons, presumably due to the film's sufficient thickness, which had been optimized. When the ITO thickness was 120 nm, however, we could still successfully shift the exciton generation zone away from the electrodes and diminish any possible quenching process at the electrodes, thereby increasing the photocurrent in real devices.

4.2 Surface Plasmonic Effects

Surface plasmons are confined electromagnetic waves propagating along the surface of a conductor [43–55]; they have many unique properties, including local field enhancement and strong light scattering, which might improve the absorption process in OPVs. Plasmonic structures for enhancing OPV performance can be

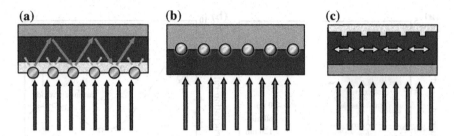

Fig. 9 Plasmonic structures for increasing the efficiency of OPVs. **a** Metal NPs at the thin film surface can scatter most incident light into the material having the higher dielectric constant, helping to confine the photons in the device. **b** Metal NPs embedded in the semiconductor materials; the induced SPs can enhance the near-field in the structure. **c** A periodic structure at the back metal electrode induces SPPs; this structure can turn the incident photons by 90° [44]

generally divided into three categories (Fig. 9) [44]. In the first case, metal nano-particles (NPs; e.g., Cu, Ag, Pt, or Au NPs), which can trigger localized surface plasmon resonance (LSPR), are placed in front of the active thin film in the OPV (Fig. 9a). In contrast to light scattering in a homogeneous medium, light will be scattered preferentially into the material having the larger permittivity. Subse-quently, through multiple and/or high-angle scattering, the overall optical path length increases, thereby improving the absorption efficiency of the OPV. Figure 9b displays the second possible structure, in which the NPs are embedded directly into the organic semiconducting layer [44]. Because the absorption is proportional to the intensity of the electromagnetic field, the enhanced near-field induced by the SPs can increase the absorption efficiency [45]. In the third case, surface plasmon polaritons (SPPs) propagating at the metal–semiconductor interface are excited by metal nanostructures (e.g., periodic arrays or gratings) (Fig. 9c) and the induced evanescent electromagnetic fields are confined near the interface. Therefore, light is turned by 90° in such a structure. Because photons are absorbed along the lateral direction, the optical length of which is several orders of magnitude longer than the thickness of the semiconductor layer, the absorption efficiency can be increased significantly.

Many remarkable plasmonic approaches have been proposed to increase the light absorption efficiency of OPVs. For example, in 2008, Kim et al. fabricated Ag NPs through pulse-current (PC) electrodeposition onto ITO substrates (Fig. 10a) [46]. This electrochemical approach was used to control the size, density, and morphology of the NP films. The authors fabricated relatively uniform Ag NPs having an average particle size of approximately 13 nm (inset to Fig. 10a). As a result, the device PCE improved from 3.05 to 3.69 % after incorporation of these Ag NPs. They attributed the increased photocurrent to the SPR effect induced by the metal NPs. Similarly, Morfa et al. deposited Ag NPs through conventional thermal evaporation on ITO-coated glass substrates; a layer of PEDOT:PSS was subsequently deposited onto the NPs [47]. Figure 10b dis-plays the device structure. The PCE of the device increased from 1.3 ± 0.2 to

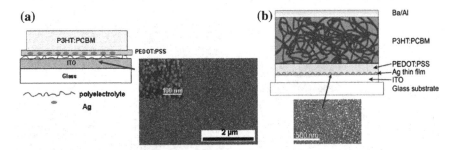

Fig. 10 Device structures of OPVs featuring Ag NPs. **a** The NPs were fabricated using an electrochemical method, which controlled their the size and density well [46]. **b** Structure of a device incorporating a thin Ag film deposited onto an ITO-coated glass substrate. Inset: Field-emission SEM micrograph of a representative 2-nm-thick Ag layer on ITO [47]

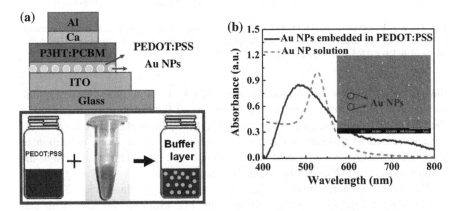

Fig. 11 **a** Structure of a device featuring Ag NPs and the method of preparation of the buffer solution containing Au NPs. **b** Absorption spectra of a Au NP solution and of Au NPs embedded in PEDOT:PSS. Inset: SEM image of a PEDOT:PSS film prepared with Au NPs blended in the matrix, revealing the uniform distribution of the Au NPs (*white dots*) in the PEDOT:PSS layer [48, 49]

2.2 ± 0.1 % after incorporating a 1- or 2-nm-thick layer of plasmon-active Ag NPs. The PCE decreased slightly upon increasing the size of the Ag NPs.

In addition to Ag NPs, gold (Au), another noble metal, is also a promising candidate for inducing LSPR phenomena. For example, our group has reported a solution-processable approach for incorporating Au NPs into OPVs; this process involved the simple blending of Au NPs and PEDOT:PSS solutions (Fig. 11a) [48, 49]. A plasmonic layer, consisting of Au NPs and PEDOT:PSS, was readily formed after spin-coating of the mixture. The size of the NPs (ca. 40 nm) was selected intentionally so that their SP peak matched the absorption wavelength of P3HT. The UV–Vis spectrum in Fig. 11b reveals that the resonance peak of the Au NPs in solution appeared near 550 nm. The standard device prepared without Au NPs exhibited a PCE of 3.57 %. The value of J_{sc} increased after incorporation of

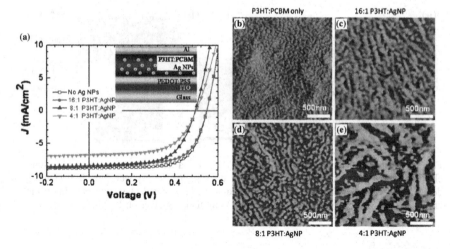

Fig. 12 **a** *J–V* curves of P3HT:PCBM solar cells incorporating Ag NPs at various concentrations; inset: device structure. **b–e** Surface morphologies of active layers containing P3HT:AgNP ratios of **b** 1:0 **c** 16:1 **d** 8:1, and **e** 4:1 [51]

the Au NPs, thereby leading to an improved PCE of 4.29 %. Two possible mechanisms might be responsible for the enhanced light absorption efficiency of the OPVs [49]. As illustrated in Fig. 9a, the optical path in the active layer might have increased as a result of the forward scattering; alternatively, excitation of the LSPR might have led to local enhancement of the electromagnetic field in the vicinity of the Au NPs, thereby increasing the absorption efficiency (Fig. 9b).

The second scenario involves the direct placement of plasmonic NPs in the semiconductor layers, as outlined in Fig. 9b. For example, in 2004 Kim and Carroll doped Ag and Au NPs directly into the photoactive layer of BHJ OPVs [50]. Although the device efficiency was improved, the authors suggested that the dominant mechanism behind the enhanced efficiency was the improved electrical conductivity. In 2010, Xue et al. also incorporated Ag NPs into the P3HT:PCBM layer; although their device efficiency did not improve (Fig. 12a) [51], they found that the mobility of the active layer increased, while the total number of extracted carriers decreased. The surface morphology in Fig. 12b–e reveals that the Ag NPs tended to phase-segregate from the organic materials, leading to the formation of a Ag NP sub-network, which was presumably responsible for the increase in mobility. On the other hand, charge trapping in the sub-network could enhance the recombination probability, thereby decreasing the degree of charge extraction.

More recently, Sha et al. developed a rigorous electrodynamic approach to investigate optical absorption in OPVs [52]. They found remarkable differences between structures in which the metal NPs had been placed at the interlayer (i.e., between the photoactive layer and the anode) and those embedded directly within the photoactive layer. Theoretical results suggested that the enhancement factors for the latter were generally greater than those of the former. In other words, direct

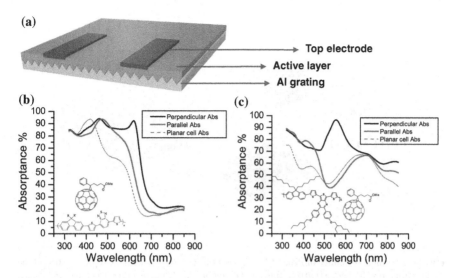

Fig. 13 **a** Periodic grating structure for propagating SSPs; grating period and height: 277 and 50 nm, respectively. **b, c** Absorption spectra of the polymer blends **b** APFO3/PCBM and **c** APFO Green5/PCBM on the grating structures, recorded while illuminating at different polarization directions [53]

contact of metal NPs with the semiconducting layer should be the most effective way to improve the absorption process in OPVs. Nevertheless, this method still encounters many problems in reality. For most of the studies reported so far, no apparent plasmonic effect has been observed, presumably because of serious phase separation between the nanostructures and the organic materials.

The third scenario is the adoption of an SPP structure (Fig. 9c). Tvingstedt et al. reported one of the earliest examples in 2007. They fabricated periodic nano-structures to induce SPPs through an Al grating; Fig. 13a presents the device structure [53]. The plasmonic grating also served as the bottom cathode; a PEDOT: PSS layer functioned as the top anode. The authors employed two different polymer blends (APFO3:PCBM and APFO Green5:PCBM) as photoactive materials. Figure 13b, c display the corresponding measured absorption spectra. When compared with the performance of the planar sample, the absorption profiles for the systems incorporating both materials were altered by the SPP nanostructures. More importantly, different spectra could be obtained when the samples were illuminated with differently directed polarized light. The authors found that the transverse electric (TE) polarized electromagnetic wave could not excite the SPPs. IPCE measurements of the device properties revealed an apparent influence of the polarization direction on the spectra, suggesting that the SPPs could indeed enhance the photocurrent.

More recently, Li et al. combined two different plasmonic nanostructures to improve the efficiencies of inverted OPVs [54]. Figure 14 displays the device structures and the chemical structures of the photoactive materials.

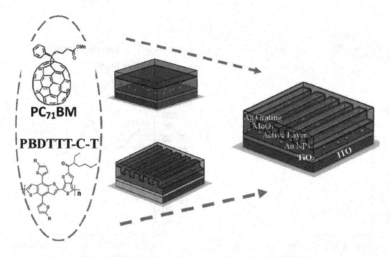

Fig. 14 Chemical structures of PBDTTT-C-T and PC$_{71}$BM (*left*). Schematic representation of the device structure: NP device (*top*), grating device (*bottom*), and dual metallic structures (*right*) [54]

Poly{[4,8-bis(2-ethylhexylthien-5-yl)benzo[1,2-*b*:4,5-*b'*]dithien-2,6-diyl]-*alt*-[2-(2'-ethylhexanoyl)thieno[3,4-*b*]thien-4,6-diyl]} (PBDTTT-C-T), a polymer having a low band gap, was used as the *p*-type polymer. The flat (control) device fabricated without any nanostructures exhibited a PCE of 7.59 ± 0.08 %. The authors used vacuum-assisted nanoimprinting at room temperature to form the nanostructure of the Ag grating. Under the optimized conditions, the PCE increased to 8.38 ± 0.20 %. Furthermore, when Au NPs were added into the active layer, the PCE increased to 8.79 ± 0.15 %. Li et al. suggested that the Au NPs offered enhanced absorption in the region 480–600 nm, whereas the Ag grating had a greater impact in the absorption regions below 400 and above 600 nm. Their study provided a general method for achieving enhanced broadband absorption [54].

Nevertheless, plasmonic-enhanced OPVs incorporating periodic nanostructures and exhibiting pronounced enhancement remain rarely reported. In addition, the one-dimensional (1-D) grating structures often exhibit high polarization-dependence, making it quite difficult to simultaneously optimize both polarization modes. More recently, two-dimensional (2-D) structures possessing higher-order symmetries have been proposed, potentially overcoming the polarization dependence [55]. In the near future, we foresee the fabrication of many more periodic nanostructures that will effectively improve device efficiencies.

5 Low-Band-Gap Materials

Organic materials usually absorb only a limited amount of the solar spectrum. For example, Fig. 15 displays the spectral response (SR) of the well-known P3HT:PCBM polymer blend. The response range is limited at approximately

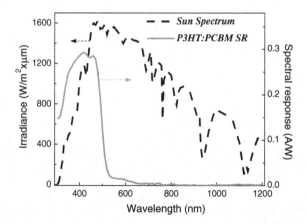

Fig. 15 Solar spectrum (*dash line*) and SR of the P3HT:PCBM polymer blend. Because of the limited absorption range of the polymeric materials, the SR was minimal beyond 650 nm

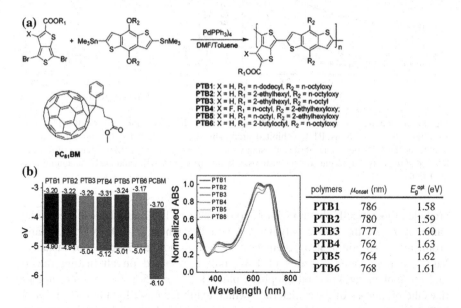

Fig. 16 **a** Molecular structures of polymers in the PTB family; **b** Onset absorptions (μ_{onset}) and relative energy levels of these polymers [58]

650 nm, corresponding to a band gap energy (E_g) of approximately 1.9 eV. In other words, the P3HT:PCBM polymer blend can harvest only approximately 22 % of all available photons from solar irradiance [56]. Therefore, to ensure the harvesting of more photons, an important task is extending the absorption region.

The development of low-band-gap (LBG) polymers is a promising approach toward harvesting long-wavelength photons from solar irradiation. Brabec et al.

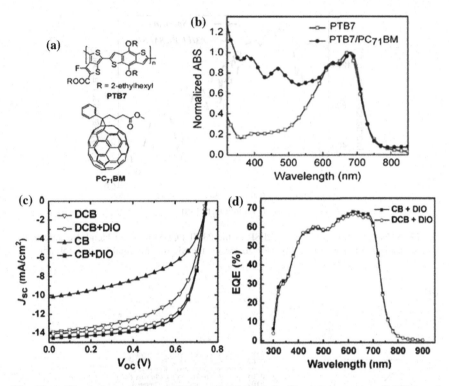

Fig. 17 **a** Chemical structures of PTB7 and PC$_{71}$BM. **b** Absorption spectra of PTB7 and PTB7:PC$_{71}$BM blend films; PTB7 exhibited strong absorbance from 550 to 750 nm. **c** J–V curves of films of the LBG polymer, fabricated using various solvent systems. **d** EQE spectra of devices fabricated using two different solvent systems: CB containing 3 % DIO and DCB containing 3 % DIO [59]

predicted theoretically that a PCE of approximately 11 % could be achieved if the value of E_g of the donor polymer were in the range 1.2–1.8 eV [57]. Many promising LBG conjugated polymers have been reported [58–61]. For instance, Yu's group prepared a series of LBG polymers based on alternating thieno [3,4-*b*]thiophene and benzodithiophene units (Fig. 16a) [58]. Figure 16b summarizes the properties of these materials, which exhibited onset light absorbencies of 762–786 nm and the relative values of E_g of 1.58–1.63 eV. After these new polymers were blended with PC$_{61}$BM to serve as active layers for OPVs, a device incorporating PTB4 exhibited the best performance, with a PCE of 5.9 %, as a result of this polymer possessing the highest hole mobility (7.7×10^{-4} cm^2 V^{-1} s^{-1}) among the tested polymers [58].

The Yu research team prepared a new PTB family member, the polymer PTB7, that provided a device exhibiting a record high PCE of 7.4 % [59]. The structure of PTB7 (Fig. 17a) featured branched side chains on its ester and benzodithiophene units to improve its solubility in organic solvents. The mobility in a film of this polymer was approximately 5.8×10^{-4} cm^2 V^{-1} s^{-1}; strong absorption occurred from 550 to 750 nm (Fig. 17b). To compensate for the absorption of PTB7, Yu et al. employed PC$_{71}$BM as the acceptor. The resulting absorption spectrum (Fig. 17b) revealed that the PTB7:PC$_{71}$BM blend absorbed broadly from 300 nm to approximately 800 nm. The authors further controlled the morphology of the thin film by altering the solvent system used to dissolve the polymer blends (Fig. 17c). The device prepared when employing 3 % 1,8-diiodoctane (DIO) in chlorobenzene (CB) as the solvent exhibited a significantly enhanced value of J_{sc} of 14.5 mA cm^{-2}, resulting in a high efficiency of 7.40 % [59].

6 Conclusion and Outlook

The research field of polymer solar cells continues to expand. Tremendous progress has been made recently on the design of new materials, the development of novel device structures, and other aspects. In this chapter, we have reviewed several important methods for enhancing device efficiency, including controlling the morphologies of polymer blends, interfacial engineering of polymer–electrode interfaces, and optical approaches toward increasing the light harvesting ability. We have also briefly noted the development of new LBG polymers. In theory, a single-junction BHJ device can achieve a maximum PCE of approximately 11 % [57]. With a recently reported PCE of approximately 9 %, we believe that single-junction OPVs with such high PCEs should appear in the near future. To achieve even higher efficiency, Dennler et al. indicated that a PCE of approximately 15 % might be possible through the development of multi-junction OPVs [62]. More recently, Janssen and Nelson suggested that limits of 20–24 % might even be reachable for single-junction OPVs based on a modified balance model; such high PCEs are similar to the highest efficiencies obtained using crystalline Si technologies [63]. The materials described herein suggest that polymer solar cells have great potential. With their additional advantageous features similar to those of plastics, including light weight and high flexibility, we foresee the commercialization of practical, large-area OPV modules in the near future.

Acknowledgment We thank the National Science Council of Taiwan and the Ministry of Education of Taiwan (through the ATU program) for financial support.

References

1. Reineke S, Lindner F, Schwartz G, Seidler N, Walzer K, Lüssem B, Leo K (2009) Nature 459:234
2. Street RA (2009) Adv Mater 21:2007
3. Capelli R, Toffanin S, Generali G, Usta H, Facchetti A, Muccini M (2010) Nat Mater 9:496
4. Clark J, Lanzani G (2010) Nat Photonics 4:438

5. R. F. Service (2011) Science 332:293
6. Kippelen B, Bredas J-L (2009) Energy Environ Sci 2:251
7. Bredas J-L, Norton JE, Cornil J, Coropceanu V (2009) Acc Chem Res 42:1691
8. Li G, Zhu R, Yang Y (2012) Nat Photonics 6:153
9. Roes AL, Alsema EA, Blok K, Patel MK (2009) Prog Photovoltaics Res Appl 17:372
10. Moliton A, Nunzi JM (2006) Polym Int 55:583
11. Padinger F, Rittberger RS, Sariciftci NS (2003) Adv Funct Mater 13:85
12. Chen FC, Lin YK, Ko CJ (2008) Appl Phys Lett 92:023307
13. Erb T, Zhokhavets U, Gobsch G, Raleva S, Stuhn B, Schilinsky P, Waldauf C, Brabec CJ (2005) Adv Funct Mater 15:1193
14. Ma W, Yang C, Gong X, Lee K, Heeger AJ (2005) Adv Funct Mater 15:1617
15. Li G, Shrotriya V, Huang JS, Yao Y, Moriarty T, Emery K, Yang Y (2005) Nat Mater 4:864
16. Chen FC, Ko CJ, Wu JL, Chen WC (2010) Sol Energy Mater Sol Cells 94:2426
17. Jin SH, Naidu BVK, Jeon HS, Park SM, Park JS, Km SC, Lee JW, Gal YS (2007) Sol Energy Mater Sol Cells 91:1187
18. Moule AJ, Meerholz K (2008) Adv Mater 20:240
19. Chen FC, Tseng HC, Ko CJ (2008) Appl Phys Lett 92:103316
20. Ko CJ, Lin YK, Chen FC (2007) Adv Mater 19:3520
21. Yip HL, Jen AKY (2012) Energy Environ Sci 5:5994
22. Li SS, Tu KH, Lin CC, Chen CW, Chhowalla M (2010) ACS Nano 4:3169
23. Brabec CJ, Shaheen SE, Winder C, Sariciftci NS, Denk P (2002) Appl Phys Lett 80:1288
24. Jiang X, Xu H, Yang L, Shi M, Wang M, Chen H (2009) Sol Energy Mater Sol Cells 93:650
25. Chen FC, Wu JL, Yang SS, Hsieh KH, Chen WC (2008) J Appl Phys 103:103721
26. Yip HL, Hau SK, Baek NS, Jen AKY (2008) Appl Phys Lett 92:193313
27. Zhang F, Ceder M, Inganäs O (2007) Adv Mater 19:1835
28. Chen FC, Chien SC (2009) J Mater Chem 19:6865
29. Chien SC, Chen FC, Chung MK, Hsu CS (2012) J Phys Chem C 116:1354
30. Zhou YH, Zhang FL, Tvingstedt K, Tian WJ, Inganäs O (2008) Appl Phys Lett 93:033302
31. Tvingstedt K, Andersson V, Zhang F, Inganas O (2007) Appl Phys Lett 91:123514
32. Tvingstedt K, Zilio SD, Inganäs O, Tormen M (2008) Opt Express 16:21608
33. Forberich J, Dennler G, Scharber MC, Hingerl K, Fromherz T, Braber CJ (2008) Thin Solid Films 516:7167
34. Peumans P, Bulovic V, Forrest SR (2000) Appl Phys Lett 76:2650
35. Tumbleston JR, Ko DH, Samulski ET, Lopez R (2009) Opt Express 17:7670
36. Ko DH, Tumbleston JR, Zhang L, Williams S, DeSimone JM, Lopez R, Samulski ET (2009) Nano Lett 9:2742
37. Kang MG, Xu T, Park HJ, Luo XG, Guo LJ (2010) Adv Mater 22:4378
38. Lindquist NC, Luhman WA, Oh SH, Holmes RJ (2008) Appl Phys Lett 93:123308
39. Kim JY, Kim SH, Lee HH, Lee K, Ma WL, Gong X, Heeger AJ (2006) Adv Mater 18:572
40. Gilot J, Barbu I, Wienk MM, Janssen RAJ (2007) Appl Phys Lett 91:113520
41. Chen FC, Wu JL, Hong Y (2010) Appl Phys Lett 96:193304
42. Andersson BV, Huang DM, Moulé AJ, Inganäs O (2009) Appl Phys Lett 94:043302
43. Maier SA, Atwater HA (2005) J Appl Phys 98:011101
44. Atwater HA, Polman A (2010) Nat Mater 9:205
45. Ferry VE, Munday JN, Atwater HA (2010) Adv Mater 22:4794
46. Kim SS, Na SI, Jo J, Kim DY, Nah YC (2008) Appl Phys Lett 93:073307
47. Morfa AJ, Rowlen KL, Reilly TH, Romero MJ, van de Lagemaat J (2008) Appl Phys Lett 92:013504
48. Chen FC, Wu JL, Lee CL, Hong Y, Kuo CH, Huang MH (2009) Appl Phys Lett 95:013305
49. Wu JL, Chen FC, Hsiao YS, Chien FC, Chen P, Kuo CH, Huang MH, Hsu CS (2011) ACS Nano 5:959
50. Kim K, Carroll DL (2005) Appl Phys Lett 87:203113
51. Xue M, Li L, Tremolet de Villers BJ, Shen H, Zhu J, Yu Z, Stieg AZ, Pei Q, Schwartz BJ, Wang KL (2011) Appl Phys Lett 98:253302

52. Sha WEI, Choy WCH, Liu YG, Chew WC (2011) Appl Phys Lett 99:113304
53. Tvingstedt K, Persson NK, Inganas O, Rahachou A, Zozoulenko IV (2007) Appl Phys Lett 91:113514
54. Li X, Choy WCH, Hou L, Xie F, Sha WEI, Ding B, Guo X, Li Y, Hou J, You J, Yang Y (2012) Adv Mater 22:3046
55. Sefunc MA, Okyay AK, Demir HV (2011) Opt Express 19:14200
56. Bundgaard E, Krebs FC (2007) Sol Energy Mater Sol Cells 91:954
57. Scharber MC, Mühlbacher D, Koppe M, Denk P, Waldauf C, Heeger AJ, Brabec CJ (2006) Adv Mater 18:789
58. Liang Y, Feng D, Wu Y, Tsai ST, Li G, Ray C, Yu L (2009) J Am Chem Soc 131:7792
59. Liang Y, Xu Z, Xia J, Tsai ST, Wu Y, Li G, Ray C, Yu L (2010) Adv Mater 22:E135
60. Chen HY, Hou J, Zhang S, Liang Y, Yang G, Yang Y, Yu L, Wu Y, Li G (2009) Nat Photonics 3:649
61. Huo L, Zhang S, Guo X, Xu F, Li Y, Hou J (2011) Angew Chem Int Ed 50:9697
62. Dennler G, Scharber MC, Ameri T, Denk P, Forberich K, Waldauf C, Brabec CJ (2008) Adv Mater 20:579
63. Janssen RAJ, Nelson J (2013) Adv Mater 25:1847. DOI: 10.1002/adma.201202873

Indium Tin Oxide-Free Polymer Solar Cells: Toward Commercial Reality

Dechan Angmo, Nieves Espinosa and Frederik Krebs

Abstract Polymer solar cell (PSC) is the latest of all photovoltaic technologies which currently lies at the brink of commercialization. The impetus for its rapid progress in the last decade has come from low-cost high throughput production possibility which in turn relies on the use of low-cost materials and vacuum-free manufacture. Indium tin oxide (ITO), the commonly used transparent conductor, imposes the majority of the cost of production of PSCs, limits flexibility, and is feared to create bottleneck in the dawning industry due to indium scarcity and the resulting large price fluctuations. As such, finding a low-cost replacement of ITO is widely identified to be very crucial for the commercial feasibility of PSCs. In this regard, a variety of nanomaterials have shown remarkable potential matching up to and sometimes even surpassing the properties of ITO. This chapter elaborates the recent developments in ITO replacement which include, but are not limited to, the use of nanomaterials such as metal nanogrids, metal nanowires, carbon nanotubes, and graphene. The use of polymers and metals as replacement to ITO is introduced as well. Finally, recent progress in large-scale experiments on ITO-free PSC modules is also presented.

1 Introduction

The impacts of global warming are increasingly becoming evident in the form of intensifying weather calamities, disappearing glaciers, and increasing water levels in the oceans. Such a scenario has the world gearing toward "green" energy technologies. Photovoltaic cells or solar cells convert sunlight directly into electricity and is one of the many classes of green technologies. Silicon-based solar

D. Angmo · N. Espinosa · F. Krebs (✉)
DTU Energy Conversion and Storage Department, Frederiksborgvej,
399 4000, Roskilde, Denmark
e-mail: frkr@dtu.dk

Z. Lin and J. Wang (eds.), *Low-cost Nanomaterials*, Green Energy and Technology,
DOI: 10.1007/978-1-4471-6473-9_8, © Springer-Verlag London 2014

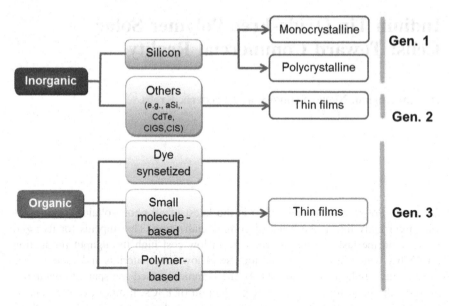

Fig. 1 A broad classification of photovoltaic technologies

cells is the most known and the oldest photovoltaic technology; there are, however, a plethora of non-silicon-based solar cells in the market or under development today. Solar cells can be generally divided into two broad categories depending on the type of the photoactive material: inorganic and organic solar cells (Fig. 1). Among the inorganic solar cells are the first generation solar cells based on silicon and the second generation solar cells based on thin films of amorphous silicon (a-Si) and on chalcogenides such as CdSe, CIGS, CdTe, etc. Organic solar cells form the third generation of photovoltaic technology. Organic solar cells can be further demarcated into three subdivisions: dye sensitized, small molecules, and polymer solar cells. Polymer solar cells (PSCs) are the most recent technology and this chapter is focused on PSCs.

The environmental impact of a photovoltaic technology (or any other green technology) can be assessed using an indicator derived from life cycle analyses, namely, energy payback time (EPBT). EPBT is the time it takes for a solar cell to generate the same amount of energy as expended in its manufacture, use, and final decommission. The lower the EPBT, the faster is the contribution to clean energy production. EPBT has been thoroughly investigated for all PV technologies and is compiled in Fig. 2 from the selected literature to present a comparison of some photovoltaic technologies [1, 2].

EPBT can simultaneously serve as an indicator for economic profitability of solar cells. The lowering of EPBT of solar cells requires either a drastic reduction in materials and manufacturing costs or a dramatic increase in the efficiency of the solar cells, or a combination thereof. Such measures ultimately reduce the cost of the electrical energy produced by the solar cells, hence, making the technology

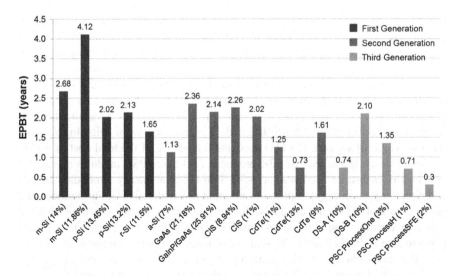

Fig. 2 Energy payback time for PV modules technologies. South Mediterranean irradiance (1,700 kWh/m²/year) and a performance ratio of 0.8 are assumed. The module efficiency is shown in brackets. DS stands for dye sensitized modules and A and B refer to the low and high values of the energy range. Process One is ITO-based PSCs, Process H is ITO-free PSCs, SFE is metal- and ITO-free PSCs

more economically competitive. Since most mature solar cells already operate at or near theoretically permissible efficiencies, the need for cost reduction in materials and processing has driven the development of solar cell technology from the first to the third generation.

The first generation of monocrystalline silicon is grown in the form of ingots via the Czochralski process that requires Si to be in molten form which is achieved at >1400 °C. The EPBT of >2 years for such cells despite its high efficiency and lifetime clearly indicates the relatively exorbitant energy investment that the first generation PV requires in comparison to other technologies (Fig. 2). The need to compensate such a slow and high energy input processing led to the development of second generation solar cells. The latter could be produced faster using techniques such as sputtering and chemical vapor deposition routes and employing cheaper materials albeit at a cost to performance.

The third generation of organic solar cells is pursued, despite the relatively lower expected efficiency and lifetime in comparison to the former generations, primarily because of their potential at drastic reduction in materials and production cost. Such a potential is envisioned accomplishable through high throughput roll-to-roll production with coating and printing at low temperatures that common plastic substrates such as PET can withstand (<140 °C). The initial experiments on large-area indium tin oxide (ITO)-based PSC modules produced by complete roll-to-roll production [3, 4] and their subsequent cost analyses [1] revealed that the use of ITO is not feasible in the low-cost production of PSCs. ITO is a commonly used transparent conductor directly adapted from other optoelectronic applications in general and inorganic solar cells in particular. It has high conductivity (sheet resistance of 10–20 Ω Υ^{-1}) and transmission (>80 %) in the visible region of the electromagnetic spectrum. However, the processing of ITO requires high preparation temperatures and vacuum-based highly energy-intensive and throughput limiting deposition techniques such as sputtering while all other components in PSCs can be coated and printed in ambient conditions. As a result, the cost footprint of ITO in PSCs is much larger than in inorganic solar cells. In fact, cost analyses [3, 5, 6] suggest that ITO accounts for >50 % of the total cost of a PSC module. Life cycle analyses of R2R-produced ITO-based PSC modules reveal that ITO on PET substrate accounts for ~90 % of the total energy (embodied energy and direct process energy) imposed by all input raw materials [1]. Furthermore, the brittle nature of ITO is not conducive for application on flexible substrates that are prone to bending and flexing and may inevitably lead to deterioration in the performance of PSCs. Lastly, the scarcity of indium resources in the world and its high demand from the display industry has created large cost fluctuations and future supply concerns. All these factors substantiate the need for finding a cost-effective alternative to ITO that ideally involves no challenges with scarcity and can be processed in a vacuum-free environment. Such an alternative would significantly improve the environmental and commercial feasibility of PSC technology.

Following a brief overview on PSCs, this chapter lays out the different alternatives of ITO that are widely researched. Each alternative is introduced along with the current development in their application in PSCs. Particular emphasis is laid on the implications of the processing choices of fabrication of ITO alternatives on the solar cell performance and suitability for low-cost upscaling. At the end of the chapter, a subsection is presented with recent progress on ITO-free roll-to-roll processed PSC modules.

2 Polymer Solar Cells: A Brief Overview

2.1 Device Structure

A polymer solar cell comprises of a thin film of the photoactive material sandwiched between two electrodes of different work functions. One of the electrodes is a transparent conductor from where light is permitted to the photoactive layer. Two different architectures are commonly used: a normal structure and an inverted structure (Fig. 3). In normal structure, holes are collected at the front electrode—the transparent conductor—while electrons are collected at the back electrode which is a lower work-function metal (usually Al) than the front electrode. In inverted structure, the reverse process takes place, that is, the electrons are collected by the transparent conductor and holes are collected by the back electrode

NORMAL STRUCTURE

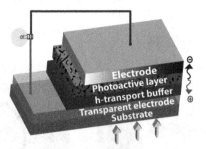

INVERTED STRUCTURES

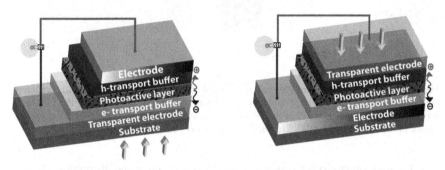

Bottom illuminated   Top illuminated

Fig. 3 Schematic illustration of device geometry used in a PSCs

such as Ag or Au which is a higher work function metal than the front electrode. Buffer layers provide charge selectively and also contribute to tuning the electric field in the device. Poly(3,4-alkenedioxythiophenes):poly(styrenesulfonate) (PEDOT:PSS) is the most commonly used hole transport layer in both structures. Zinc oxide (ZnO) is often used as electron selective buffer layer in inverted structure while LiF is the most commonly employed as buffer layers in normal devices. PEDOT:PSS also functions as a surface planarization on ITO or other electrodes. Inverted structure is most popular when studying large-scale low-cost processing of PSCs while normal structure is mostly used in the development of PSCs, for example, in the evaluation of new polymers and other materials.

2.2 Mechanism of Current Generation in PSCs

The operational mechanism of a PSC that dictates the conversion of sunlight into electricity involves four steps (Fig. 4):

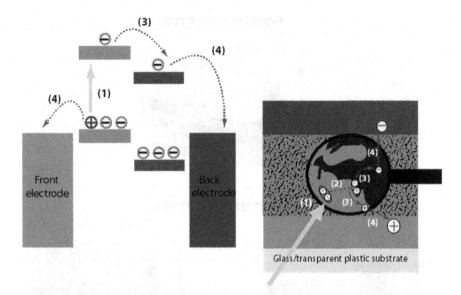

Fig. 4 Energy level diagram of PSC in normal architecture (*left*) and a schematic illustration of bulk heterojunction (BHJ) morphophology. Superimposed in both diagrams are the four steps involved in the current generation in a PSC: (*1*) Exciton generation (*2*) Exciton diffusion (not shown in the band diagram) (*3*) Exciton dissociation (*4*) Charge transport

(1) Exciton generation: Exciton generation occurs upon the absorption of incident light (photon) having an energy equal or higher than the bandgap of the photoactive polymer. The band gap of the polymer is characterized by the energy difference between the lowest unoccupied molecular orbital (LUMO) and the highest occupied molecular orbital (HOMO) of the polymer. The lower the band gap, the higher is the amount of exciton generated. Hence, exhaustive research is being carried out in tailoring low-band gap-conjugated polymers.

(2) Exciton diffusion: The exciton thus generated in the photoactive polymer has high binding energy which does not dissociate at room temperature unlike inorganic solar cells such as silicon solar cell. An acceptor molecular provides the energy impetus for exciton dissociation. To achieve this, an exciton must diffuse to a donor: acceptor interface. The optimum distance of the exciton to a donor: acceptor interface must be similar to the exciton diffusion length in conjugated polymers, which is in the order of 10–20 nm [7]. This is realized by the intermixing of donor and acceptor materials during processing that result in a bulk heterojunction (BHJ) morphology in the deposited film. BHJ is characterized by interpenetrating network of donor and acceptor domains. The difference in the HOMO of the donor polymer and the LUMO of the acceptor molecule largely determines the open circuit voltage in a PSC. Hence, conjugated polymers having deeper HOMO levels, apart from having low-band gap, are preferable.

(3) Exciton dissociation: At the donor: acceptor interface, the exciton dissociates forming free charge carriers.

(4) Charge transport: Once free charge carriers are generated, they are transported to their respective electrodes according to the electric field in the device. The holes and electrons travel through the percolated network of donor and acceptor, respectively. Once collected by the electrodes, they are channeled into the external circuit.

Poly(3-hexylthiophene-2,5-diyl) (P3HT) and [6,6]-phenyl-C_{61}-butyric acid methyl ester (PCBM) are extensively studied and commonly used donor polymer and acceptor molecule, respectively. A comprehensive overview of all photoactive polymers and acceptor molecules could be found elsewhere [8, 9]. Similarly, several reviews could be consulted for deeper understanding of device physics of PSCs [7, 10, 11].

2.3 Photovoltaic Characterization

Like inorganic solar cells, the photovoltaic properties of a PSC are described with four key parameters derived from a current–voltage (*IV*) curve (Fig. 5). These parameters are open circuit voltage (V_{OC}), short circuit current density (J_{SC}), fill factor (*FF*), and power conversion efficiency (PCE). PCE is the most common parameter used as a figure of merit to describe the performance of any solar cell. It indicates the percentage conversion of power density received by a solar cell from the incident light into electrical power. From the *IV* curve, PCE can be deduced as follows:

$$PCE(\%) = \frac{P_{max}}{P_{in}} \times 100 = \frac{I_{sc} \times V_{oc} \times FF}{area \times P_{in}} \times 100 \qquad (1)$$

FF describes the "squareness" of a JV curve and is given by the ratio of ($I_{max} \times V_{max}$)/($I_{SC} \times V_{OC}$). *FF* denotes the extent of internal loses of generated current in a solar cell and is affected by high series -and low shunt- resistance. Shunt resistance is calculated from the inverse of the slope at the J_{sc} point in the IV curve. Low-shunt resistance or "shunting" is a result of manufacturing defects where the positive and negative electrodes within the device are not well isolated (for example, intercalation of spikes in the topology of bottom electrode into the top electrode). Such defects result in current leakage and excessive current leakage may lead to lowering of V_{oc} and can even cause "short circuit" rendering the solar cell nonfunctional. Series resistance, on the other hand, is characterized by the inverse slope at the V_{oc} point in the IV curve. Series resistance is a result of recombination at the material interfaces, defects, poor BHJ morphology, lack of percolation in the donor: acceptor network, the contact resistance, and the sheet resistances of the electrodes. Comprehensive information on the device physics of PSCs can be found elsewhere [10].

Fig. 5 Current–voltage
curve of a solar cell in dark
and under illumination

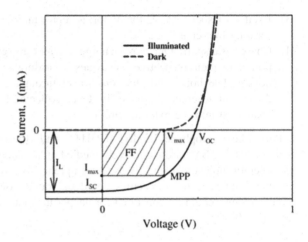

3 Alternatives to ITO

Indium tin oxide is a ubiquitous material in most optoelectronic devices and as
such the efforts in finding an alternative to ITO is exhaustive. In general, the
alternatives of ITO could be categorized into four broad material groups: (1)
nanomaterials; (2) polymers; (3) metals; and (4) metal oxides. Nanomaterials can
further be classified into carbon nanotubes, graphene, metal nanowires, and metal
nanogrids. These material groups are not mutually exclusive and are often used in
some combination with each other. Particularly, Poly(3,4-ethylenedioxythio-
phene):poly(styrenesulfonate) or PEDOT:PSS often used in combination with
many of these alternatives. Figure 6 summaries material categories that are
investigated as alternatives to ITO.

When evaluating the efficacy of transparent conductors for application in solar
cells, two property parameters are of prime importance: optical transparency and
sheet resistance (R_{sh}). However, these parameters are often in competition with
one another and an improvement in one often requires a sacrifice in the other.

Commercially available ITO usually has an R_{sh} of 10 Ω Υ^{-1} and a visible
transmittance of >80 %. Such a property of ITO is used as a benchmark against
which alternatives are being developed and assessed. A figure of merit relationship
that enables property comparison between different transparent conductors is the
ratio between DC conductivity (σ) and absorptivity (α) which is given by [12];

$$\frac{\sigma}{\alpha} = \frac{\frac{1}{t \cdot R_{sh}}}{\frac{-\ln(1-A)}{t}} \tag{2}$$

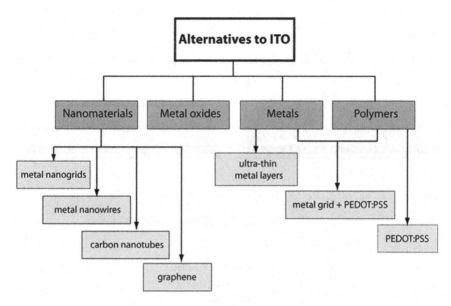

Fig. 6 Flowchart illustrating the classification of alternatives to ITO

where R_{sh} is the sheet resistance, A is the absorbance, and t is the film thickness. For thin films, this equation equates to $\approx \frac{G_{sh}}{A}$; where G_{sh} is the sheet conductance. Alternatively, $\frac{\sigma}{\alpha}$ for thin films such as inorganic oxides in terms of transmittance T and reflectance R can be written as [13];

$$\frac{\sigma}{\alpha} = -\{R_{sh}\, In(T+R)\}^{-1} \tag{3}$$

where T is the total visible transmittance and R is the total visible reflectance. Note that these equations do not take into account the percolation nature of nanowires and carbon nanotubes.

For an ITO film with an R_{sh} of 6 Ω Υ^{-1} and an absorption coefficient of 0.04, figure of merit ratio is 4. According to Eq. (1), a $\frac{\sigma}{\alpha} \geq 1$ is found necessary for minimum power losses in various upscaling geometries of thin film solar cells shown in Fig. 7 [12]. This corresponds to a T of 90 % and a R_{sh} of 10 Ω Υ^{-1}. Values lower than this would result in precipitous loses in the efficiency of monolithically integrated modules than in single cells (Fig. 7). Monolithically integrated modules are used in thin films inorganic and in PSCs, and are the more cost-effectively produced designs that require no post processing assembly. Single cells, on the other hand, demands labor-intensive post processing assembly and are used in the first generation solar cells wherein silicon wafers are manually interconnected to produce a module.

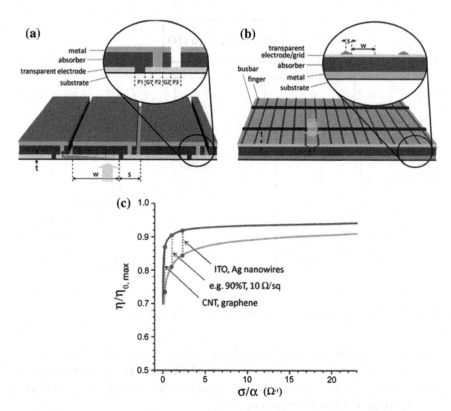

Fig. 7 Schematic illustration of two upscaling geometries of solar cells: **a** a cross-section of a monolithically integrated thin film module where width *w* is the active area that contributes to the power generation and the interconnect area of width *s* is lost area. The *dashed lines* show the path of current which is injected into the transparent electrode and then driven laterally to one edge of the device where contact to the back metal electrode of the adjacent device is made. **b** Cross-section of a standard cell (6 inch^2) using metalgrid/TC. These individual cells are stacked adjacent to each other and serially connected in a long string. Area under the grid lines is lost area and photogenerated current between the lines is conducted laterally over a short distance by the TC film. **c** Maximum fraction of nominal efficiency as a function of the TC material figure of merit for monolithic integration (*lower green line*) and for standard integration (*upper blue line*). For monolithic integrations, achieving high module efficiencies will require $\sigma/\alpha > 1$. Reproduced from Ref. [12] with permission. © 2013 RSC Publishing

3.1 Nanomaterials

3.1.1 Metal Nanogrids

Metal nanogrids are characterized by an array of periodic nanoscale metal grid lines. For metal nanowire to be effective replacement of ITO, two design considerations are important: (1) the grid lines width should be subwavelength to provide sufficient visible transparency; and (2) the period of the mesh should be submicrometer to ensure uniformity of charge collection in any device [14]. Such a periodic array of metal grid lines enables light scattering and coupling, therefore, enhancing photon flux to the photoactive material [15]. Unlike in ITO where transparency and conductivity are competing parameters, conductivity in nano metal grids can be improved by increasing the line height without increasing the width of the grid lines and therefore avoiding decrease in transmittance due to shading from the grid lines. The relationship of the R_{sh} of a periodic array of metal grid to physical attributes of the grid lines can be deducted from Kirchoff's rule. Assuming a square wire network with $N \times N$ wires, R_{sh} is given by;

$$R_{sh} = \frac{N}{N+1} \frac{\rho L}{wh} \tag{3}$$

where ρ is the wire resistivity, L is the wire length, w is the wire width, and h is the wire height [16].

Optical simulations based on a finite-difference frequency-domain (FDFD) method have indicated that a Ag metal nanogrid structure having a period of 400 nm, grid height of 100 nm, and a grid width of 40 nm, an optical transmission of >80 % is achievable in both 1- and 2-dimensional network structure with a R_{sh} of 1.6 $\Omega \, \Upsilon^{-1}$ [17]. Experimentally, metal nanopatterns based on Ag, Cu, and Au with a line width of 70 nm, height of 80 nm, and a period of 700 nm, a transmittance above 70 % is observed over the entire visible spectrum at a R_{sh} of ~ 10 $\Omega \, \Upsilon^{-1}$ [18]. These metal nanopatterns based on Ag and Cu when incorporated in laboratory scale PSCs have resulted in similar performance to ITO-based reference devices [18, 19]. Figure 8 shows an example of reported properties of metal nanogrids.

Although metal nanowire grids show comparable properties to ITO, its upscaling is not cost feasible in the processing of large-area PSCs. Processing of nanopattern can be carried out using a variety of lithography techniques [21]. One such method based on nanoimprinting lithography (NIL) is utilized the printing of metal nanogrids for application in organic solar cells and OLEDs [18, 22]. However, such a technique requires multiple processing steps including evaporation in the preparation of resist template as well as in the subsequent processing of the electrode metal grid. Figure 9 shows an example of the processing steps utilized in the preparation of metal grids for application in PSCs and OLEDs [19, 20]. More information on NIL can be found elsewhere [23]. Recently, an R2R line and vacuum-free methods have been successfully applied in preparation of the, however, evaporation of the final metal is still needed (Fig. 10) [24, 25].

3.1.2 Metal Nanowires

The low-cost processing limitations of metal nanogrids led to the development of metal nanowires (NW) as a potential transparent electrode. Metal NWs offer the possibility of solution processing and a dispersed random network of metal NWs

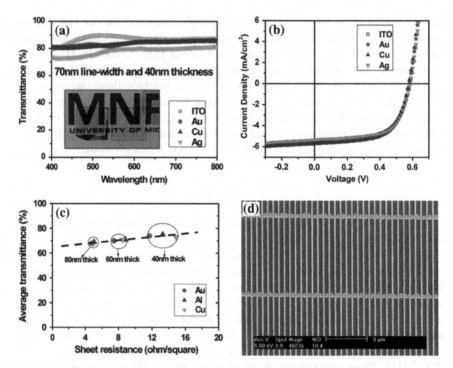

Fig. 8 a Optical transmittance of nanogrids (line width 70 nm and period 700 nm) made different metals; **b** IV characteristic of PSCs incorporating metal nanogrid transparent electrode with the structure: transparent grid/PEDOT:PSS/P3HT:PCBM/LiF/Al. **c** Average transmittance versus sheet resistance of metal grids with a line width of 120 nm. **d** Scanning electron microscopy image of an as-fabricated metal nanogrid electrode with a line width of 70 nm and a period of 700 nm. © 2013. John Wiley and Sons and IEEE. Reprinted, with permission from Refs. [18, 20, 26]

can exhibit transmission and conductivity even superior to ITO. Figure 11 shows a comparison the relationship between transmittance and R_{sh} in ITO, metal nanogrids, and metal nanowires. In metal oxide transparent conductors such as ITO, there is an inherent trade off between R_{sh} and transparency which is determined by film thickness. While higher film thicknesses allow lower R_{sh}, however, at a significant loss of transparency. Calculations on ITO have shown that the optical transparency of ITO suffers greatly with increasing film thicknesses particularly below R_{sh} of 15 Ω Υ^{-1} [27]. In case of metal nanogrids, we have earlier noted that the R_{sh} can be improved at no cost to optical transparency by increasing grid height and maintaining the line width. Experimental results on metal nanowires indicate that a random network film of Ag nanowires exhibit comparable properties to either ITO or metal grids. At the same time, the solution processing possibility of metal nanowire marks its competitive edge over ITO and nanogrids in the low-cost processing.

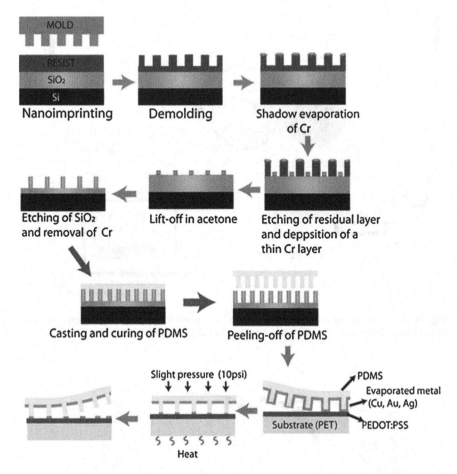

Fig. 9 Schematic illustration of the fabrication procedure of the *narrow line* width mold and subsequent steps in the preparation of metal nanogrids. Color of the arrows indicates three stages involved: NIL stage in preperation of mold (*magenta*), preparation of PDMS stamp (*green*), and metal transfer to the final substrate (*blue*). © 2013. John Wiley and Sons and IEEE. Adapted, with permission from Refs. [18, 20, 26]

A number of factors influence the R_{sh} of random network of NWs: wire length, wire resistance, wire-to-wire contact (junction) resistance, and wire density. At very low density, the effect of the junction resistance is more pronounced. This is because of significantly greater number of parallel NW connections present at higher densities, hence the lowest resistance determines the overall resistance of the film [27]. Higher densities, however, require a trade-off with optical transparency. Nevertheless, a random network of silver nanowires scatters a significant amount of transmitted light which can result in improved photocurrent generation when incorporated in solar cells. Reports suggest that as much of 20 % of transmitted light is scattered at $>10°$ in Ag nanowire network [27].

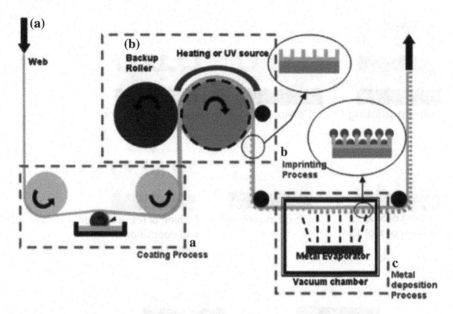

Fig. 10 Schematic of the roll-to-roll nanoimprinting process for preparation of metal nanogrids. © 2013 John Wiley and Sons. Reprinted, with permission from Ref. [25]

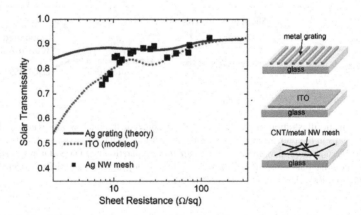

Fig. 11 Solar photon flux-weighted transmissivity versus sheet resistance: Ag gratings (*blue line*), data is collected from finite element modeling with grating period is 400 nm, the Ag line width is 40 nm, and varying thickness to achieve different sheet resistance; ITO (*red dotted line*), the data are computed based on optical constants for e-beam-deposited ITO acquired using spectroscopic ellipsometry; Ag nanowire (*black square*),the data are obtained experimentally. © 2013 American Chemical Society. Adapted, with permission from Ref. [27]

The percolation theory predicts that for a random network of conducting sticks, the percolation threshold dramatically decreases as the length of sticks increases as given in Eq. (4) [28]. Therefore, longer and thinner AgNWs are preferable for optimizing transparency and R_{sh} of a random of NWs. Processing methods are sought to produce nanowires of longer length. The use of electro spinning, for example, could produce ultra-long metal wires (nanofibers). Such nanofibers of Cu films have been demonstrated to exhibit similar R_{sh} and transmittance as ITO films:

$$N_c = \frac{1}{\pi} \left(\frac{4.236}{L} \right)^2 \tag{4}$$

where N_c is the percolation threshold and L is length of the sticks [29].

Nanowire films, however, have a drawback. When incorporated in solar cells in the normal device architecture (Fig. 3), the high roughness leads to interpenetration of the nanowires to the counter electrodes leading to large dark current leakage or even short circuit. The roughness of the NW film can exceed the layer thickness of the overlying photoactive materials (50–200 nm). Several processing strategies have been explored to passivate the surface roughness of NW films. First, an alternate inverted structure, the top illuminated inverted device in Fig. 3, is often employed in which NWs are laminated onto PEDOT:PSS layer. Such a structure alleviates the problem of shunting due to roughness although complete elimination is seldom accomplished. With the use of a short pulse high voltage to burn the remaining shunts, a working module with the structure substrate/Ag film/Cs$_2$O3/P3HT:PCBM/PEDOT:PSS/AgNW has shown a PCE of 2.5 %; J_{sc}: 10.59 mA/cm^2; V_{oc}: 0.51 V, and FF: 46 % on a device area of 2 mm^2 [30]. In contrast, AgNW transparent conductors in a P3HT:PCBM-based solar cell in a normal architecture showed a PCE of 1.1 % with significantly lower V_{oc} than ITO-based reference devices [31]. Second, in a bottom illuminated inverted device, the problem of roughness can be circumvented by increasing the thickness of the buffer layer. Increasing TiO$_2$ thickness in a device structure substrate/AgNW/TiO$_2$/P3HT:PCBM/Mo$_2$O$_3$/Ag has shown a PCE of 3.42 %; J_{sc}:10.1 mA/cm^2; V_{oc} of 0.56 V; and a FF: 61.1 % [32]. Alternatively, by using a composite electrode of AgNW, sol gel TiO$_2$, and PEDOT:PSS in a normal structure AgNW/TiO$_2$/PEDOT:PSS/P3HT:PCBM/Ca/Al, JV properties of the cell were found similar to cells on commercial ITO substrates (Fig. 12) as well as it was found that TiO$_2$ and PEDOT:PSS bind the AgNWs and also result in strong adhesion to the substrate [33]. Such buffer layers function as a smoothening layer on top of rough AgNW film, improve packing of AgNW and their adhesion to substrate, tune work function of the electrode, and provide charge selectivity.

AgNW films provide superior mechanical flexibility than its ITO counterparts and can withstand severe bending under numerous bending cycles. For example, tests have shown that a AgNW (L_{avg}.6.5 μm; diameter 85 nm) film at a nanowire density of 79 mg m^{-2} can withstand 1,000 bend cycles without any change in R_{sh} whereas an ITO substrate catastrophically fails after a 160 bend cycles [34]. Similarly, AgNW wire films on PET could withstand a bending angle of 160° without significant change in the R_{sh} of the film while ITO-PET when subjected to bending angle of 60° showed a third order of magnitude decrease in the R_{sh} (Fig. 12) [35].

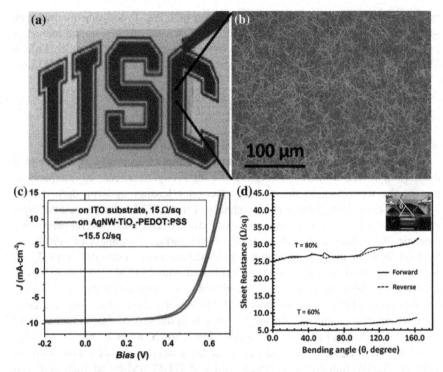

Fig. 12 **a** Photograph of a silver nanowire film deposited on a ∼2 × 2 cm PET substrate with T 80 % and R_{sh} 25 Ω Υ^{-1}; **b** SEM image of the percolating nanowire distribution of the sample shown in **a**; **c** IV curve of P3HT:PCBM cells on ITO or AgNW/TiO$_2$/PEDOT:PSS substrates; **d** A plot of R_{sh} versus bending angle of two Ag nanowire films deposited on PET with transmittance values of 60 and 80 %. The *solid lines* represent the change in sheet resistance during bending and the *dotted lines* represent the change in sheet resistance during unbending. The Ag nanowire film remains conductive even under severe bending. (**a, b**, and **d**) reprinted from Ref. [35] © IOP Publishing, 2013. Reproduced with permission from IOP Publishing. All rights reserved. **c** © 2013 Amercian Chemical Society. Reprinted, with permission from Ref. [33]

Research on metal nanowires is very recent. So far, studies on NW films have affirmed its applicability as an effective replacement to ITO. Currently, research on metal nanowires is slowly gearing toward finding large-scale compatible techniques. In early reports, metal NW films were processed by drop casting a suspension of surfactant-assisted dispersed AgNW, followed by annealing [27, 30]. Such a processing method is not scalable and cannot generate uniform films over large area. As such, these films were less than a cm^2 of area. Spray-coating is a scalable technique that has been recently employed in the preparation of large-area AgNW transparent substrates [36, 37]. A composite AgNW and PEDOT:PSS, both spray coated consecutively have demonstrated low roughness and a R_{sh} of 10 Ω Υ^{-1} with a mean optical transmittance of 85 % without any annealing steps [37].

Fig. 13 High intensity pulse laser sintering of as-deposited AgNW films. Cross-sectional SEM images of AgNW films on PET substrates **a** before and **b** and **c** after HIPL sintering with light intensities of 1.14 and 2.33 J cm^{-2}, respectively. © 2013 RSC Publishing.Reprinted, with permission from Ref. [41]

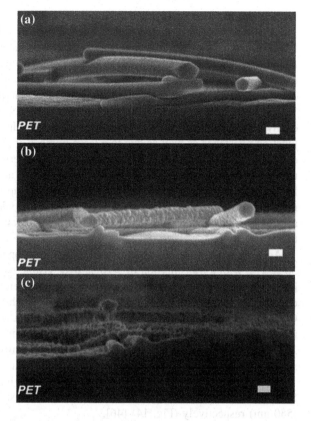

Other methods include Meyer rod coating in which a metal nanowire suspension is placed on the substrate and a Meyer rod is pulled over it. Using this method, an 8 × 8 inch substrate has been prepared [38].

Although the solution-based nature of nanowire inks opens wide opportunities in low-cost processing on large scales using various printing and coating techniques as described in Ref. [39]; however, the early reports have identified some challenges that ought to be solved before large-scale processing could be carried out. For example, the poor adhesion AgNW on substrates, the large junction resistance, the removal of surfactants present in the ink at temperatures that flexible substrates such as PET can withstand (<140 °C) and the rough topology of the resulting film. Very recently, the use of photonic sintering methods such as Direct-Write Pulse Laser Sintering [40] and High Intensity Pulse Light Sintering (HIPS) [41] has achieved successful results at simultaneously eliminating most challenges associated with AgNW films without sacrificing the properties of the films (Fig. 13).

3.1.3 Carbon Nanotubes

It was not too long after the first report on carbon nanotubes in 1991 by Sumio Iljima that the excellent mechanical, electrical, and optical properties of carbon nanotubes were identified. Theoretically, single-wall carbon nanotubes (SWCNTs) can have a current conductivity up to $1-3 \times 10^6$ S m^{-1} and mobility of 100,000 cm^2 V^{-1} s^{-1} [42, 43]. In films, a random network of CNTs possess far lower conductivity and mobility than a single tube primarily due to the presence of high junction resistance as well as due to a large number of parameters that affect CNT conductivity and mobility such as purity, lattice perfection, bundle size, wall number, metal/semiconductor ratio, diameter, length, and doping level. Nonetheless, conductivity up to 6,600 SCM^{-1} and mobility up to 10 cm^2 V^{-2} is experimentally observed in films [42]. As a result, a wide variety of advanced applications have been envisioned for CNTs; transparent conductors being one of them.

An early report noted that a 50 nm thin film of p-doped SWNT film has a R_{sh} of 30 Ω Υ^{-1} with a transmission of >70 % over visible region of the light spectrum [44]. However, accomplishing such results in subsequent reports has not been easy due to the large number of parameters involved. Much like nanowires, the R_{sh} of CNT films is dominated by junction resistance. The use of doping by acid treatments has been shown to cause a threefold decrease in junction resistance and a 30 % increase in the nanotube conductivity when compared to pristine untreated samples [45] (Fig. 14). For example, acid treatment of CNTs has resulted in improvement in R_{sh} of spray-coated SWCNT film from 110 Ω Υ^{-1} (pristine SWCNT) to 37 Ω Υ^{-1} (doped SWCNT) with transmittance of 78 and 76 % (550 nm) respectively (Fig. 14) [46].

Apart from the large junction resistance, the roughness of the CNTs thin films and their adhesion to the substrates has been equally impeding factors to their efficacy in organic solar cells. Such issues are similar to those observed in nanowire-based transparent conductors as well. Shunts due to roughness of the SWCNT surface is circumvented, for example, by either using thick active layer (0.5–1 μm thick) or using a planarization layer such as PEDOT:PSS, or both [48, 49]. Other approaches include, for example, the application of pressure upon transferring of dispersed named with a PDMS stamp [50]. Using such a method, a low roughness value of 10 nm over a scan area of 25 μm^2 was observed in an untreated SWCNT-based film. When the film is used in the fabrication of PSCs, a PCE of 2.5 % was attained. With the use of doped SWCNT, such a transfer method will likely yield similar performance to ITO (PCE 3 %).

The earlier reports were generally proof of principle studies wherein upscaling compatibility of processing was not the central concern. With the promising results shown by these proof of principle studies, recently, reports on large-scale compatible processing have emerged. Spray coating is recently investigated in the preparation of large-area CNT transparent film on PET substrate [46, 47, 51, 52]. Generally, CNTs are processed from suspensions in a solvent. High van der Waals forces render the nanotubes very susceptible to bundling. De-bundling is achieved

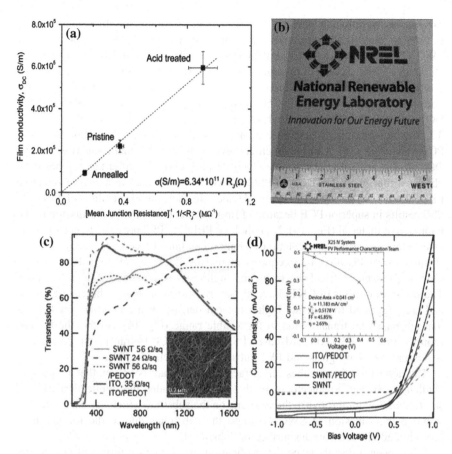

Fig. 14 **a** Plot of measured film conductivity as a function of the inverse of mean junction resistance; **b** Ultrasonically spray deposited large-area (6 × 6 inch²) SWCNTs film;. **c** Comparison of transmission spectra for ITO and SWCNT (with and without PEDOT:PSS). Inset shows TEM image of ultrasonically sprayed; and **d** Light (*solid lines*) and dark (*dashed lines*) JV curves for devices on SWNT and ITO transparent electrodes, with and without PEDOT as an HTL. Inset shows NREL-certified light IV curve for a BHJ device on a SWNT electrode without an HTL. Reprinted permissions **a** © 2013 American Chemical Society. Reprinted, with permissions from Ref. [45]; **c** © 2013 John Wiley and Sons. Reprinted, with permission from Ref. [47]; and **b**, **d** © 2013 American Institute of Physics. Reprinted, with permission from Ref. [46]

with the use of ultrasonication of the nanotubes dissolved in a solvent. The surfactant and/or acid treatments are generally used in addition to ultrasonication to achieve stable dispersions. However, when these suspensions are deposited as films, the surfactants are hard to eliminate and require high-temperature thermal treatment, which is not compatible to flexible substrate such as PET. Such challenges are constantly being addressed with the rapid progress in processing of carbon nanotube films. For example, in one method, SWCNTs are dispersed in aqueous solvents with high molecular weight (~90,000 MW) cellulous–sodium carboxymethylcellulose (CMC)—as a dispersing agent and an ultrasonic spraying is used to deposit the film over large area (6 × 6 inch). Once deposited, the SWCNTs are doped by exposing the film to nitric acid which simultaneously eliminates CMC. Such a film is found to be highly homogenous (rms roughness of 3 nm scanned over 100 μm^2 area) and have resulted in a R_{sh} of 60 Ω Υ^{-1} at a transmission of 60–70 % (at 550 nm). PSCs with the structure SWCNT/PEDOT:PSS/P3HT:PCBM/Ca/Al fabricated on such a film have resulted in a PCE of 3.1 % which is close to reference device with ITO substrate that demonstrated a PCE of 3.6 % [46]. Unlike in other devices where the main function of PEDOT:PSS is providing charge selective transportation as a buffer layer, it has been observed that in SWCNT-based PSC devices does not require a charge selective PEDOT:PSS buffer layer in normal devices. In fact, the elimination of PEDOT:PSS results in superior PCE because of improvement in optical transmission to the photoactive material (Fig. 14). Nonetheless, PEDOT: PSS provides a planarization layer suppressing the intercalation of nanotubes into the counter electrodes. The elimination of PEDOT: PSS requires increasing thickness of the active layer in order to prevent electrical shorts due to roughness in the surface of CNT film [47]. The transparent films of CNTs have higher transmission beyond the visible region and are suggested to generate higher current density than theoretical predictions that are based on transmission in the visible range (Fig. 14). So far, the highest reported PCE of P3HT:PCBM-based PSCs with SWCNTs transparent conductors is 3.6 and 2.6 % on glass and PET substrates, respectively [53].

The adhesion of nanotubes on surface of a substrate is yet another processing challenge. Recent advancement in this regard includes the application of 1 % solution of 3-aminopropyltriethoxy silane on the substrate which has been found to improve the adhesion of SWCNTs onto the substrates due to the formation of cross-linked siloxane on the surface of substrates.

It has been a decade since the application of CNTs as transparent conductors were recognized, however progress has been slow primarily owing to the challenges in processing. There is a long way before SWCNTs transparent conductors can make their way into R2R processing. Reports thus far have demonstrated proof of principle and the task ahead is to demonstrate a robust processing technique(s) that is feasible for application in the processing of PSCs and is easily scalable.

3.1.4 Graphene

Graphene is a two-dimensional material consisting of a monolayer of sp^2-hybridized carbon atoms resulting in a hexagonal arrangement. A monolayer of graphene theoretically would exhibit a reflectance of less than 0.1 % and an absorbance of 2.3 %, therefore resulting in a theoretical transmission limit of 97.7 %. However, graphene is a zero band gap semiconductor and undoped graphene sheet has high R_{sh} values of 6 kΩ Υ^{-1} [54, 55]. Doping is an efficient method that allows tailoring of various properties of graphene and can be carried out following various chemical and electrical routes [56]. So far, a R_{sh} of 30 Ω Υ^{-1}

and a corresponding transmission of 90 % has been observed in doped graphene sheets, thus making graphene potentially suitable as transparent conductors [57]. Apart from R_{sh} and transmission, several other properties such as the high chemical and thermal stability, high charge carrier mobility (200 cm^2 V^{-1} s^{-1}), high current carrying capacity (3 × 10^8 A cm^{-2}), high stretchability, and low-contact resistance with organic materials render graphene a very favorable alternative to ITO [58, 59].

In principle, a monolayer of graphene possesses ballistic charge transport due to delocalization of electrons over the complete sheet, however, in practice defects are introduced during growth and processing of graphene. Such defects, for example, lattice defects, grain boundaries, and oxidative traps due to functionalization result in high R_{sh} of graphene [54, 60]. Such a challenge is reflected in its application as transparent conductors in PSCs. The earlier reports on graphene as transparent conductors were adopted in dye sensitized and small molecule solar cells, however, the performance of such devices were limited (PCEs of <1 %) largely due to the high R_{sh} of the films which is often in the kΩ range [60, 61]. The doping of graphene with AuCl$_3$ is reported to reduce R_{sh} of graphene by 77 % with only 2 % decrease in transmission [62]. As such, an improvement of PCE of PSCs from 1.36 % for undoped graphene to 1.63 % for AuCl$_3$-doped graphene film has been observed [63]. With high quality CVD-grown multilayer (15 layers) graphene, a PCE of 2.60 % has been observed in a P3HT:PCBM-based PSCs. A strong dependence of the photovoltaic properties of PSCs on the growth temperature of high quality CVD-grown graphene has been observed [64]. In addition to the high R_{sh}, the poor wetting properties of graphene is yet another hindrance to their application in PSCs. Graphene is hydrophobic and requires functionalization (e.g., UV/ozone treatment or acid treatment) to improve its wetting properties. Such functionalization in turn creates defects in graphene sheets increasing its R_{sh} and therefore limiting the final performance of a solar cell. Noncovalent functionalization improves wetting while maintaining the structural integrity of graphene. Several routes for noncovalent functionalization have been proposed. For example, a noncovalent functionalization of graphene with self-assembled pyrene butanoic acid succidymidyl ester (PBASE) is observed to improve the wetting properties of graphene toward PEDOT:PSS, thereby, resulting in greater than two-fold increase in PCE [65]. Other methods include deposition of a thin (20 Å) layer of MoO$_3$ over graphene which has been seen to improve the wetting properties of graphene toward PEDOT:PSS as well as tune work function of graphene electrodes [66]. Using MoO$_3$ in PSC in normal architecture (doped graphene/MoO$_3$/PEDOT:PSS/ P3HT:PCBM/LiF/Al), a significant improvement in the device was observed and resulted in a PCE of 2.5 %. In comparison, ITO-based equivalent cells had a PCE of 3 %.

These preliminary investigations of graphene transparent conductors in PSCs are mere proof- of -concept studies. Most graphene transparent conductor film are reported to have R_{sh} values seldom lower than 100 Ω Υ^{-1} and are often investigated on very small area devices (often less than a cm^2). Such high R_{sh} may not be critical in small area devices but will prove detrimental to photovoltaic properties upon upscaling. The foray of graphene into large-scale processing will require significant advancement in large-scale processing.

Currently, high quality graphene is either micromechanically cleaved or grown by chemical vapor deposition; both of which are not low cost and large-scale compatible. Defects are more prominent in graphene films processed by solution-based methods such as liquid-phase cleaving with ultrasonication or by the reduction of graphene oxide. Although these techniques provide lower cost alternatives to processing of graphene, however, graphene produced by such methods exhibit poor properties with R_{sh} in the $k\Omega \, \Upsilon^{-1}$ range due to structural defects and poor interlayer contact as a result of vigorous exfoliation and reduction processes [67]. Several reviews on the properties and processing of graphene are present elsewhere [59, 68, 69] and a recent review elaborates on the application of graphene as electrodes in electrical and optical devices [70].

3.2 Transparent Conductor Oxides

Transparent conductor oxides (TCO) are semiconductor materials composed of binary and ternary oxides containing one or two metallic elements. They have a wide optical band gap of >3 eV making them optically transparent in the visible range. The doping of intrinsic semiconductor oxides with metallic elements in a non-stoichiometric composition results in increasing conductivity without degrading their optical properties. There is a wide variety of transparent conductor oxides such as ZnO, In_2O_3 SnO_2, CdO; ternary compounds like Zn_2SnO_4, $ZnSnO_3$, $Zn_2In_2O_5$, $Zn_3In_2O_6$, In_2SnO_4, $CdSnO_3$; and multicomponent oxides such as Sn: In_2O_2 (ITO) and $F:SnO_2$ [13, 71]. Of these, apart from ITO (Sn: In_2O_3), AZO and GZO (Al-and Ga- doped ZnO, respectively) have optical transparency and conductivity similar to ITO and therefore exhibit a figure of merit ratio $\sigma/\alpha > 1 \, \Omega^{-1}$ [13]. AZO is the best candidate for replacement of ITO because of its nontoxicity, inexpensive materials, and low resistivity.

Transparent conductive materials can be prepared using a wide variety of thin film deposition techniques, through physical vapor deposition methods such as evaporation, magnetron sputtering, molecular beam epitaxy; and through chemical vapor deposition (CVD) techniques such as high-temperature CVD, metal–organic CVD (MOCVD), atomic layer deposition. Such methods are, however, not suitable for high throughput production of PSCs. Liquid-based deposition methods are also employed in the processing of TCO thin films, for example, sol-gel, and chemical bath deposition. However, films produced by these methods have far inferior conductivity. For example, the reported resistivity of sol gel-produced films ranges from 7×10^{-4} to $10 \, \Omega$ cm whereas sputtered films have reported resistivity as low as $1 \times 10^{-4} \, \Omega$ cm [72]. The low resistivity is achieved by sintering at high temperature under vacuum. Furthermore, TCO are brittle materials and hence flexibility restricting. As such, it is unlikely that TCO will be explored for very large-scale high throughput production of PSCs. They are, however, more attractive in other optoelectronic applications such as flat panel displays and inorganic solar cells. In PSCs, their application has been limited to buffer layers that are solution processed or to the use of acceptor material in hybrid solar cells.

3.3 Ultrathin Metal Films

Prior to the discovery and the subsequent dominance of ITO as the material of choice for transparent conductors over the last four decades [73], very thin metals usually evaporated were used as semitransparent conductors in optoelectronics. With the advent of ITO that exhibited far superior properties, these thin semi-transparent metals were rapidly replaced by ITO. However, the economic and physical incompatibility of ITO in low-cost applications particularly has led to revisiting ultrathin metals as a replacement to ITO.

In thin metal films, the surface scattering of free charge carriers causes an inverse relationship between film resistivity and thickness. As a result, there is a threshold below which further reduction in thickness leads to dramatic increases in R_{sh} of the metal film. This critical threshold thickness is observed to lie between 5 and 10 nm in commonly employed electrode metals such as Al, Au, and Ag as predicted by Fuchs-Sondheimer (FS)-Mayadas-Shatzes (MS) model and also experimentally observed [74, 75]. The FS-MS model accounts for surface scattering as well as scattering from the grain boundaries. At a thickness of 10 nm, a transmission of the Ag film is observed to be ~ 60 % (Fig. 15) [74, 75]. Metals also exhibit very high reflection. Calculations have shown that while absorption of 10 nm thick silver is merely 3 %, reflections can amount to 63 % [76]. A dielectric capping layer, similar to antireflection coating, can suppress this reflection and induce a desired interference pattern enhancing the amount of photons reaching the photoactive layer in a solar cell. Such a capping layer can be either organic or inorganic dielectric compounds materials [77]. Coating of a metal electrode with aluminum hydroxiquinoline (Alq_3) in small molecular-based solar cell is reported to cause an improvement in photocurrent up to 50 % [77, 78].

A very few scientific reports are available on thin metal films as transparent conductor in the fabrication of PSCs [74, 79, 80]. Nonetheless, ultrathin metal layer has been adopted as transparent conductors in PSCs in both top-and bottom-illuminated inverted architectures. In bottom illuminated solar cell with the structure substrate/ultrathin Au /ZnO/P3HT:PCBM/PEDOT:PSS/100 nm Au, a PCE of 2.52 % is observed. Such a PCE is nonetheless lower than the ITO-based reference devices (PCE: 3.53 %). The lower PCE is mainly attributed to the lower J_{sc} which in turn is caused by the reduced incoupling of light into the device. In the same structure when top illuminated realized by using 10 nm thick Au top electrode and 100 nm Au bottom electrode, the lowest PCE 1.75 % is observed and was attributed due to optical transmission loss due to the presence of PEDOT:PSS layer [79]. The top illuminated inverted structure has been explored in other studies and has resulted in poor performance in comparison to that of reference

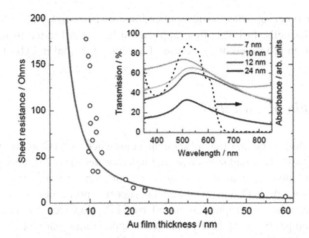

Fig. 15 Thickness dependence of sheet resistance in ultra-thin gold films determined by simulation based on FS-MS model (*solid line*) and experimentally verified (*dots*). Inset shows visible transmission of ultra-thin Au films with various thicknesses. The *dashed line* is absorbance of P3HT:PCBM. © 2013 Elsevier. Reprinted, with permission from Ref. [74]

devices. In such a structure, controlling the metal film properties on different substrates is not an easy task. Film morphology is affected by the type of substrate and its roughness, the surface treatment of the substrate, and the deposition conditions [75]. Without any morphological control, it is unlikely that small amounts of Au or Ag can be deposited on PEDOT:PSS can result in continuous films. Several authors have noticed that Ag has a tendency to coalesce when deposited on organic layers. Incorporating processing strategies to achieve continuous film formation of Ag or Au on organic layer may reduce the loss in performance. Such strategies involve deposition of a work function compatible low-surface energy metals. For example, MgAg can form very flat semitransparent metal films at <5 nm thickness while Ag forms island [81, 82].

Ultrathin metal films have been deposited by evaporation. However, recently a solution processed semitransparent Ag electrode was reported having a R_{sh} of 5 Ω Υ^{-1} and a corresponding transmission of 30 % at 550 nm along with a roughness of 2 nm. When incorporated in the fabrication of a P3HT: PCBM-based solar cell, a PCE: 1.6 %; FF: ~ 60 %; V_{oc}: 0.51 V; and J_{sc}:5.3 mA/cm^2 was observed. The low J_{sc} is attributed to lower incoupling of light [83].

3.4 Polymers

The most important advantage of polymeric transparent conductors is that they can be solution processed and are therefore readily processed in an R2R setup using the plethora of coating and printing techniques available [39]. Poly(3,4-ethylenedioxythiophene):poly(styrenesulfonate) or PEDOT:PSS is the most widely

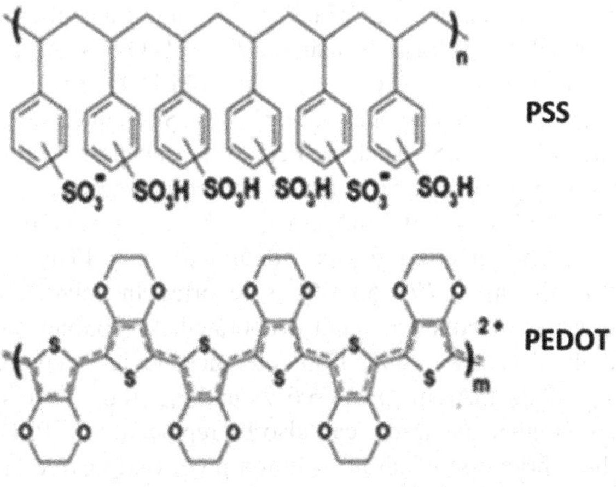

Fig. 16 Chemical structure
of PEDOT:PSS

used polymeric salt for this application where PEDOT is a conjugated polymer in its oxidized state carrying a positive charge and PSS is a polymer with deprotonated sulfonyl groups carrying a negative charge (Fig. 16). PSS is added to EDOT during polymerization as a charge-balancing counter ion and to improve the inherently low solubility of PEDOT in aqueous medium. Several in-depth reviews on PEDOT:PSS in general are present [84, 85].

PEDOT:PSS was initially and still widely used in antistatic coating where R_{sh} is not as critical as when used in PSCs. In earlier reports on PEDOT:PSS as transparent conductor films, a typical conductivity values 1–10 S cm^{-1} (R_{sh} of 10^5 Ω Υ^{-1}) was observed [86, 87]. Such conductivity values were three orders of magnitude lower than that of ITO (>4,000 S cm^{-1}) at similar transmission (80 %). Several approaches were subsequently employed to increase the inherently low conductivity of PEDOT:PSS and many have been successful. Some of these methods include the addition of high-boiling temperature polar compounds such as diethylene glycol [88]; ethylene glycol [89, 90]; sorbitol [91, 92]; dimethylsulfoxide (DMSO) [88, 93, 94]; glycerol [91, 93, 95]; by different chemistry methods such as controlling synthetic conditions, fundamental alteration of the polymer back bone, and by functionalizing the backbone with substituent side groups [87]. Currently, highly conductive formulations of PEDOT:PSS are commercially available, for example, from Heraeus with the latest generations Clevios^{TM} PH500 and PH1000 having a conductivity of 300 and 850 S cm^{-1}, respectively. This is in stark contrast to previous generations such as H.C. Stark Baytron P variants having R_{sh} of 10^5 Ω Υ^{-1} or 1–10 Scm^{-1}.

Like TCO, the conductivity of PEDOT:PSS increases with increasing film thickness reaching saturation at a finite film thickness [85, 96]. However, increasing thickness leads to decreasing optical transmittance. As a result, a trade off between transmission and conductivity is required. Hence, particularly in earlier reports prior to the commercial availability of highly conductive formulations of PEDOT:PSS, such a tradeoff between transmission and conductivity resulted in poor power conversion efficiencies of ITO-free PSCs and modules.

In some cases, PEDOT:PSS having transmission in the range 10–30 % (350–600 nm) for a comparable R_{sh} to ITO (10–20 Ω Υ^{-1}) has been reported [97]. As a result, often PSC devices-based PEDOT: PSS delivered lower PCE than ITO-based reference devices [89, 91, 98]. Subsequently, the use of highly conductive (hc) PEDOT:PSS such as PH500 as transparent conductor has resulted in comparable performance to ITO-based control devices. A maximum PCE of 3.27 %, V_{oc}: 0.63 V, J_{sc}: 9.7 mA cm^{-2}, FF: 53.5 % have been reported [99].

In the inverted device configuration, a PCE of 3.08 %, V_{oc}: 0.61 V; J_{sc}: 9.1 mA cm^{-2}; FF: 53.3 % is reported in solar cells based on hcPEDOT:PSS transparent conductor and evaporated Ag counter electrode. Such a PCE, nonetheless, is lower than ITO reference devices having an average PCE: 4.20 %, J_{sc}: 10.25 mA/cm^2, FF: 66.6 %, and V_{oc}: 0.62 V. In such a device, the evaporated Ag counter electrode, can also be replaced with PEDOT:PSS, therefore allowing the fabrication of an all-solution-processed device [100]. Initially, such a device showed poor rectification with a performance of PCE: 0.47 %, V_{oc}: 0.31 V, J_{sc}: 5.94 mA cm^{-2}, and FF: 27.7 % [100]. However, later reports observed good rectification with improved ZnO buffer films deposited by atomic layer deposition [101]. With a new generation low-band gap polymer, a PCE of 2.69 % was observed in a polymer-based solar cell processed with a roll-coated hcPEDOT:PSS as the front electrode in combination with PEDOT:PSS/Ag as the back electrode [102]. Nonetheless, all such devices had lower PCE than ITO-based reference devices. However, PEDOT:PSS electrodes render improve mechanical flexibility than ITO electrodes where PCE showed a 92 % retention in PEDOT:PSS-based PSCs while only 50 % of the initial PCE was retained in ITO-based electrode after 300 bend cycle [100].

3.5 Metal grid/PEDOT:PSS Composite Electrode

Despite the development of highly conductive formulations of PEDOT: PSS, its R_{sh} values (10^2–10^3 Ω Υ^{-1}) still remains significantly higher than ITO (10–60 Ω Υ^{-1}) as a result of which stand-alone PEDOT:PSS front electrodes have yielded lower PCE than ITO-based control devices. By physically reinforcing PEDOT:PSS with metal grids, conductivity of PEDOT:PSS can be significantly improved. Such a composite electrode when used in place of ITO in normal structure solar cells is observed to cause a threefold decrease in series resistance (R_s) from > 1 kΩ in cells with only PEDOT:PSS as electrode to 400 Ω with composite electrodes, ultimately resulting in a threefold increase in J_{sc} [98]. The challenge for using such a composite electrode is in the optimization required between shadow losses due to the metal grids and the resistive losses due to the resistance of the combined PEDOT:PSS/metal electrode. Depending on the R_{sh} of PEDOT:PSS films, completely different configurations of metal grid design are required in the optimization of the cells [103]. Overall, the surface coverage of the metal grids should be as small as possible so as to minimize loss of incoming

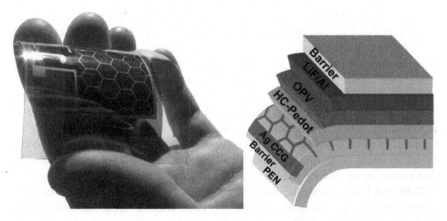

Fig. 17 A photograph of ITO-free PSC (2×2 cm^2) with screen-printed Ag grid and highly conductive PEDOT: PSS as front electrode and evaporated Al back electrode (*left*). The device comprised of a layer stack schematically shown alongside (*right*). These devices had similar photovoltaic performance to ITO. © 2013 Elsevier. Reprinted, from with permission from Ref. [106]

radiation reaching the photoactive layer. Conductivity can be increasing by increasing line height while maintaining minimum shading losses [104]. In general, for a given PEDOT:PSS, the optimized geometry of the metal grid that results in minimum shading fractional power losses is simply given by the empirical relation ($W / W + S$) where W is the grid width and S is the grid separation [103, 105].

With rigorous optimizations of grid design and PEDOT:PSS thickness, the composite metal grid and PEDOT:PSS transparent conductor layers have resulted in a performance similar to ITO-based solar cells in both normal and inverted device geometries [103, 104, 106] (Fig. 17). Some methods used for deposition of metal grids in laboratory cells are lithography [103, 107, 108]; thermal evaporation through shadow masks [109–111]; sputtering in combination with photolithography for patterning [112]; microfluidic deposition and nanoimprinting methods [104]; precision-weaved metalized polymer fabric electrodes [113]; and printing methods such as screen printing [3]; inkjet printing [107, 114, 115]; and flexographic printing [115]. Among all these methods, currently only the printing methods are readily adoptable in a fast large-scale roll-to-roll processing of low-cost PSC. These methods have been experimentally demonstrated as well.

In a normal structure, the use of screen-printed metal grids in combination with highly conductive PEDOT:PSS on flexible substrates has shown superior PCE to equivalent ITO-based cells [106]. Similarly, the use of inkjet-printed metal grid as well as embedded grids in flexible substrates has also resulted in similar results with higher reproducibility [107, 114].

Top illuminated inverted structures have been also adopted in the demonstration of ITO-free large-area PSC modules in a structure: substrate/metal/buffer layer/photoactive layer/PEDOT:PSS/Ag grid. The PEDOT:PSS/Ag grid forms the front electrode and the metal layer on the substrate forms as the back electrode. A range of metals has been applied as back electrodes, for example, Ag, Al/Cr, and Cu/Ti [116–118]. The choice of metal for back electrodes can allow tuning of the work function of the electrode allowing higher voltage extraction and therefore higher PCE. This is observed in small devices when Ag grid is deposited by evaporation which in turn allows PEDOT:PSS layer to be made very thin, ultimately allowing superior light transmission into the device. When Ag grid is deposited by ambient processing technique such as screen printing, a higher thickness of PEDOT:PSS because the solvents from screen printing formulation of Ag ink is observed to diffuse through the underlying PEDOT:PSS layer into the photoactive layer, thereby destroying it. Consequently, the higher thickness of PEDOT:PSS leads to poor transmission to the PAL and thereby very poor current. In early reports on R2R produced screen-printed grids, a transmission as low as 30 % is reported [116–118].

The most recent advancement in ITO-free R2R produced modules were made in bottom illuminated inverted structures. A vacuum-free all R2R-processed PSCs employing high conductive PEDOT:PSS/metal grid as a front electrode and PEDOT:PSS/metal grid as the back electrode in an inverted structure was recently demonstrated [115]. Metal grids were printed by three R2R methods: R2R thermal imprinting of embedded grids, R2R inkjet printing, and R2R flexographic printing. R2R flexographic-printed and R2R-embedded grids delivered similar albeit unprecedented performance for fully R2R processed vacuum-free large-area ITO-free cells (active area 6 cm^2) under ambient conditions with devices based on flexo grids having a *PCE*: 1.82 %, J_{sc}: 7.1 mA cm^{-2}, V_{oc}: 5.1 V, *FF*: 51.2 %, and embedded grids with *PCE*: 1.92 %, J_{sc}: 7.06 mA cm^{-2}, V_{oc}: 0.50 V, *FF*: 54.6 %. The raised topography and relatively poor conductivities in the R2R inkjet-printed silver grids resulted in significantly lower PCE attributed to the lower *FF* and J_{sc} caused as a result of shunt paths: *PCE* 0.75 %, J_{sc}: 4.27 mA cm^{-2}, V_{oc}: 0.50 V, and *FF*: 35.1 %. Among all these techniques, flexographic printing emerged to be the favorable low-cost technique presenting no topography issues and the need for multiple R2R steps as required in the embedded grids. Flexographic printing has been recently adopted in producing large-area modules that are produced by all R2R processing and have delivered PCE comparable to ITO-based devices. This is discussed in the next section.

4 Roll-to-Roll Processing of ITO-Free Polymer Solar Cell Modules

To date, almost all polymer solar cells reported in the literature are prepared using a combination of two laboratory techniques: spin coating and metal evaporation. However, neither of these techniques can be expected to share a future with PSCs, where only processes that rely on flexible substrates and the absence of vacuum steps are expected.

Solution-based processing is an appealing alternative to vacuum-based deposition approaches. The straightforward comparative advantages of solution methods include atmospheric pressure processing, which requires significantly lower capital equipment costs, suitability for large-area and flexible substrates, higher throughput, and the combination of more efficient materials usage and lower temperature processing. These approaches can also be readily adapted for simultaneously patterning the materials while coating or printing, which eliminates the need for additional processing steps.

An enormous palette of film-forming techniques has been investigated for processing one or more of the layers in a PSC, which usually is comprised of five or six layers. Some of these film-forming techniques investigated are slot die coating, gravure coating, knife-over-edge coating, offset coating, spray coating, and printing techniques such as inkjet printing, pad printing, and screen printing. A complete review on roll-to-roll processing can be found elsewhere [39, 119].

Recently, fully functional large area highly flexible ITO-free modules, reaching a total area of 180 cm^2 were reported [120]. These modules were fabricated on a 60 μm thick barrier substrate on which a flexographic printed silver grid in combination with printed hcPEDOT: PSS formed the transparent electrode. Such a composite front electrode exhibits a transmittance of >60 % in the visible region and printed silver shows a sheet resistance of >1 Ω Υ^{-1}. The complete device stack comprised of Ag grid/hcPEDOT:PSS/ZnO/P3HT:PCBM/PEDOT:PSS/Ag is known as the *IOne* process. The R2R processing carried out in the preparation of IOne stack is shown in Fig. 18. Such devices were found to be highly scalable; single cells of 6 cm^2 demonstrated a *FF* of 51 % while up-scaled modules with an area of 180 cm^2 had a *FF* of 55 % (Fig. 19). These modules were subjected to several accelerated lifetime testing conditions and were found to be rather stable for more than 1,000 h under different operational and storage conditions [120], thereby attesting to >1 year of lifetime.

The *IOne* process has been further adopted in the processing of large arrays of serially interconnected modules, all accomplished in the printing and coating processing [122]. Such a structure can ideally be used to produce infinitely interconnected solar cell modules on an infinitely long flexible substrate and be readily installed for power generation. Such a concept has been demonstrated on an 80 m foil comprising of 16,000 serially connected cells delivering a voltage of 8.12 KV under outdoor conditions. Characterization under overcast conditions at 138 Wm^{-2} showed a PCE of 1.6 %, V_{oc}: 874 V, I_{sc}: 5.49 mA cm^{-2}, and *FF*: 50 % on an area of 11 meters.

5 Summary and Future Outlook

Polymer solar cells present an attractive technology by the sole virtue of its low-cost processing potential. Research efforts in low-cost processing until now have shown that very low energy payback time (0.30 years) can be obtained with PSCs which inevitably would allow the technology can be within the economical reach

Fig. 18 Photographs of the stepwise R2R printing and coating processes in fabrication of the modules: **a** flexography printing of Ag grid; **b–d** slot die coating of hcPEDOT:PSS, P3HT:PCBM, and PEDOT:PSS, respectively; **e** flat-bed screen printing of Ag paste; and **f** final module after step **e**. © 2013 Elsevier. Reprinted, with permission from Ref. [120]

of mass consumers. ITO currently remains the biggest hurdle in reaching this objective and finding a cost-effective replacement to ITO that can be solution processed in a vacuum-free environment is crucial for PSCs to witness a commercial reality. While all but ITO layers in a PSC stack are solution based that can be processed on a roll-to-roll coating and printing line at fast web speeds (>10 m min^{-1}), the processing of a patterned ITO substrate requires multistep including sputtering and etching, ultimately incurring more than 50 % of the cost

(a)   **(b)**

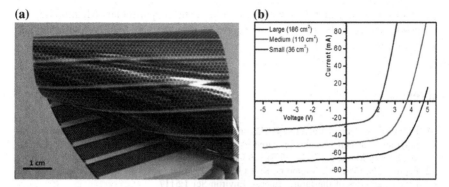

Fig. 19 **a** An R2R processed highly flexible *IOne* PSC modules with an area >100 cm^2 where front electrode is metal grid/PEDOT:PSS composite; **b** IV characteristic of the module shown in **a** with different sizes depicting scalability of the structure. © 2013 Elsevier. Adapted, with permission from Ref. [120]

Fig. 20 A summary of reported properties of TCO alternatives. *Solid lines* are fit according to some figure of merit equations. The *dotted rectangle* shows the target region for transparent conductive electrode application. © 2013 Nature Publishing group. Reprinted, with permission from Ref. [123]

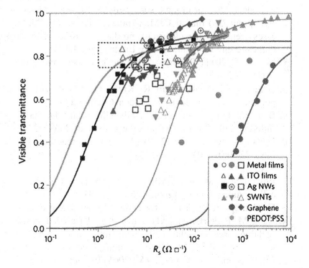

of a PSC module. Hence, finding an alternative is crucial for the successful commercialization of PSCs.

While most of the alternatives presented in this review have the potential to fulfill such requirements (comparative summary of TCE properties of all alternatives is shown in Fig. 20), several processing issues have impeded their progress from laboratory to large-scale production. Particularly, nanomaterials including CNTs, graphene, and metal nanowires have shown remarkable properties that even surpass the benchmark that ITO has set.

With the rapid progress in processing of such nanomaterials being reported, it is not too long before they are commercially utilized as transparent conductors for all optoelectronics in general and organic solar cells in particular. Meanwhile, the composite PEDOT:PSS/metal grid electrode remains the readily available and upscalable alternative to ITO.

References

1. Espinosa N, García-Valverde R, Urbina A, Krebs FC (2011) A life cycle analysis of polymer solar cell modules prepared using roll-to-roll methods under ambient conditions. Sol Energy Mater Sol Cells 95(5):1293–1302
2. Espinosa N, Hösel M, Angmo D, Krebs FC (2012) Solar cells with one-day energy payback for the factories of the future. Energy Environ Sci 1:5117
3. Krebs FC, Tromholt T, Jørgensen M (2010) Upscaling of polymer solar cell fabrication using full roll-to-roll processing. Nanoscale 2(6):873–886
4. Krebs FC, Gevorgyan SA, Alstrup J (2009) A roll-to-roll process to flexible polymer solar cells: model studies, manufacture and operational stability studies. J Mater Chem 19(30):5442–5451
5. Emmott CJM, Urbina A, Nelson J (2012) Environmental and economic assessment of ITO-free electrodes for organic solar cells. Sol Energy Mater Sol Cells 97:14–21
6. Azzopardi B, Emmott CJM, Urbina A, Krebs FC, Mutale J, Nelson J (2011) Economic assessment of solar electricity production from organic-based photovoltaic modules in a domestic environment. Energy Environ Sci 4:3741
7. Nelson J (2011) Polymer: fullerene bulk heterojunction solar cells. Mater Today 2011 14(10):462–470
8. Helgesen M, Søndergaard R, Krebs FC (2010) Advanced materials and processes for polymer solar cell devices. J Mater Chem 20(1):36
9. Brabec CJ, Gowrisanker S, Halls JJM, Laird D, Jia S, Williams SP (2010) Polymer: fullerene bulk-heterojunction solar cells. Adv Mater 22(34):3839–3856
10. Deibel C, Dyakonov V (2010) Polymer-fullerene bulk heterojunction solar cells. Rep Prog Phys 73(9):096401
11. XXX
12. Rowell MW, McGehee MD (2011) Transparent electrode requirements for thin film solar cell modules. Energy Environ Sci 4(1):131
13. Gordon R (2000) Criteria for choosing transparent conductors. MRS Bull 25(8):52–57
14. Kang M, Park HJ, Ahn SH, Xu T, Guo LJ (2010) Toward low-cost, high-efficiency, and scalable organic solar cells with transparent metal electrode and improved domain morphology. IEEE J Sel Top Quantum Electron 16(6):1807–1820
15. Liu C, Kamaev V, Vardeny ZV (2005) Efficiency enhancement of an organic light-emitting diode with a cathode forming two-dimensional periodic hole array. Appl Phys Lett 86(14):143501
16. van de Groep J, Spinelli P, Polman A (2012) Transparent conducting silver nanowire networks. Nano Lett 12(6):3138
17. Catrysse PB, Fan S (2010) Nanopatterned metallic films for use as transparent conductive electrodes in optoelectronic devices. Nano Lett 10(8):2944
18. Kang M, Kim MM, Kim J, Guo LJ (2008) Organic solar cells using nanoimprinted transparent metal electrodes. Adv Mater 20(23):4624
19. Kang M, Joon Park H, Hyun Ahn S, Jay Guo L (2010) Transparent Cu nanowire mesh electrode on flexible substrates fabricated by transfer printing and its application in organic solar cells. Sol Energy Mater Sol Cells 94(6):1179–1184
20. Kang M, Park HJ, Ahn SH, Xu T, Guo LJ (2010) Toward low-cost, high-efficiency, and scalable organic solar cells with transparent metal electrode and improved domain morphology. IEEE J Sel Top Quantum Electron 16(6):1807

21. Byeon K, Lee H (2012) Recent progress in direct patterning technologies based on nano-imprint lithography. Eur Phys J Appl Phys 59(1):10001
22. Kang M, Guo LJ (2007) Semitransparent Cu electrode on a flexible substrate and its application in organic light emitting diodes. J Vac Sci Technol, B 25(6):2637
23. Guo LJ (2007) Nanoimprint lithography: methods and material requirements. Adv Mater 19(4):495
24. Ahn SH, Guo LJ (2009) Large-area roll-to-roll and roll-to-plate nanoimprint lithography: a step toward high-throughput application of continuous nanoimprinting. ACS Nano 3(8):2304
25. Ahn SH, Guo LJ (2008) High-speed roll-to-roll nanoimprint lithography on flexible plastic substrates. Adv Mater 20(11):2044
26. Kang M, Guo LJ (2007) Nanoimprinted semitransparent metal electrodes and their application in organic light-emitting diodes. Adv Mater 19(10):1391
27. Lee J, Connor ST, Cui Y, Peumans P (2008) Solution-processed metal nanowire mesh transparent electrodes. Nano Lett 8(2):689
28. Hu L, Wu H, Cui Y (2011) Metal nanogrids, nanowires, and nanofibers for transparent electrodes. MRS Bull 36(10):760
29. Wu H, Hu L, Rowell MW, Kong D, Cha JJ, McDonough JR et al (2010) Electrospun metal nanofiber webs as high-performance transparent electrode. Nano Lett 10(10):4242
30. Gaynor W, Lee J, Peumans P (2010) Fully solution-processed inverted polymer solar cells with laminated nanowire electrodes. ACS Nano 4(1):30–34
31. Yang L, Zhang T, Zhou H, Price SC, Wiley BJ, You W (2011) Solution-processed flexible polymer solar cells with silver nanowire electrodes. Acs Appl Mater Interfaces 3(10):4075
32. Leem D, Edwards A, Faist M, Nelson J, Bradley DDC, de Mello JC (2011) Efficient organic solar cells with solution-processed silver nanowire electrodes. Adv Mater 23(38):4371
33. Zhu R, Chung C, Cha KC, Yang W, Zheng YB, Zhou H et al (2011) Fused silver nanowires with metal oxide nanoparticles and organic polymers for highly transparent conductors. ACS Nano 5(12):9877–9882
34. De S, Higgins TM, Lyons PE, Doherty EM, Nirmalraj PN, Blau WJ et al (2009) Silver nanowire networks as flexible, transparent, conducting films: extremely high DC to optical conductivity ratios. ACS Nano 3(7):1767
35. Madaria AR, Kumar A, Zhou C (2011) Large scale, highly conductive and patterned transparent films of silver nanowires on arbitrary substrates and their application in touch screens. Nanotechnology 22(24):245201
36. Scardaci V, Coull R, Rickard D, Coleman JN (2011) Spray deposition of highly transparent, low resistance networks of silver nanowires over large areas. Small 7(8):2621–2628
37. Choi DY, Kang HW, Sung HJ, Kim SS (2013) Annealing-free, flexible silver nanowire-polymer composite electrodes via a continuous two-step spray-coating method. Nanoscale 5:977–983
38. Hu L, Kim HS, Lee J, Peumans P, Cui Y (2010) Scalable coating and properties of transparent, flexible, silver nanowire electrodes. ACS Nano 4(5):2955
39. Søndergaard R, Hösel M, Angmo D, Larsen-Olsen TT, Krebs FC (2012) Roll-to-roll fabrication of polymer solar cells. Mater Today 15(1–2):36–49
40. Spechler JA, Arnold CB (2012) Direct-write pulsed laser processed silver nanowire networks for transparent conducting electrodes. Appl Phys A-Mater Sci Process 108(1):25
41. Jiu J, Nogi M, Sugahara T, Tokuno T, Araki T, Komoda N et al (2012) Strongly adhesive and flexible transparent silver nanowire conductive films fabricated with a high-intensity pulsed light technique. J Mater Chem 22(44):23561
42. Hecht DS, Hu L, Irvin G (2011) Emerging transparent electrodes based on thin films of carbon nanotubes, graphene, and metallic nanostructures. Adv Mater 23(13):1482
43. Shim BS, Tang Z, Morabito MP, Agarwal A, Hong H, Kotov NA (2007) Integration of conductivity transparency, and mechanical strength into highly homogeneous layer-by-layer composites of single-walled carbon nanotubes for optoelectronics. Chem Mater 19(23):5467

44. Wu ZC, Chen ZH, Du X, Logan JM, Sippel J, Nikolou M et al (2004) Transparent, conductive carbon nanotube films. Science 305(5688):1273

45. Nirmalraj PN, Lyons PE, De S, Coleman JN, Boland JJ (2009) Electrical connectivity in single-walled carbon nanotube networks. Nano Lett 9(11):3890

46. Tenent RC, Barnes TM, Bergeson JD, Ferguson AJ, To B, Gedvilas LM et al (2009) Ultrasmooth, large-area, high-uniformity, conductive transparent single-walled-carbon-nanotube films for photovoltaics produced by ultrasonic spraying. Adv Mater 21(31):3210

47. Barnes TM, Bergeson JD, Tenent RC, Larsen BA, Teeter G, Jones KM et al (2010) Carbon nanotube network electrodes enabling efficient organic solar cells without a hole transport layer. Appl Phys Lett 96(24):243309

48. Pasquier AD, Unalan HE, Kanwal A, Miller S, Chhowalla M (2005) Conducting and transparent single-wall carbon nanotube electrodes for polymer-fullerene solar cells. Appl Phys Lett 87(20):1–3

49. van de Lagemaat J, Barnes TM, Rumbles G, Shaheen SE, Coutts TJ, Weeks C et al (2006) Organic solar cells with carbon nanotubes replacing In2O3: Sn as the transparent electrode. Appl Phys Lett 88(23):233503

50. Rowell MW, Topinka MA, McGehee MD, Prall H, Dennler G, Sariciftci NS et al (2006) Organic solar cells with carbon nanotube network electrodes. Appl Phys Lett 88(23):233506

51. Li Z, Kandel HR, Dervishi E, Saini V, Xu Y, Biris AR et al (2008) Comparative study on different carbon nanotube materials in terms of transparent conductive coatings. Langmuir 24(6):2655–2662

52. Kim S, Yim J, Wang X, Bradley DDC, Lee S, de Mello JC (2010) Spin- and spray-deposited single-walled carbon-nanotube electrodes for organic solar cells. Adv Funct Mater 20(14):2310–2316

53. Kim S, Wang X, Yim JH, Tsoi WC, Kim J, Lee S et al (2012) Efficient organic solar cells based on spray-patterned single wall carbon nanotube electrodes. J Photonics Eng 2(1):021010–021011

54. Geim AK, Novoselov KS (2007) The rise of graphene. Nat Mater 6(3):183

55. Bonaccorso F, Sun Z, Hasan T, Ferrari AC (2010) Graphene photonics and optoelectronics. Nat Photonics 4(9):611–622

56. Lv R, Terrones M (2012) Towards new graphene materials: doped graphene sheets and nanoribbons. Mater Lett 78:209–218

57. Bae S, Kim H, Lee Y, Xu X, Park J, Zheng Y et al (2010) Roll-to-roll production of 30-inch graphene films for transparent electrodes. Nat Nanotechnol 5(8):574–578

58. Kang J, Hwang S, Kim JH, Kim MH, Ryu J, Seo SJ et al (2012) Efficient transfer of large-area graphene films onto rigid substrates by hot pressing. ACS Nano 6(6):5360

59. Park H, Brown PR, Buloyic V, Kong J (2012) Graphene as transparent conducting electrodes in organic photovoltaics: studies in graphene morphology, hole transporting layers, and counter electrodes. Nano Lett 12(1):133

60. Wu J, Becerril HA, Bao Z, Liu Z, Chen Y, Peumans P (2008) Organic solar cells with solution-processed graphene transparent electrodes. Appl Phys Lett 92(26):263302

61. Wang X, Zhi L, Muellen K (2008) Transparent, conductive graphene electrodes for dye-sensitized solar cells. Nano Lett 8(1):323

62. Kim KK, Reina A, Shi Y, Park H, Li L, Lee YH et al (2010) Enhancing the conductivity of transparent graphene films via doping. Nanotechnology 21(28):285205

63. Park H, Rowehl JA, Kim KK, Bulovic V, Kong J (2010) Doped graphene electrodes for organic solar cells. Nanotechnology 21(50):505204

64. Choe M, Lee BH, Jo G, Park J, Park W, Lee S et al (2010) Efficient bulk-heterojunction photovoltaic cells with transparent multi-layer graphene electrodes. Org Electron 11(11):1864

65. Wang Y, Chen X, Zhong Y, Zhu F, Loh KP (2009) Large area, continuous, few-layered graphene as anodes in organic photovoltaic devices. Appl Phys Lett 95(6):063302

66. Wang Y, Tong SW, Xu XF, Oezyilmaz B, Loh KP (2011) Interface engineering of layer-by-layer stacked graphene anodes for high-performance organic solar cells. Adv Mater 23(13):1514–1518
67. Kim KS, Zhao Y, Jang H, Lee SY, Kim JM, Kim KS et al (2009) Large-scale pattern growth of graphene films for stretchable transparent electrodes. Nature 457(7230):706
68. Pang S, Hernandez Y, Feng X, Muellen K (2011) Graphene as transparent electrode material for organic electronics. Adv Mater 23(25):2779
69. Huang X, Yin Z, Wu S, Qi X, He Q, Zhang Q et al (2011) Graphene-based materials: synthesis, characterization, properties, and applications. Small 7(14):1876
70. Jo G, Choe M, Lee S, Park W, Kahng YH, Lee T (2012) The application of graphene as electrodes in electrical and optical devices. Nanotechnology 23(11):112001
71. Castañeda L (2011) Present status of the development and application of transparent conductors oxide thin solid films. Mater Sci Appl 2:1233
72. Zhou H, Yi D, Yu Z, Xiao L, Li J (2007) Preparation of aluminum doped zinc oxide films and the study of their microstructure, electrical and optical properties. Thin Solid Films 515(17):6909–6914
73. Kumar A, Zhou C (2010) The race to replace tin-doped indium oxide: which material will win? ACS Nano 4(1):11
74. Wilken S, Hoffmann T, von Hauff E, Borchert H, Parisi J (2012) ITO-free inverted polymer/fullerene solar cells: Interface effects and comparison of different semi-transparent front contacts. Sol Energy Mater Sol Cells 96(1):141–147
75. O'Connor B, Haughn C, An K, Pipe KP, Shtein M (2008) Transparent and conductive electrodes based on unpatterned, thin metal films. Appl Phys Lett 93(22):223304
76. Koeppe R, Hoeglinger D, Troshin PA, Lyubovskaya RN, Razumov VF, Sariciftci NS (2009) Organic solar cells with semitransparent metal back contacts for power window applications. Chemsuschem 2(4):309
77. O'Connor B, An KH, Pipe KP, Zhao Y, Shtein M (2006) Enhanced optical field intensity distribution in organic photovoltaic devices using external coatings. Appl Phys Lett 89(23):233502
78. Meiss J, Furno M, Pfuetzner S, Leo K, Riede M (2010) Selective absorption enhancement in organic solar cells using light incoupling layers. J Appl Phys 107(5):053117
79. Ajuria J, Etxebarria I, Cambarau W, Munecas U, Tena-Zaera R, Carlos Jimeno J et al (2011) Inverted ITO-free organic solar cells based on p and n semiconducting oxides. New designs for integration in tandem cells, top or bottom detecting devices, and photovoltaic windows. Energy Environ Sci 4(2):453–458
80. Al-Ibrahim M, Sensfuss S, Uziel J, Ecke G, Ambacher O (2005) Comparison of normal and inverse poly(3-hexylthiophene)/fullerene solar cell architectures. Sol Energy Mater Sol Cells 85(2):277–283
81. Meiss J, Riede MK, Leo K (2009) Optimizing the morphology of metal multilayer films for indium tin oxide (ITO)-free inverted organic solar cells. J Appl Phys 105(6):063108
82. Oyamada T, Sugawara Y, Terao Y, Sasabe H, Adachi C (2007) Top light-harvesting organic solar cell using ultrathin Ag/MgAg layer as anode. Japan J Appl Phys Part 1-Regular Papers Brief Communications & Review Papers 2007;46(4A):1734
83. Angmo D, Hösel M, Krebs FC (2012) All solution processing of ITO-free organic solar cell modules directly on barrier foil. Sol Energy Mater Sol Cells 107:329–336
84. Kirchmeyer S, Reuter K (2005) Scientific importance, properties and growing applications of poly(3,4-ethylenedioxythiophene). J Mater Chem 15(21):2077
85. Levermore PA, Chen L, Wang X, Das R, Bradley DDC (2007) Highly conductive poly(3,4-ethylenedioxythiophene) films by vapor phase polymerization for application in efficient organic light-emitting diodes. Adv Mater 19(17):2379
86. Groenendaal BL, Jonas F, Freitag D, Pielartzik H, Reynolds JR (2000) Poly(3,4-ethylenedioxythiophene) and its derivatives: past, present, and future. Adv Mater 12(7):481

87. Ha YH, Nikolov N, Pollack SK, Mastrangelo J, Martin BD, Shashidhar R (2004) Towards a transparent, highly conductive poly(3,4-ethylenedioxythiophene). Adv Funct Mater 14(6):615

88. Hsiao Y, Whang W, Chen C, Chen Y (2008) High-conductivity poly(3,4-ethylenedioxythiophene): poly(styrene sulfonate) film for use in ITO-free polymer solar cells. J Mater Chem 18(48):5948–5955

89. Ouyang BY, Chi CW, Chen FC, Xi QF, Yang Y (2005) High-conductivity poly (3,4-ethylenedioxythiophene): poly(styrene sulfonate) film and its application in polymer optoelectronic devices. Adv Funct Mater 15(2):203

90. Chang Y, Wang L, Su W (2008) Polymer solar cells with poly(3,4-ethylenedioxythiophene) as transparent anode. Org Electron 9(6):968

91. Zhang FL, Johansson M, Andersson MR, Hummelen JC, Inganas O (2002) Polymer photovoltaic cells with conducting polymer anodes. Adv Mater 14(9):662

92. Admassie S, Zhang FL, Manoj AG, Svensson M, Andersson MR, Ingaas O (2006) A polymer photodiode using vapour-phase polymerized PEDOT as an anode. Sol Energy Mater Sol Cells 90(2):133

93. Ahlswede E, Hanisch J, Powalla M (2007) Comparative study of the influence of LiF, NaF, and KF on the performance of polymer bulk heterojunction solar cells. Appl Phys Lett 90(16):163504

94. Do H, Reinhard M, Vogeler H, Puetz A, Klein MFG, Schabel W et al (2009) Polymeric anodes from poly(3,4-ethylenedioxythiophene): poly(styrenesulfonate) for 3.5 % efficient organic solar cells. Thin Solid Films 517(20):5900

95. Huang T, Huang C, Su Y, Chen Y, Fang J, Wen T (2010) Extension of active region in crossbar-type polymer solar photovoltaics induced by highly conductive PEDOT:PSS buffer layer. J Vac Sci Technol, B 28(4):702

96. Zhang XG, Butler WH (1995) Conductivity of metallic-films and multilayers. Phys Rev B 51(15):10085

97. Winther-Jensen B, Krebs FC (2006) High-conductivity large-area semi-transparent electrodes for polymer photovoltaics by silk screen printing and vapour-phase deposition. Sol Energy Mater Sol Cells 90(2):123

98. Aernouts T, Vanlaeke P, Geens W, Poortmans J, Heremans P, Borghs S et al (2004) Printable anodes for flexible organic solar cell modules. Thin Solid Films 451:22

99. Na S, Kim S, Jo J, Kim D (2008) Efficient and flexible ITO-free organic solar cells using highly conductive polymer anodes. Adv Mater 20(21):4061

100. Hau SK, Yip H, Zou J, Jen AK (2009) Indium tin oxide-free semi-transparent inverted polymer solar cells using conducting polymer as both bottom and top electrodes. Org Electron 10(7):1401

101. Zhou Y, Cheun H, Choi S, Potscavage WJ Jr, Fuentes-Hernandez C, Kippelen B (2010) Indium tin oxide-free and metal-free semitransparent organic solar cells. Appl Phys Lett 97(15):153304

102. Larsen-Olsen TT, Machui F, Lechene B, Berny S, Angmo D, Søndergaard R et al (2012) Round-robin studies as a method for testing and validating high-efficiency ITO-free polymer solar cells based on roll-to-roll-coated highly conductive and transparent flexible substrates. Adv Energy Mater 2(9):1091–1094

103. Tvingstedt K, Inganas O (2007) Electrode grids for ITO-free organic photovoltaic devices. Adv Mater 19(19):2893

104. Kang M, Kim M, Kim J, Guo LJ (2008) Organic solar cells using nanoimprinted transparent metal electrodes. Adv Mater 20(23):4408

105. Cheknane A (2011) Optimal design of electrode grids dimensions for ITO-free organic photovoltaic devices. Prog Photovoltaics 19(2):155–159

106. Galagan Y, Rubingh JJM, Andriessen R, Fan C, Blom PWM, Veenstra SC et al (2011) ITO-free flexible organic solar cells with printed current collecting grids. Sol Energy Mater Sol Cells 95(5):1339–1343

107. Galagan Y, Zimmermann B, Coenen EWC, Jørgensen M, Tanenbaum DM, Krebs FC et al (2012) Current collecting grids for ITO-free solar cells. Adv Energy Mater 2(1):103–110
108. Zou J, Yip H, Hau SK, Jen AK (2010) Metal grid/conducting polymer hybrid transparent electrode for inverted polymer solar cells. Appl Phys Lett 96(20):203301
109. Zimmermann B, Glatthaar M, Niggemann M, Riede MK, Hinsch A, Gombert A (2007) ITO-free wrap through organic solar cells: a module concept for cost-efficient reel-to-reel production. Sol Energy Mater Sol Cells 91(5):374–378
110. Glatthaar M, Niggemann M, Zimmermann B, Lewer P, Riede M, Hinsch A et al (2005) Organic solar cells using inverted layer sequence. Thin Solid Films 491(1–2):298–300
111. Zimmermann B, Schleiermacher HF, Niggemann M, Wuerfel U (2011) ITO-free flexible inverted organic solar cell modules with high fill factor prepared by slot die coating. Sol Energy Mater Sol Cells 95(7):1587–1589
112. Choi S, Potscavage WJ Jr, Kippelen B (2010) ITO-free large-area organic solar cells. Opt Express 18(19):A458–A466
113. Kylberg W, de Castro FA, Chabrecek P, Sonderegger U, Chu BT, Nüesch F et al (2011) Woven electrodes for flexible organic photovoltaic cells. Adv Mater 23(8):1015–1019
114. Galagan Y, Coenen EWC, Sabik S, Gorter HH, Barink M, Veenstra SC et al (2012) Evaluation of ink-jet printed current collecting grids and busbars for ITO-free organic solar cells. Sol Energy Mater Sol Cells 104:32–38
115. Yu J, Kim I, Kim J, Jo J, Larsen-Olsen TT, Søndergaard RR et al (2012) Silver front electrode grids for ITO-free all printed polymer solar cells with embedded and raised topographies, prepared by thermal imprint, flexographic and inkjet roll-to-roll processes. Nanoscale 4:6032
116. Krebs FC (2009) All solution roll-to-roll processed polymer solar cells free from indium-tin-oxide and vacuum coating steps. Org Electron 10(5):761–768
117. Manceau M, Angmo D, Jørgensen M, Krebs FC (2011) ITO-free flexible polymer solar cells: from small model devices to roll-to-roll processed large modules. Org Electron 12(4):566–574
118. Krebs FC (2009) Roll-to-roll fabrication of monolithic large-area polymer solar cells free from indium-tin-oxide. Sol Energy Mater Sol Cells 93(9):1636–1641
119. Krebs FC (2009) Fabrication and processing of polymer solar cells: a review of printing and coating techniques. Sol Energy Mater Sol Cells 93(4):394–412
120. Angmo D, Gevorgyan SA, Larsen-Olsen TT, Søndergaard R, Hösel M, Jørgensen M et al (2013) Scalability and stability of very thin, roll-to-roll processed, large-area, indium-tin-oxide free polymer solar cells. Org Electron 14(3):984–994
121. Larsen-Olsen TT, Søndergaard RR, Norrman K, Jørgensen M, Krebs FC (2012) All printed transparent electrodes through an electrical switching mechanism: a convincing alternative to indium-tin-oxide, silver and vacuum. Energy Environ Sci 15:36–49
122. Sommer-Larsen P, Jørgensen M, Søndergaard RR, Hösel M, Krebs FC (2013) It is all in the pattern–high-efficiency power extraction from polymer solar cells through high-voltage serial connection. Energy Technol 1(1):15–19
123. Ellmer K (2012) Past achievements and future challenges in the development of optically transparent electrodes. Nat Photonics 6(12):808

Low-Cost Fabrication of Organic Photovoltaics and Polymer LEDs

Hongseok Youn, Hyunsoo Kim and L. Jay Guo

Abstract Polymer light-emitting diodes (PLEDs) and organic photovoltaics (OPVs) are considered as next generation electronics due to the low-cost, flexibility, and lightweight features. However, there are challenges such as large-area processing technologies, film coating quality, and long-term stability toward scalable and low-cost polymer electronics. This chapter deals with the various scalable processing methods and evaluates the coating performance as well as electrical performances in polymer electronics fabricated by the solution processes. Special attention on coating instability is elaborated in the context of important material components in PLEDs and OPVs. Proper coating techniques can be chosen by considering the thickness requirement of each functional layer with good reproducibility. Additionally, we will evaluate mechanical/optical characteristics of the polymer anode for ITO-free electrodes; and introduce the metal mesh in combination with conductive polymers as the ITO-free transparent electrode for large area applications.

1 Introduction to Roll-Coating Techniques

Polymer electronics have great potential toward large-scale, lightweight, and low-cost devices. Moreover, since polymers can be dissolved in common solvents, polymer electronics can be fabricated using scalable solution processes such as roll-to-roll, inkjet, blade, and spray coating methods. In particular, the polymer solar cells are considered as next generation and sustainable photovoltaic devices. They are attractive as additional power sources for mobile devices and could become ubiquitous in the near future. Moreover, the flexible polymer solar cells could also be applied for building integrated photovoltaics (BIPV) such as walls,

H. Youn · H. Kim · L. J. Guo (✉)
Department of Electrical Engineering and Computer Science, University of Michigan,
Ann Arbor, MI 48109, USA
e-mail: guo@umich.edu

Z. Lin and J. Wang (eds.), *Low-cost Nanomaterials*, Green Energy and Technology, 227
DOI: 10.1007/978-1-4471-6473-9_9, © Springer-Verlag London 2014

Fig. 1 Various future applications of the Polymer Solar Cells with a large-scale and flexible advantages. (The images captured at http://www.solarmer.com/products.html, Copyright © Solarmer Energy Inc)

roofs, and windows as shown in Fig. 1. The manufacturing processes of polymer solar cells are relatively environmentally friendly due to low-emission and waste free features. On the other hand, Polymer LEDs are strong contenders for low-cost and flexible solid-state lighting.

Since the pioneering work of Tang [1], the OLEDs field has witnessed rapid progress. The introduction of triplet phosphorescence materials by Forrest and Thompson [2] has enabled commercialization of small molecule-based OLEDs. In the OPVs area, after the introduction of the bi-layer heterojunction structure by Tang in 1986, the efficient photo-induced charge transfer between conjugated polymer and fullerene in the bulk heterojunction (BHJ) structure was reported by Heeger et al. in 1992 [3]. The power conversion efficiency of the polymer (PBHJ) solar cells has now increased to over 10 % by employing the low band gap polymer [4].

To enable large manufacturing, however, there are few reports on high efficiency OPVs devices fabricated by practical processing. There are some considerations for developing fully printable solar cells. First, the materials should be soluble in common solvents and stable in the air environment. Second, the device structure should be simple and the film thickness should not be too thin considering the variability in film thickness in large area fabrication. If the device structure contains many layers, the limitations and compatibility of solvents for each functional layer should be considered due to the dissolution or intermixing problems with the underlying layers. Thinner layers, e.g., of only a few nanometers thickness, are very challenging to realize uniform film thickness in the large area devices. Third, the cost of the process and that of materials and substrates should not be too expensive. For example, material utilization is low in both vacuum evaporation and spin-coating deposition methods, and the former also

Table 1 The material costs in the polymer solar cells

	Spin casting ($/inch^2)	Slot/Blade coating ($/inch^2)	Slot/Blade coating (ITO-free) ($/inch^2)	Slot/Blade coating (ITO-free and evaporation-free) ($/inch^2)
ITO	0.20	0.20	–	–
Ink (PEDOT, Active, etc.)	0.060 + 1.200	0.003 + 0.060	0.003 + 0.060	0.003 + 0.060 + 0.002 (Ag ink)
Aluminum	0.104	0.104	0.104	–
Total cost (inch^2)	1.564	0.367	0.167	0.065

requires expensive vacuum facility. Thus, to meet the requirements of throughput and material utilization, various practical and efficient coating methods such as roll-to-roll, blade coating, spray coating, and slot-die coating are being developed and reported. Here, we give an estimate of the material cost for making OPVs, which was calculated per inch^2-area device. Because there is little waste of solution in the blade/slot coating process, the volume of the solution consumption is 20-fold less than that of spin coating. For instance, the blade coating having an ink supplying nozzle utilizes 30 μl solution for a $2'' \times 3''$ size substrate. On the other hand, spin coating requires more than 500 μl to cover the same area substrate, clearly not suitable for large-scale manufacturing process considering the high material cost of electronic materials. Another consideration is the ITO transparent conductor commonly used for organic electronic devices: even though it has good transparency and high conductivity, its cost has increased significantly due to the worldwide demand and production of LCD and OLED displays, and touch panels along with the fact that the indium is a relatively scarce element on the planet earth. In fact, ITO coated glass accounts for 12 % (spin coating) and 54 % (slot coating) of the whole device cost as shown in Table 1. Moreover, the flexible ITO-PET is more expensive than ITO-glass and occupies 35 % in an electronics device application [5]. To achieve 1$/W target in the polymer solar cells, the cost of transparent conductors and the active materials during the process should be reduced.

1.1 Various Printing/Coating Technologies for Polymer Electronics

Among conventional printing/coating technologies, the roll-to-roll printing, gravure, flexography, and offset printing offer great productivity compared with other manufacturing technologies (Table 2). However, whether these conventional printing processes are appropriate for manufacturing polymer electronics should be considered. For instance, screen printing is a simple printing process but it is

Table 2 The printing parameters and features in conventional printing technologies

	Inkjet	Screen	Gravure	Flexo.	Offset
Resolution (μm)	16–50	30	20–75	75	30–50
Thickness (nm)	<100	>1000	<50 nm	<50 nm	<1000
Viscosity (mPa s)	<20	500–50000	50–200	50–500	20–100 k
Coating speed (m²/s)	0.1	<10	60	10	20
Polymer ink	○	○	○	○	×

limited by the requirement of high ink viscosity. Inkjet printing is useful for fine patterning and thinner film using less viscous solution, but due to its serial process nature it is not suitable for large area film coating. In terms of large area coating, roll-to-roll coating has more flexibility to be adapted for the various coating area and to obtain varied film thickness.

1.2 General Coating Characteristics in Coating Methods: Thickness Control

This section introduces and compares general coating processes such as spin coating, knife coating, and slot-die coating. Since film thickness and uniformity are more important factors for electronics devices than traditional coating industries, we will examine some equations to get a clear picture of the key parameters affecting the film thickness. We aim at understanding the general coating characteristics for the practical coating process.

1.2.1 Spin Coating

Spin coating is one of the most popular coating methods as it is simple and easy to realize the intended film thickness. The solution dispensed on the substrate can be spread by a centrifugal force and the wet film is thinned down during the spin process. The film thickness depends on the viscosity and concentration of the solution. However, the final thickness of the film is mainly controlled by the spin speed, and thinner film can be achieved at higher spin speed.

In the following equations h_w is the wet film thickness, x: rate of the solids contacting on the substrate, ω: angular speed, ρ: solution concentration, C: constant for the coating gas in the coating chamber, μ: viscosity of the solution. The wet film thickness can be expressed as follows:

$$h_w = x\left(\frac{e}{2(1-x)K}\right)^{1/3} \tag{1}$$

where $e = C\sqrt{\omega}$, and $K = \frac{\rho\omega^2}{3\mu}$.

When we consider the evaporation of the solvent during the coating process, the final film thickness can be expressed as follows:

$$h_f = x h_w$$

and finally [6],

$$h_f \propto x \left(\frac{\mu}{\rho\omega}\right)^{1/2} \tag{2}$$

Therefore, the film thickness can be precisely controlled by the parameters in Eq. (2). However, very small amounts of the solution remain on the substrate to form the film during the spinning process. The rest of the solution is totally wasted. As a result, material utilization in the spin-coating process is commonly too low, sometimes less than 5 % of the solution volume. Therefore, spin coating is not considered as a practical manufacturing process.

1.2.2 Blade (Knife-Edge) Coating

The blade-coating method is commonly used in various coating industries and generally referred as knife-edge coating. The amount of solution (metered or not metered) is dispensed ahead of the blade, and the blade moves with a given speed as shown in Fig. 2. Contrary to spin coating where the film thickness is inversely proportion to the spin rate, the film thickness in the blade coating increases with the blade speed (V_{blade}). The film thickness in blade coating is also related to viscosity of the solution μ, surface tension of the solution, and curvature radius of the downstream meniscus R [7].

$$h = 1.34 \left(\frac{\mu V_{blade}}{\sigma}\right)^{2/3} R \tag{3}$$

The downstream meniscus can be expressed with pressure difference between P_1 and P_2.

$$h = 1.34 \left(\frac{\mu V_{blade}}{\sigma}\right)^{2/3} (P_2 - P_1) \tag{4}$$

Thus, the film thickness in the blade coating increases with blade speed, because as the blade moves faster, it has less chance to remove the solution on the substrate with the other parameters fixed. The volume of the moving bead will be smaller in the high-speed blade coating. Therefore, a large amount of solution will remain on the substrate after the blade and result in a thicker film. In the same

Fig. 2 The schematic
diagram of the blade-coating
process

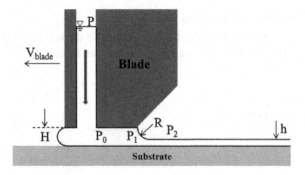

manner, if the surface tension is small, a large amount of volume will stay after the coating. This phenomenon is also known in the dip-coating process, where the typical film thickness equation is expressed as [8],

$$h = 0.946 \left(\frac{\mu V_{\text{blade}}}{\sigma} \right)^{1/6} \left(\frac{\mu V_{\text{blade}}}{\rho g} \right)^{1/2} \tag{5}$$

If the density of the solution ρ, viscosity μ, and surface tension σ are treated as constants, we can lump them together as a coating constant K. Similar treatment can be applied to blade coating. Therefore in both cases the film thickness versus blade speed can be expressed simply as

$$h \cong K(V_{\text{blade}})^{2/3} \tag{6}$$

In the actual blade-coating experiment, we found that the film thickness of P3HT:PC$_{61}$BM blend film used as the active layer (substrate size: 2″ × 3″) in the polymer solar cells is well matched with theoretical values as shown in Fig. 3. The solution consisted of 20 mg P3HT and 16 mg PC$_{61}$BM dissolved in 1.2 mg dichlorobenzene.

1.2.3 Slot (Die) Coating

If there is no pumping pressure to deliver the ink from the reservoir to slot nozzle, the thickness relation is the same as that of blade coating. However, conventional slot coating uses the static pump to extrude the solution. Therefore, the thickness depends on the volume-rate of the pumped solution. The thickness can be expressed as follows:

$$h_w \simeq \frac{\dot{V}}{W V_{\text{slot}}} \tag{7}$$

$\dot{V}$: volume flow rate of the pumping, W: coating width of the slot-die, V_{slot}: coating speed.

Fig. 3 The relationship between the various film thicknesses of the P3HT:PC$_{61}$BM layer and the blade speeds

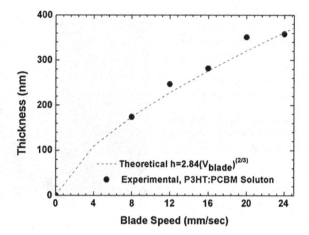

1.3 Coating Instabilities

In typical roll-to-roll coating process such as gravure coating, the surface of the printing rolls has many engraved cells. A picture of the gravure coating system is shown in Fig. 8. The ink in the engraved cell on the roll surface is transferred to substrate in the roll-to-roll coating, while there is direct contact between the gravure roll and the substrate (in comparison with blade or die coating where a fluid layer exists between the two rolls, and later will be referred as non-contact coating process). The direct contact mechanism of the ink transfer causes printing/coating instability, such as in the form of streak patterns in the film as shown in Fig. 6. The small amount of the ink in the gravure cell moves like an extensional viscous flow during the ink transfer. The extended viscous fluid causes irregular streak patterns. An example is shown in Fig. 4 for the emissive polymer solution/material. This coating instability can be explained by the capillary number, C_a. This dimensionless number is defined by the surface energy and viscosity of the solution.

$$C_a = \frac{\mu V}{\sigma} \tag{8}$$

μ: the viscosity of fluid, V: the flow rate (roll speed), σ: the surface tension. Typically, larger C_a will cause coating instabilities, such as ribbing, cascading, and coating mist, due to the cavitation.

The viscosity of the solution increases for higher molecular weight material. For instance, the emissive polymer in the light-emitting diodes has relatively high molecular weight around 1,000,000. Therefore, the coating problem can be more severe. Generally, viscosity also increases with the solution concentration, which

Fig. 4 The image of the irregular coating streaks (*yellow patterns*) with the thermally deformed aluminum cathode (metallic *bubble-like patterns*) due to the leakage current. The aluminum cathode was peeled out from the polymer light emissive layer after the device test

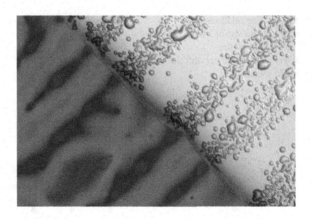

leads to higher capillary number and instability. Reducing the solution concentration, the film quality can be improved as shown in Fig. 5. To improve the film coating quality, the capillary number should be controlled by, e.g., reduction of coating speed or viscosity. However, the desired film thickness may not be achievable. Therefore, the contact coatings causing the extensional viscous flow of the polymer solution between the roll and substrate have more serious instability issues. The instability analysis can also apply to the coating processes such as blade coating where the blade does not have direct physical contact with the substrate (and hence will be referred to as non-contact coating). Because the blade or the slot-die coatings do not cause extensional viscose flow, they are frequently more suitable choices.

In the case of low-band gap polymer semiconductor (iI-T3) [9] dissolved into dichlorobenzene, because the solution has very low surface tension, the capillary number depends highly on the coating speed. Coating meniscus can be broken due to low cohesive energy of the solution under fast coating condition, showing as irregular patterns due to instability. On the other hand, the coating quality of the low-band gap polymer will be improved by reducing the coating speed as shown in Fig. 6.

There is another important parameter to determine the coating instability. It is the leveling time for the solvent in the wet film to be evaporated after the coating process. If the solvent evaporates relatively slowly, the film can be self-leveled under the surface tension, and the irregular coating pattern resulted from the instability will disappear by the flow itself.

The wet film thickness containing the instability can be considered as a simple wave with amplitude δ and characteristics wavelength $(2\pi/q)$ as follow (Fig. 7):

$$e = e_0 + \delta e \cos(qx) \tag{9}$$

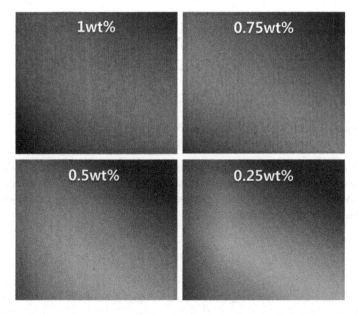

Fig. 5 The irregular coating patterns in the emissive material due to instability with respect to the solution concentration, 1, 0.75, 0.5 and 0.25 wt%. The lower concentration improves the film quality, the irregular pattern disappear in the case of 0.25 w%. The other coating condition are 80 m/min, 200 LPI (line per inch), 22 µm of the gravure cell depth

Fig. 6 The irregular coating patterns in iI-T3:PC$_{71}$BM blend PV material due to instability with respect to the blade-coating speed, 20, 16, 12 and 8 mm/s. Clearly lower speed removes instability and leads to uniform film

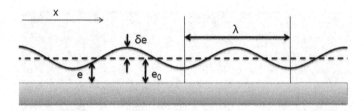

Fig. 7 The ideal modeling of the coating ripples of the wet film solution

The film thickness variation, expressed as δ, will be reduced by the self-leveling flow. We estimate the leveling time τ. First the fluid speed (v) can be deduced from the 1-dimensional Navier–Stokes equation. Then the flow rate Q can be deduced from Young–Laplace equation:

$$-\frac{\partial p}{\partial x} + \mu \frac{\partial^2 v}{\partial z^2} = 0 \tag{10}$$

$$Q \approx \frac{e_0}{\mu} \sigma q^3 \delta e \sin(qx) \tag{11}$$

Considering the volume conservation,

$$\frac{\partial Q}{\partial x} = -\mu \frac{\partial e}{\partial t} \tag{12}$$

$$\frac{\mathrm{d}\delta e}{\mathrm{d}t} = -\frac{\delta e}{\tau} \tag{13}$$

Solving the differential equation yields the leveling time as follows:

$$\tau = \frac{\mu \lambda^4}{\sigma e_0^3} \tag{14}$$

If leveling time is smaller than evaporation time, the pattern will disappear due to the leveling flow. However, if the leveling time is too long or evaporation time too fast, the irregular pattern will remain. The leveling time depends on the viscosity and surface tension of the solution. For instance, when the viscosity of the given polymer solution is 100 mPa s, the surface tension is 25 m N/m, the characteristic pitch of the stripes is 500 μm, the wet film thickness is 10 μm, and the leveling time will be 250 s. To remove the irregular patterns, the solvent having higher boiling temperature should be considered. By choosing a solvent that can evaporate slowly beyond the leveling time, flat and uniform films can be produced (Fig. 8).

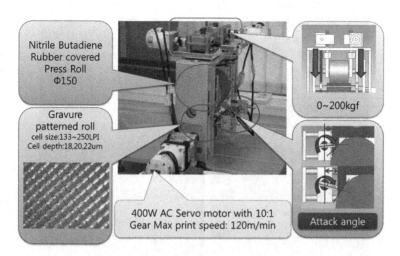

Nitrile Butadiene
Rubber covered
Press Roll
Φ150

Gravure
patterned roll
cell size:133~250LPI
Cell depth:18,20,22um

0~200kgf

400W AC Servo motor with 10:1
Gear Max print speed: 120m/min

Attack angle

Fig. 8 The gravure coating system and its coating parts

Uniform film is important to ensure proper device function. Example, the streak patterns of the emissive material shown in Fig. 9 can cause electrical shorting problems under the bias voltage. This unwanted streak pattern can be controlled by adjusting the printing speed and ink viscosity. The viscosity of the solution can be reduced by diluting the solution as discussed earlier for the emissive polymer. For another example, the commonly used conductive polymer PEDOT:PSS solution is also difficult to form a uniform film on the polymer substrate as shown in Fig. 10. In this case, since the solution is water-born and hydrophilic solution, it is difficult to coat uniformly. The irregular pattern of the PEDOT:PSS layer can be controlled by adding isopropyl alcohol (IPA) solution to reduce viscosity. Dilution with IPA has better effect than that of adding a surfactant. However, it is hard to obtain an appropriate film thickness by one time coating using a lower concentration solution. Therefore, in the case of using the viscous solution, the film quality is sensitive to the coating instability caused by ink transfer in the contact coating method. Due to the basic problems of the contact coating process, non-contact coating methods will be introduced. Slot-die coating as a typical non-contact coating is a good manufacturing tool for the large scale solar cells.

2 Characteristics of Roll-to-Roll Coated Functional Layers Toward the Large Scale Devices

This section describes the roll-to-roll/blade-coating process with applications in large area OPV and OLED devices. We discuss the requirements of the functional layers (e.g., transparent electrode, hole extraction layer, charge separation layer,

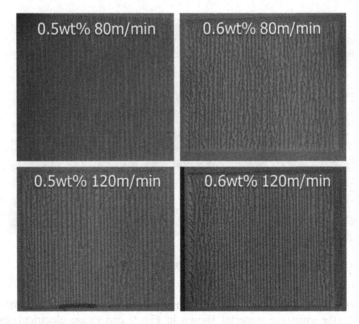

Fig. 9 The irregular streak patterns of the emissive layer in PLEDs with respect to the different concentrations and coating speeds

electron transport layer, and electron extraction layer) in terms of the coating performances such as coating thickness, uniformity, and electrical properties.

2.1 Roll-to-Roll and Blade-Coating System as the Noncontact-Coating Methods

Among the various printing processes to make large area polymer PVs and PLEDs, the roll-to-roll coating system is one of the most favorable candidates. By adding a facile doctor blade to the roll-to-roll system, a simple, noncontact coating platform can be constructed, where the blade does not contact the substrate directly but via a liquid layer. As compared with the slot-die system, it is a valuable tool for fabrication of large-sized thin films.

Since the conventional blade-coating system has advantages such as easy operation, simple structure, and low cost, it has been widely used in not only laboratory but also industrial scale. Previously, there have been efforts to make organic electronics using simple blade coating due to these advantages. However, there is a serious problem in the conventional blade-coating process: because the ink supplying is not homogeneous, the layer thickness varies: the initial layer thickness was frequently thicker than the final layer thickness. In addition, the ink supply in the slot-die coater relies on the use of external pumping, which cannot be

Fig. 10 The control of the irregular streak patterns by adding surfactant (Dynol 604, Air product inc.) and isopropyl alcohol (IPA)

thin enough, e.g., for the electron injection layer using the low-viscous solutions. To address the nonuniform film thickness problem, a pre-metering mechanism, or nozzle system should be added like that used in a slot coating. In order to achieve better film uniformity for much thinner organic layers such as the electron injection layer in PLED and to solve the fundamental ink supplying problem, we attach a simple slide glass to the blade surface to meter the solution. The whole system was developed with 8″-based rollers and it is driven by an AC servo-motor. As shown in Fig. 11 the solution contained in the slit between the glass plate and the blade flows and created a meniscus in front of the glass, followed by a homogeneous laminar flow at the back of the blade. This fine laminar flow is the key to uniformity and better film quality. The control variables of the film thickness are the blade gap, the slit gap, the blade speed, the ink concentration, and the surface energy of the substrate, and the slit.

The blade-slit coating method utilizes only the gravity of the solution for the ink supply, the flow rate can be minimized effectively. The estimated minimal thickness of the blade-slit coating is around a few nanometers. The minimal wet film thickness in the slot-die system is around 1 μm, and the estimated dry film thicknesses are few tens of nanometers. Because the device performance sensitively depends on the film thickness in the organic electronics, the thickness of each layer should be controlled

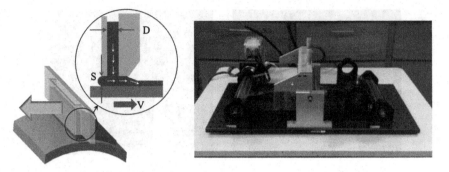

Fig. 11 The schematic diagram of blade-slot coating showing the ink supply and coating flows; and a 8″ based roll-to-roll coating system

fairly accurately to optimize the device performance. When designing the various device structures by the solution coating processes, special consideration should be given to whether each layer thickness can be realized by the practical coating method such as the blade/roll-to-roll coating equipment.

2.2 Conventional Device Structure in the PLEDs

We now discuss the roll-to-roll fabrication of PLEDs. A typical PLEDs is indium tin oxide (ITO)/Hole transporting layer/Emissive layer/electron transporting layer/ metal cathode, poly(3,4-ethylenedioxythiophene)poly-(styrenesulfonate) (PEDOT: PSS) conductive polymer is a commonly used hole transporting layer. Phenyl substituted poly(para-phenylene vinylene)—known as "Super Yellow," (will simply be referred as S-Y, Merck, PDY-132) will be used as yellow light-emitting layer. Zinc oxide (ZnO) nanoparticle (NP) is used as electron transporting layer, and Al as cathode, poly(ethylene oxide) (PEO), and tetra-n-butylammonium tetrafluoborate (TBABF$_4$) in acetonitrile (ionic solution) is used to induce a dipole layer to adjust the work function of the cathode, as shown in Fig. 12 [10].

2.3 Flexible and Transparent Electrode for Large-Scale PLEDs

2.3.1 Optical and Electrical Property of the Roll-to-Roll Coated Polymer Electrode

Since indium tin oxide (ITO) has good optical and electrical properties, it has been commonly used as a transparent electrode in displays and solar cells. However, the cost of the ITO is expensive and mechanical bending strength is poor due to its

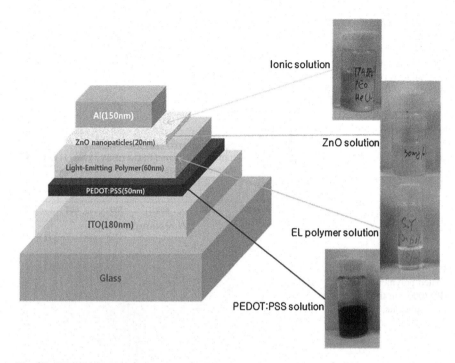

Fig. 12 Illustration of the multilayer structure and the solutions of the functional layers

brittleness. Moreover, the flexible ITO sputtered on PET substrate has less transmittance and conductivity than that of the ITO-glass. There are many candidates for the alternative transparent electrodes such as metal mesh, silver nanowire [11], graphene [12], carbon nanotubes [13], and conducting polymer [14]. The conducting polymer of PEDOT:PSS fabricated by roll-to-roll coating offers satisfactory optical transparency and electrical conductivity (Fig. 13) as polymer anode. For example, a 200 nm thick polymer anode (PEDOT:PSS, H.C. Stock PH1000) can be coated by roll-to-roll blade-slit coating method over large substrate at one time without wasting of the solution as shown in Fig. 14. In comparison, spin coating method requires three time coatings and wasting lots of solutions to cast the same 200 nm thick polymer anode. The sheet resistance of the polymer anode made by roll-to-roll coating is around 40 Ω/sq, when the thickness is 250 nm. The transmittance including PET substrate of the polymer anode is 74 % at 550 nm wavelength. For a thinner, 120 nm thick polymer anode, the sheet resistance is 110 Ω/sq and the transmittance is around 85 %.

Fig. 13 Comparison of transmittance between ITO-PET and R2R coated PH1000 (42 and 112 Ω/sq) in the visible wavelength range. Reproduced from [15] with permission of Wiley-VCH Verlag GmbH & Co. KGaA © 2013

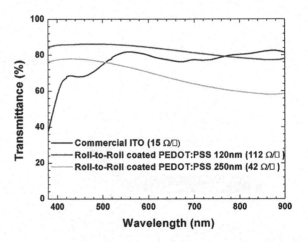

Fig. 14 The roll-to-roll coated polymer anode (PH1000) on the 8″ flexible PET substrate

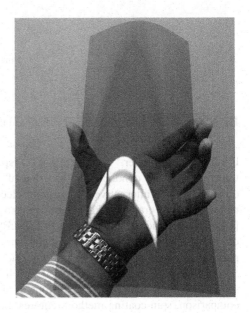

2.3.2 The Mechanical and Stability of the Transparent Polymer Electrode

The polymer electrode has better bending strength than ITO-PET as shown in Table 3. The variation of the sheet resistance of the film is less than ITO-PET. The ITO-PET is vulnerable to the bending stress: extensive cracks are developed if the ITO surface is subject to just one time excessive bending stress. Thus in the case of PLED, we could not observe the light emission from the device after the several bending cycles.

Table 3 The sheet resistance change of conducting polymer anode and ITO-PET following the change in bending angle

	0°	30°	60°	90°	120°	150°	180°
ITO PET (Ω/sq)	13.9	17.6	26.1	40.3	52.6	68.2	309.3
Polymer anode (Ω/sq)	68.8	69.0	69.0	69.1	69.2	69.5	69.7

The sheet resistance of ITO is significantly increased when the bending angle is increased. In contrast, the conducting polymer shows almost the same sheet resistance. Reproduced from [15] with permission of Wiley-VCH Verlag GmbH & Co. KGaA © 2013

2.3.3 Coating Evaluations of Functional Layers in the PLEDs

For large-scale devices, the thickness of the blade-only coating is nonuniform in the coating direction. Accordingly, the layer thickness in the initial state is usually greater than that in the final state for a large-scale device. Thickness variation is a serious problem associated with PLEDs, because difference in the functional layer thickness, in particular the very thin electron injection layer causes variation in the luminance in different areas. The final thinner area will be brighter than that of initial thicker area as shown in Fig. 15. Furthermore, an excessively thin area can even cause electrical shorts in the PLED layers.

Coating of the active layers in PLEDs using the blade-slit coating system can effectively address the above issues. The coating mechanism is similar to the slot-die coating. The main difference between the two is the ink supply. Organic electronic devices such as OPVs, OTFTs, and OLEDs have thin layers (from a few nanometers to a few hundreds of nanometers). Thus, it requires smaller feeding capacity of the solution than what typical slot-die coater delivers. Especially in the PLEDs illustrated above, the ZnO NP layer/Ionic complex layer as the electron transport/electron injection layer is even thinner: the total thickness of the two layers is only from 15 to 30 nm. Therefore, it requires less and homogeneous amount of ink supply. In this effort, we aimed at fabricating not only the hole injection layer, emissive layer, and electron transport layer but also the much thinner electron injection layer using the new blade-slit coating method. The blade-slit coating system we developed does not employ external pumping system, but utilizes only natural gravity and surface tension of the solution to flow out from the capillary to the surface of the substrate, which can effectively reduce the flow rate and the wet film thickness.

The following is a summary of the fabrication process. Each layer was fabricated at a temperature of 45 °C on the hot plate under ambient air conditions. The blade-slit speed was 15 mm/s. Sputtered ITO glass (15 Ω/sq) was cleaned beforehand by ultrasonic treatment in pure water, acetone, and IPA. It was then subjected to a UV-ozone treatment for approximately 1 h. A PEDOT:PSS layer (40 nm) was blade-slit coated onto the ITO glass, where the slit gap between the blade and the slide glass was 70 μm. The yellow light-emitting polymer (S-Y) dissolved in toluene at 0.6 wt% was then blade-slit coated (approximately 75 nm), with the slit gap of 210 μm. The ZnO NP layer (approximately 30 nm) was

Fig. 15 The light-emitting images of the blade-coated devices, non-uniformly coated by the only blade coating (*left*) and uniformly coated layer utilizing the slit nozzle

blade-slit coated onto the emissive layer, with a slit gap of 210 μm. Finally, the ionic solution was blade-slit coated onto the ZnO NP layer, with the slit gap of 350 μm. The ZnO NPs dispersed solution, dissolved in 1-butanol at a concentration of 30 mg/mL, was synthesized according to the method described by Beek et al. [24] The ZnO NPs appeared rather monodispersed with an average size of approximately 5 nm. The ZnO NPs layer was thin and porous, which allows the ionic solution containing TBABF$_4$ and PEO to permeate into the porous ZnO NP layer. The aluminium cathode (100 nm) was thermally evaporated under 2×10^{-6} torr in this demonstration, but could be replaced by the blade-coated Ag paste. The film uniformity of the emissive layer is easily verified by UV-light exposure as shown in Fig. 16. However, the other layer should be measured by surface profiler or SEM image. The amount of the solution for each layer to be coated is only 25 μL in area of 50 mm × 50 mm. The whole volume of the slit space is 550 μL. In addition, it is inexpensive and easy to change and clean the nozzle for the different solutions. The blade-slit coating employs transparent slide glass for the slit capillary, which is quite useful to observe the fluids flow in the slit nozzle.

2.3.4 Uniformity of the Functional Layers in PLEDs Coated by Blade-Slit Coating

As stated, the blade-slit coating improved the coating uniformity in the moving direction. The measured standard deviations of the blade-only coated PEDOT:PSS (hole injection layer) and Super Yellow (yellow light-emitting polymer, structure shown in Fig. 17) layers were about 5.7 and 5.7 nm, respectively, representing a nonuniformity of 7.9 and 9.1 %, respectively, of film thickness. These were calculated from thickness data measured at ten given positions (2 columns and 5 rows, each spaced 25 and 12 mm, respectively) in the area of the coated layer

Fig. 16 The UV-light irradiation for evaluation of the coating quality in the emissive layer, *left* films are spin-coated and *right* films are blade-coated films

Fig. 17 Molecular structure of the *yellow* light-emitting phenyl substituted poly(phenylene) vinylene

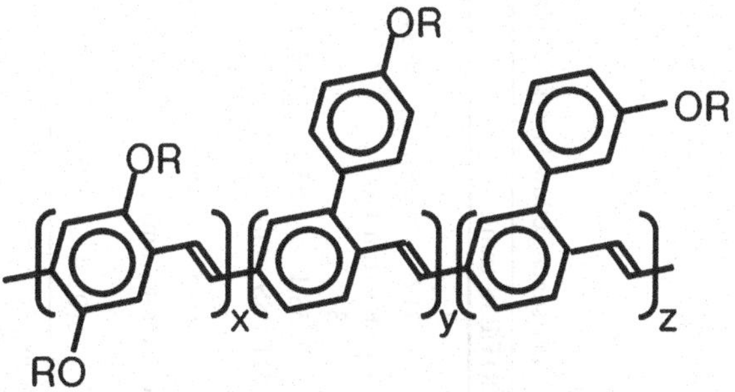

(substrate area: 70 mm × 80 mm). The lack of the uniformity of the blade-only coated layers created serious nonuniform luminance when the device was driven electrically. In comparison, in the blade-slit coated layer the standard deviation was only 0.68 and 2.3 nm for the PEDOT:PSS and emissive layer S-Y. So non-uniformity is only 2.1 and 2.2 % of their respective layer thicknesses (Table 4). It exhibited dramatic improvement in film uniformity compared to the blade-only coating. The electrical performances of PLED devices made of polymer anode are discussed in Sect. 3.

Table 4 Film uniformity of PEDOT:PSS and super yellow layers for blade-only coating versus blade-slit coating

Blade-only		1	2	3	4	5	6	7	8	9	10	Avg[a] (nm)	SD[b] (nm)	U[c] (%)
	Hole injection layer, PEDOT:PSS (nm)													
	Pos[d]	1	2	3	4	5	6	7	8	9	10			
	Thk[e]	73.1	79.3	71.6	76.7	75.0	78.0	73.3	61.1	63.2	73.1	72.4	5.7	7.9
	Light-emitting layer, super yellow (nm)													
	Pos[d]	1	2	3	4	5	6	7	8	9	10			
	Thk[e]	71.3	55.8	58.1	55.9	60.2	71.4	60.9	64.7	68.9	60.2	62.7	5.7	9.1
Blade slit	Hole injection layer, PEDOT:PSS (nm)													
	Pos[d]	1	2	3	4	5	6	7	8	9	10			
	Thk[e]	32.9	32.4	32.2	33.3	31.3	33.7	32.6	32.0	31.9	32.0	32.4	0.68	2.1
	Light-emitting layer, super yellow (nm)													
	Pos[d]	1	2	3	4	5	6	7	8	9	10			
	Thk[e]	102.2	100.0	102.9	105.4	104.3	102.2	100.2	100.1	106.1	105.7	102.9	2.3	2.2

Reproduced from [16] with permission from Elsevier © 2012

[a] Avg: average thickness of ten points (nm)

[b] SD: standard deviation (nm)

[c] U: non-uniformity (%): (SD/Avg)

[d] Pos: measured position

[e] Thk: thickness

Table 5 Thickness variations of layers for blade-slit coating method with respect to gap distance between the blade and substrate

Super yellow		ZnO nanoparticle	
Blade gap (μm)	Film thickness (nm)	Blade gap (μm)	Film thickness (nm)
30	60.0	45	15.3
35	75.0	55	20.4
40	91.0	65	30.0

Reproduced from [16] with permission from Elsevier © 2012

2.3.5 Solution Processable Electron Transport/Injection Layer

Low-work-function metals or compounds such as LiF, CsF, NaF, Cs_2CO_3, Ca, Ba, and Mg have often been used as electron injection/extraction layer [17–23]. In spite of the use of lithium fluoride and aluminium bi-layer cathodes, these alkali or alkaline-earth metal-halides as electron injection materials are still sensitive to oxygen and moisture, resulting in degradation of the OLEDs and consequently shorting their lifetimes. The ZnO NP layer used here is not only a stable electron injection layer but can also be fabricated by the solution coating process. In particular, the interface dipole formed by the $TBABF_4$ and PEO ionic liquid significantly lowers the electron injection barrier. Therefore, the ZnO NP and the ionic interlayer are appropriate for the solution process [10]. Because the ZnO NP layers were too thin to evaluate by the surface profiler, the coating thickness, and uniformity of the film were investigated with scanning electron microscopy (SEM). The film thickness of the ZnO NP layer was again controlled by the blade gap distance. The thickness of the ZnO NP layer, using a 30 mg/mL ZnO solution, was increased from 15.3 to 30.0 nm when the blade gap was increased from 45 to 65 μm. See Table 5. Surprisingly, the very thin ZnO NP layers can also be controllably coated using our blade-slit coating system.

2.4 Experimental Characterization of Polymer Light-Emitting Diodes

The device structures of the PLEDs are shown in Fig. 12. The device performance characteristics of the J (current density), V (voltage), L (luminance) curves were obtained from a Minolta CS-100 luminance meter and a Keithley 2400 source meter. The measurement process was conducted in ambient air conditions without encapsulation.

The control PLED devices were also fabricated by the spin coating process because it is easier to do on a small scale and also offers high efficiency. For a more accurate and fair comparison of blade-slit coating and spin coating, the layer thicknesses of the two devices were kept similar. As a result, the turn-on-voltages

Fig. 18 The image of uniform luminance of the large-area device (the substrate size was 50 mm × 100 mm) fabricated by the blade-slit coating. Reproduced from [16] with permission from Elsevier © 2012

and the maximum luminance voltage of the two devices were almost the same, as shown in Fig. 19a, indicating the blade-slit coating method can realize the same coating quality as that of spin coating but easily covers larger areas. In addition, Fig. 19b compares the performance (in terms of luminescence intensity) between the spin-coated devices and blade-slit coated devices. Though the blade-slit coated device was made in the air environment, the luminous efficiency of the blade-slit coated PLED reached 5.26 cd/A. The luminous efficiency of the spin-coated device fabricated in the Nitrogen-filled glove-box was 6.30 cd/A. These comparable results support the advantage of the blade-slit coating method, which is easier and less wasteful for materials for large-sized devices (100 mm × 50 mm) as shown in Fig. 18. Furthermore, the luminous efficiency of the PLEDs using the ZnO and ionic surfactant (6.30 cd/A) are over two-fold higher than that using Ca transporting layer (3.01 cd/A) as shown in Fig. 19b. The turn-on voltages of the two devices are almost the same as 2.2 V as shown in Fig. 19a. Since the peak emission wavelength of the S-Y is 560 nm as shown in the EL-spectra in Fig. 19c, the optical band gap of the S-Y is calculated at 2.2 eV, the same as turn-on voltage. This means that the devices have good electrical contact with Al cathode. In addition, it reveals that the ammonium ions can be effectively incorporated into the electron injection layer which lowers the electron injection barrier.

2.4.1 Roll-to-Roll Processed PLEDs

In this section, we show that the polymer anode (see Sect. 2.3) is superior to the ITO-PET in making flexible PLEDs and resulting in higher PLEDs performance. To evaluate the performance between the two devices, the two type devices were fabricated using the same process, and with similar film thicknesses. As a result, the turn-on-voltages and voltages at the maximum luminance of the two types of devices were very similar (Fig. 20a). It implies that the thickness of each layer in the two devices is about the same. The maximum luminance of ITO anode PLEDs was greater than that of polymer anode PLEDs. However, the maximum

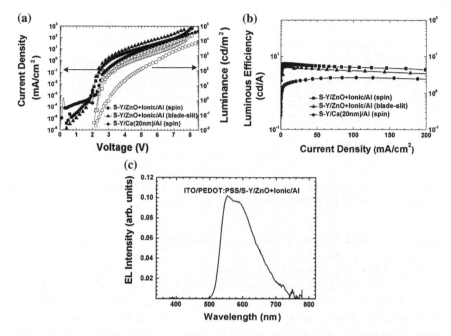

Fig. 19 **a** Current density, voltage, and luminance (J-V-L) characteristics with the spin-coated, the blade-slit coated, and the spin-coated Ca (20 nm)/Al devices. **b** Luminous efficiency of the spin-coated, the blade-slit coated, and the spin-coated Ca (20 nm)/Al devices. **c** The electroluminescence spectra at a voltage of 5 V [16]

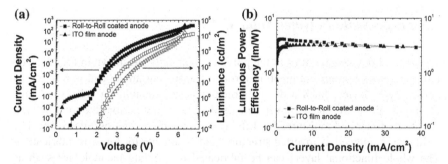

Fig. 20 The performance comparison of the ITO anode PLED and polymer anode PLED. **a** Current density, voltage, and luminance (J-V-L) characteristics with the ITO anode PLED and polymer anode PLED. **b** Luminous power efficiency of the ITO anode PLED and polymer anode PLED. Reproduced from [15] with permission of Wiley-VCH Verlag GmbH & Co. KGaA © 2013

luminance is only of secondary importance because commercialized devices are not driven at the extreme condition such as the maximum luminance. In this case, the PLED with polymer anode shows high performance in the range of commercial

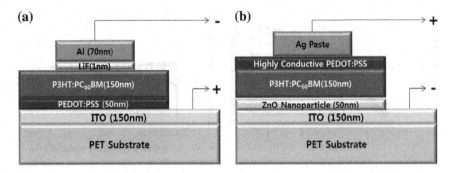

Fig. 21 The schematic diagrams of the two device structures, **a** bottom illuminated conventional structure and **b** bottom illuminated inverted structure

operation, about 2–6 V. In addition, the leakage current of the polymer anode PLEDs was lower than that of ITO anode PLEDs. That is because the polymer anode of PLEDs had improved surface roughness. The luminous power efficiency was 4.13 lm/W for PLED with polymer anode and 3.21 lm/W for that on ITO as shown in Fig. 20b. Consequently, the PLED made with polymer anode marked 28 % enhanced performance. This is because the polymer anode represents a higher transparency than the ITO around the wavelength of 560 nm, which is the emission peak in the EL-spectra of the SY. Furthermore, greatly suppressed roughness was important to enhance the device efficiency because it reduces leakage current between the anode and cathode.

2.5 Large-Scale Polymer Solar Cells

Two typical device structures of the polymer solar cells are shown in Fig. 21. The left one shows bottom illumination through transparent anode and is referred as normal type device. Such a device structure is not suitable for roll-to-roll processing as both LiF and Al layer will need vacuum deposition due to their high sensitivity to air and moisture, and LiF layer is extremely thin, of order 1 nm. The right one is refereed as inverted structure, and is suitable for low-cost fabrication. The whole functional layers can be fabricated by printing technologies such as roll-to-roll coating and screen printing.

2.5.1 Uniformity of the Functional Layers in OPVs by Blade Coating

The coating of PEDOT:PSS and ZnO NP layers as a hole extraction layer and electron transport layer in OPVs is the same as the PLEDs described in the previous section. The film uniformity by the blade-slit coating show better quality for the commonly used P3HT:PC$_{61}$BM (20 mg P3HT and 16 mg PC$_{61}$BM in 1.2 g

Table 6 Film uniformity of P3HT:PC$_{61}$BM blend film and iI-T3and PC$_{71}$BMblend film layers by spin-coating versus blade-slit coating

Spin	P3HT:PC$_{61}$BM									Avg[a]	SD[b]	U[c]
	Pos[d] 1	2	3	4	5	6	7	8	9	(nm)	(nm)	(%)
	Thk[e] 350.2	278.8	327.3	337.7	265.7	318.4	328.3	277.8	317.1	311.3	28.1	9.0
Blade	P3HT:PC$_{61}$BM											
	Pos[d] 1	2	3	4	5	6	7	8	9			
	Thk[e] 247.1	249.2	252.5	266.1	212.6	236.1	264.7	253.1	287.9	252.1	19.7	7.8
	Il-T3(Low-band gap polymer):PC$_{71}$BM											
	Pos[d] 1	2	3	4	5	6	7	8	9			
	Thk[e] 114.2	105.6	125.1	99.2	102.2	118.2	95.3	103.2	110.2	108.1	9.1	8.4

[a] *Avg*: average thickness of nine points (nm)
[b] *SD*: standard deviation (nm)
[c] *U*: nonuniformity (%): (SD/Avg)
[d] *Pos*: measured position
[e] *Thk*: thickness

Fig. 22 P3HT:PC$_{60}$BM and low band gap Il-T3:PC$_{71}$BM blend films by blade-slit coating method and measuring points by surface profiler

dichlorobenzens) as described in Table 6. As another example, uniform film of a low band gap polymer iL-T3 (10 mg iI-T3 [9] and 15 mg PC$_{71}$BM in 1.0 g dichlorobenzens) coated by the blade-slit coating is also shown in Table 6 and Fig. 22. The thickness data were measured at nine given positions shown in Fig. 22 over the coated area of 50 mm × 75 mm. Furthermore, the film quality by the blade-slit coating is better than that by spin-coating as shown in Fig. 23. Since the surface quality of the film can affect the device performance such as shunt

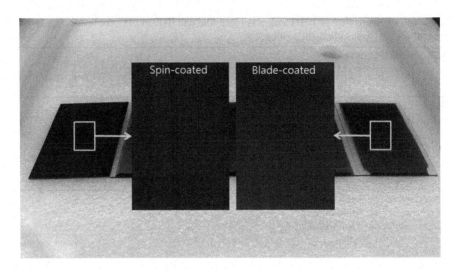

Fig. 23 The coating quality images of the blade-coated layers versus spin-coated layer. Defects can be seen in the latter

Table 7 The relationship between the blade speed and thickness of the P3HT:PCBM solution

Blade speed (mm/s)	8	12	16	20	24
Thickness (nm)	174.0	247.1	281.8	351.0	368.9

resistance and the fill factor, the overall power conversion efficiency will depend on the surface morphology. The film thickness can be controlled by modifying the coating speed. As discussed in the previous section, the film thickness is well matched by the theoretical relationship between the blade-coating speed and the film thickness as shown in Table 7. Higher speed leads to thicker film as shown in Fig. 24.

2.5.2 Device Performance in the Roll-to-Roll Processed Conventional Structure in Solar Cells

Conventional OPVs with the structure of ITO-glass/PEDOT:PSS/P3HT:PC$_{61}$BM/LiF/Al was fabricated as a control device. Among the functional layers, inorganic layers such as ITO, LiF, and Al cannot be coated by roll-to-roll process. However, the ITO can be replaced by polymer anode on PET substrate. The polymer anode was coated by roll-to-roll coating method. Low conductive PEDOT:PSS (AI4083) as a hole extraction layer and P3HT:PC$_{61}$BM donor-acceptor blend as polymer active layer were then roll-to-roll coated at air atmosphere. The layer thicknesses were 50 and 200 nm respectively. The LiF (1 nm) and Al (80 nm) were thermally evaporated in the vacuum chamber at a base pressure of 10^{-6} mbar.

Fig. 24 The images of the blade-coated layers on the glass substrate (50 mm × 75 mm) with respect to the different coating speeds

Table 8 The performance of OPVs fabricated by the roll-to-roll coating

	Jsc (mA/cm^2)	Voc (V)	FF (%)	PCE (%)	Rs (Ω cm^2)	Rp (Ω cm^2)
50 Ω/sq	−13.63	0.59	36.71	2.26	30.00	174.06
150 Ω/sq	−10.59	0.59	42.66	3.43	20.07	196.25

Polymer anodes with two different thicknesses were used to make the OPVs. The device utilizing thinner polymer anode has 50 % higher series resistance than the thicker anode (200 nm) as shown in Table 8. The power conversion efficiency of OPVs using thicker anode was 3.43 %, while that using thinner anode (100 nm) showed less performance due to the series resistance even though it has similar shunt resistance (Fig. 25). The impact on the device efficiency for large-scale devices are discussed in detail in Sect. 2.5.6.

2.5.3 Device Performance of the Inverted OPV Structure

The typical inverted OPV structure is ITO-glass/interfacial layer/P3HT:PCBM/ MoO3/Ag. We will examine the electrical performance for the different size devices. To fabricate the inverted OPVs, first ITO-glass was pre-cleaned by washing with DI water, acetone, and IPA, followed by O_2 plasma treated with 100 W for 1 min. The PEIE as the interfacial layer [27] and P3HT:PCBM layer was blade-slit coated except for the small area (1 mm diameter) control device which was spin-coated in the air atmosphere. The amino-groups in the PEIE polymer form an interfacial dipole between ITO and active layer, which can lower the work function of the ITO and collectively act as the cathode of the solar cell. The target layer thickness of the active layer was 250 nm, similar to that in the small area control device. The 6 nm MoO_3 and then 80 nm Ag were thermally evaporated under 10^{-6} mbar base pressure. From Table 9, the power conversion

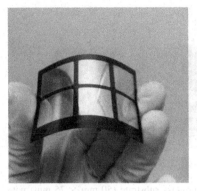
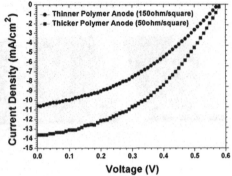

Fig. 25 The J-V characteristics of the roll-to-roll processed devices utilizing different thickness polymer anodes

Table 9 The device performances with respect to the device size which was fabricated by the roll-to-roll coating

	Jsc (A/cm^2)	Voc (V)	FF (%)	PCE (%)	Rs (Ωcm^2)	Rp (Ωcm^2)
0.00785 cm^2	11.57	0.58	53.64	3.60	8.66	468.9
4 cm^2	10.79	0.58	36.94	2.31	23.59	154.62
6 cm^2	10.66	0.58	38.94	2.41	22.87	181.56
12 cm^2	10.01	0.55	29.08	1.60	33.87	79.07

efficiency is reduced when the device size is increased. This is the result of a number of factors, including reduced fill factor, increased series resistance, and reduced shunt resistance. The blade-coated device showed better efficiency than the spin-coated device. The spin coated samples have lots of voids and aggregated particles in the polymer film as shown in Fig. 26. The shunt resistance of the spin coated device in the Table 10 is significantly reduced in device sized in 6 cm^2. This causes the lower fill factor and then lower efficiency.

2.5.4 Non-conventional Fabrication Methods Toward All-Solution Processed Polymer Solar Cells

As mentioned earlier materials such as ITO, MoO$_3$, and Ag have to be replaced with soluble materials to enable all-solution processes. Here we aim to replace the layers by polymer cathode, PEDOT:PSS and Ag nanoparticle ink, respectively, all in solution phase. To realize all-solution processable OPVs using these materials, the work function of the polymer anode should be modified with interfacial layer to create interface dipole and lower the work function [10]. Among the various materials, ammonium/amino-based organic surfactant is appropriate for the solution process such as roll-to-roll process. In addition, the PEDOT:PSS should has a

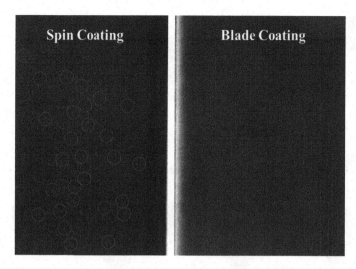

Fig. 26 The coating quality images of the blade-coated layers versus spin-coated layer. The circles indicate the voids on the P3HT:PC$_{61}$BM layer. The images were captured with X5 magnification rate

Table 10 The device performances of the blade-coated device and spin-coated device

	Jsc (A/cm^2)	Voc (V)	FF (%)	PCE (%)	Rs (Ω cm^2)	Rp (Ω cm^2)
Blade, 6 cm^2	10.66	0.58	38.94	2.41	22.87	181.56
Spin, 6 cm^2	11.21	0.50	35.27	1.98	22.17	95.06

good coat-ability on the hydrophobic polymer active layer. Moreover, the Ag nanoparticle ink should not affect the underlying layers during the coating process. Unfortunately, because the PEDOT:PSS is water-based polymer solution, it is unable to be coated on the polymer active layer. In addition, the Ag nanoparticle ink is easy to infiltrate and contaminate the underlying layer. To avoid these problems, we used nonconventional methods such as lamination or transfer technique to make the OPVs: the two layers will be coated on a separate substrate at first, and then the dried layers will be transferred to be laminated on the surface of the polymer active layer. After lamination, the contact between the two parts such as PEDOT:PSS and active polymer layer are not perfect. Thus, thermal annealing can improve the contact between the layers.

2.5.5 All-Solution Processed Lamination Technique

As a proof-of-concept experiment, the ITO-PET and ITO-glass were used to verify the feasibility of the lamination technique as an alternative process. The brief fabrication process is as follow (Fig. 27). The ZnO nanoparticle solution was

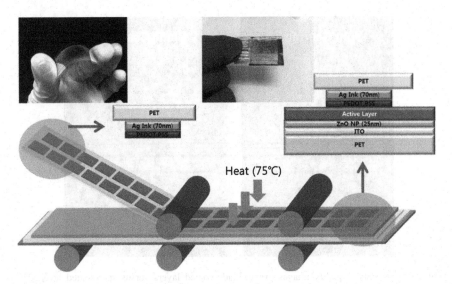

Fig. 27 Schematic diagram of the ideal lamination process and embedded images of the patterned Silver layer (*left*) and Laminated device (*right*)

spin-coated on the ITO-PET or ITO-glass. The P3HT:PC$_{61}$BM was spin-coated on the ZnO nanoparticle layer. On a separate PET substrate, the Ag nanoparticle ink was spin-coated, and then the PEDOT:PSS was spin-coated over the silver film. Finally, the two parts were laminated together on the hot plate at 130 °C for 5 min. The Ag electrode was patterned previously to have a role of the anode. Figure 35 shows the proposed continuous process to produce large area OPVs by using lamination technique. Because the metal film could not infiltrate through the voids or defects in the active layer, the lamination process is helpful to maintain the high shunt resistance in the large-scale devices. Moreover, the PET substrate acts as a barrier film preventing the oxygen and moisture from diffusing into the Ag layer and the active polymer layer.

The power conversion efficiency of the laminated device is 2.76 % using the ITO glass in the small area device (1 mm diameter, Fig. 28) as summarized in Table 11. The device performance of the ITO-glass is better than the ITO-PET, because the sheet resistance of the ITO-glass was 10 Ω/sq and the ITO-PET was 20 Ω/sq.

For larger device low-band gap polymer (iI-T3) was used. The ZnO nanoparticle layer and polymer active layer was blade-coated to 30 and 100 nm thickness respectively. The device performance is promising for the large area devices (2 and 3 cm^2). Generally, when the device size is increased, shunt resistance is decreased. The shunt resistance of the lamination processed device is higher than that of the evaporated device. In the large-size device, the efficiency of the optimized device was 2.27 % for 3 cm^2 devices as shown in Table 12. The series resistance was reduced after the thermal annealing (Table 14), due to improved contact at the laminated interface.

Fig. 28 The device performance of the laminated devices using ITO-glass and ITO-PET

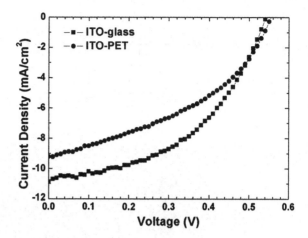

Table 11 The device performances fabricated by the lamination process

Device	Jsc (mA/cm^2)	Voc (V)	FF (%)	PCE (%)
ITO-PET	9.21	0.56	40.95	2.11
ITO-glass	10.87	0.55	46.1	2.76

2.5.6 Influence of Series Resistance in Large Area Solar Cells

We have already seen the impact of series resistance in large area device to the achievable OPV efficiency. Practical large area solar cell devices suffer more from resistive losses, which should be minimized to maintain the performance of the devices. We will quantify the effect of series resistance, R_S by computing the resistive power loss. In small area devices, organic semiconducting active layers contribute more to the resistive power loss than other factors. However, as the size of the devices increases, the R_S of the transparent electrodes become the main factor of the resistive loss [17]. The total resistive power loss per unit area is given by

$$P_R = \frac{R_S (J_{max}A)^2}{A} = R_S A J_{max}^2, \tag{15}$$

where P_R is the total resistive power loss per unit area, R_S is the series resistance in the device, J_{max} the current density, and A the area of the devices. R_S in organic solar cell contains the resistances of the anode, active layer, contacts, and cathode:

$$R_S = R_{anode} + R_{active} + R_{contacts} + R_{cathode}, \tag{16}$$

where R_{active} and $R_{contacts}$ are the series resistances of active layer and contacts of each layer, respectively. When we assume the conventional structure of organic solar cell, cathode side is the reflective metal and the anode side is the transparent

Table 12 The performances in the large-scale device fabricated by the lamination process

	Jsc (A/cm^2)	Voc (V)	FF (%)	PCE (%)	Rs (Ω cm^2)	Rp (Ω cm^2)
Non	3.62	0.73	37.59	0.99	93.26	516.62
100 °C, 5 min	6.58	0.71	44.94	1.50	34.53	506.63
100 °C, 10 min	7.04	0.69	46.71	2.27	21.59	371.70

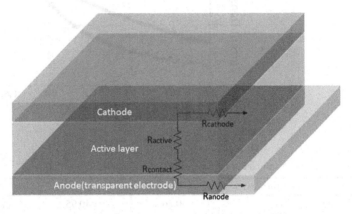

Fig. 29 Description of series resistance components inside conventional OPV cells

electrode. R_{active} and R_{contacts} does not increase with area scaling since they are vertical components in the cell which have the same carrier travel distance regardless of cell area. Furthermore, the metal cathode has negligible series resistance compared to transparent anode electrode such as ITO, with typical conductivity ratio of $\sim 100{:}1$ [18]. Therefore, the main factor that determines the resistive loss with increasing device size is the resistance of transparent electrode. Figure 29 shows each component inside the cell graphically.

The effect of the series resistance can be studied in more detail with the non-ideal equivalent circuit model, which is widely acceptable for both inorganic solar cells and organic solar cells [19]. With the parasitic resistances included, the diode equation becomes

$$J(V) = J_0 \left(e^{\frac{q(V - J(V)R_S A)}{nkT}} - 1 \right) + \frac{V - J(V)R_S A}{R_{sh}A} - J_{\text{ph}}, \qquad (17)$$

where J_0 is the reverse saturation current, V is the applied voltage, J is the current density of the cell, n is the diode ideality factor, k is Boltzmann's constant, T is temperature, R_S is series resistance, R_{sh} is shunt resistance and J_{ph} is the photocurrent generated by the cell. The parameters J_0, n, R_S, R_{sh}, and J_{ph} can be obtained by fitting the diode model to the experimental J–V data from actual OPV devices under illuminated and dark conditions. Servaites et al. [20] showed the impact of R_S on the performance of large area P3HT:PCBM BHJ OPVs using ITO anode.

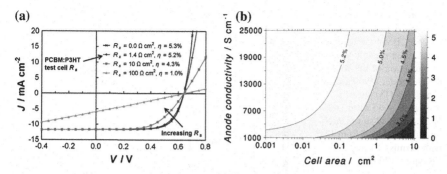

Fig. 30 **a** The effect of R_S variation on projected J–V characteristics for the P3HT:PCBM test cell. **b** The effect of anode conductivity and cell area on cell power conversion efficiency for the P3HT:PCBM test cell. Reproduced from [20] with permission of Wiley-VCH Verlag GmbH & Co. KGaA © 2010

Figure 30 shows the effect of R_S variation on the J–V curves for the P3HT:PCBM test cell. In the graph, R_S is the only parameter that changes and all the other parameters in Eq. (16) are kept constant. The results show that the resistive losses due to R_S remain negligible with cell areas under ~ 0.1 cm². However, as cell area increases over 0.2 cm², the efficiency starts to drop significantly due to steep increase of anode series resistance. Figure 30b shows the relationship between anode conductivity and cell area on power conversion efficiency of the P3HT:PCBM device, illustrating the importance of the anode conductivity as the cell size increases.

2.5.7 Design Strategy of Transparent Electrode for Low Series Resistance

To reduce the series resistance of the transparent electrode of the large area OPVs, metal grid patterns can be added to the anode side to reduce the resistive loss. Deliberate design of electrode geometry is imperative. Several strategies have been employed to reduce the negative effect of the series resistance on OPVs including bus bar design [17], ring type design [21], stripe type design [22], and deep trench type metal grating design with conductive polymer [23]. Stripe design of the large area OPVs is the most popular type due to its simplicity and adaptability to roll-to-roll process. The resistive losses from relatively high resistance of the transparent electrode can be minimized by reducing the travel distance of the photo-generated charges at the transparent electrode. The simplest way to achieve this is by putting at least one side of metal lines as close as possible to the regions of photocurrent generation. In other words, the width of the PV cell stripe should be kept narrow with the contacts put on the long sides while the length of the stripe does not matter as long as each layer of the solar cell device has no defects or discontinuities. The effects of the width of the stripe in organic solar cells were

Fig. 31 Calculated power conversion efficiency of a P3HT:PCBM-based single *rectangular* organic solar cell as a function of the width of the electrode. Two cases are shown, an ITO sheet resistance of 15 Ω/sq (*solid line*) and 60 Ω/sq (*dashed line*). Reproduced from [24] with permission from Elsevier © 2007

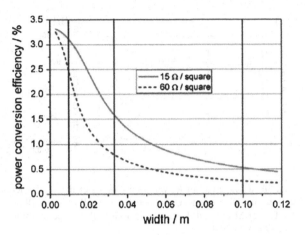

investigated by Lungenschmied et al. [24]. They calculated theoretical power conversion efficiencies of a P3HT:PCBM BHJ cells with two cases of ITO sheet resistance, 15 and 60 Ω/sq. As shown in Fig. 31, the rapid decrease of the power conversion efficiency with increasing width of the rectangular solar cell can be observed while the length of the device kept same, indicating that the performance of stripe typed large area OPV devices are mainly affected by the width of the stripes as a result of the series resistance of the transparent electrode.

Another effective way to reduce the R_S of transparent electrode is the combination of metal mesh and PEDOT:PSS to serve as transparent electrode [25]. By increasing the metal width and reducing the mesh period, the sheet resistance can be made smaller than the conventional ITO electrode. A trade-off between the optical transmittance and electrical conductivity still exists, though much smaller than that of ITO [25]. This trade-off can be further compensated by embedding high aspect-ratio metal mesh into conductive polymer, which eliminates the low optical transmittance problem without sacrificing the conductivity. Kuang et al. [23] fabricated a nanoscale metallic grating with optical transparency over 80 % and low electrical resistance under 2.4 Ω/sq. To make the structure, a polyurethane (PU) grating structure was prepared for oblique metal deposition and argon ion milling process to get the metal layers on the PU sidewalls. The typical sheet resistance of the patterned structures with gold and silver were 9.6 and 3.2 Ω/sq, respectively, which are lower than the sheet resistance of ITO for typical applications of OPVs and OLEDs.

2.5.8 Fine Metal Mesh Design for Transparent Electrode

Previously, we demonstrated normal type OPVs by using fine transparent metal mesh electrodes (TME) made by nanoimprinting [25] and transfer-printing [26] techniques. Recently, we fabricated TME by conventional photolithography

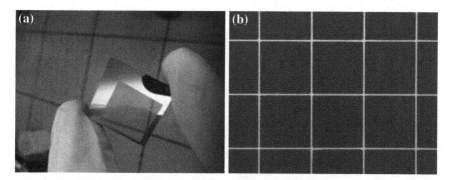

Fig. 32 **a** The fabricated transparent metal electrode on glass. **b** The optical microscope image of the fabricated transparent silver electrode. The line width of silver is 5 μm and the distance between lines is 150 μm

Fig. 33 UV–Vis transmittance spectra of bare glass, ITO glass, and the transparent Ag, Al mesh electrodes

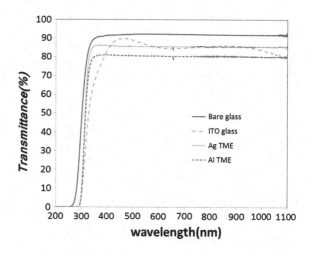

technique and successfully fabricated inverted type OPVs with the TMEs using aluminum and silver with 150 μm cell size and 5 μm line width (Fig. 32). The thickness of the patterned TMEs for both metals was 50 nm. Figure 33 shows the UV–vis transmittance spectra from 200 to 1100 nm wavelength range of the TME. The transmittance of Ag TME substrate was 85 %, which is comparable to ITO glass. The transmittance of Al TME was 80 %, which is slightly lower than ITO glass and Ag TME substrate. Inverted solar cells were fabricated with structure shown in Fig. 34. On the TME substrates, PEDOT:PSS was spun-coated to planarize the TME patterns. Polyethylenimine ethoxylated (PEIE) solution was coated subsequently to reduce the work-function of the PEDOT:PSS for it to function as a cathode for extracting electrons [27]. Both P3HT:PC$_{61}$BM and a low band gap polymer PIDT-phanQ:PC$_{71}$BM [28] bulk heterojunction system were

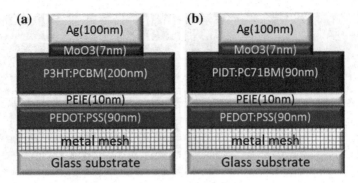

Fig. 34 The device structure of the inverted TME solar cell for **a** P3HT:PCBM bulk heterojunction. **b** PIDT-phanQ:PCBM bulk heterojunction

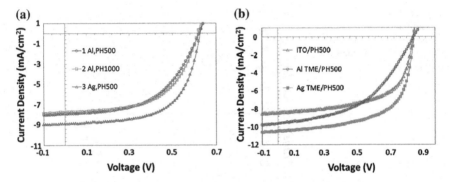

Fig. 35 J-V characteristics of organic solar cells fabricated using different transparent metal electrodes. **a** P3HT:PCBM BHJ inverted cells with Al TME/PH500, Al TME/PH1000, and Ag TME/PH500. **b** PIDT-phanQ:PC71BM BHJ inverted cells with ITO glass/PH500, Al TME/PH500 and Ag TME/PH500

used to fabricate the solar cells. MoO_3 and silver were deposited sequentially for the anode formation.

To compare the performance of the TME substrates for organic solar cell application two PEDOT:PSS materials with different conductivity, pristine PH500 $\sigma = 0.2$ S/cm) and PH1000 ($\sigma = 0.4$ S/cm) purchased from HC Starck, were used to compare the performance of Al and Ag TME electrodes. The current density versus voltage characteristics are shown in Fig. 35. The figures of merits are summarized in Table 13. When we compare the P3HT:PC_{61}BM BHJ device with Al TME and Ag TME (Table 13), the OPV with Ag TME showed superior performance than Al TME due to the higher conductivity of Ag. In Table 14, the Ag TME-based OPV devices were compared with ITO based OPV devices for the PIDT-phanQ:PC_{71}BM BHJ system. The OPV with Ag TME showed the best performance with higher current density and fill factor compared to the OPV with

Table 13 Device characteristics of solar cells fabricated by transparent metal mesh of (a) P3HT:PCBM BHJ inverted cell

P3HT:PCBM	Jsc (mA/cm^2)	Voc (V)	FF (%)	PCE (%)	Rs (Ω cm^2)	Rsh (Ω cm^2)
Al TME/PH500	7.90	0.63	52.46	2.62	19.52	3588.17
Al TME/PH1000	8.20	0.63	56.22	2.89	11.85	767.65
Ag TME/PH500	9.01	0.64	61.57	3.53	6.29	1932.42

Table 14 Device characteristics of solar cells fabricated by transparent metal mesh of PIDT-phanQ:PC71BM BHJ inverted cell

PIDT-phanQ:PC$_{71}$BM	Jsc (mA/cm^2)	Voc (V)	FF (%)	PCE (%)	Rs (Ω cm^2)	Rsh (Ω cm^2)
ITO/PH500	8.53	0.87	59.62	4.41	5.95	1320.41
Al TME/PH500	9.66	0.88	46.90	3.99	32.10	591.74
Ag TME/PH500	10.54	0.88	61.49	5.68	4.14	1927.38

Al TME and ITO electrode. One of the reasons is that Ag TME with PH500 show the lowest R_S. Once again these results show the importance of R_S in determining the OPV overall efficiency.

3 Conclusions and Discussion

This chapter described various coating methods that can be used for large area PLED and OPVs. The coating mechanisms and instability issues in practical solution process have been examined. Using the various coating process such as roll-to-roll coating, blade coating, blade-slit coating, and spin coating, the coating performance was compared in terms of the film uniformity and controlling the coating instabilities. The blade and roll-to-roll coating are more favorable for the large-scale device in terms of the device performance and the manufacturing cost. Metal mesh structures in combination with conductive polymers was introduced as the ITO-free transparent electrode. Finally, nonconventional process such as lamination and transfer techniques was used to improve the shunt resistance and the series resistance in the large-scale and vacuum-free processed devices. However, several other issues remained to be overcome such as the lifetime stability and environmentally friendly solvents, and materials costs for the realization toward $1/W OPVs. The polymer electronics are growing fields, more research efforts are needed to improve the device performance, lifetime, process-ability, and large area manufacturing, which will bring the bright future of the polymer electronics to reality.

References

1. Tang CW (1986) Two-layer organic photovoltaic cell. Appl Phys Lett 48:183–185
2. Baldo MA, O'Brien DF, Thompson ME, Forrest SR (1999) Excitonic singlet-triplet ratio in a semiconducting organic thin film. Phys Rev B 60:14422–14428
3. Yu G, Gao J, Hummelen JC, Wudl F, Heeger AJ (1995) Polymer photovoltaic cells: enhanced efficiencies via a network of internal donor-acceptor heterojunctions. Science 270:1789–1791
4. He Z, Zhong C, Su S, Xu M, Wu H, Cao Y (2012) Enhanced power-conversion efficiency in polymer solar cells using an inverted device structure. Nat Photon 6:591–595
5. Espinosa N, García-Valverdea R, Krebs FC (2011) Life-cycle analysis of product integrated polymer solar cells. Energy Environ Sci 4:1547–1557
6. Hall DB, Underhill P, Torkelson JM (1998) Spin coating of thin and ultrathin polymer films. Polym Eng Sci 38:2040–2045
7. Weinstein SJ, Ruschak KJ (2004) Coating flows. Annu Rev Fluid Mech 36:29–53
8. Landau L, Levich B (1942) Dragging of a liquid by a moving plate. Acta Physicochim URSS 17:42–54
9. Stalder R, Grand C, Subbiah J, So F, Reynolds JR (2012) An isoindigo and dithieno [3,2-b:2′,3′-d]silole copolymer for polymer solar cells. Polym Chem 3:89–92
10. Youn H, Yang M (2010) Solution-processed polymer light-emitting diodes utilizing a ZnO/organic ionic interlayer with Al cathode. Appl Phys Lett 97:243302
11. Yu Z, Zhang Q, Chen LLQ, Niu X, Liu J, Pei Q (2011) Highly flexible silver nanowire electrodes for shape-memory polymer light-emitting diodes. Adv Mater 23:664–668
12. Han T-H, Lee Y, Choi M-R, Woo S-H, Bae S-H, Hong BH, Ahn J-H, Lee T-W (2011) Extremely efficient flexible organic light-emitting diodes with modified graphene anode. Nat Photon 6:105–110
13. Li J, Hu L, Wang L, Zhou Y, Grüner G, Marks TJ (2006) Nano Lett 6:2472–2477
14. Lee BH, Park SH, Back H, Lee K (2011) Novel film-casting method for high-performance flexible polymer electrodes. Adv Mater 21:287–493
15. Shin S, Yang M, Guo LJ, Youn H (2013) Noble roll-to-roll cohesive, coated, flexible, high-efficiency polymer light-emitting diodes utilizing ITO-free polymer anodes. Small 9:1–9
16. Youn H, Jeon K, Shin S, Yang M (2012) All-solution blade-slit coated polymer light-emitting diodes. Org Electron 13:1470–1478
17. Choi S, Potscavage WJ, Kippelen B (2009) Area-scaling of organic solar cells. J Appl Phys 106:054507
18. Yang Y, Jin S, Medvedeva JE, Ireland JR, Metz AW, Ni J, Hersam MC, Freeman AJ, Marks TJ (2006) CdO as the archetypical transparent conducting oxide. Systematics of dopant ionic radius and electronic structure effects on charge transport and band structure. JACS 127:8796–8804
19. Cheknane A, Hilal HS, Djeffal F, Benyoucef B, Charles J-P (2008) An equivalent circuit approach to organic solar cell modelling. Microelectron J 39:1173–1180
20. Servaites JD, Yeganeh S, Marks TJ, Ratner MA (2010) Efficiency Enhancement in organic photovoltaic cells: consequences of optimizing series resistance. Adv Funct Mater 20:97–104
21. Krebs FC, Jørgensen M, Norrman K, Hagemann O, Alstrup J, Nielsen TD, Fyenbo J, Larsen K, Kristensen J (2009) A complete process for production of flexible large area polymer solar cells entirely using screen printing—first public demonstration. Sol Energy Mater Sol Cells 93:422–441
22. Angmo D, Hösel M, Krebs FC (2012) All solution processing of ITO-free organic solar cell modules directly on barrier foil. Sol Energy Mater Sol Cells 107:329–336
23. Kuang P, Park J-M, Leung W, Mahadevapuram RC, Nalwa KS, Kim T-G, Chaudhary S, Ho K-M, Constant K (2011) A new architecture for transparent electrodes: relieving the trade-off between electrical conductivity and optical transmittance. Adv Mater 23:2469–2473

24. Lungenschmied C, Dennler G, Neugebauer H, Sariciftci SN, Glatthaar M, Meyer T, Meyer A (2007) Flexible, long-lived, large-area, organic solar cells. Sol Energy Mater Sol Cells 91:379–384
25. Kang M-G, Kim M-S, Kim J, Guo LJ (2008) Organic solar cells using nanoimprinted transparent metal electrodes. Adv Mater 20:4408–4413
26. Kang M-G, Park HJ, Ahn S-H, Guo LJ (2010) Transparent Cu nanowire mesh electrode on a flexible substrate fabricated by simple transfer printing and its application in organic solar cell. Sol Energy Mater Sol Cells 94:1179–1184
27. Zhou Y, Fuentes-Hernandez C, Shim J, Meyer J, Giordano AJ, Li H, Winget P et al (2012) A universal method to produce low–work function electrodes for organic electronics. Science 336:327–332
28. Zhang Y, Zou J, Yip H-L, Chen K-S, Zeigler DF, Sun Y, Jen AK-Y (2011) Indacenodithiophene and quinoxaline-based conjugated polymers for highly efficient polymer solar cells. Chem Mater 23:2289–2291

Low-Cost Nanomaterials
for Photoelectrochemical Water Splitting

Gongming Wang, Xihong Lu and Yat Li

Abstract Hydrogen represents a clean and high gravimetric energy density chemical fuel that could potentially replace fossil fuels and natural gas in electricity generation and powering vehicles. Central to the success of hydrogen technology and economy, the sustainability, efficiency and cost of hydrogen generation are the major factors. Industrial hydrogen is currently obtained from steam methane reforming and water-gas shift reaction, however, this method still relays on fossil fuels. Therefore, it is important to develop efficient, low-cost, and scalable method to produce hydrogen in a sustainable manner. Photoelectrochemical (PEC) water splitting to produce hydrogen is one of most promising and sustainable approaches. The development of low-cost and efficient nanostructured photoelectrodes is the key to achieve this goal. In this chapter, we will give a brief background on PEC water splitting and review the recent advancement of developing low-cost nanostructured photoelectrodes.

1 Background of Photoelectrochemical Water Splitting

Photogeneration of hydrogen by water splitting has attracted a lot of attentions, because water is the most abundant source of hydrogen carrier on the earth [23, 25, 42, 65, 89, 92, 96, 105]. However, water splitting is known to be an uphill reaction:

$$H_2O(l) \rightarrow H_2(g) + 1/2O_2(g) \quad \Delta G = \sim 237.2 \text{ KJ/mol.}$$

This process can be achieved via electrolysis with applied external potentials and a minimal potential of 1.23 V is needed to drive this reaction. Due to the

G. Wang · X. Lu · Y. Li (✉)
Department of Chemistry and Biochemistry, University of California,
Santa Cruz, CA 96064, USA
e-mail: yatli@ucsc.edu

Z. Lin and J. Wang (eds.), *Low-cost Nanomaterials*, Green Energy and Technology,   267
DOI: 10.1007/978-1-4471-6473-9_10, © Springer-Verlag London 2014

overpotential on anode and cathode electrodes, practical potential needed to drive water splitting is larger than 1.8 V. A promising solution is using renewable energy to split water, for example photoelectrolysis [11]. There are two approaches to utilize solar energy for electrolysis. For instance, electrolyzer can be coupled with photovoltaic (PV) cells, which provide photovoltage for water electrolysis [11]. However, the relative high cost of PV cells and electrolyzer could limit its practical application. Alternatively, water splitting can be achieved in a photo-electrochemical (PEC) cell [11]. PEC cells are consisted of semiconductor pho-toelectrodes for harvesting solar light. The photogenerated electrons and holes can reduce and oxidize water to produce hydrogen and oxygen gas, respectively. In order to compete with PV/electrolyzer system, the key of this approach is to use low-cost semiconductor materials as photoelectrodes.

The concept of PEC water splitting was first demonstrated by Honda and Fujishima in 1972 [17]. Figure 1 shows the schematic diagram of PEC cell that consists of an n type semiconductor as photoanode and counter electrode [23]. Upon light illumination, photoexcited electron-hole pairs will be generated in the photoelectrode. The electron-hole pairs will be separated due to the band bending at the electrolyte/semiconductor interface and/or the application of external bias. The holes and electrons will oxidize and reduce water to produce oxygen and hydrogen, respectively. The reaction can be described by the following equations:

$$\text{Photanode: } H_2O + 2h^+ \rightarrow 2H^+ + 1/2O_2 \quad E^o_{ox} = -1.23\,V$$
$$\text{Cathode: } 2H^+ + 2e^- \rightarrow H_2 \quad\quad E^o_{red} = 0\,V.$$

According to the Nernst equation, a minimum energy of 1.23 V is needed to drive water electrolysis. Therefore, in order to achieve hydrogen evolution and oxygen evolution reactions simultaneously under light illumination, the band-gap energy (E_g) of the semiconductor photoelectrode must be larger than 1.23 eV and the band-edge potentials (conduction band and valence band) should straddle the electrochemical potentials of $E^o(H^+/H_2)$ and $E^o(O_2/H_2O)$. Moreover, a favorable semiconductor photoelectrode should able to absorb a significant portion of solar light, and has low kinetic overpotential for water oxidation and reduction, good corrosion resistivity, and electrochemical stability in aqueous solution [92].

In order to properly evaluate the performance of photoelectrodes, it is critical to define the parameters to characterize solar to hydrogen (STH) conversion effi-ciency. The PEC properties of photoelectrodes are typically studied in a three electrode electrochemical cell system, which consists of a working electrode (photoelectrode), a reference electrode, and a counter electrode. For the reference electrode, the internationally accepted primary reference is the standard hydrogen electrode (SHE) or reversible hydrogen electrode (RHE), which has all compo-nents at unit activity [(Pt/H$_2$ (a = 1)/H$^+$ (a = 1, aqueous)]. However, this kind of reference electrode is impractical, and therefore other reference electrodes such as saturated Ag/AgCl or saturated calomel electrode (SCE) are normally used in the measurement. In this case, the experimentally obtained potentials should be

Fig. 1 Principle of operation of photoelectrochemical water splitting cells based on *n* type semiconductors. Reproduced with permission from [23]

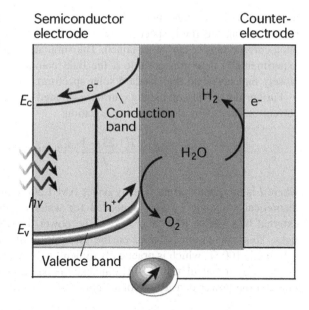

converted to the potential versus RHE for calculating the solar conversion efficiency. The conversion equation is expressed as followed [36]:

$$E_{RHE} = E_{Ag/AgCl} + 0.059\,pH + E^o_{Ag/AgCl}$$

where E_{RHE} is the potential against RHE; $E_{Ag/AgCl}$ is the applied potential versus Ag/AgCl reference electrode, pH is the pH value of the electrolyte solution and $E^o_{Ag/AgCl}$ is the potential of reference electrode (Ag/AgCl) versus RHE electrode. Platinum (Pt) structures (e.g., wire, plate, mesh) are typically used as counter electrode, because Pt has the lowest overpotential for hydrogen evolution. The most common parameter to evaluate the PEC performance of photoelectrodes is the photocurrent density obtained under standard one sun illumination (AM 1.5G, 100 mW/cm². However, the photocurrent density can vary based on the power and the model of solar simulator and filters used for the experiment. Alternatively, the electrode performance can also be evaluated by incident photon to current conversion efficiency (IPCE). IPCE measurement presents the photoactivity of the electrode as a function of monochromatic incident light wavelength. It can be calculated with the following equation:

$$IPCE = \frac{(1240\,eV \cdot nm) \times (I\,\mu A/cm^2)}{(\lambda\,nm) \times (J\,\mu W/cm^2)} \times 100\,\%$$

where I is the photocurrent density obtained under the illumination of the monochromatic light, λ is the wavelength of the monochromatic light, J is the power density of the monochromatic light. The equation shows that the IPCE measurement is independent of the power and model of the light source. Moreover, by integrating the IPCE spectrum with the standard solar spectrum, a simulated photocurrent density can be obtained. The simulated value should be close to the experimentally determined value if the light coming from the solar simulator is closely matched with the standard solar spectrum.

Furthermore, STH efficiency can be estimated based on the simulated photo-current density using the following equation:

$$\eta = \frac{I(1.23 - V_{\text{bias}})}{J_{\text{light}}} \times 100\,\%$$

where I is the photocurrent density under AM 1.5G light illumination; 1.23 is the theoretical value of voltage is needed for water splitting; V_{bias} is the applied external bias versus RHE and J_{light} is the power density of the AM 1.5G white light. This calculation is based on the assumption of faradic efficiency of water splitting is 100 %, which is practically difficult to be achieved. Therefore, the most precise way to calculate the STH efficiency is based on the hydrogen production rate and the power density of solar light.

1.1 Low-Cost Metal Oxide Nanomaterials: Promising Photoelectrodes for Water Splitting

Nanostructured electrodes have potential advantages over their bulk counterparts. Nanomaterials provide not only extremely large surface area for oxidation and reduction reactions, but also a short diffusion length for minority carriers that can improve the charge separation efficiency and reduce the loss through electron-hole recombination. Moreover, nanostructured electrodes could increase light scattering and absorption, compared to planar structure. With such potentials, nanostructured materials could open up new opportunities in improving the performance of photoelectrodes.

A number of semiconductor nanomaterials have been studied as photoelectrodes for PEC water splitting, including metal oxides, metal chalcogenides, and III-V compounds. Figure 2 shows the band edge positions of common semiconductors in aqueous electrolyte with pH 1 [23]. However, not a single semiconductor material can meet all the technical requirements discussed above. For example, metal chalcogenides such as CdS and CdSe are photocorrosive. Hole scavengers such as Na_2S and Na_2SO_3 have to be added to protect the photoelectrode from self-oxidation [28, 50, 94, 102, 103]. Likewise, electrochemical instability is also a major concern for III-V semiconductor materials [112]. Additionally, the cost of III-V materials such as GaN ($1,900 for 2 inch bulk GaN wafer) and GaP ($50/g) is relatively high, which could limit their applications [87, 93]. In terms of chemical stability and materials cost, metal oxides hold great promise as photoelectrode materials for PEC water splitting. Metal oxides such as TiO_2 ($2–3/Kg), ZnO ($1–3/Kg),

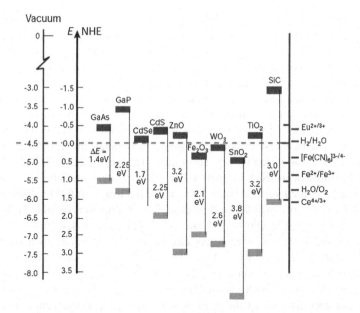

Fig. 2 Band edge positions of several semiconductors in contact with aqueous electrolyte at pH 1 with energy scale versus vacuum level and the normal hydrogen electrode (NHE). Reproduced with permission from [23]

WO_3 ($40–60/Kg), and Fe_2O_3 (less than $1/Kg) have been extensively studied [23, 96]. In this chapter, we will focus on the synthesis with low-cost techniques and discuss their application in PEC water splitting.

2 Low-Cost Synthetic Methods

During the last two decades, a number of methods have been demonstrated for the synthesis of metal oxide nanomaterials. Among these methods, hydrothermal and solvothermal synthesis, sol–gel, electrochemical deposition, and anodization method show great promise as low-cost and scalable approaches to prepare nanomaterials for photoelectrodes, as these methods require simple equipment and mild synthetic conditions. In this section, we are going to give a brief overview of these low-cost synthetic methods.

2.1 Hydrothermal and Solvothermal Method

Hydrothermal synthesis is one of the most extensively used approaches to prepare metal oxide nanomaterials such as TiO_2 [97, 101], WO_3 [33, 85, 99], ZnO [24, 102, 111] and Fe_2O_3 [47, 48]. Hydrothermal synthesis is typically performed in a closed system such as autoclave filled with precursor solution. By heating the autoclave, the chemical reaction can be carried out under controlled temperature (typically <300 °C) and pressure. The reaction temperature, pressure, and concentration of reactants play important roles in determining the material properties of the products such as dimension, morphology, and crystal phase. For example, Greene et al. developed a facile seed-mediated hydrothermal method for the preparation of vertically aligned ZnO nanowire arrays [24]. As shown in Fig. 3, the entire 4-inch wafer can be covered with a highly uniform and densely packed array of ZnO nanowires. Importantly, this is a general strategy to grow ZnO nanowire arrays on virtually any substrates. The seed-mediated hydrothermal method has also been used to grow other metal oxide materials. For example, Hoang et al. prepared vertical rutile TiO_2 nanowire arrays on FTO substrates by a simple seed-assistant hydrothermal method. These single-crystalline TiO_2 nanowires have average diameter of ~ 5 nm and length up to ~ 4.4 μm [31]. Grimes et al. fabricated WO_3 nanowire arrays directly on FTO substrate by a seed-mediated solvothermal method [85]. WO_3 nanowires and nanoflakes with hexagonal and monoclinic structure could be readily obtained by adjusting the precursor composition. Moreover, hydrothermal methods have also been used to grow metal oxide nanomaterials without the need of seed layer [33]. For instance, Lee et al. synthesized WO_3 nanocrystals using a hydrothermal process as a precursor and studied the effect of annealing temperature on their morphology and photocatalytic performance [33]. Zhong et al. utilized a solvothermal method to synthesize 3D flower-like hematite nanostructure by using an ethylene glycol-mediated self-assembly process [125]. In comparison to other gas-phase synthetic methods such as chemical vapor deposition, atomic layer deposition, solution-based hydrothermal methods are simpler, cheaper, and more scalable, which offer significant advantages for large scale production of photoelectrodes for water splitting.

2.2 Sol–gel Methods

Sol–gel methods typically involve either dip-coating or drop-coating colloidal precursors directly onto conductive substrates, followed by calcination at various temperatures. They are well-developed growth techniques and are known for convenience and environmental friendliness. For example, the sol–gel spin-coating process provides a simplified fabrication route for nanolayers, as it eliminates the need of vapor-phase deposition equipment.

Various metal oxide thin films with different morphologies, have been prepared using sol–gel techniques. Laberty-Robert et al. fabricated transparent α-Fe_2O_3 mesoporous films via a template-directed sol–gel method coupled with the dip-coating approach, followed by thermally annealing at various temperatures from 350 to 750 °C in air [26]. The crystallite size is about 14 nm at 400 °C and becomes two times larger at 500 °C (30 nm) due to the thermal aggregation. It has been found that the heat treatment has an obvious effect on the optical property of

Fig. 3 ZnO nanowire array on a 4-inch (ca. 10 cm) silicon wafer. At the center is a photograph of a coated wafer, surrounded by SEM images of the array at different locations and magnifications. These images are representative of the entire surface. Scale bars, clockwise from *upper left* 2, 1 mm. 500 and 200 nm. Reproduced with permission from [24]

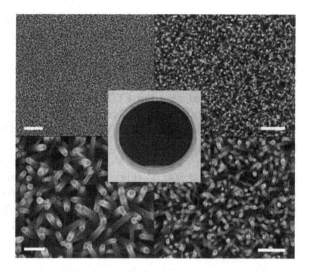

the film. Mesoporous films heat-treated below 300 °C appeared yellow, compared to the red films obtained after annealing at higher temperatures. Prochazka et al. synthesized the exceptionally dense TiO_2 films via dip-coating from a sol containing poly(hexafluorobutyl methacrylate) as the structure-directing agent [72]. The TiO_2 films can be deposited on glass, F-doped SnO_2, and crystalline silicon (111) faces. Noriyuki et al. fabricated $ZnO/Zn_{0.85}Mg_{0.15}O$ heterostructure thin films by sol–gel spin-coating method on sapphire (0001) substrate [66]. The bottom ZnO layer was prepared from zinc acetate dissolved in 2-methoxyethanol. And, monoethanolamine was added to the solution until reaching the same molar ratio with Zn ions. Its solution was deposited onto sapphire substrate by spin-coating process, and then annealed in air at 280 °C for 1 min to remove the solvent. Likewise, Elfanaoui et al. used a very simple and low-cost sol–gel synthesis technique to prepare granular TiO_2 film [12]. Thin films of TiO_2 were deposited by the method of spin coating in air at room temperature, on corning glass as substrate. The films were dried at 100 °C for 1 h and annealed at 300, 350, and 400 °C in air during 1 h. They investigated the effect of the number of coatings and the annealing temperature on the films as well.

In comparison to sol–gel spin-coating techniques, doctor-blading of thin films is an even more simple and convenient method for creating ordered and mesoporous nanostructures [1, 34, 75, 83]. Films that are 10 μm thick can be fabricated with relative ease using this method. For example, Kabre et al. prepared a Tin-doped TiO_2 photoanodes with a thickness of 2 mm and a resistance of 8 Ω/sq on a 2.5 × 2.5 cm of FTO glass [34]. A TiO_2 slurry was prepared by grinding 2 g of Degussa P25 TiO_2 powder in 3 mL of 10 % acetic acid for 30 min, followed by addition of 1 drop of an aqueous 10 % Triton-X solution. The slurry was then deposited on the conductive side of the glass plate using the doctor blade technique with a mask made of 3 M tape applied to three sides of the plate. After air drying for 20 min, the coated glass was subjected to a sintering process. The resulting mesoporous semiconducting thin films were utilized as a photoanode in dye-sensitized solar cells (DSSC). Furthermore, Sivula et al. synthesized mesoporous hematite (Fe_2O_3) photoanodes on FTO substrate by doctor-blading and reported their enhanced PEC properties [83]. The Fe_2O_3 colloid solution containing the porogen was coated onto the FTO substrate via doctor-blading with a 40 μm invisible tape (3 M) as a spacer. The original size of the Fe_2O_3 nanoparticles was about 10 nm before annealing (Fig. 4a). After annealed at 400 °C, the film became porous with necked particles of 30 nm (Fig. 4b). Porous films with larger feature size of the necked particles ca. 60 and 75 nm were obtained after subsequent sintering at 700 and 800 °C (Figs. 4c and d), respectively. These porous Fe_2O_3 films yield water-splitting photocurrents of 0.56 mA/cm^2 under standard conditions (AM 1.5G 100 mW/cm^2, 1.23 V vs. reversible hydrogen electrode, RHE) and over 1.0 mA/cm^2 before the dark current onset (1.55 V vs. RHE).

2.3 Electrochemical Deposition Methods

Electrochemical deposition is a powerful approach for the fabrication of nanomaterials on conductive substrates due to its simplicity, ease of scale-up, low cost, and environmental friendliness [55, 56, 58]. Nanomaterials grown directly on conductive substrates can facilitate the transport of electrons and electrical signals, which will broaden their applications in sensors, solar cells, electronic and electrochemical devices [18]. Moreover, the morphology and composition could be easily tuned by adjusting the electrochemical deposition parameters such as applied potential, current density, temperature, electrolyte, substrate, etc.

The growth of various metal oxides, with different morphologies such as nanowires, nanorod arrays, nanodendritic structures, and thin films, using electrochemical deposition have been achieved. For instance, Mao et al. have used an electrochemical deposition method to deposit iron into AAO template channels and subsequently removed the AAO template using a NaOH solution, followed by thermal oxidation converting the as-prepared iron nanorod arrays into Fe_2O_3 [58]. Lu et al. recently reported the controllable growth of ZnO nanostructures such as nanowire arrays, hierarchical architectures on FTO glass substrates using electrochemical deposition [55, 56]. The morphology and size of ZnO nanostructures can be readily controlled by adjusting the electrochemical deposition parameters such as reaction temperature and the concentration of electrolyte [55, 56]. Vertically aligned and single-crystalline ZnO nanowire with an average diameter of ∼200 nm and length of ∼2 μm were grown directly on a FTO glass substrate by cathodic electrochemical deposition [56]. Furthermore, Gan et al. recently synthesized vertically aligned In_2O_3 nanorod arrays on FTO substrate via a template- and surfactant-free electrochemical deposition method from aqueous solution [18]. Moreover, Mao et al. synthesized three-dimensional (3D) hierarchical Cu_2O stars on FTO substrates using a rapid and facile electrochemical deposition approach

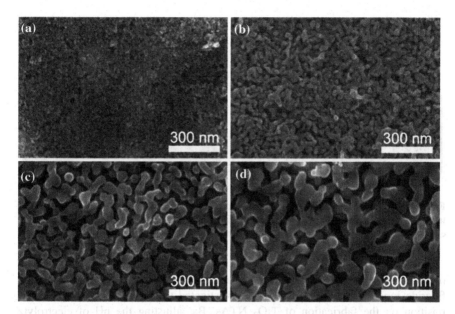

Fig. 4 2 Scanning electron micrographs of mesoporous hematite films prepared on $SiO_2/F:SnO_2$ substrates after different heat treatments: **a** As deposited with porogen **b** After 10 h at 400 °C, and **c** After 20 min at 700 °C and **d** 800 °C. Reproduced with permission from [83]

from aqueous solutions at room temperature [58]. The multifaced 3D structure act as a light-scattering layer, which can significantly improves the photocurrent by reducing the loss via surface light reflection.

2.4 Electrochemical Anodization Method

Electrochemical anodization of metal foils is a simple and convenient method for creating ordered and self-oriented porous metal oxide nanostructures perpendicular to the substrate with controllable pore size and length, especially for nanotube arrays (NTAs) [21, 60, 70, 71, 104]. Anodization of metal foils is usually conducted in a conventional two-electrode electrochemical cell with a platinum foil as cathode at a constant potential. A schematic diagram for the anodization of metal foils is illustrated in Fig. 5.

A type example is electrochemical anodization of Ti foil to prepare vertically and highly ordered TiO_2 NTAs. In 2001, Grimes and co-workers first reported the formation of TiO_2 NTAs via electrochemical anodization of Ti foil in a hydrofluoric electrolyte [21]. Further studies focused on the precise control the morphology of TiO_2 NTAs such as pore size, length, and wall thickness. Electrolyte composition plays a critical role in determining the morphology and structure of TiO_2 NTAs. So far, TiO_2 NTAs with various diameter and length are easily

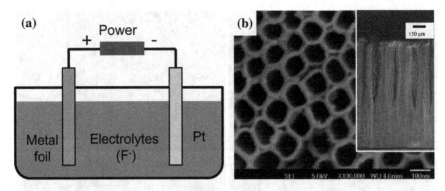

Fig. 5 a A schematic diagram of a two-electrode electrochemical cell for anodization. **b** SEM images of TiO$_2$ NTAs prepared by potentiostatic anodization of Ti foils in an ethylene glycol electrolyte containing NH$_4$F and H$_2$O. Reproduced with permission from [70]

obtained by anodic oxidization in aqueous or organic electrolytes containing fluoride ions. For example, Cai et al. investigated the effect of electrolyte composition on the fabrication of TiO$_2$ NTAs. By adjusting the pH of electrolyte using different additives, TiO$_2$ NTAs with different lengths ranging from 300 nm to 6.4 μm were formed in an aqueous electrolyte containing potassium fluoride [4]. Schmuki et al. synthesized TiO$_2$ NTAs with a length of up to 4 μm in a neutral fluoride solution containing phosphate [20]. Due to the high chemical dissolution rate in an aqueous solution containing fluoride ions, the length of the prepared TiO$_2$ NTAs was short. In comparison to aqueous electrolytes, much longer TiO$_2$ NTAs can be formed in polar organic electrolytes due to low chemical dissolution rate resulting from low water content. The most commonly used organic electrolyte is ethylene glycol [77], glycol [57], acetic acid [91], formamide (FA) [81], and dimethylsulfoxide (DMSO) [116]. By using organic electrolyte, the length of TiO$_2$ NTAs could be extended to up to 100 μm under the proper anodization conditions in organic electrolyte [70, 71]. Grimes and co-workers have widely explored the organic electrolytes in preparing the TiO$_2$ NTAs and have made significant progress. Recently, they prepared highly ordered TiO$_2$ NTAs of over 1,000 μm in length and aspect ratio about 10,000 by potentiostatic anodization of Ti foils in an ethylene glycol electrolyte containing NH$_4$F and H$_2$O [70]. The length and the wall thickness of the TiO$_2$ NTAs were readily controlled by adjusting the electrochemical parameters such as the anodization duration, the composition and temperature of the electrolyte, and the anodization voltage. Besides TO$_2$ nanotube, other metal oxide nanotube and nanoporous nanostructures have developed such as Fe$_2$O$_3$ [39, 109], WO$_3$ [118, 120], Nb$_2$O$_5$ [19], and Ta$_2$O$_5$ [15, 40] via the same electrochemical anodization method. These research works have demonstrated electrochemical anodization to be an effective approach to fabricate high surface area metal oxide nanostructures.

3 Low-Cost Nanomaterials for PEC Water Splitting

Metal oxide semiconductors such as TiO_2, ZnO, Fe_2O_3 and WO_3 are the most common materials used for photocatalytic and PEC water splitting, due to their excellent chemical stability, low cost, and suitable band edge positions; however, each metal oxide has its own limitations as photoelectrode for PEC water splitting. Various techniques have been developed, such as morphology engineering, element doping, and surface modification in order to solve these limitations. In this section, we will review the recent research progress on these popular and low-cost metal oxide nanomaterials for PEC water splitting and the strategies have been used to solve their limitations.

3.1 TiO₂ and ZnO Nanomaterials for PEC Water Splitting

Since the first demonstration of PEC water splitting on TiO_2 by Honda and Fujishima in 1972 [17], TiO_2 has been widely studied as photocatalyst [2, 49] and photoelectrodes [101, 117] for water splitting. In comparison to bulk materials, nanostructures could provide larger surface area and further facilitate charge separation at the interface between semiconductor and electrolyte [42]. Various nanostructured TiO_2 such as nanoparticle films [14, 110], nanowire arrays [16, 101, 108], branched nanowire arrays [7], and nanotube arrays [43, 127] have been developed and implemented as photoelectrodes for water splitting. For instance, Shankar et al. used electrochemical anodization method to synthesize TiO_2 nanotube arrays and applied them for PEC water splitting [80]. They systematically studied the effects of the anodization voltages, times, nanotube lengths, and annealing temperatures on the PEC performance [70, 71, 80]. Rutile TiO_2 nanowire arrays grown on FTO glass have also been used for PEC water splitting [16, 101]. Compared to the polycrystalline TiO_2 nanotube arrays, the single crystal nanowires could have better charge transport property. Besides, the rutile TiO_2 has a relative smaller band-gap of 3.0 eV than anatase TiO_2 (3.2 eV), which allows it to utilize longer wavelength light in the solar spectrum. The large band-gap energy is still the major limitation for TiO_2 materials for solar energy conversion. The theoretical solar energy conversion efficiency for rutile TiO_2 should be ~ 2.5 %, depends on its band-gap energy [62]. However, the reported STH conversion efficiencies were much lower than this theoretical value. For example, ~ 0.7 % of STH conversion efficiency was obtained on pristine TiO_2 nanowire arrays [16]. It suggests that the conversion efficiency of TiO_2 is also limited by another factor, which is believed to be the rapid electron-hole recombination.

To reduce the electron-hole recombination loss, the charge separation and collection efficiencies should be improved. Wang et al. reported that hydrogen thermal treatment improves charge transport of TiO_2 nanowire arrays by controlled incorporation of oxygen vacancies [101]. Figure 6 shows the PEC and electrochemical properties of TiO_2 before and after hydrogen treatment. The photocurrent densities of TiO_2 nanowire were significantly increased in the entire potential windows after hydrogen treatment (Fig. 6a). IPCE studies further showed that the increased photocurrent density was mainly due to the enhanced photoactivities in the UV region (Fig. 6b). The results suggested that the enhanced photocurrent was due to the improved charge collection efficiency and hydrogen treatment did not narrow the band-gap of TiO_2. STH conversion efficiency was calculated to be around 1.1 %, by integrating IPCE spectra with standard solar spectrum shown in Fig. 6c; which was the highest efficiency obtained on TiO_2 nanostructure. Mott-Schottky studies confirmed the carrier densities of hydrogen treated TiO_2 were significantly enhanced by orders of magnitude (Fig. 6d). The enhanced PEC performance was attributed to increased oxygen vacancies and hydrogen impurities, which work as shallow donors for TiO_2. The increased donor density could facilitate charge transport in the semiconductor and charge separation at the interface between semiconductor and substrate.

Furthermore, other chemical modification methods such as element doping [22, 37, 64] and sensitization [28, 103] have been used to increase visible light photoactivity of TiO_2. Element doping is a typical approach to extend light absorption spectrum of TiO_2 into visible light region. Various impurity elements such as Fe [13, 86], C [37, 64, 68], P [22], and N [9, 31, 61] have been reported to increase visible light absorption of TiO_2. For instance, Hoang et al. reported nitrogen-doped TiO_2 nanowire arrays and used them for PEC water oxidation [31]. Nitrogen doping was achieved by nitridation of TiO_2 in ammonia gas flow at temperature of 500 °C. Nitrogen-doped TiO_2 nanowire photoanode was light yellow in color and IPCE studies showed that it exhibited obvious photoactivity in the visible region from 400 to 500 nm (Fig. 7). The visible light photoactivity is attributed to the introduction of N impurity states in the electronic band structure of TiO_2, which narrows its band-gap. Hoang et al. further demonstrated that hydrogenation of nitrogen-doped TiO_2 could further increase their performance of TiO_2 for PEC water splitting, due to a synergistic interaction between nitrogen dopant and the oxygen vacancy (Ti^{3+}) in the TiO_2 [30].

Sensitization is another common method used to increase visible light photoactivity of large band-gap semiconductor metal oxides. For example, TiO_2 sensitized with small band-gap semiconductors such as CdS and CdSe have been reported [28, 50, 103]. As shown in Fig. 8, upon light illuminated, electron and hole pairs will be generated in both TiO_2 and the sensitizer CdSe quantum dot. By forming a type II heterojunction between TiO_2 and the sensitizer, the photogenerated electrons in the sensitizer can efficiently transfer to TiO_2 (Fig. 8) In this case, the sensitized TiO_2 composite structure can utilize visible solar light. However, the major drawback of this approach is the instability of chalcogenides. Hole scavengers such as sodium sulfide should be added into the electrolyte solution to avoid the self-oxidation of CdS and CdSe, because the oxidation potential of CdSe and CdS is much lower than water oxidation [28, 103]. As a result, although the excited electrons are used for water reduction to produce hydrogen, the overall reaction is no longer water splitting.

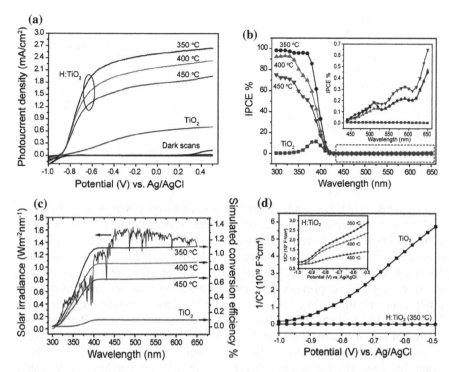

Fig. 6 **a** Linear sweeps of pristine TiO_2 and hydrogen treated TiO_2 (denoted as H-TiO_2) nanowire arrays in 1.0 M NaOH aqueous solution. **b** IPCE spectra of TiO_2 and H-TiO_2 at different temperatures. **c** Integrating solar to hydrogen efficiencies of TiO_2 and H-TiO_2 nanowire arrays. **d** Mott-schottky plots of TiO_2 and H-TiO_2 collected under dark condition at the frequency of 10,000 Hz. Reproduced with permission from [101]

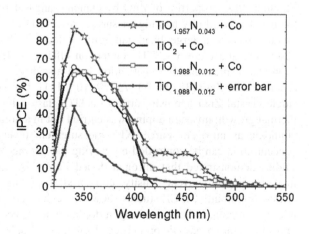

Fig. 7 IPCE spectra of TiO_2 and nitrogen doped TiO_2 with/without cobalt catalyst modification, collected at the potential of 1.4 V versus RHE. Reproduced with permission from [31]

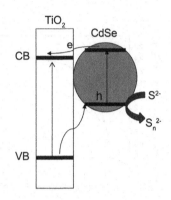

Fig. 8 The schematic illustration of the band structure of CdSe sensitized TiO$_2$. The photoexcited hole in CdS transfer to the interface between semiconductor and electrolyte to oxidize S^{2-} to S$_n^{2-}$; the excited electron is injected into TiO$_2$

Zinc oxide (ZnO) has similar band-gap and band-edge positions as TiO$_2$, which is also a suitable semiconductor material for PEC water splitting [76, 88, 107, 111, 126]. Although the chemical stability of ZnO is not as good as TiO$_2$, due to the dissolution in basic and acidic solution, the electron mobility of ZnO is typically 10–100 folds higher than that of TiO$_2$ [27, 111], which is favorable for charge transport and PEC application. Nanostructured ZnO such as nanoparticle film [107], nanowire film [111], 3D nanotree film [88], and nanotube arrays [8] have been developed and studied for PEC water splitting. For instance, Wolcott et al. fabricated nanostructured ZnO thin film using three different deposition strategies: normal pulsed laser deposition, pulse laser oblique-angle deposition and electron-beam glancing-angle deposition [107]. They found that the thin film morphologies played significant effect on their PEC properties. Normal pulsed laser deposition produced densely packed particle films with ∼200 nm grain sizes; oblique-angle deposition produced nanoplatelets with a fish scale morphology thin film; while glancing-angle deposition generated a highly porous, interconnected nanoparticle network. The highly porous thin film by glancing-angle deposition exhibited the best PEC water splitting performance, among these three kinds of ZnO thin films. The high PEC properties of ZnO by glancing-angle deposition were attributed to the better crystallinity and higher surface area [107].

ZnO is also a large band-gap semiconductor, which limits the solar light absorption and therefore STH conversion efficiency. Element doping and small band-gap semiconductor sensitization have also been used to increase the light absorption of ZnO [38, 44, 63, 73, 79, 102, 111]. For example, Yang et al. grew single crystal ZnO nanowire arrays on ITO glass using a seed-mediated hydro-thermal growth; nitrogen doping was carried out in a home-built CVD system with ammonia as nitrogen source. XPS measurement revealed that nitrogen dopant concentration can be controlled by varying the doping treatment time [111]. PEC studies demonstrated that nitrogen doped ZnO show more than one magnitude increase in photocurrent density, compared to pristine undoped ZnO nanowire arrays. Importantly, IPCE studies shows the enhancement of photoactivity in vis-ible light region, indicating nitrogen doping is an effective method to modify the optical and photocatalytic properties of ZnO. Besides, Wang et al. demonstrated to use quantum dot sensitization method to further increase the visible light absorption of ZnO [102]. They designed a novel double-sided CdS and CdSe quantum dot co-sensitized ZnO nanowire arrays and used them for PEC hydrogen evolution. The double-sided design represents a simple analog of tandem cell structure to maximize light absorption. This double-sided architecture enabled direct interaction between quantum dot and nanowire, which could improve charge collection efficiency, compared to single sided co-sensitized structure. A maximum photocurrent density of ~ 12 mA/cm^2 at 0.4 V versus Ag/AgCl was obtained on the double-sided quantum dot co-sensitized ZnO nanowire arrayed photoanode [102].

Although enormous efforts have been devoted to develop more efficient ZnO- and TiO$_2$-based photoelectrodes, the overall efficiency is still limited by relatively poor visible light absorption due to their large band-gap energies. Therefore, it is still necessary and important to develop new method to modify their optical and electronic properties and improve their performance for PEC water splitting.

3.2 WO₃ Nanomaterials for PEC Water Oxidation

In comparison to TiO$_2$ and ZnO, WO$_3$ has a relatively small band-gap (2.5–2.8 eV) that allows it to capture about 12 % of the solar spectrum and is more favorable for visible light absorption [62]. In addition, WO$_3$ also has a moderate hole diffusion length (~ 150 nm) and good electron transport property. These features make it a very promising photoanode material for PEC water oxidation [51, 53]. Considerable efforts have been devoted to develop various WO$_3$ films as photoanodes for water oxidation [10, 29, 54, 78, 84, 85, 99], since Hodes et al. first demonstrated the possibility of using WO$_3$ as a photoanode for water splitting [32]. However, WO$_3$ also has its own limitations. First, the energy level of its conduction band is more positive than the potential for water reduction. As a result, water splitting can be only achieved with the application of additional external bias. Second, WO$_3$ is an indirect band-gap semiconductor that requires a relatively thick film for increasing light absorption. However, a thick layer of active electrode material usually cause significant electron–hole recombination loss and decrease photoactivity of WO$_3$ [99]. Nanostructured materials have large surface area and short carrier diffusion distance, which could potentially address this issue. Recently, various nanostrcutured WO$_3$ such as nanowire [85], nanoflake [74] and nanoparticle film [78] have been reported for PEC water oxidation. For instance, Cristineo et al. synthesized the wormlike WO$_3$ nanostructured photoanodes on tungsten foils using potentiostatic anodization method in high dielectric constant organic solvents [10]. The highest photocurrent density obtained was approximately 3.5 mA/cm^2 at 1.5 V (vs. SCE) in 1.0 M H$_2$SO$_4$ solution under an incident power of 150 mW/cm^2 and was more than 9 mA cm^{-2} under incident light power of 370 mW/cm^2. Qin et al. fabricated vertically aligned WO$_3$ nanoparticulate flakes that have both high surface area and sufficient film thickness for good light absorption, using a simple surfactant-free hydrothermal growth in a mixture of concentrated nitric acid and hydrochloric acid, followed by sintering at 500 °C in air [74]. Recently, Chen et al. prepared large area and high-quality WO$_3$ photonic crystal photoanodes with inverse opal structure [6]. This 3D-photonic crystal design is used to enhance light absorption and further increase IPCE of WO$_3$ photoanodes. The photonic stop-bands of these WO$_3$ photoanodes can be tuned experimentally by variation of the pore sizes of inverse opal structures. It was found that when the red-edge of the photonic stop-band of WO$_3$ inverse opals overlapped with the electronic absorption edge of WO$_3$ at Eg = 2.6–2.8 eV, a maximum of 100 % increase in photocurrent intensity was observed under visible light irradiation ($\lambda > 400$ nm), in comparison with a disordered porous WO$_3$ photoanode [6]. When the red-edge of the stop-band was tuned well within the electronic absorption range of WO$_3$, noticeable but less amplitude of enhancement in the photocurrent intensity was observed. It was further shown that the spectral region with a selective IPCE enhancement of the WO$_3$ inverse opals exhibited a blue-shift in wavelength under off-normal incidence of light, in agreement with the calculated stop-band edge locations. (Fig. 9) The enhancement was attributed to a longer photon–matter interaction length as a result of the slow-light effect at the photonic stop-band edge, thus leading to a remarkable improvement in the light-harvesting efficiency. This method can provide a potential and promising approach to effectively utilize solar energy for visible light responsive photoanodes.

Additionally, WO$_3$ is known to be thermodynamically unstable in electrolyte solution with pH > 4 due to OH$^-$-induced chemical dissolution and photocorrosion induced by the peroxo species created during water oxidation [106]. In order to suppress the formation of peroxo species, oxygen evolution catalyst has been used to modify the surface of WO$_3$ [52, 78]. For example, Seabold et al. demonstrated the photo-oxidation reactions and photostability of electrochemically prepared WO$_3$ can be improved by coating with Co–Pi OEC catalyst (Fig. 10) [78]. When a bare WO$_3$ photoanode was used, a significant portion of the photogenerated holes (ca. 39 %) were used to form peroxo species and dissolve WO$_3$; therefore, the accumulated peroxo species on the WO$_3$ surface result in a gradual loss of photoactivity. In contrast, when a thick Co–Pi OEC layer (>1.5 µm) was deposited on the WO$_3$ electrode, most photon-generated holes were used for O$_2$ production and suppress the formation of peroxo species. The suppression of the peroxide species not only increases O$_2$ production, but also improves the long-term photostability of WO$_3$. Alternatively, Wang et al. also demonstrated that oxygen-deficient tungsten oxides are resistive to the peroxo species-induced corrosion [99]. They showed that the photostability and photoactivity of WO$_3$ for water oxidation can be simultaneously enhanced by self-doping with oxygen vacancies via hydrogen treatment. The improved photoactivity was believed to be due to the increased donor densities, which could facilitate charge transport. Importantly, the photocurrent of hydrogen treated WO$_3$ was stabilized for at least 7 h without substantial loss, while pristine WO$_3$ lost 80 % of initial photocurrent in the first two hours. The results proved that hydrogen treatment is an effective method to improve and stabilize the photoactivity of WO$_3$.

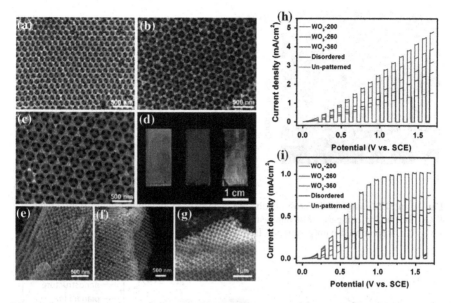

Fig. 9 a–c SEM top view of WO$_3$ inverse opals with 200, 260 and 360 nm pore size. **d** photograph of the inverse opal WO$_3$-200, WO$_3$-260 and WO$_{3-}$360. **e–g** SEM images of the cross-section view of WO$_3$-200, WO$_3$-260 and WO$_{3-}$360, respectively. **h–i** linear sweep voltammograms of all the WO$_3$ electrodes under chopped UV light illumination and visible light illumination. Reproduced with permission from [6]

3.3 α-Fe₂O₃ Nanomaterials for PEC Water Oxidation

Hematite (α-Fe$_2$O$_3$), as one of the most prevalent metal oxides on earth, has attracted extensive attentions for PEC water splitting, due to its favorable band-gap, natural abundance, low cost, and superior chemical stability in aqueous solution [35, 45, 59, 82, 89, 105]. However, hematite has a few severe drawbacks that limits its application as photoelectrode. First, the conduction band of hematite is more positive than the water reduction potential, as a result, a large external bias is needed to drive water splitting. Furthermore, hematite has a very short excited state lifetime (~ 10 ps) and a short hole diffusion distance (2–4 nm), which cause significant loss via electron-hole recombination [47, 48, 100]. In this regard, very thin layer of hematite is required to reduce the diffusion length for minority carrier and facilitate charge separation. Nanostructured materials open up new opportunity to address this issue by providing extremely large surface area, as well as short hole diffusion distance. For instance, Lin et al. reported the fabrication of ultrathin hematite thin films on TiSi$_2$ nanonet using atomic layer deposition technique (Fig. 11) [46]. This core shell structure has several advantages over bulk structure. The TiSi$_2$ scaffold provides large surface area and the TiSi$_2$ is also highly conductive, which is beneficial for charge transport. Besides, the ultrathin hematite

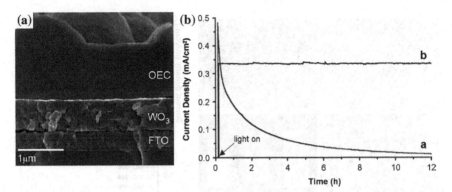

Fig. 10 a SEM image of the cross-section of Co–Pi oxygen evolution catalyst (OEC) modified WO$_3$ film. **b** The comparison of photocurrent density versus time of WO$_3$ and OEC modified WO$_3$ film. Reproduced with permission from [78]

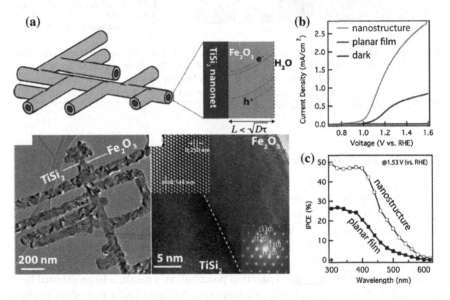

Fig. 11 a Schematic illustration of the TiSi$_2$/Fe$_2$O$_3$ core shell nanonet structure and corresponded TEM studies. **b** Linear sweep voltammograms of planar hematite film and TiSi$_2$/Fe$_2$O$_3$ nanonet film collected in 1.0 M NaOH solution, under 100 mW/cm^2 AM 1.5G simulated solar light. **c** The comparison of the external quantum efficiency (IPCE) with/without TiSi$_2$ nanonet structure. Reproduced with permission from [46]

film offers short hole diffusion distance to minimize electron-hole recombination loss. Figure 11 shows the PEC performance of α-Fe$_2$O$_3$-TiSi$_2$ core shell nanonets and α-Fe$_2$O$_3$ planar structure. The photocurrent density of α-Fe$_2$O$_3$-TiSi$_2$ nanonets is three times larger than that of planar α-Fe$_2$O$_3$ and the IPCE spectra is almost the same as the absorbed photon conversion efficiency (APCE), indicating the benefits of the nanostructure in the charge collection and transport [46].

Although the ultrathin nanostructure can improve the charge separation, hematite is an indirect band-gap material that requires a thick layer of active material for increasing light absorption. Nevertheless, thick hematite films will lead to significant electron-hole recombination loss. Therefore, it is also important to modify the intrinsic electronic property of hematite, with a goal of increasing charge transport and suppressing the electron-hole recombination. Element doping is the most common method to modify the electronic property of hematite. A number of element dopants including titanium [100, 119], silicon [5, 36], and tin [41, 48] have been studied for hematite. These dopants function as electron donors and can significantly enhance the electrical conductivity of hematite by increasing its donor density. Gratzel and co-workers developed silicon-doped hematite by spray pyrolysis and atmospheric pressure chemical vapor deposition (APCVD) [5, 36]. These silicon-doped hematite achieved the best photocurrent density of 2.3 mA/cm^2 at 1.23 V versus Ag/AgCl, without the need of catalyst modification [36]. Recently, Sn doping has also been demonstrated to be an effective dopant for hematite. Upon high temperature annealing, Sn can diffuse into hematite film from fluorine-doped tin oxide (FTO) glass substrate. Sn-doped hematite can also be achieved by intentionally mixing Sn precursor with Fe precursor during hydrothermal synthesis of hematite [48]. Sn doping treatment could activate hematite nanowires for PEC water splitting. Moreover, Wang et al. also reported that Ti-doped hematite can be achieved by a drop-annealing method using titanium butoxide as dopant precursor [100]. Electrochemical impedance studies showed that the donor density of Ti-doped hematite was significantly enhanced by orders of magnitude, indicating the element doping is a good strategy to increase the electrical conductivity of hematite.

Alternative to the incorporation of extrinsic dopants into hematite, Ling et al. also demonstrated a new method to improve the electrical conductivity of hematite by creating intrinsic defects such as oxygen vacancies [47]. Oxygen vacancies (Fe^{2+}) function as a shallow donor for hematite, playing a similar role as extrinsic dopants such as silicon, titanium, and tin. Ling et al. annealed FeOOH nanowires in an oxygen-deficient condition to create the oxygen vacancies [47]. They found that this method could activate hematite nanowires at relatively low activation treatment of 550 °C [48]. A maximum photocurrent density of 3.37 mA/cm^2 at 1.5 V versus RHE, was obtained for the oxygen-deficient hematite nanowire photoanode. Mott-Schottky studies showed that the carrier density was improved by an order of magnitude, as a result of the creation of oxygen vacancies (Fig. 12). This work demonstrated a simple and effective method to prepare highly photoactive hematite nanowire for PEC water splitting application, at relatively low activation temperature without the need of extrinsic dopants.

Furthermore, the large overpotential for water oxidation is another major issue limiting the performance of hematite for PEC water splitting. Water oxidation catalyst modification has been proved to be an effective method to suppress water oxidation overpotential. A variety of catalysts such as Co–Pi [3, 122–124], CoO$_x$ [36], NiO$_x$ [98] and IrO$_x$ [90] have been previously reported. Co–Pi is the most

Fig. 12 **a** Linear sweep voltammograms collected for air annealed hematite (A-hematite) and hematite annealed in N_2 (N-hematite) at a scan rate of 10 mV/s in 1.0 M NaOH aqueous solution, under 100 mW/cm^2 AM 1.5G simulated solar light. **b** Mott-Schottky plots measured for A-hematite and N-hematite nanowire films. Reproduced with permission from [47]

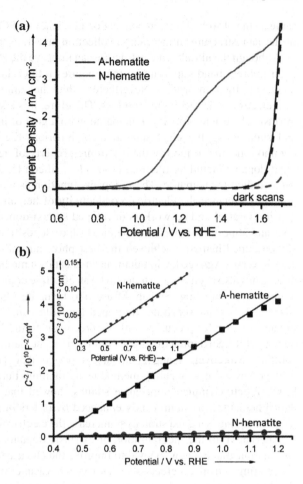

common water oxidation catalyst. Zhong et al. deposited Co–Pi on silicon doped hematite film by electrochemical deposition method [122–124]. The catalyst modified hematite showed a negative shift of photocurrent onset potential, indicating the suppression of water oxidation overpotential. Iridium oxide is another promising water oxidation catalyst, although the cost is relatively higher than Co–Pi catalyst. Tilley et al. used IrO_2 nanoparticle to modify silicon hematite and studied their PEC performance (Fig. 13). A dramatic shift in the onset photocurrent potential was observed from 1.0 to 0.8 V versus RHE and plateau photocurrent density was enhanced from 3.45 to 3.75 mA/cm^2, which achieved a new benchmark photocurrent for hematite photoanode for PEC water splitting [90].

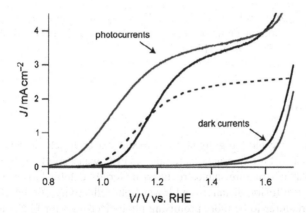

Fig. 13 Photoelectrochemical performance of the unmodified hematite photoanode and the same anode modified with IrO_2 nanoparticles under AM 1.5G 100 mW/cm^2 simulated sunlight. The *dashed line* is the photocurrent for the former state-of-the-art hematite photoanode. Reproduced with permission from [90]

3.4 Other Low-Cost Nanomaterials for PEC Water Oxidation

In addition to these binary metal oxides, ternary metal oxides such as $SrTiO_3$ [67, 117] and $BiVO_4$ [113–115, 121] have also been studied for PEC water oxidation. Ternary metal oxides could provide new opportunities in developing photoelectrode, as their electronic bands are formed by atomic orbitals from more than one element and the modulation of the stoichiometric ratio of the elements could finely tune the potentials of valence and conduction bands as well as the band-gap energy [95]. $SrTiO_3$ has favorable band-edge positions that straddled the water oxidation and reduction potentials. However, the large band-gap $SrTiO_3$ (3.75 eV) limits its efficiency for water oxidation. Zhang et al. used a hydrothermal method to fabricate $SrTiO_3$ nanocrystals modified TiO_2 nanotube arrays for PEC water splitting [117]. They found that the incorporation of $SrTiO_3$ could shift the flat band potential of TiO_2 to a more negative value, and thus, improve its PEC performance. $BiVO_4$ has a favorable band-gap of 2.3 eV for efficient absorption of solar light. Furthermore, its conduction band is close to 0 V versus RHE at pH = 0, as a result of the overlap of empty Bi 6p orbitals with anti-bonding V 3d-O 2p states, which can reduce the need for external bias for PEC water splitting. However, charge transport and interfacial charge transfer have been found to be key limiting factors for PEC water splitting. A number of methods have been demonstrated to improve their charge transport efficiency and further improve their PEC performance. For example, Mo- and W-doped $BiVO_4$ have been demonstrated to increase the solar energy conversion efficiency for PEC water splitting by improving their carrier densities with W and

Mo as *n* type dopants [69, 113–115, 121]. Recently, Wang et al. also demonstrated that hydrogen treatment increased the carrier density of $BiVO_4$ and improved its photoactivity for PEC water oxidation.

4 Conclusion

In this chapter, we have highlighted the recent research achievements in developing low-cost metal oxide nanomaterials for PEC water splitting. In comparison to bulk or planar structure, nanostructured materials exhibit larger surface area, shorter carrier diffusion distance, and lower light reflectivity; all these advantages make nanostructures to be more promising photoelectrodes for PEC water spitting application. A number of strategies, such as morphology engineering, element doping, sensitization, and chemical modifications have been developed to address the intrinsic limitations of metal oxides in light absorption, charge separation, and transport. Although significant process has been made in the past decades, there are still several outstanding challenges remain in order to commercialize the photoelectrodes for PEC water spitting. The cost, efficiency, and long-time stability are still the most important issues that need to be further improved. To date, the overall efficiency of metal oxides for PEC water splitting is still relatively low (<2 %); and there is not a single semiconductor material could meets all the requirements for a high efficient photoelectrode. A possible solution to the problem is to develop new nanomaterials and device architectures. It is promising to develop semiconductor alloys such as ternary or quaternary metal oxide nanostructures, which allow the band-gap and band-edge positions to be engineerable for maximizing the efficiency for water splitting. Additionally, it is equally important to develop simple and low-cost synthetic methods to fabricate these nanostructures with controlled optical and electronic properties.

References

1. Ahn S, Kim K, Cho A, Gwak J, Yun JH, Shin K, Ahn S, Yoon K (2012) CuInSe$_2$ (CIS) thin films prepared from amorphous Cu-In-Se nanoparticle precursors for solar cell application. Acs Appl Mater Inter 4(3):1530–1536
2. Alfano OM, Bahnemann D, Cassano AE, Dillert R, Goslich R (2000) Photocatalysis in water environments using artificial and solar light. Catal Today 58(2–3):199–230
3. Barroso M, Cowan AJ, Pendlebury SR, Gratzel M, Klug DR, Durrant JR (2011) The role of cobalt phosphate in enhancing the photocatalytic activity of alpha-Fe$_2$O$_3$ toward water oxidation. J Am Chem Soc 133(38):14868–14871
4. Cai QY, Paulose M, Varghese OK, Grimes CA (2005) The effect of electrolyte composition on the fabrication of self-organized titanium oxide nanotube arrays by anodic oxidation. J Mater Res 20(1):230–236

5. Cesar I, Sivula K, Kay A, Zboril R, Graetzel M (2009) Influence of feature size, film thickness, and silicon doping on the performance of nanostructured hematite photoanodes for solar water splitting. J Phys Chem C 113(2):772–782

6. Chen X, Ye J, Ouyang S, Kako T, Li Z, Zou Z (2011) Enhanced incident photon-to-electron conversion efficiency of tungsten trioxide photoanodes based on 3D-photonic crystal design. ACS Nano 5(6):4310–4318

7. Cho IS, Chen ZB, Forman AJ, Kim DR, Rao PM, Jaramillo TF, Zheng XL (2011) Branched TiO_2 nanorods for photoelectrochemical hydrogen production. Nano Lett 11(11):4978–4984

8. Chouhan N, Yeh CL, Hu SF, Liu RS, Chang WS, Chen KH (2011) Photocatalytic CdSe QDs-decorated ZnO nanotubes: an effective photoelectrode for splitting water. Chem Commun 47(12):3493–3495

9. Cong Y, Zhang JL, Chen F, Anpo M (2007) Synthesis and characterization of nitrogen-doped TiO_2 nanophotocatalyst with high visible light activity. J Phys Chem C 111(19):6976–6982

10. Cristino V, Caramori S, Argazzi R, Meda L, Marra GL, Bignozzi CA (2011) Efficient photoelectrochemical water splitting by anodically grown WO_3 electrodes. Langmuir 27(11):7276–7284

11. Currao A (2007) Photoelectrochemical water splitting. Chimia 61(12):815–819

12. Elfanaoui A, Elhamri E, Boulkaddat L, Ihlal A, Bouabid K, Laanab L, Taleb A, Portier X (2011) Optical and structural properties of TiO_2 thin films prepared by sol-gel spin coating. Int J Hydrogen Energ 36(6):4130–4133

13. Farhangi N, Chowdhury RR, Medina-Gonzalez Y, Ray MB, Charpentier PA (2011) Visible light active Fe doped TiO_2 nanowires grown on graphene using supercritical CO_2. Appl Catal B-Environ 110:25–32

14. Fei HH, Yang YC, Rogow DL, Fan XJ, Oliver SRJ (2010) Polymer-templated nanospider TiO_2 thin films for efficient photoelectrochemical water splitting. Acs Appl Mater Inter 2(4):974–979

15. Feng XJ, LaTempa TJ, Basham JI, Mor GK, Varghese OK, Grimes CA (2010) Ta_3N_5 nanotube arrays for visible light water photoelectrolysis. Nano Lett 10(3):948–952

16. Feng XJ, Shankar K, Varghese OK, Paulose M, Latempa TJ, Grimes CA (2008) Vertically aligned single crystal TiO_2 nanowire arrays grown directly on transparent conducting oxide coated glass: synthesis details and applications. Nano Lett 8(11):3781–3786

17. Fujishima A, Honda K (1972) Electrochemical photolysis of water at a semiconductor electrode. Nature 238(5358):37–38

18. Gan JY, Lu XH, Zhai T, Zhao YF, Xie SL, Mao YC, Zhang YL, Yang YY, Tong YX (2011) Vertically aligned In_2O_3 nanorods on FTO substrates for photoelectrochemical applications. J Mater Chem 21(38):14685–14692

19. Ghicov A, Aldabergenova S, Tsuchiya H, Schmuki P (2006) TiO_2-Nb_2O_5 nanotubes with electrochemically tunable morphologies. Angew Chem-Int Edit 45(42):6993–6996

20. Ghicov A, Tsuchiya H, Macak JM, Schmuki P (2005) Titanium oxide nanotubes prepared in phosphate electrolytes. Electrochem Commun 7(5):505–509

21. Gong D, Grimes CA, Varghese OK, Hu WC, Singh RS, Chen Z, Dickey EC (2001) Titanium oxide nanotube arrays prepared by anodic oxidation. J Mater Res 16(12):3331–3334

22. Gopal NO, Lo HH, Ke TF, Lee CH, Chou CC, Wu JD, Sheu SC, Ke SC (2012) Visible light active phosphorus-doped TiO_2 nanoparticles: an EPR evidence for the enhanced charge separation. J Phys Chem C 116(30):16191–16197

23. Gratzel M (2001) Photoelectrochemical cells. Nature 414(6861):338–344

24. Greene LE, Law M, Goldberger J, Kim F, Johnson JC, Zhang YF, Saykally RJ, Yang PD (2003) Low-temperature wafer-scale production of ZnO nanowire arrays. Angew Chem-Int Edit 42(26):3031–3034

25. Gust D, Moore TA, Moore AL (2009) Solar Fuels via Artificial Photosynthesis. Acc Chem Res 42(12):1890–1898

26. Hamd W, Cobo S, Fize J, Baldinozzi G, Schwartz W, Reymermier M, Pereira A, Fontecave M, Artero V, Laberty-Robert C, Sanchez C (2012) Mesoporous alpha-Fe_2O_3 thin films synthesized via the sol-gel process for light-driven water oxidation. Phys Chem Chem Phys 14(38):13224–13232

27. Hendry E, Koeberg M, O'Regan B, Bonn M (2006) Local field effects on electron transport in nanostructured TiO_2 revealed by terahertz spectroscopy. Nano Lett 6(4):755–759

28. Hensel J, Wang GM, Li Y, Zhang JZ (2010) Synergistic effect of CdSe quantum dot sensitization and nitrogen doping of TiO_2 nanostructures for photoelectrochemical solar hydrogen generation. Nano Lett 10(2):478–483

29. Hill JC, Choi KS (2012) Effect of electrolytes on the selectivity and stability of n-type WO_3 photoelectrodes for use in solar water oxidation. J Phys Chem C 116(14):7612–7620

30. Hoang S, Berglund SP, Hahn NT, Bard AJ, Mullins CB (2012) Enhancing visible light photo-oxidation of water with TiO_2 nanowire arrays via cotreatment with H_2 and NH_3: synergistic effects between Ti^{3+} and N. J Am Chem Soc 134(8):3659–3662

31. Hoang S, Guo SW, Hahn NT, Bard AJ, Mullins CB (2012) Visible light driven photoelectrochemical water oxidation on nitrogen-modified TiO_2 nanowires. Nano Lett 12(1):26–32

32. Hodes G, Cahen D, Manassen J (1976) Tungsten trioxide as a photoanode for a photoelectrochemical cell (PEC). Nature 260(5549):312–313

33. Hong SJ, Jun H, Borse PH, Lee JS (2009) Size effects of WO_3 nanocrystals for photooxidation of water in particulate suspension and photoelectrochemical film systems. Int J Hydrogen Energ 34(8):3234–3242

34. Kabre J, LeSuer RJ (2012) Modeling diffusion of tin into the mesoporous titanium dioxide layer of a dye-sensitized solar cell photoanode. J Phys Chem C 116(34):18327–18333

35. Katz MJ, Riha SC, Jeong NC, Martinson ABF, Farha OK, Hupp JT (2012) Toward solar fuels: water splitting with sunlight and "rust"? Coord Chem Rev 256(21–22):2521–2529

36. Kay A, Cesar I, Gratzel M (2006) New benchmark for water photooxidation by nanostructured alpha-Fe_2O_3 films. J Am Chem Soc 128(49):15714–15721

37. Khan SUM, Al-Shahry M, Ingler WB (2002) Efficient photochemical water splitting by a chemically modified n-TiO_2 2. Science 297(5590):2243–2245

38. Kim H, Seol M, Lee J, Yong K (2011) Highly efficient photoelectrochemical hydrogen generation using hierarchical ZnO/WO_x nanowires cosensitized with CdSe/CdS. J Phys Chem C 115(51):25429–25436

39. LaTempa TJ, Feng XJ, Paulose M, Grimes CA (2009) Temperature-dependent growth of self-assembled hematite (alpha-Fe_2O_3) nanotube arrays: rapid electrochemical synthesis and photoelectrochemical properties. J Phys Chem C 113(36):16293–16298

40. Lee KY, Schmuki P (2011) Highly ordered nanoporous Ta_2O_5 formed by anodization of Ta at high temperatures in a glycerol/phosphate electrolyte. Electrochem Commun 13(6):542–545

41. Li LS, Yu YH, Meng F, Tan YZ, Hamers RJ, Jin S (2012) Facile solution synthesis of alpha-FeF3 center dot 3H_2O nanowires and their conversion to alpha-Fe_2O_3 nanowires for photoelectrochemical application. Nano Lett 12(2):724–731

42. Li Y, Zhang JZ (2010) Hydrogen generation from photoelectrochemical water splitting based on nanomaterials. Laser Photon Rev 4(4):517–528

43. Liang SZ, He JF, Sun ZH, Liu QH, Jiang Y, Cheng H, He B, Xie Z, Wei SQ (2012) Improving photoelectrochemical water splitting activity of TiO_2 nanotube arrays by tuning geometrical parameters. J Phys Chem C 116(16):9049–9053

44. Lin YG, Hsu YK, Chen YC, Chen LC, Chen SY, Chen KH (2012) Visible-light-driven photocatalytic carbon-doped porous ZnO nanoarchitectures for solar water-splitting. Nanoscale 4(20):6515–6519

45. Lin YJ, Xu Y, Mayer MT, Simpson ZI, McMahon G, Zhou S, Wang DW (2012) Growth of p-type hematite by atomic layer deposition and its utilization for improved solar water splitting. J Am Chem Soc 134(12):5508–5511

46. Lin YJ, Zhou S, Sheehan SW, Wang DW (2011) Nanonet-based hematite heteronanostructures for efficient solar water splitting. J Am Chem Soc 133(8):2398–2401

47. Ling Y, Wang G, Reddy J, Wang C, Zhang JZ, Li Y (2012) The influence of oxygen content on the thermal activation of hematite nanowires. Angew Chem Int Ed Engl 51(17):4074–4079

48. Ling YC, Wang GM, Wheeler DA, Zhang JZ, Li Y (2011) Sn-doped hematite nanostructures for photoelectrochemical water splitting. Nano Lett 11(5):2119–2125

49. Linsebigler AL, Lu GQ, Yates JT (1995) Photocatalysis on TiO$_2$ surfaces- principles, mechanisms and selected results. Chem Rev 95(3):735–758

50. Liu LP, Wang GM, Li Y, Li YD, Zhang JZ (2011) CdSe quantum dot-sensitized Au/TiO$_2$ hybrid mesoporous films and their enhanced photoelectrochemical performance. Nano Res 4(3):249–258

51. Liu R, Lin Y, Chou LY, Sheehan SW, He W, Zhang F, Hou HJM, Wang D (2011) Water splitting by tungsten oxide prepared by atomic layer deposition and decorated with an oxygen-evolving catalyst. Angew Chem 123(2):519–522

52. Liu R, Lin YJ, Chou LY, Sheehan SW, He WS, Zhang F, Hou HJM, Wang DW (2011) Water splitting by tungsten oxide prepared by atomic layer deposition and decorated with an oxygen-evolving catalyst. Angew Chem-Int Edit 50(2):499–502

53. Liu X, Wang F, Wang Q (2012) Nanostructure-based WO$_3$ photoanodes for photoelectrochemical water splitting. Phys Chem Chem Phys 14(22):7894–7911

54. Liu X, Wang FY, Wang Q (2012) Nanostructure-based WO$_3$ photoanodes for photoelectrochemical water splitting. Phys Chem Chem Phys 14(22):7894–7911

55. Lu XH, Wang D, Li GR, Su CY, Kuang DB, Tong YX (2009) Controllable electrochemical synthesis of hierarchical ZnO nanostructures on FTO glass. J Phys Chem C 113(31):13574–13582

56. Lu XH, Wang GM, Xie SL, Shi JY, Li W, Tong YX, Li Y (2012) Efficient photocatalytic hydrogen evolution over hydrogenated ZnO nanorod arrays. Chem Commun 48(62):7717–7719

57. Macak JM, Schmuki P (2006) Anodic growth of self-organized anodic TiO$_2$ nanotubes in viscous electrolytes. Electrochim Acta 52(3):1258–1264

58. Mao YC, He JT, Sun XF, Li W, Lu XH, Gan JY, Liu ZQ, Gong L, Chen J, Liu P, Tong YX (2012) Electrochemical synthesis of hierarchical Cu$_2$O stars with enhanced photoelectrochemical properties. Electrochim Acta 62:1–7

59. Mayer MT, Du C, Wang DW (2012) Hematite/Si nanowire dual-absorber system for photoelectrochemical water splitting at low applied potentials. J Am Chem Soc 134(30):12406–12409

60. Mohamed AE, Rohani S (2011) Modified TiO$_2$ nanotube arrays (TNTAs): progressive strategies towards visible light responsive photoanode, a review. Energy Environ Sci 4(4):1065–1086

61. Mrowetz M, Balcerski W, Colussi AJ, Hoffmann MR (2004) Oxidative power of nitrogen-doped TiO$_2$ photocatalysts under visible illumination. J Phys Chem B 108(45):17269–17273

62. Murphy AB, Barnes PRF, Randeniya LK, Plumb IC, Grey IE, Horne MD, Glasscock JA (2006) Efficiency of solar water splitting using semiconductor electrodes. Int J Hydrogen Energ 31(14):1999–2017

63. Myung Y, Jang DM, Sung TK, Sohn YJ, Jung GB, Cho YJ, Kim HS, Park J (2010) Composition-tuned ZnO-CdSSe core-shell nanowire arrays. ACS Nano 4(7):3789–3800

64. Neville EM, Mattle MJ, Loughrey D, Rajesh B, Rahman M, MacElroy JMD, Sullivan JA, Thampi KR (2012) Carbon-doped TiO$_2$ and carbon, tungsten-codoped TiO$_2$ through sol-gel processes in the presence of melamine borate: reflections through photocatalysis. J Phys Chem C 116(31):16511–16521

65. Nocera DG (2012) The artificial leaf. Acc Chem Res 45(5):767–776

66. Noriyuki H, Toru K, Tomoe H, Kenji K, Hiroshi H (2011) Fabrication of ZnO/Zn$_{1-x}$Mg$_x$O heterostructure thin films by sol-gel spin-coating method. Phys Status Solidi C 8 (2):2511–2515

67. Ohko Y, Saitoh S, Tatsuma T, Fujishima A (2002) Photoelectrochemical anticorrosion effect of SrTiO$_3$ for carbon steel. Electrochem Solid State Lett 5(2):B9–B12

68. Park JH, Kim S, Bard AJ (2006) Novel carbon-doped TiO$_2$ nanotube arrays with high aspect ratios for efficient solar water splitting. Nano Lett 6(1):24–28

69. Parmar KPS, Kang HJ, Bist A, Dua P, Jang JS, Lee JS (2012) Photocatalytic and photoelectrochemical water oxidation over metal-doped monoclinic BiVO$_4$ photoanodes. ChemSusChem 5(10):1926–1934

70. Paulose M, Prakasam HE, Varghese OK, Peng L, Popat KC, Mor GK, Desai TA, Grimes CA (2007) TiO$_2$ nanotube arrays of 1000 mu m length by anodization of titanium foil: phenol red diffusion. J Phys Chem C 111(41):14992–14997

71. Paulose M, Shankar K, Yoriya S, Prakasam HE, Varghese OK, Mor GK, LaTempa TJ, Fitzgerald A, Grimes CA (2006) Anodic growth of highly ordered TiO$_2$ nanotube arrays to 134 microm in length. J Phys Chem B 110(33):16179–16184

72. Prochazka J, Kavan L, Zukalova M, Janda P, Jirkovsky J, Zivcova Z (2012) Dense TiO2 films grown by sol–gel dip coating on glass, F-doped SnO2, and silicon substrates. J. Mater. Res. 28(3): 385–393

73. Qi XP, She GW, Liu YY, Mu LX, Shi WS (2012) Electrochemical synthesis of CdS/ZnO nanotube arrays with excellent photoelectrochemical properties. Chem Commun 48(2):242–244

74. Qin DD, Tao CL, Friesen SA, Wang TH, Varghese OK, Bao NZ, Yang ZY, Mallouk TE, Grimes CA (2012) Dense layers of vertically oriented WO$_3$ crystals as anodes for photoelectrochemical water oxidation. Chem Commun 48(5):729–731

75. Qiu YC, Chen W, Yang SH (2010) Double-layered photoanodes from variable-size anatase TiO$_2$ nanospindles: a candidate for high-efficiency dye-sensitized solar cells. Angew Chem-Int Edit 49(21):3675–3679

76. Qiu YC, Yan KY, Deng H, Yang SH (2012) Secondary branching and nitrogen doping of ZnO nanotetrapods: building a highly active network for photoelectrochemical water splitting. Nano Lett 12(1):407–413

77. Raja KS, Gandhi T, Misra M (2007) Effect of water content of ethylene glycol as electrolyte for synthesis of ordered titania nanotubes. Electrochem Commun 9(5):1069–1076

78. Seabold JA, Choi KS (2011) Effect of a cobalt-based oxygen evolution catalyst on the stability and the selectivity of photo-oxidation reactions of a WO$_3$ photoanode. Chem Mat 23(5):1105–1112

79. Seol M, Kim H, Kim W, Yong K (2010) Highly efficient photoelectrochemical hydrogen generation using a ZnO nanowire array and a CdSe/CdS co-sensitizer. Electrochem Commun 12(10):1416–1418

80. Shankar K, Basham JI, Allam NK, Varghese OK, Mor GK, Feng XJ, Paulose M, Seabold JA, Choi KS, Grimes CA (2009) Recent advances in the use of TiO$_2$ nanotube and nanowire arrays for oxidative photoelectrochemistry. J Phys Chem C 113(16):6327–6359

81. Shankar K, Mor GK, Fitzgerald A, Grimes CA (2007) Cation effect on the electrochemical formation of very high aspect ratio TiO$_2$ nanotube arrays in formamide—Water mixtures. J Phys Chem C 111(1):21–26

82. Sivula K, Le Formal F, Gratzel M (2011) Solar Water Splitting: Progress Using Hematite (alpha-Fe$_2$O$_3$) Photoelectrodes. ChemSusChem 4(4):432–449

83. Sivula K, Zboril R, Le Formal F, Robert R, Weidenkaff A, Tucek J, Frydrych J, Gratzel M (2010) Photoelectrochemical water splitting with mesoporous hematite prepared by a solution-based colloidal approach. J Am Chem Soc 132(21):7436–7444

84. Su J, Guo L, Bao N, Grimes CA (2011) Nanostructured WO$_3$/BiVO$_4$ heterojunction films for efficient photoelectrochemical water splitting. Nano Lett 11(5):1928–1933

85. Su JZ, Feng XJ, Sloppy JD, Guo LJ, Grimes CA (2011) Vertically aligned WO$_3$ nanowire arrays grown directly on transparent conducting oxide coated glass: synthesis and photoelectrochemical properties. Nano Lett 11(1):203–208

86. Su R, Bechstein R, Kibsgaard J, Vang RT, Besenbacher F (2012) High-quality Fe-doped TiO2 films with superior visible-light performance. J Mater Chem 22(45):23755–23758

87. Sun JW, Liu C, Yang PD (2011) Surfactant-free, large-scale, solution-liquid-solid growth of gallium phosphide nanowires and their use for visible-light-driven hydrogen production from water reduction. J Am Chem Soc 133(48):19306–19309

88. Sun K, Jing Y, Li C, Zhang XF, Aguinaldo R, Kargar A, Madsen K, Banu K, Zhou YC, Bando Y, Liu ZW, Wang DL (2012) 3D branched nanowire heterojunction photoelectrodes for high-efficiency solar water splitting and H_2 generation. Nanoscale 4(5):1515–1521

89. Tachibana Y, Vayssieres L, Durrant JR (2012) Artificial photosynthesis for solar water-splitting. Nat Photonics 6(8):511–518

90. Tilley SD, Cornuz M, Sivula K, Gratzel M (2010) Light-induced water splitting with hematite: improved nanostructure and iridium oxide catalysis. Angew Chem-Int Edit 49(36):6405–6408

91. Tsuchiya H, Macak JM, Taveira L, Balaur E, Ghicov A, Sirotna K, Schmuki P (2005) Self-organized TiO_2 nanotubes prepared in ammonium fluoride containing acetic acid electrolytes. Electrochem Commun 7(6):576–580

92. Walter MG, Warren EL, McKone JR, Boettcher SW, Mi QX, Santori EA, Lewis NS (2010) Solar water splitting cells. Chem Rev 110(11):6446–6473

93. Wang DF, Pierre A, Kibria MG, Cui K, Han XG, Bevan KH, Guo H, Paradis S, Hakima AR, Mi ZT (2011) Wafer-level photocatalytic water splitting on GaN nanowire arrays grown by molecular beam epitaxy. Nano Lett 11(6):2353–2357

94. Wang G, Li Y (2013) Nickel catalyst boosts solar hydrogen generation of CdSe nanocrystals. ChemCatChem 5(6):1294–1295

95. Wang G, Ling Y, Lu XH, Qian F, Tong YX, Zhang JZ, Lordi V, Leao CR, Li Y (2013) Computational and photoelectrochemical study of hydrogenated bismuth vanadate. J Phys Chem C 117(21):10957–10964

96. Wang GM, Ling YC, Li Y (2012) Oxygen-deficient metal oxide nanostructures for photoelectrochemical water oxidation and other applications. Nanoscale 4(21):6682–6691

97. Wang GM, Ling YC, Lu XH, Wang HY, Qian F, Tong YX, Li Y (2012) Solar driven hydrogen releasing from urea and human urine. Energy Environ Sci 5(8):8215–8219

98. Wang GM, Ling YC, Lu XH, Zhai T, Qian F, Tong YX, Li Y (2013) A mechanistic study into the catalytic effect of $Ni(OH)_2$ on hematite for photoelectrochemical water oxidation. Nanoscale 5(10):4129–4133

99. Wang GM, Ling YC, Wang HY, Yang XY, Wang CC, Zhang JZ, Li Y (2012) Hydrogen-treated WO_3 nanoflakes show enhanced photostability. Energy Environ Sci 5(3):6180–6187

100. Wang GM, Ling YC, Wheeler DA, George KEN, Horsley K, Heske C, Zhang JZ, Li Y (2011) Facile synthesis of highly photoactive alpha-Fe_2O_3-based films for water oxidation. Nano Lett 11(8):3503–3509

101. Wang GM, Wang HY, Ling YC, Tang YC, Yang XY, Fitzmorris RC, Wang CC, Zhang JZ, Li Y (2011) Hydrogen-treated TiO_2 nanowire arrays for photoelectrochemical water splitting. Nano Lett 11(7):3026–3033

102. Wang GM, Yang XY, Qian F, Zhang JZ, Li Y (2010) Double-sided CdS and CdSe quantum dot Co-sensitized ZnO nanowire arrays for photoelectrochemical hydrogen generation. Nano Lett 10(3):1088–1092

103. Wang HY, Wang GM, Ling YC, Lepert M, Wang CC, Zhang JZ, Li Y (2012) Photoelectrochemical study of oxygen deficient TiO_2 nanowire arrays with CdS quantum dot sensitization. Nanoscale 4(5):1463–1466

104. Wei W, Lee K, Shaw S, Schmuki P (2012) Anodic formation of high aspect ratio, self-ordered Nb_2O_5 nanotubes. Chem Commun 48(35):4244–4246

105. Wheeler DA, Wang GM, Ling YC, Li Y, Zhang JZ (2012) Nanostructured hematite: synthesis, characterization, charge carrier dynamics, and photoelectrochemical properties. Energy Environ Sci 5(5):6682–6702

106. Widenkvist E, Quinlan RA, Holloway BC, Grennberg H, Jansson U (2008) Synthesis of nanostructured tungsten oxide thin films. Cryst Growth Des 8(10):3750–3753

107. Wolcott A, Smith WA, Kuykendall TR, Zhao YP, Zhang JZ (2009) Photoelectrochemical study of nanostructured ZnO thin films for hydrogen generation from water splitting. Adv Funct Mater 19(12):1849–1856

108. Wolcott A, Smith WA, Kuykendall TR, Zhao YP, Zhang JZ (2009) Photoelectrochemical water splitting using dense and aligned TiO$_2$ nanorod arrays. Small 5(1):104–111

109. Xie KY, Li J, Lai YQ, Lu W, Zhang ZA, Liu YX, Zhou LM, Huang HT (2011) Highly ordered iron oxide nanotube arrays as electrodes for electrochemical energy storage. Electrochem Commun 13(6):657–660

110. Xu YL, He Y, Cao XD, Zhong DJ, Jia JP (2008) TiO$_2$/Ti rotating disk photoelectrocatalytic (PEC) reactor: a combination of highly effective thin-film PEC and conventional PEC processes on a single electrode. Environ Sci Technol 42(7):2612–2617

111. Yang XY, Wolcott A, Wang GM, Sobo A, Fitzmorris RC, Qian F, Zhang JZ, Li Y (2009) Nitrogen-doped ZnO nanowire arrays for photoelectrochemical water splitting. Nano Lett 9(6):2331–2336

112. Yang Y, Ling YC, Wang GM, Lu XH, Tong YX, Li Y (2013) Growth of gallium nitride and indium nitride nanowires on conductive and flexible carbon cloth substrates. Nanoscale 5(5):1820–1824

113. Ye H, Lee J, Jang JS, Bard AJ (2010) Rapid screening of BiVO$_4$-based photocatalysts by scanning electrochemical microscopy (SECM) and studies of their photoelectrochemical properties. J Phys Chem C 114(31):13322–13328

114. Ye H, Park HS, Bard AJ (2010) Screening of electrocatalysts for photoelectrochemical water oxidation on W-doped BiVO$_4$ photocatalysts by scanning electrochemical microscopy. J Phys Chem C 115(25):12464–12470

115. Yin WJ, Wei SH, Al-Jassim MM, Turner J, Yan YF (2011) Doping properties of monoclinic BiVO$_4$ studied by first-principles density-functional theory. Phys Rev B 83(15):11

116. Yoriya S, Paulose M, Varghese OK, Mor GK, Grimes CA (2007) Fabrication of vertically oriented TiO$_2$ nanotube arrays using dimethyl sulfoxide electrolytes. J Phys Chem C 111(37):13770–13776

117. Zhang J, Bang JH, Tang CC, Kamat PV (2010) Tailored TiO$_2$-SrTiO$_3$ heterostructure nanotube arrays for improved photoelectrochemical performance. ACS Nano 4(1):387–395

118. Zhang J, Wang XL, Xia XH, Gu CD, Zhao ZJ, Tu JP (2010) Enhanced electrochromic performance of macroporous WO$_3$ films formed by anodic oxidation of DC-sputtered tungsten layers. Electrochim Acta 55(23):6953–6958

119. Zhang ML, Luo WJ, Li ZS, Yu T, Zou ZG (2010) Improved photoelectrochemical responses of Si and Ti codoped alpha-Fe$_2$O$_3$ photoanode films. Appl Phys Lett 97(4):3

120. Zheng HD, Sadek AZ, Latham K, Kalantar-Zadeh K (2009) Nanoporous WO$_3$ from anodized RF sputtered tungsten thin films. Electrochem Commun 11(4):768–771

121. Zhong DK, Choi S, Gamelin DR (2011) Near-complete suppression of surface recombination in solar photoelectrolysis by "Co-Pi" catalyst-modified W:BiVO$_4$. J Am Chem Soc 133(45):18370–18377

122. Zhong DK, Cornuz M, Sivula K, Graetzel M, Gamelin DR (2011) Photo-assisted electrodeposition of cobalt-phosphate (Co-Pi) catalyst on hematite photoanodes for solar water oxidation. Energy Environ Sci 4(5):1759–1764

123. Zhong DK, Gamelin DR (2011) Photoelectrochemical water oxidation by cobalt catalyst ("Co-Pi")/alpha-Fe$_2$O$_3$ composite photoanodes: oxygen evolution and resolution of a kinetic bottleneck. J Am Chem Soc 132(12):4202–4207

124. Zhong DK, Sun JW, Inumaru H, Gamelin DR (2009) Solar water oxidation by composite catalyst/alpha-Fe$_2$O$_3$ photoanodes. J Am Chem Soc 131(17):6086–6087

125. Zhong LS, Hu JS, Liang HP, Cao AM, Song WG, Wan LJ (2006) Self-assembled 3D flowerlike iron oxide nanostructures and their application in water treatment. Adv Mater 18(18):2426–2431

126. Zhong M, Li YB, Yamada I, Delaunay JJ (2012) ZnO-ZnGa$_2$O$_4$ core-shell nanowire array for stable photoelectrochemical water splitting. Nanoscale 4(5):1509–1514
127. Zhu W, Liu X, Liu HQ, Tong DL, Yang JY, Peng JY (2010) Coaxial heterogeneous structure of TiO$_2$ nanotube arrays with CdS as a superthin coating synthesized via modified electrochemical atomic layer deposition. J Am Chem Soc 132(36):12619–12626

Magnesium and Doped Magnesium Nanostructured Materials for Hydrogen Storage

Daniel J. Shissler, Sarah J. Fredrick, Max B. Braun and Amy L. Prieto

Abstract Hydrogen is an attractive fuel for many applications because of its high energy density as molecular hydrogen, as well as the clean exhaust produced when burned with oxygen. One significant challenge to the widespread adoption of hydrogen, for mobile applications in particular, is the inability to efficiently store large amounts of readily accessible hydrogen in small volumes at ambient temperature and pressure. This chapter describes the current research on one particularly interesting candidate for hydrogen storage, nanostructured magnesium. The synthetic methods currently used to control the size and shape of nanostructured magnesium are described, as are the measured kinetics of hydrogen storage, the modeling used to explain the observed kinetics, and theoretical models that can be used to guide experimental efforts.

1 Introduction

Hydrogen is a very attractive fuel for many applications because it is the most abundant element on earth (although less than 1 % is present as molecular hydrogen), the gravimetric energy density is three times higher than liquid hydrocarbons (142 MJ/kg versus 47 MJ/kg), and when burnt in oxygen the only exhaust is water. One significant hurdle to the widespread adoption of hydrogen-burning vehicles, however, is the development of new materials that can absorb and desorb large amounts of hydrogen safely at low pressures and ambient temperatures. The Department of Energy has set ambitious goals for the capacity, cycle life, and delivery pressures required to make hydrogen a viable fuel for mobile applications (Table 1). There are no current materials that can meet these goals to date.

D. J. Shissler · S. J. Fredrick · M. B. Braun · A. L. Prieto (✉)
Chemistry Department, Colorado State University, Campus Delivery 1872,
Fort Collins, CO 80523, USA
e-mail: alprieto@lamar.colostate.edu

Z. Lin and J. Wang (eds.), *Low-cost Nanomaterials*, Green Energy and Technology,
DOI: 10.1007/978-1-4471-6473-9_11, © Springer-Verlag London 2014

Table 1 Targets set by the Department of Energy (released in 2009)

Storage parameters	Units	2010	2015	Ultimate
System gravimetric capacity	kWh kg^{-1}	1.5	1.8	2.5
	kg H$_2$ kg^{-1} system	0.045	0.055	0.075
System volumetric capacity	kWh L^{-1}	0.9	1.3	2.3
	kg H$_2$ L^{-1} system	0.028	0.040	0.070
Min/max delivery temperature	°C	−40/85	−40/85	−40/95–105
Cycle life (1/4 tank to full)	Cycles	1,000	1,500	1,500
Min/max delivery pressure for fuel cell	Bar (abs)	5/12	5/12	3/12
Min/max delivery pressure for internal combustion engine	Bar (abs)	35/100	35/100	35/100
Onboard reversible system efficiency	%	90	90	90
System fill rate	kg H$_2$ min^{-1}	1.2	1.5	2.0

Used with permission by Chen 2012

Current research in the storage of hydrogen is focused on several areas, including the synthesis of new materials with high porosity and controllable functionality, [1] the adsorption of H$_2$ in zeolites and carbonaceous materials including carbon nanotubes, [2] and the modification of existing metal hydride systems [3]. A comparison of a range of materials as a function of their volumetric and gravimetric hydrogen storage is shown in Fig. 1. Among these research areas, metal hydrides containing light elements such as MgH$_2$ offer high density storage at low cost.

The significant advantage to light metal hydrides such as MgH$_2$ is the potentially high hydrogen capacity in small volumes at ambient pressures and temperatures. However, essentially all of the target compounds are plagued by slow kinetics for the hydrogenation and dehydrogenation reactions, which means that they are currently cycled at elevated temperatures (typically near 300 °C) and low/high pressure for the dehydrogenation/hydrogenation reactions. The compounds that can cycle at or near room temperature are hindered by low mass density for hydrogen (such as Pd, which converts to PdH$_{0.6}$) [3]. A fundamental understanding of the structure and properties of complex hydrides would be extremely helpful in decoupling the challenges in terms of thermodynamic stability of the relevant phases as well as the kinetics. Without an understanding of the interplay among structure, bonding, and diffusion in these materials, optimization to improve the kinetics at lower temperatures and ambient pressures relies on trial and error. For that reason a portion of this chapter is dedicated to theoretical predictions that are guiding current research.

A reasonable argument can be made that *only* light hydrogen storage materials with high gravimetric capacities will be able to achieve the DOE targets. For that reason, Mg, doped Mg, and Mg alloys are promising materials for hydrogen storage

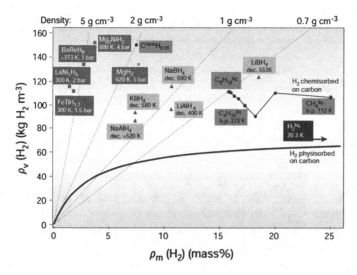

Fig. 1 A comparison of a range of materials as a function of stored hydrogen per volume and per mass. Reproduced with permission from Schlapbach 2001

due to their high theoretical hydrogen storage capacities (e.g., 7.6 wt.% for MgH_2) [3, 4]. *Pure Mg particles can store 1 kg of H_2 in 6.9 L of solid metal hydride.* However, bulk Mg is less than ideal as a hydrogen storage material due to the slow kinetics and high temperatures required for hydrogen absorption/desorption.

One common approach to mitigating the sluggish kinetics for bulk magnesium is to increase the surface area to volume ratio of the material. The reaction of bulk Mg and molecular hydrogen first involves the adsorption of molecular hydrogen onto the surface of the magnesium, the breaking of the hydrogen–hydrogen bond, and then the diffusion of the atomic hydrogen into the magnesium. Once a critical concentration is reached, MgH_2 nucleates and grows (shown in Fig 2). One can imagine that by reducing the physical dimensions of the material to the nanoscale, the diffusion length required for hydrogen to diffuse by solid-state diffusion from the surface of the material to the center is dramatically reduced, which can reduce the time necessary for hydrogenation (and the same effect should be true for the reverse reaction, removing hydrogen from the hydride).

Recent studies have shown that hydrogen storage properties can be greatly improved by preparing these materials with nanocrystalline grain sizes, often achieved by mechanical ball-milling, and/or by alloying Mg with other metals [3–5]. Previous studies have shown that reducing the particle size of metal hydrides such as Mg_2Ni significantly increases the kinetics of hydrogen absorption [4]. The increase in the kinetics for hydrogen absorption may be a result of a decrease in the diffusion length for hydrogen due to the small sizes of the particles (as mentioned previously), but the lack of control over the size and homogeneity of the samples has made it difficult to determine this conclusively. The synthetic methods used (such as ball milling [4] and hydrogen plasma–metal reactions [6]) can produce samples that are

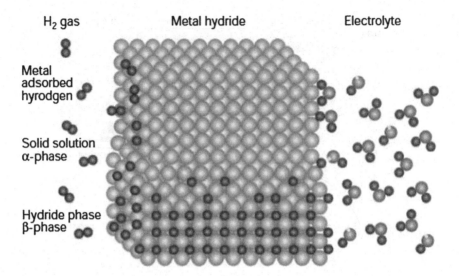

Fig. 2 Schematic model of a metal structure with H atoms in the interstices between the metal atoms, and H₂ molecules at the surface. Hydrogen atoms are from physisorbed hydrogen molecules on the *left-hand side* and from the dissociation of water molecules on the *right-hand side*. Reproduced with permission from Schlapbach 2001

inhomogeneous, poorly crystalline, and with ill-defined particle size. An important but unanswered question is still whether or not the increased kinetics observed for nanostructured materials are due purely to increased surface area, or if grain boundaries and trace impurities or catalysts are important for the hydridation reaction.

This chapter will provide an overview of the synthesis of nanostructured magnesium-based materials, the resulting kinetics of these materials (and how these kinetics are currently modeled) as well as theoretical predictions of what can be obtained using different approaches toward magnesium. For the purposes of brevity, we have focused on nanostructured magnesium, as well as doped magnesium, but we have not included a discussion of composite materials where magnesium is not the major component. The references contained herein are meant to be as comprehensive as possible, to provide the reader with reasonable resources with which to read further about this fascinating material.

2 Synthesis

In order to optimize the performance of any material, a deep understanding of the effects of the synthesis of the material on the resulting properties is imperative. For hydrogen storage materials in particular, it is important to fully understand the sorption kinetics for these systems and how the kinetics depend on the quality of the material as made. The challenge with magnesium-based materials in particular is that elemental magnesium is very easily oxidized. Any synthetic method that is designed to produce zero valent magnesium will need to be carried out under inert conditions, and if the precursors are common magnesium salts, under fairly reducing conditions. For the purpose of studying the effects of parameters such as particle size and the incorporation of dopants (common variables that are tuned for magnesium as a hydrogen storage material), a synthesis that allows for fine control and manipulation of these variables is desired. The general goals of modern synthetic methods for magnesium are to control the size and morphology of the particles and/or to controllably incorporate dopants.

2.1 Undoped Magnesium Nanocrystal Syntheses

There are two general strategies for synthesizing magnesium for hydrogen storage. The first is to start with bulk elemental magnesium, and then either ball mill the bulk material to reduce the grain size, or to use vapor transport reactions to create nanostructures. For the purposes of clarity, a distinction must be made between nanostructured magnesium and nanocrystalline magnesium; while one can be both nanostructured (wherein the physical features of the particles are on the nanoscale) and nanocrystalline (where the grain sizes composing the particles are on the nanoscale), the other may be nanocrystalline but microstructured (Fig. 3) [7, 8]. This is an important distinction because different synthetic methods are required to control the physical dimensions of the particles versus the physical dimensions of the grains contained within a particle. The second general method for synthesizing magnesium for hydrogen storage is to start with magnesium precursors where magnesium is present in the +2 oxidation state, and then reduce the precursor in solution (either chemically or electrochemically, see Fig. 4) to form neutral magnesium.

The most prevalent synthesis in the current literature for reducing the grain sizes of bulk magnesium to micro or nanocrystalline grains is that of mechanical milling, or ball milling [7, 9]. In addition to standard ball milling, high energy mechanical milling (under either high pressure or high temperature) has been used by many for the synthesis of magnesium nanocrystals [10]. This technique can be utilized under hydrogen pressure to directly make MgH_2 [11–14] rather than making the pure metal and hydriding it post-synthesis. Mechanochemical milling has also been used to produce magnesium nanocrystals in a similar fashion [14]. Another mechanical method for producing nanocrystalline magnesium is the method of cold rolling and cold forging to reduce grain sizes [15]. Although these types of mechanical milling are effective in reducing grain size and incorporating dopant materials (to be discussed in Sect. 2.2), they offer little control over nanocrystal size distribution, shape, and impurity phases (Fig. 3b).

Plasma reactions [16, 17] have been demonstrated to be effective in producing nanoscale magnesium; however, impurity phases such as oxides can also be a problem in these systems. Gas-phase condensation of metallic magnesium into

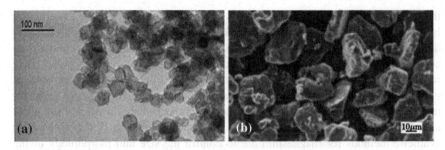

Fig. 3 Comparison of **a** TEM micrograph of nanostructured magnesium synthesized by gas phase condensation reaction and **b** SEM micrograph of nanocrystalline magnesium microstructures. Reproduced with permission from Zaluska 1999 and Friedrichs 2007

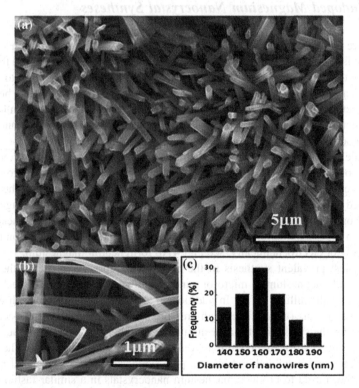

Fig. 4 SEM image **a** of the nanowires obtained by electrochemical reduction of EtMgCl **b** high resolution SEM and **c** size distribution of the diameters of the nanowires obtained by this method. Figure used with permission from Viyannalage 2012

nanocrystalline powders has been shown to be effective and have the advantage of in situ hydriding, [8, 18] but the high temperatures and/or pressures required for this synthesis make it less than ideal. Alternatively, vapor transport has been used to grow magnesium nanowires also used for hydrogen storage studies [19].

Direct decomposition of Grignard reagents in the presence of hydrogen was shown to give pure magnesium hydride that can be used for dehydrogenation studies, however purity varied among samples [20]. A recent example of a solvated metal atom dispersion method followed by digestive ripening with organic surfactants produced small (2–4 nm in diameter), monodisperse nanocrystals [21]. This synthesis is arguably the most effective at making ultrasmall clusters of magnesium that are expected to show the fastest kinetics at lower temperatures as well as improved thermodynamics.

The solution-phase synthesis of nanoparticles has emerged as an elegant and rational method which can yield a high level of control over size [22], shape [23, 24], surface properties [25], and connectivity [26]. Synthetic methods such as electrochemical reduction from solution result in small particles exhibiting improved kinetics, but are not amenable to the controllable addition of catalyst particles [27]. Mg nanoparticles synthesized from solution have been tested for hydrogen storage properties, but the early preparation methods were not designed to control particle size, and the average size and surface composition of these nanoparticles were not investigated, therefore, not providing a good description of how much particle diameter can influence the material's properties [28]. Mg nanowires have been synthesized in high yield, and show enhanced kinetics for hydrogen sorption [29]. As mentioned previously, ball-milled Mg nanoparticles show even faster kinetics, although there is a large size distribution in these samples [30]. A solution-phase synthesis would provide a route toward narrow size dispersion, high quality nanoparticles.

In a push toward lower energy methods for producing nanoscale magnesium, there have been several recent reports of solution synthesis methods. The low reduction potential of magnesium has allowed for electrochemical reduction to produce magnesium nanocrystals out of THF [31]. A similar synthesis has produced both nanoparticles and nanowires, [32] and there is no indication that impurities were a byproduct. Direct chemical reduction of magnesium salts to magnesium metal in glyme at near-room temperature has been shown [33]. This method has also demonstrated control over particle size with corresponding enhanced kinetics. The powder X-ray diffraction characterization of this material showed pure, crystalline magnesium, with significant peak broadening indicative of the small size (Fig. 5). One final example of a solution method not only reduces magnesium salts to magnesium metal at room temperature, but also encapsulates the nanoparticles in a polymer matrix, which keeps the highly pyrophoric magnesium air stable for weeks in air (Fig. 6) [34].

2.2 Doped Magnesium Nanocrystal Synthesis

In addition to reducing particle size, dopants can be incorporated into the pure magnesium nanocrystals to enhance the poor sorption kinetics of magnesium. This chapter will not specifically address magnesium-based composites such as

Fig. 5 X-ray diffraction patterns of magnesium nanoparticles with varying sizes. Reproduced with permission from Norberg 2011

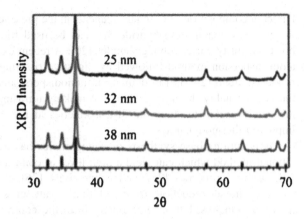

MgH_2–$LiAlH_4$, but rather will focus on dopants in small percentages (10 % or less). Information on composites can be found elsewhere [35–38]. A variety of dopants can be incorporated through many of the previously mentioned synthesis methods. Several papers have been written on a series of transition metals [5, 39–43] or transition metal oxides [44–46] while others have focused on specific elemental dopants, the most common being Al, [47, 48] Si, [47, 49] Ti, [50, 51] V and its corresponding oxides, [52, 53] Cr, [54] Fe, [50, 55–57] Ni, [17, 50, 58–63] Ge, [64] Nb and corresponding oxides [52, 54, 57, 60, 65–68], and Pd [59]. More discussion of these dopants and their effects on sorption kinetics will be discussed in Sects. 3 and 4.

3 Evaluating the Kinetics of Hydrogen Sorption and Desorption

There are several different methods by which the kinetics of hydrogenation and dehydrogenation can be determined. Non-isothermal methods, also known as Temperature Programmed Desorption (TPD), involve the use of either Thermogravimetric Analysis (TGA) or Differential Scanning Calorimetry (DSC). Isothermal approaches include the measurements of Gravimetric-Composition-Isotherms (GCI) and Pressure-Composition-Isotherms (PCI's). GCI measure the change in mass of a sample under constant temperature and pressure conditions. Conversely, a Sievert's apparatus is used with dynamic pressure and constant temperature to measure PCI's [69, 70]. This is a volumetric method which determines the amount of hydrogen absorbed or desorbed by a sample through the ideal gas law, $PV = nRT$, by measuring the change in pressure in a system with known volume and temperature [71].

A downside of TPD methods is that they can only be used to analyze desorption of hydrogen from a material, not the adsorption. However, since the measurements are done with a TGA or DSC instrument, common facility instruments,

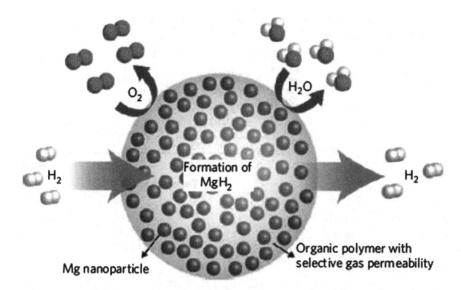

Fig. 6 Mg/polymer composite hydrogen storage material. Here Mg nanoparticles are encapsulated in PMMA, a gas-permeable polymer. Reproduced with permission from Jeon 2011

TPD methods are employable without an apparatus specifically built for hydrogenation/dehydrogenation purposes. Also, the data analysis is more flexible than that of the isothermal methods, i.e., PCI and GCI.

3.1 Non-isothermal Methods

When using TGA or DSC, the hydrogen desorption kinetics of different materials can be qualitatively characterized by comparing the initial onset of transformation in the materials. For example, in TGA, a characteristic plot of weight percent versus temperature (Fig. 7) [72] displays an onset of weight loss, corresponding to loss of hydrogen at different temperatures, for an assortment of materials. This onset can be observed at lower temperature for the various catalyzed materials than for that of the non-catalyzed MgH_2. Similarly in DSC, where a quantitative endotherm or exotherm signifies a phase change as different temperatures are scanned, it can be seen that hydrogen desorption for the nickel-catalyzed sample occurs at a much lower temperature with respect to that of the non-catalyzed sample (Fig. 8) [40].

For the purposes of hydrogen storage, the better performing material will exhibit an onset of desorption at a lower temperature. However, due to the variation within experimental procedure and instrument calibration, if the performance of a material is to be compared to that of the other literature, an activation energy for desorption should be determined. The activation energy can be calculated by analyzing TGA and DSC data with one of the two following equations:

Fig. 7 TGA measurements
on magnesium hydride with
and without different
nanocomposites. Reproduced
with permission from Sabitu
2012

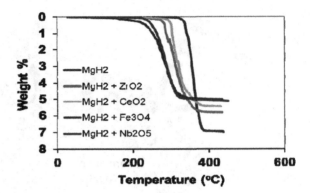

$$\ln \beta = -\frac{0.457 E_a}{RT} - 2.315 - \log \left[\frac{R}{AE_a} \int_0^\alpha \frac{d\alpha}{f(\alpha)} \right] \qquad (1)$$

$$\ln \frac{\beta}{T^2} = -\frac{E_a}{RT} - \ln \left[\frac{E_a}{RA} \int_0^\alpha \frac{d\alpha}{f(\alpha)} \right] \qquad (2)$$

where β is the ramp rate, E_a is the activation energy, R is the gas constant, and A is the pre-exponential factor of the Arrhenius equation. The kinetic function, $f(\alpha)$, is dependent on the fraction of converted material at a given time over the maximum converted material [73]. These two variables will be discussed further in later sections. Determining the activation energy with Eq. (1) is known as the Ozawa–Flynn–Wall (OFW) method [74, 75]. Alternatively, the Kissinger–Akahira–Sunose (KAS) method can be used to find the activation energy with Eq. (2) [76]. By varying the ramp rates of several TGA or DSC scans, a linear relationship can be observed by plotting $\ln(\beta)$ versus $1/T$ or $\ln(\beta/T^2)$ versus $1/T$, depending on which method is used. The slope of the plot can then be used in correlation with Eq. (1) or (2) to calculate the activation energy. Different data sets are obtained in TGA than with DSC. As can be seen in the TGA data shown in Fig. 9 [74], different fractions of completion can be chosen prior to analysis. From here, the different values of ramp rates and corresponding temperatures can be used to form, for the purpose of repetition, several linear plots. Alternately with DSC, the various ramp rates are plotted with the temperature at which the reaction rate is greatest (Fig. 8) [40]. There has been much debate about which method, OFW or KAS, should be utilized due to the accuracy of the calculated value. However, the discrepancies between the methods are small enough that a semiquantitative comparison of activation energies can be done to compare the kinetics [73, 77]. The advantage of using these methods to study kinetics of desorption is that an activation energy may be calculated without the difficulty of fitting the desorption curve to an appropriate kinetic function, $f(\alpha)$. However, this is not the case for the PCI or GCI methods, which have the added advantage of being able to study both the hydrogen adsorption and desorption kinetics.

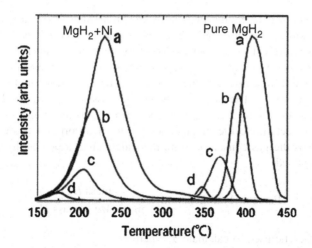

Fig. 8 Thermal desorption mass spectra of hydrogen under various heating rates 1, 5, 10, 20 °C/min (**a**), (**b**), (**c**), and (**d**), respectively, for the pure MgH$_2$ milled for 15 min at 200 rpm and the 2 mol% Ni catalyzed MgH$_2$ composite prepared by milling for 15 min at 200 rpm. The intensities of the longitudinal axes indicate the amount of hydrogen desorption per unit time. Adapted with permission from Hanada 2005

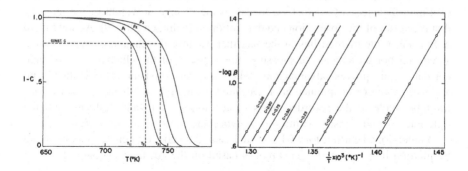

Fig. 9 Residual fraction versus temperature for three different rates of heating (*left*). Log (β) versus 1/T for different degrees of conversion, C. Adapted with permission from Flynn 1966

3.2 Isothermal Methods

Both GCI and PCI techniques require a reaction chamber that can be evacuated and highly pressurized, as well as a sample holder that can be thermally regulated. In order to calculate the activation energies using these methods, kinetics rate constants are used in association with the Arrhenius equation. However, to find the kinetic rate constant, the sorption curves must be fit to the modeled function, $f(\alpha)$, of the appropriate kinetic mechanism. As will be explained, the GCI and PCI methods differ by the manner in which the fraction of completion, α, is determined.

The method of determining PCIs, also known as Sievert's method, requires a chamber of known volume [69]. When a sample is exposed to hydrogen, the change in pressure can be accounted for absorption of hydrogen. Alternatively, when a hydrogenated sample is exposed to vacuum, or a pressure lower than the desorption equilibrium pressure, the pressure change can be attributed to desorption of hydrogen [78]. The recorded change in pressure, P, with the known volume, V, and temperature, T, of the chamber, can be applied to the ideal gas law, $PV = nRT$, to determine the molar amounts, n, of absorption or desorption of hydrogen (R is the gas constant). From this calculation, a percent hydrogenation can be determined:

$$\text{Percent Hydrogenation} = \frac{\text{mass } H_2}{\text{mass } Mg + \text{mass } H_2} \tag{3}$$

which can then be used to calculate α, where

$$\alpha = \frac{H_2 \text{ weight\%}}{\text{Maximum } H_2 \text{ weight\%}} \tag{4}$$

which can then be used in kinetic fittings.

Using a similar apparatus to the one used in Sievert's method, with the addition of a microbalance, a GCI method measures the change in weight of the material that is a result of the absorption and desorption of hydrogen [20, 79]. Another main difference in the method lies in the fact that because the change of mass is measured and not the change of pressure, the pressure of the chamber can be held constant. This proves advantageous when a kinetic constant is calculated and applied, for reasons explained later. Despite this mentioned advantage, corrections must be made to the recorded mass to account for buoyancy effects from the differences in air density within the chamber and the surroundings [79].

As mentioned earlier, for PCI and GCI data, the activation energy can be found by plotting ln k versus $1/T$. This plot is based on the Arrhenius equation:

$$\ln(k) = \frac{-E_a}{R} \frac{1}{T} + \ln(A) \tag{5}$$

Before this can be done, however, the kinetic rate constant has to be deduced for various temperatures.

The kinetic rate constant is found by fitting the experimental data with different model equations, $f(\alpha)$ (Table 2) [80]. Each function has a different rate-determining step, which is mentioned in the description in Table 2. The best function will produce a plot that is linear, with the slope that is equal to the rate constant, k (Fig. 10) [79, 80]. However, because of the high surface area-to-volume ratio of nanoscaled materials, fitting kinetic data to kinetic models based on bulk materials proves arduous, if not impossible [81, 82]. The two most common models used are the Johnson–Mehl–Avrami (JMA) model of Nucleation and Growth (NG) [83–86]

Table 2 Kinetic equations used for fitting experimental sorption data

Model equation	Description
$\alpha = kt$	Surface-controlled reaction with chemisorption being the RDS
$[-\ln(1-\alpha)]^{1/3} = kt$	JMA nucleation and growth: three-dimensional growth of existing nuclei with constant interface velocity
$[-\ln(1-\alpha)]^{1/2} = kt$	JMA nucleation and growth: two-dimensional growth of existing nuclei with constant interface velocity
$1 - (1-\alpha)^{1/3} = kt$	CV: three-dimensional growth with constant interface velocity
$1 - (1-\alpha)^{1/2} = kt$	CV: two-dimensional growth with constant interface velocity
$1 - \left(\frac{2\alpha}{3}\right) - (1-\alpha)^{2/3} = kt$	CV: three-dimensional growth with decreasing interface velocity

Adapted with permission from Barkhordarian 2006

Fig. 10 Different kinetic curves based on the equations in Table 1 for magnesium catalyzed with 1 mol% Nb_2O_5 and milled for 100 h (T = 100 °C). Reproduced with permission from Barkhordarian 2006

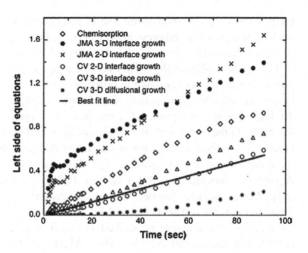

and Contracting Volume (CV) [87] model (Fig. 11). Both models make assumptions that may not be valid during growth of differing phases of nanoparticles. For instance, the CV model makes the assumption that nucleation of the secondary phases are instantaneous (the nucleation of MgH_2, for example) and that the core/shell morphology is quickly reached. As a result of the model not accounting for the initial nucleation and growth, the fitting will likely have discrepancies for low α values. In contrast, the fitting for JMA will likely have discrepancies for the high α values. This is because the JMA model does not account for the overlap of the growing phase that is likely to occur once, most of the material is transformed to MgH_2 [70, 88]. Due to the many assumptions made by these models, the experimental data rarely fit well for a whole data set [70]. However, even though an entirely acceptable fit may not be made, the nature of the mechanism can be inferred through characterization of a portion of the full data set [70, 81].

As mentioned earlier, the GCI method can be performed at constant pressure, which is advantageous because pressure changes add even more complexity to isothermal measurements. In addition to the difficulty of fitting the data, the PCI

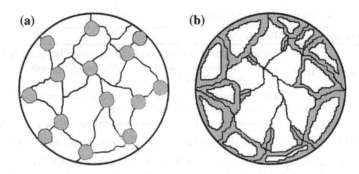

Fig. 11 Schematic picture of phase growth according to **a** JMA and **b** CV models. The *dark areas* represent the transformed phase. Reproduced with permission from Barkhordarian 2006

method calls for data correction due to the kinetic behavior being dependent on both the temperature and pressure. It has been shown that as the change in pressure is measured, which is how the data is collected, the driving force for the exchange of hydrogen is altered [89]. This driving force is dependent on the equilibrium pressure, which is defined as the pressure at which the system must be above or below, in order for hydrogenation or dehydrogenation to proceed. The change in driving force is not significant if the process is performed with a pressure considerably higher or lower than the equilibrium pressure, depending on which process is monitored. However, if this is not the case, the deviation of the driving force can be corrected by plotting $F(\alpha)/F(P)$ versus t, where $F(P)$ is the pressure dependence function [68, 89]. Common $F(P)$ functions can be found in Table 3 [89]. Proper methods for determining the pressure dependence function can be found in the work of Sanchez et al. [90]. P is the measured pressure at time t, P_{br} is the reaction bed pressure (described in detail in reference) [89], and P_{eq} is the equilibrium pressure of the $Mg + H_2 \rightarrow MgH_2$ reaction. P_{eq} values can be determined by performing the pressure sweep method [91] or estimated by utilizing values reported in the literature [68, 92]. Again, the slope of the new plot, $F(\alpha)/F(P)$ versus t, is the kinetic rate constant.

Quantitative characterization of the sorption kinetics of nanoscaled materials has proven difficult if not unattainable. However, with the methods that have been developed thus far, at least qualitative comparison of catalytic ability is possible. As will be evident in the next section, the lack of activation energy values reported in the literature makes the comparison of hydrogenation and dehydrogenation kinetics elusive, and with respect to mechanism, arduous to fully understand.

4 Doping Studies

As mentioned earlier in Sect. 2.2, in addition to reducing the grain size of magnesium-based hydrogen storage materials, researchers have attempted alloying or "doping" Mg nanostructures with small amounts, less than $\sim 10\,\%$, of various

Table 3 Different pressure dependence functions, $F(P)$, used by investigators

$F(P)$	$F(P)$
$\frac{\lvert P_{eq}-P \rvert}{P_0}; P_0 = 1$ atm	$\left[\frac{P_{eq}-P}{P_0}\right]^n; P_0 = 1$ atm
$Ln\left(\frac{P_{eq}}{P}\right)$	$Ln\left(\frac{P}{P_{eq}}\right)$
P	$\sqrt{P}$
$\frac{(P_{eq}-P)}{P_{eq}}$	$\frac{(P_{eq}-P_b)}{P_{eq}}$
$P^{1/2} - P_{eq}^{1/2}$ for bulk	$P_0^{1/2} - P_{eq}^{1/2}; P_0 = $ initial pressure
$P_{eq} - P$	$P - P_{eq}$
$F\left(\frac{P_{eq}}{P}\right)^n, F\left(\frac{P}{P_{eq}}\right)^n$	$P_{eq} - P$ for surface control

Reproduced with permission from Ron 1999
P system pressure; P_b reaction bed pressure; P_{eq} desorption/absorption equilibrium pressure; Not all forms of $F(P)$ that have been suggested are presented here

metals and metal oxides as catalysts to improve kinetic performance. Understanding the physical nature of the dopants in the magnesium requires an understanding of the structure, size, and shape of the nanoparticles. Using techniques such as x-ray diffraction (XRD), x-ray photoelectron spectroscopy (XPS), scanning electron microscopy (SEM), transmission electron microscopy (TEM) among others, researchers have attempted to understand the physical nature of these dopants. Having said that, the following sections will mainly concern themselves with understanding the kinetic and mechanistic abilities of the different dopants. Although direct comparison of the results is inherently difficult because of differences in the apparatuses, methods and calculations, and kinetic modeling (or lack thereof), the effects of additional chemical species on sorption kinetics are undoubtedly apparent. In light of the inability to directly compare the results of different doping papers with respect to each other, this section will attempt to highlight the findings in this area.

4.1 Dopant Effects on Hydrogen Sorption Kinetics

One of the earliest reports on the effects of alloying (~ 15–30 % Cu) on the hydrogen storage properties of sub-bulk, micron-sized, magnesium was in 1979 when Karty et al. studied the hydrogen sorption kinetics of a Mg/Mg$_2$Cu microstructured eutectic alloy via a pressure sweep method. Using the Johnson–Mehl–Avrami–Kolmogorov equation, an adapted version of the JMA equation discussed earlier, they were able to obtain suggestive evidence that the alloying of Mg with Cu reduced both the hydrogen desorption and adsorption activation energies with respect to measured activation for the pure Mg system. They propose that the Mg$_2$Cu's role in decreasing these activation energies is through offering an alternative pathway for diffusion of hydrogen into and out of the magnesium [91].

Twenty years later, the next notable study, from Liang et al., compared the sorption kinetics of MgH_2 that was ball milled with different transition metals (MgH_2 5 % at. TM where TM = Ti, V, Mn, Fe, and Ni) [5]. Using the CV model equation (Table 2), they were able to plot the desorption kinetics curves at different temperatures and obtain activation energies for the desorption (Fig. 12) [5]. They were, 62.3, 67.6, 71.1, 88.1, and 104.6 kJ mol^{-1} for the MgH_2—vanadium, iron, titanium, nickel, and manganese nanocomposites, respectively. These values are much lower than the calculated pure balled milled MgH_2 desorption activation energy of 120 kJ mol^{-1}. They did not calculate the adsorption activation energies; however, they report increased rate of absorption compared to pure ball-milled MgH_2, for all nanocomposites at 10 MPa pressure and temperatures of 473, 373, and 303 K. A mechanism of catalysis was not specified, but they ruled out changes in the Mg–H bond thermodynamics by measuring PCIs for all systems [5, 68].

Oelerich and coworkers investigated a series of MgH_2/Me_xO_y nanocomposites that were synthesized through high energy ball milling ($Me_xO_y = Sc_2O_3$, TiO_2, V_2O_5, Cr_2O_3, Mn_2O_3, Fe_3O_4, CuO, Al_2O_3, and SiO_2). Only the transition metal oxides acted as catalysts in the hydrogen adsorption/desorption processes at ~573 K. As a result, the authors suggest that the ability of a metal to take on different oxidation states may play a part in the role of the catalyst. The best performing catalysts were Fe_3O_4 and Cr_2O_3 which allowed for rapid adsorption at room temperature and pressure of 8.4 bar, and desorption at ~473 K and pressure of 0 bar [45, 46, 53]. The team also compared the catalytic activity of V, V_2O_5, VC, and VN. They found that the vanadium already bonded with oxygen, nitrogen, or carbon resulted in the best performance, again supporting the theory that the oxidation state plays an important role. Additionally, they found that exposure of the MgH_2–V to small amounts of oxygen increased the catalytic activity of the vanadium. This further evidences that oxygen, or more specifically the oxidation state of vanadium is also playing a role in the catalytic mechanism [53]. Not long after, Barkhodarian et al. compared alloying NbO, NbO_2, and Nb_2O_5 with those researched by Oelerich et al. (Fig. 13) [80]. They found that 0.5 mole% Nb_2O_5 was a superior catalyst in both adsorption and desorption of hydrogen, with a tentative (based on JMA) lowered desorption activation energy of 62 kJ mol^{-1}. In their numerous papers on the subject, Barkhordarian et al. present thorough discussion on the differences in the kinetic models that can be used. While using the different approaches discussed in Sect. 3 to analyze their kinetics, they ultimately conclude that the mechanism of the catalytic process is unclear, but they agree with Oelerich and coworkers [45, 46, 53] that the ability of the transition metals to take on different electronic states must play a major role. It was proposed that the transition metal goes through electronic exchange reactions with the hydrogen molecule, which could accelerate the overall reaction. They claim there are five distinct steps in the reaction of metals with hydrogen: physisorption, chemisorption, surface penetration, diffusion, and hydride formation. The slowest of these steps will be the one that causes slow kinetics; they found that the rate limiting step changes with varying amounts of their catalyst, Nb_2O_5 [80]. Tentatively, the different transition metal catalysts were concluded to influence both the chemisorption and diffusion steps.

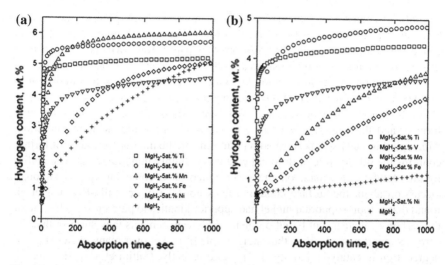

Fig. 12 Hydrogen adsorption curve of Mg–TM composites at **a** 273 K and **b** 373 K. Adapted with permission from Liang 1999

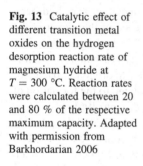

Fig. 13 Catalytic effect of different transition metal oxides on the hydrogen desorption reaction rate of magnesium hydride at $T = 300$ °C. Reaction rates were calculated between 20 and 80 % of the respective maximum capacity. Adapted with permission from Barkhordarian 2006

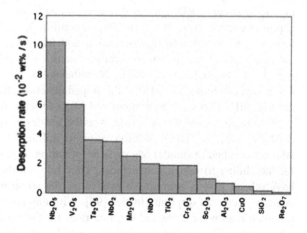

Around the same time, Hanada et al. also performed kinetic studies on nano-composites of MgH_2 with iron, cobalt, copper, nickel, and Nb_2O_5 [80, 93–95]. They found that each of these catalysts improves kinetics, with nickel and Nb_2O_5 performing the best at 2 and 1 mol%, respectively. Using the JMA equation to model their kinetics and TDMS, they were able to effectively apply the Kissinger method [76] to estimate the activation energies to be 323 ± 40, 94 ± 3, and 71 ± 3 kJ $molH_2^{-1}$ for the non-catalyzed Mg–H_2, nickel-catalyzed, and Nb_2O_5 catalyzed hydrogen desorption reaction, respectively. The difference in the activation energy of the pure MgH_2 desorption process compared to that found by Liang et al. [5] highlights the difficulty in comparing the capabilities of different catalysts to improve sorption kinetics. As with others, Hanada et al. conclude the absolute mechanism of catalysis is elusive; however, they propose the surface reaction is a first order process and that must take into account certain surface conditions prior to modeling [40, 66]. In their paper, Kalisvaart et al. describe a neutron reflectometry experiment that studies the effects of vanadium and chromium doping on Mg hydrogenation kinetics. It was found that with pure thin film Mg, the deuterium forms an MgD_2 blocking layer on the surface of the film, which slows the kinetics of hydrogenation. With the addition of catalysts, no such layer formed, and the deuterium was able to diffuse into the film without inhibition. Although this study was performed on thin films in the micrometer range, it further confirms the surface role of different catalysts is highly important [54]. The ability for the catalyst to remain on the surface may also come into play. Although discussion about the surface condition does not take place, Lillo-Rodenas et al. found that carbon-supported nickel had superior kinetics to pure nickel. The role of the carbon is unclear, but based on their DSC measurements they surmise that the carbon support may stabilize the catalytic role of nickel over many cycles (i.e., no degradation in catalytic activity) [61]. More recently, Callini et al. performed an in-depth investigation on magnesium nanoparticles decorated with nickel by thermal evaporation. The nanoparticles were exposed to oxygen in order to form a protective MgO layer, which was suggested not to hinder diffusion of H_2. Using in situ powder XRD during different simultaneous annealing and hydrogen exposure experiments, they were able to confirm the presence of the nickel in the form of Mg_2Ni prior to hydrogenation and Mg_2NiH_4 after hydrogenation. The increased rate of the hydrogen sorption kinetics was clearly attributed to the formation of Mg_2Ni from Mg_2NiH_4. Nonetheless, a lowered hydrogen capacity of 4.4 % was a definite downside of this doped material, attributed to the formation of the Mg_2NiH_4 (3.6 wt.% hydrogen) and MgO (0 wt.% hydrogen) [96].

With regards to MgO formation, Ares-Fernandez et al. investigated the possible catalytic effect of MgO addition prior to milling. They found that the MgO addition resulted in smaller Mg particle formation, referring to MgO as a lubricant for the sliding tribological interfaces. They rule out the possibility that MgO, as have others, [97] is acting as a catalyst in the sorption processes, confirming it actually decreases sorption kinetics [98]. With this confirmation, they suggest that the main reason for increased sorption rates for the particles is a result of particle size; contrarily, a recent theoretical paper by Shevlin et al. has proposed that nanostructuring does not decrease dehydrogenation enthalpies unless the nanocluster contains less than three magnesium atoms, much smaller than those reported by Ares-Fernandez et al. This particular paper will be discussed more in the next section [99]. Another 2013 paper written by Ma et al. formulates a niobium gateway model to explain the superior catalytic ability of Nb_2O_5 in MgH_2 dehydrogenation. The gateway model begins by NbH_2 (confirmed by XRD of a 50 wt.% XRD) decomposing rapidly because of its thermodynamic instability. This is followed by formation of Nb, which then instigates diffusion of hydrogen to form an NbH_x solid solution. This solid solution allows for the flow of hydrogen from MgH_2 to the outside of the particle until MgH_2 is exhausted. They rule out the possibility of NbO acting as a catalyst by investigating Nb_2O_5, Nb, and

NbO-doped samples. They found that both Nb_2O_5 and Nb formed NbH_2, which allows for the Nb gateway to take place; however, the NbO-doped samples did not yield the same kinetic rates or NbH_2 species. This study is one of the more recent studies that straightforwardly attempts to characterize the mechanism of catalysis for doped Magnesium. However, as mentioned earlier, most of the niobium phases confirmed to be present in this study were confirmed by XRD of a 50 wt.% dopant sample, much more than that used in most kinetic studies. Nonetheless, this paper gives insight into possible mechanisms for catalysis; further proposing the hypothesis that destabilization of the hydride phase occurs through bond formation between the catalyst and hydrogen itself [96]. Many other groups have worked with doping magnesium with nickel and Nb_2O_5 as well as their effects on the hydrogen sorption kinetics. For example, many of them have tried investigating the physical nature with high resolution micrographs (Fig. 14), [60, 67] and the others can be sought out for more information than already mentioned [13, 61, 62, 65, 67]. As with nickel and Nb_2O_5, more studies have since been conducted in order to further investigate the effects of adding various metals and metal composites to Mg, e.g., Ti, Fe, Ge, Si, Al, and Fe, however, very little new insight has been given into the nature and mechanism of these catalysts besides reconfirmation of their respective abilities to improve sorption kinetics [17, 48, 50, 51, 55, 57, 64, 100–104].

4.2 Theoretical Modeling Studies on Nature of Dopants

In attempt to investigate and characterize the kinetics and mechanism of the hydrogen sorption reactions, many groups have tried using different theoretical approaches to elucidate the nature of the reactions, both catalyzed and non-catalyzed. Song et al. carried out theoretical calculations to explore the effects of alloying Mg with Cu, Ni, Al, Nb, Fe, and Ti. Using the full-potential linearized augmented plane-wave method [based on density functional theory (DFT)], they found that alloying the magnesium should destabilize the MgH_2 by weakening the Mg–H bond, with increasing effect from Ti, Fe, Nb, Al, Ni to Cu [105]. This is contrary to what Liang et al. had previously reported [5], however, the calculations were not necessarily taking into account particle size [105]. Similarly, Cleri and coworkers did extensive electronic structure and DFT calculations on the effects of doping with small amounts of Cu, Ni, Fe, Ti, Zr, Pd, Co, Al, Cr, and Nb. Comparing their results with experiment, they were able to establish a trend, with a few exceptions [39]. Additionally, Moser et al. studied the stability of Mg-transition metal hydrides, where TM = Ti, Zr, Hf, V, Nb, and Ta. Their DFT calculations suggest the H is more strongly bonded to the transition metal, causing a stepwise dehydrogenation process [41]. Kelkar et al. also carried out a computational study on the effects of pure and aluminum and silicon-doped alpha-, gamma-, and beta-MgH_2 hydrogen sorption kinetics. They did this using ab initio plane wave pseudopotential method based on DFT, similar to Song et al. [47, 105].

Fig. 14 *Top* Bright field
TEM micrograph suggesting
that most of the Nb$_2$O$_5$
particles are embedded in the
MgH$_2$ matrix (*darker areas*).
Bottom TEM micrograph of
the MgH$_2$-Ni nanostructured
composite with (*1*) being
87.5 % Ni and (2) being only
5.2 % Ni by EDX.
Reproduced with permission
from Porcu 2008 and Hanada
2008

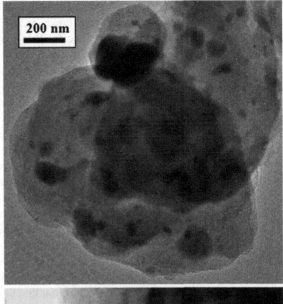

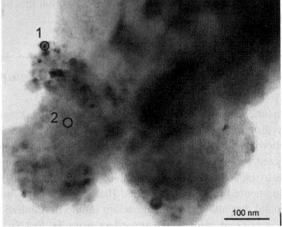

More recently Zeng et al. used a similar method to characterize another group of
3d transition metals, with a few different considerations taken into account [43].

As stated earlier, most of the above stated theoretical papers do not take into
account the effects of nanostructuring. More recently, Shevlin et al. used DFT and
many different exchange-correlation functionals, i.e., LDA, PBE, and PBESol, to
investigate the thermodynamics of nanostructuring and transition metal doping.
They found that unless the magnesium cluster contains less than three magnesium
atoms, it will have a higher dehydrogenation enthalpy than the bulk. Thus, for all
the experiments discussed thus far, all increased performance through nanostructuring is purely kinetic. However, they did find that nickel doping is highly
effective in decreasing both dehydrogenation thermodynamics and kinetics [99].

In summary, the effects of adding small amounts of dopants have been highlighted and briefly discussed. The mechanism and nature of these catalysts is still relatively unclear, however, with the development of new kinetic models, and the incorporation of theoretical analysis, a deeper understanding of these catalysts is on the horizon. Clearly, the catalysts play an important role in multiple steps of the hydrogen sorption process, and with more research, proper utilization of these catalysts may make the practicality of Mg nanostructures for hydrogen storage, a potentially realizable goal.

5 Conclusions and Outlook

Bulk Mg itself is not likely to be a viable hydrogen storage material due to the slow kinetics at ambient temperatures and pressures. However, there are strong indications from experiments as well as theory that nanostructured magnesium particles doped with low percentages of other elements spanning the periodic table exhibit enhanced kinetics at lower temperature, but efficient hydrogenation and dehydrogenations at room temperature is a goal that has still not been achieved. Significant challenges in the field are structurally characterizing the doped materials effectively, so that an iterative loop between synthesis, structural characterization, evaluation of kinetics, and modeling can be achieved.

References

1. Rosi NL, Eckert J, Eddadoudi M, Vodak DT, Kim J, O'Keeffe M, Yaghi O (2003) Science 300:1127
2. Hirscher M, Becher M, Haluska M, Dettlaff-Weglikowska U, Quintel A, Duesberg GS, Choi YM, Downes P, Hulman M, Roth S, Stepanek I, Bernier P (2001) Appl Phys A 72:129
3. Schlapbach L, Zuttel A (2001) Nature 414:253
4. Zaluska A, Zaluski L, Strom-Olsen JO (2001) Appl Phys A 72:157
5. Liang G, Huot J, Boily S, Van Neste A, Schulz R (1999) J Alloy Compd 292:247
6. Shao H, Xu H, Wang Y, Li X (2004) Nanotechnology 15:269
7. Zaluska A, Zaluski L, Strom-Olsen JO (1999) J Alloy Compd 288:217
8. Friedrichs O, Kolodziejczyk L, Sanchez-Lopez JC, Lopez-Cartes C, Fernandez A (2007) J Alloys Compd 434–435:721
9. Huhn PA, Dornheim M, Klassen T, Bormann R (2005) J Alloys Compd 404–406:499
10. Imamura H, Masanari K, Kusuhara M, Katsumoto H, Sumi T, Sakata Y (2005) J Alloys Compd 386:211
11. Huot J, Tremblay ML, Schulz R (2003) J Alloys Compd 356–357:603
12. Doppiu S, Schultz L, Gutfleisch O (2007) J Alloys Compd 427:204
13. Varin RA, Czujko T, Chiu C, Wronski Z (2006) J Alloys Compd 424:356
14. Paskevicius M, Sheppard DA, Buckley CE (2010) J Am Chem Soc 132:5077
15. Leiva DR, Floriano R, Huot J, Jorge AM, Bolfarini C, Kiminami CS, Ishikawa TT, Botta WJ (2011) J Alloys Compd 509:S444
16. Shao H, Liu T, Wang Y, Xu H, Li X (2008) J Alloys Compd 465:527

17. Zou J, Sun H, Zeng X, Ji G, Ding W (2012) J Nanomater 2012:592147
18. Zhu C, Akiyama T (2012) Cryst Growth Des 12:4043
19. Li W, Li C, Ma H, Chen J (2007) J Am Chem Soc 129:6710
20. Setijadi EJ, Boyer C, Aguey-Zinsou KF (2012) Phys Chem Chem Phys 14:11386
21. Kalidindi SB, Jagirdar BR (2009) Inorg Chem 48:4524
22. Redl FX, Cho K-S, Murray CB, O'Brien S (2003) Nature 423:968
23. Peng X, Manna L, Yang W, Wickham J, Scher E, Kadavanich A, Alivisatos AP (2000) Nature 404:59
24. Puntes VF, Krishnan KM, Alivisatos AP (2001) Science 291:2115
25. Ouyang M, Awschalom DD (1074) Science 2003:301
26. Milliron DJ, Hughes SM, Cui Y, Manna L, Li J, Wang L-W, Alivisatos AP (2004) Nature 430:190
27. Aguey-Zinsou KF, Ares-Fernandez J-R (2008) Chem Mater 20:376
28. Bogdanovic B, Spliethoff B (1987) Int J Hydrogen Energy 12:863
29. Li WY, Li CS, Ma H, Chen J (2007) J Am Chem Soc 129:6710
30. Huot J, Liang G, Schulz R (2001) Appl Phys A-Mater Sci Process 72:187
31. Aguey-Zinsou KF, Ares-Fernandez JR (2008) Chem Mater 20:376
32. Viyannalage L, Lee V, Dennis RV, Kapoor D, Haines CD, Banerjee S (2012) Chem Commun 48:5169
33. Norberg NS, Arthur TS, Fredrick SJ, Prieto AL (2011) J Am Chem Soc 133:10679
34. Jeon KJ, Moon HR, Ruminski AM, Jiang B, Kisielowski C, Bardhan R, Urban JJ (2011) Nat Mater 10:286
35. Aguey-Zinsou KF, Ares-Fernandez JR (2010) Energy Environ Sci 3:526
36. Sakintuna B, Lamari-Darkrim F, Hirscher M (2007) Int J Hydrogen Energy 32:1121
37. Wagner LK, Majzoub EH, Allendorf MD, Grossman JC (2012) Phys Chem Chem Phys 14:6611
38. Yang J, Sudik A, Wolverton C, Siegel DJ (2010) Chem Soc Rev 39:656
39. Cleri F, Celino M, Montone A, Bonetti E, Pasquini L (2007) In: Uskokovic DP, Milonjic SK, Rakovic DI (eds) Research trends in contemporary materials science, vol 555. p 349
40. Hanada N, Ichikawa T, Fujii H (2005) J Phys Chem B 109:7188
41. Moser D, Bull DJ, Sato T, Noreus D, Kyoi D, Sakai T, Kitamura N, Yusa H, Taniguchi T, Kalisvaart WP, Notten P (2009) J Mater Chem 19:8150
42. Shang CX, Bououdina M, Song Y, Guo ZX (2004) Int J Hydrogen Energy 29:73
43. Zeng XQ, Cheng LF, Zou JX, Ding WJ, Tian HY, Buckley C (2012) J Appl Phys 2012:111
44. Hanada N, Ichikawa T, Isobe S, Nakagawa T, Tokoyoda K, Honma T, Fujii H, Kojima Y (2009) J Phys Chem C 113:13450
45. Oelerich W, Klassen T, Bormann R (2001) Adv Eng Mater 3:487
46. Oelerich W, Klassen T, Bormann R (2001) J Alloy Compd 315:237
47. Kelkar T, Pal S (2009) J Mater Chem 19:4348
48. Liu T, Qin C, Zhang T, Cao Y, Zhu M, Li X (2012) J Mater Chem 22:19831
49. Vajo JJ, Mertens F, Ahn CC, Bowman RC, Fultz B (2004) J Phys Chem B 108:13977
50. Shahi RR, Tiwari AP, Shaz MA, Srivastava ON (2013) Int J Hydrogen Energy 38:2778
51. Lu J, Choi YJ, Fang ZZ, Sohn HY, Ronnebro E (2010) J Am Chem Soc 132:6616
52. Schimmel HG, Huot J, Chapon LC, Tichelaar FD, Mulder FM (2005) J Am Chem Soc 127:14348
53. Oelerich W, Klassen T, Bormann R (2001) J Alloy Compd 322:L5
54. Kalisvaart P, Luber E, Fritzsche H, Mitlin D (2011) Chem Commun 47:4294
55. Asselli AAC, Leiva DR, Jorge AM, Ishikawa TT, Botta WJ (2012) J Alloy Compd 536:S250
56. Montone A, Aurora A, Gattia DM, Antisari MV (2012) Catalysts 2:400
57. Yavari AR, Vaughan G, de Castro FR, Georgarakis K, Jorge AM, Nuta I, Botta WJ (2012) J Alloy Compd 540:57
58. Bogdanovic B, Hofmann H, Neuy A, Reiser A, Schlichte K, Spliethoff B, Wessel S (1999) J Alloy Compd 292:57

59. Gutfleisch O, Dal TS, Herrich M, Handstein A, Pratt A (2005) J Alloys Compd 404–406:413
60. Hanada N, Hirotoshi E, Ichikawa T, Akiba E, Fujii H (2008) J Alloy Compd 450:395
61. Lillo-Rodenas MA, Aguey-Zinsou KF, Cazorla-Amoros D, Linares-Solano A, Guo ZX (2008) J Phys Chem C 112:5984
62. Varin RA, Czujko T, Wasmund EB, Wronski ZS (2007) J Alloys Compd 432:217
63. Xie L, Liu Y, Zhang X, Qu J, Wang Y, Li X (2009) J Alloys Compd 482:388
64. Walker GS, Abbas M, Grant DM, Udeh C (2011) Chem Commun 47:8001
65. da Conceicao MOT, Brum MC, dos Santos DS, Dias ML (2013) J Alloy Compd 550:179
66. Hanada N, Ichikawa T, Hino S, Fujii H (2006) J Alloy Compd 420:46
67. Porcu M, Petford-Long AK, Sykes JM (2008) J Alloys Compd 453:341
68. Liang G, Huot J, Boily S, Schulz R (2000) J Alloy Compd 305:239
69. Blach TP, Gray EM (2007) J Alloy Compd 446–447:692
70. Mintz MH, Zeiri Y (1995) J Alloy Compd 216:159
71. Lee Y-W, Clemens BM, Gross KJ (2008) J Alloy Compd 452:410
72. Sabitu ST, Goudy AJ (2012) Metals 2:219
73. Markmaitree T, Ren R, Shaw LL (2006) J Phys Chem B 110:20710
74. Flynn JH, Wall LA (1966) J Polym Sci Part B: Polym Lett 4:323
75. Ozawa T (1970) J Therm Anal Calorim 2:301
76. Kissinger HE (1957) Anal Chem 29:1702
77. Li C-R, Tang T (1999) J Mater Sci 34:3467
78. Karty A, Grunzweig-Genossar J, Rudman PS (1979) J Appl Phys 50:7200
79. Stander CM (1977) J Inorg Nucl Chem 39:221
80. Barkhordarian G, Klassen T, Bormann R (2006) J Alloy Compd 407:249
81. Zhdanov VP (2008) Surf Rev Lett 15:605
82. Berube V, Radtke G, Dresselhaus M, Chen G (2007) Int J Energy Res 31:637
83. Avrami M (1939) J Chem Phys 7:1103
84. Johnson WA, Mehl RF (1939) Trans Am Inst Min Metall Eng 135:416
85. Avrami M (1940) J Chem Phys 8:212
86. Avrami M (1941) J Chem Phys 9:177
87. Jacobs PWM, Tompkins FC (1955) Chemistry of the solid state. Butterworths, London
88. Tran NE, Lambrakos SG, Imam MA (2006) J Alloy Compd 407:240
89. Ron M (1999) J Alloy Compd 283:178
90. Fernandez JF, Sanchez CR (2002) J Alloy Compd 340:189
91. Karty A, Grunzweiggenossar J, Rudman PS (1979) J Appl Phys 50:7200
92. Reilly JJ, Wiswall RH (1967) Inorg Chem 6:2220
93. Barkhordarian G, Klassen T, Bormann R (2006) J Phys Chem B 110:11020
94. Barkhordarian G, Klassen T, Bormann R (2003) Scripta Mater 49:213
95. Barkhordarian G, Klassen T, Bormann R (2004) J Alloy Compd 364:242
96. Callini E, Pasquini L, Jensen TR, Bonetti E (2013) Int J Hydrogen Energy 38:12207
97. Ma T, Ma S, Isobe Y, Wang N, Hashimoto S, Ohnuki T (2013) J Phys Chem C 117:10302
98. Ares Fernandez J-R, Ares Fernández K-F, Aguey Zinsou J-R, Ares F (2012) Catalysts 2:330
99. Shevlin SA, Guo ZX (2013) J Phys Chem C 117:10883
100. Fu Y, Groll M, Mertz R, Kulenovic R (2008) J Alloys Compd 460:607
101. Cermak J, Kral L (2012) J Power Sources 214:208
102. Liu J, Song XP, Pei P, Chen GL (2009) J Alloys Compd 486:338
103. Niaz NA, Hussain ST, Nasir S, Ahmad I (2012) Key Eng Mater 510–511:371
104. Shahi RR, Yadav TP, Shaz MA, Srivastva ON (2010) Int J Hydrogen Energy 35:238
105. Song Y, Guo ZX, Yang R (2004) Phys Rev B 2004:69

1D Pd-Based Nanomaterials as Efficient Electrocatalysts for Fuel Cells

Yizhong Lu and Wei Chen

Abstract Since the first experiment conducted by William Grove in 1839, fuel cell, a device that converts the chemical energy stored in fuels into electricity through electrochemical reactions with oxygen or other oxidizing agents, has attracted worldwide attention in the past few decades. However, despite extensive research progress, the widespread commercialization of fuel cells is still a big challenge partly because of the low catalytic performance and high-cost of the Pt-based electrocatalysts. In addition, the hydrogen storage is another critical issue for the commercialization of hydrogen-powered fuel cells. Among the metal catalysts, Pd has been found to be a promising alternative because of its excellent catalytic properties and lower cost than Pt. Moreover, Pd-based materials exhibit high hydrogen storage capabilities. In this chapter, we summarize recent progress in the synthesis of one-dimensional (1D) Pd-based nanomaterials and their applications as electrocatalysts on both anodic and cathodic sides of fuel cells, and their applications in hydrogen storage. We demonstrated here that various 1D Pd-based nanomaterials, such as nanorods, nanowires, and nanotubes have been successfully prepared through different synthetic routes. The nanostructured 1D Pd-based materials exhibit high catalytic performance for electrooxidation of small organic molecules and oxygen reduction reaction (ORR). Moreover, high capacities for hydrogen storage have also been reported with 1D Pd-based nanomaterials.

Y. Lu · W. Chen (✉)
State Key Laboratory of Electroanalytical Chemistry, Changchun Institute of Applied Chemistry, Chinese Academy of Sciences, Changchun 130022, Jilin, China
e-mail: weichen@ciac.ac.cn

Y. Lu
University of Chinese Academy of Sciences, Beijing 100039, China

Z. Lin and J. Wang (eds.), *Low-cost Nanomaterials*, Green Energy and Technology, DOI: 10.1007/978-1-4471-6473-9_12, © Springer-Verlag London 2014

1 Introduction

Due to the increasing worldwide energy demand and environmental concerns, much effort has been devoted to the seeking for efficient and clean energy sources to replace the traditional fossil fuels such as gasoline and diesel [1]. Fuel cell, a device that converts the chemical energy stored in fuels into electricity through electrochemical reactions efficiently without pollution, has been attracting increasing research interest as a new power source for portable applications due to their high energy-conversion efficiency and relatively low operating temperature [2, 3]. Based on the operating temperature and the type of used electrolyte, fuel cells are usually classified into phosphoric-acid fuel cells (PAFCs), solid oxide fuel cells (SOFCs), alkaline fuel cells (AFCs), molten-carbonate fuel cells (MCFCs), polymer exchange membrane fuel cells (PEMFCs), and direct-methanol fuel cells (DMFCs) [4]. While all types of fuel cells work on the similar principle: hydrogen or other fuels oxidation at anode and oxygen reduction at cathode. Among different types of fuel cells, the PEMFCs and DMFCs are especially promising for automotive and portable electronic applications owing to the low operation temperatures [5, 6]. It should be noted that despite extensive research progress, there are still many scientific and technological challenges to realize the widespread commercialization of fuel cells. For instance, the reactions on both anode and cathode need electrocatalysts with high catalytic performance. To improve the efficiency and durability, and to reduce the cost of fuel cells, the conventional Pt catalysts have to be replaced by novel nanostructured electrocatalysts with high electrocatalytic activity, high stability, and low-cost. For fuel cells, platinum has been regarded as the best electrocatalyst because of the highest electrocatalytic activity among the metal catalysts for electrooxidation of small organic fuels and for oxygen reduction [7]. However, with platinum as an anode catalyst, its surface is usually heavily poisoned by CO intermediates produced during the oxidation of organic fuels, resulting in the lowering of catalytic performance. On the other hand, the state-of-the-art cathode electrocatalysts, Pt nanoparticles (2–5 nm) supported on amorphous carbon materials (Pt/C), usually suffer from poor durability caused by the fast and significant loss of platinum electrochemical surface area (ECSA) over time during fuel cell operation. Moreover, the high price and the limited global supply of Pt largely drive up the cost of fuel cells. To reduce the cost and minimize the self-poisoning of catalysts, Pt-based electrocatalysts alloyed with other transition metals (Fe, Co, Ni, Cu, Mn, Ir) with controlled surface composition and structures have been extensively studied in recent years [8–17]. Compared to pure platinum, Pt-based electrocatalysts exhibit higher performance and a reduced sensitivity toward CO poisoning because of the so-called bifunctional mechanism [18–20] or ligand effect [21, 22]. For instance, Pt-Ru alloys have shown enhanced electrocatalytic activity for fuel cell anode reactions and the enhancement could be well explained by the ligand effect and the bifunctional mechanism [23, 24]. Based on the ligand effect, the presence of ruthenium was proposed to alter the electronic properties of Pt, resulting in less strongly CO adsorption on the catalyst surface. While according to the bifunctional mechanism, ruthenium provides oxygen-containing species, which could then oxidize CO on the adjacent platinum sites at more negative potentials than pure platinum. Due to the large surface area, various Pt-based bimetallic nanostructures have been synthesized and applied to both anode and cathode catalysts. For example, Xia and co-workers [25] synthesized Pd–Pt bimetallic nanodendrites, which exhibited enhanced catalytic activity for oxygen electro-reduction. Recently, Stamenkovic et al. [12] demonstrated that extended single crystal surfaces of $Pt_3Ni(111)$ exhibit an enhanced ORR activity that is 10-fold higher than the corresponding Pt(111) surface and 90-fold more active than the current state-of-the-art Pt/C catalysts for PEMFCs. Such a remarkable activity was attributed to the unusual electronic structure (d-band center position) and arrangement of surface atoms in near-surface region. For the anode reaction, Yang et al. [10] have found that the (100) facet-terminated Pt_3Co nanocubes are much more active than the Pt nanocubes due to the weaker and slower CO adsorption.

In recent years, to further reduce the loading of Pt, intensive studies have focused on low-Pt or non-Pt electrocatalysts [26–37]. Among the studied non-platinum electrocatalysts, Pd-based nanomaterials have become hot research topics as electrocatalysts in fuel cells because of their similar intrinsic properties to platinum, such as the valence band structure and lattice parameters [38, 39]. More significantly, compared to Pt, Pd is cheaper and exhibits higher electrochemical stability. Recent research progress in Pd-based catalysts has revealed that Pd can catalyze the oxidation of formic acid and alcohols at the anode of polymer electrolyte membrane fuel cells (PEMFCs) with greater tolerance to CO than Pt catalysts and comparable activity toward the cathode oxygen reduction [40–50].

For catalysts on nanoscale, their catalytic efficiency, selectivity and reaction durability are highly dependent on the size, shape, composition, and surface structure [8, 26, 41, 51–54]. Similarly to Pt-based nanoalloys, Pd-based alloy nanomaterials have also shown enhanced electrocatalytic activities compared to pure Pd catalyst due to the synergistic effect or the modulation effect [55, 56]. Recent investigations have found that alloying Pd with some transition metals, such as Au [57], Ag [58], Fe [59], Co [60], Ti [55, 61], Ni [62], Sn [63], etc., is an effective way to improve the catalytic activity of Pd metal. At the same time, some of Pd alloys exhibited comparable activity and stability to those of Pt catalysts.

Among various Pd or Pd-based nanomaterials, one-dimensional (1D) nanostructured electrocatalysts, such as nanowires (NWs) [64–72], nanorods (NRs) [44, 73], nanoleaves (NLs) [74], and nanotubes (NTs) [75–80], have attracted more and more attention in recent years due to their unique structures and high surface area. From a structural perspective, 1D Pd nanomaterials possess largely pristine surfaces with long segments of crystalline planes, as compared with their corresponding 0D morphologies. Additionally, the anisotropic growth of 1D structured materials typically results in the preferential surface display of low energy crystal facets in order to minimize the surface energy of the systems [81]. In the previous studies on 1D Pd nanostructures, high resolution transmission electron microscopy (HRTEM) and selected-area-electron diffraction patterns (SAED) showed that the surfaces of the Pd nanowires are enclosed by {111} and {200} planes, suggesting the <110> growth direction of the nanowires [82]. It is well known that the catalytic activities of metal nanomaterials are strongly dependent on their exposed surface facets. Earlier studies have shown that the Pd(100) plane exhibits the highest catalytic activity for formic acid oxygen among the three low-index faces Pd(111), Pd(100), and Pd(110) [83, 84]. On the other hand, the single-crystalline 1D nanomaterials with structural anisotropy contain fewer surface defect sites in comparison with the corresponding zero-dimensional nanoparticles. When used as electrocatalysts, the electron transport properties of catalyst materials are of importance for the electrochemical reactions. Compared to metal nanoparticles, 1D nanomaterials can provide more efficient electron transfer, thus lowering the electronic resistance and enhancing the fuel oxidation or Oxygen reduction reactions [76, 85]. Meanwhile, for nanoparticle-based electrocatalysts, the electrochemical scanning process may lead to the particle aggregation, dissolution, and Oswald ripening, which is one of the biggest challenges in designing efficient fuel cell electrocatalysts. However, for 1D nanomaterials, the asymmetric structure nature can effectively prevent their structure destruction from aggregation, dissolution, and ripening [75, 86, 87]. From above, the large surface area, high electrochemical stability, efficient electron transport, and excellent CO-tolerance of 1D Pd-based nanomaterials would be highly advantageous for their applications in fuel cells as anode and cathode catalysts.

Among the fuels fed to the anode of fuel cells, hydrogen is considered to be one of the most promising clean energy carriers due to its light weight, high energy density, and no harmful chemical by-products from its combustion. Hydrogen can be generated from renewable sources, such as water splitting via photolysis [88–92]. For hydrogen powered fuel cells, despite prodigious efforts, how to develop a safe and easy storage method is still a remaining significant challenge for the widespread application of hydrogen as the fuel of choice in mobile transportation. The efficient and safe storage of hydrogen is crucial for promoting the "hydrogen economy". The safety and cost issues of conventional hydrogen storage as compressed gas or liquid largely limit the practical use of these methods. Other developed methods for hydrogen storage by physical adsorption on materials with large surface area or formation of chemical bonds have been proved to be efficient, convenient, and safe approaches [93]. Among the hydrogen storage materials, Pd-based nanomaterials exhibited excellent storage capacity and their storage properties are strongly related to the morphologies, composition, and size of the nanostructures [94].

In this chapter, we highlight the recent progress of one-dimensional Pd-based nanomaterials, including the synthetic techniques and their application in fuel cells as electrocatalysts on both anode and cathode sides. With different preparation routes, various structured nanomaterials with different surface morphologies can be realized. Since the composition and surface structure of Pd-based alloy materials play decisive roles in determining their electrocatalytic properties, the structure and properties of 1D Pd materials can be manipulated by changing synthetic conditions to meet the scientific and technological demands of fuel cell catalysts. In addition, the application of one-dimensional Pd-based nanomaterials in the hydrogen storage is also discussed here due to their importance in fuel cells. Finally, a brief conclusion and an outlook will be given on the future research directions of 1D Pd-based nanomaterial as efficient electrocatalysts and hydrogen storage materials.

2 One-Dimensional Pd-Based Nanomaterials as Effective Cathode Electrocatalysts

From the previous studies, the ORR is considered to proceed by the efficient four-electron pathway or the less efficient two-step pathway, depending on the electrolyte and the catalytic activity of the cathode materials [95].

A. Direct 4-electron pathway:

In alkaline solutions

$$O_2 + 2H_2O + 4e^- \rightarrow 4OH^- \tag{1}$$

Acid solutions.

$$O_2 + 4H^+ + 4e^- \rightarrow 2H_2O \tag{2}$$

B. Peroxide pathway:

In alkaline solutions

$$O_2 + H_2O + 2e^- \rightarrow HO_2^- + OH^- \tag{3}$$

followed by either the further reduction reaction

$$HO_2^- + H_2O + 2e^- \rightarrow 3OH^- \tag{4}$$

or the decomposition reaction

$$2HO_2^- \rightarrow 2OH^- + O_2 \tag{5}$$

In acid solutions

$$O_2 + 2H^+ + 2e^- \rightarrow H_2O_2 \tag{6}$$

followed by either

$$H_2O_2 + 2H^+ + 2e^- \rightarrow 2H_2O \tag{7}$$

or

$$2H_2O_2 \rightarrow 2H_2O + O_2 \tag{8}$$

On the basis of the mechanism of ORR, O_2 molecules are firstly adsorbed on the surface of catalysts, and then electrochemically reduced either directly to water or indirectly to intermediate of H_2O_2. According to the previous studies on the Pd and Pt-catalyzed ORR, the adsorption energy (AE) of O adsorption can serve as a good descriptor for the catalytic activity of the catalyst surface toward the ORR [56, 96]. Therefore, it would be of great interest to develop a robust and practical Pd or Pd-based catalysts with low AE toward O species. Recently, Abruna et al. [44] succeeded in tailoring the morphology of the deposited Pd from nanoparticles to nanorods by simply adjusting the precursor concentration in the electrochemical deposition of Pd. They found that the surface-specific activity of Pd nanorods (Pd NRs) toward ORR is not only higher than that of Pd nanoparticles (Pd NPs), but also becomes comparable to that of bulk Pt catalysts under fuel cell operating potentials. It can be seen from Fig. 1a, b that, under the experimental conditions of 1×10^{-5} M $PdCl_2$ precursor and 3000 s electrochemical deposition time, only Pd nanoparticles were formed with the size range from 5 to 10 nm. However, when the $PdCl_2$ concentration increased to 3×10^{-4} M and deposition time decreased to 100 s, uniform Pd nanorods can be produced with an average diameter of 5 nm and an aspect ratio of ~ 8. By comparing the electrocatalytic activities of the formed Pd NPs, Pd NRs, and bulk Pt for ORR as shown in Fig. 1c, the half-wave potential ($E_{1/2}$) obtained from Pd NRs shifts positively by 85 mV compared to that of Pd NPs and approaches that of bulk Pt catalyst. Figure 1d demonstrates that the kinetic current density (j_K) of the Pd NRs approaches that of bulk Pt and is 10-fold higher than that of the Pd NPs at +0.85 V (a practical operating potential of a PEMFC cathode). The superior ORR activity of Pd NRs was attributed to the exceptionally weak interaction between the exposed Pd(110) facet of Pd NRs and the adsorbed O atoms, which was confirmed by the CO stripping experiments and density functional theory (DFT) calculations. This study indicated that 1D Pd nanostructures could be an efficient and cost-effective cathodic electrocatalysts for PEMFCs.

By taking advantage of the higher activity offered by certain platinum alloys, Yan and coworkers [75] reported that unsupported PtPd nanotubes (PtPdNTs) exhibited much enhanced mass activity toward ORR as fuel cell cathode electrocatalysts. In the study, Pt and PtPd nanotubes were prepared through a galvanic replacement reaction between the pre-synthesized silver nanorods and Pt or Pd precursors. Figure 2 shows the SEM and TEM images of the as-synthesized PtPd NTs. It can be clearly seen that the PtPd NTs display uniform diameter (45 nm), wall thickness (7 nm), and length (10 μm). The ORR activity and durability test are shown in Fig. 3. From Fig. 3b, it can be seen that the half-wave potential of the PtPdNTs (0.851 V) is higher than that of the synthesized Pt nanotubes (PtNTs) (0.837 V), platinum black (0.817 V), and Pt/C (0.828 V). The shift to more positive potentials suggests clearly that the overpotential of ORR can be effectively reduced by using the bimetallic system. Moreover, the calculated mass- and specific-activity of the PtPd nanotubes are 1.4 and 1.8 times higher than those of the platinum black, respectively. Figure 3a shows the ECSA change with number of cyclic voltammetry (CV) cycles at three electrocatalysts. After 1000 cycles,

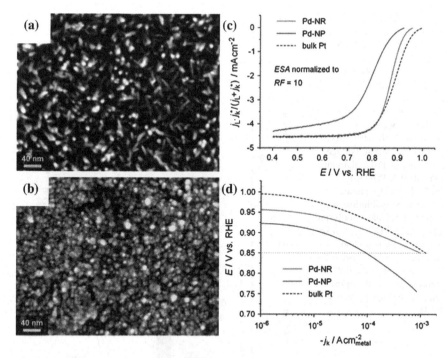

Fig. 1 SEM images of two typical morphologies of Pd obtained by adjusting the precursor concentration and deposition time: **a** Pd nanorods (Pd-NRs), 3×10^{-4} M PdCl$_2$, 100 s; **b** Pd nanoparticles (Pd-NPs), 1×10^{-5} M PdCl$_2$, 3000 s. ORR data for Pd-NR, Pd-NP, and bulk Pt, obtained in an O$_2$ saturated 0.1 M HClO$_4$ solution at a rotation rate of 900 rpm and a scanning rate of 10 mV s^{-1}. **c** Normalized ORR profiles and **d** Tafel plots. The diffusion limited current (j_L) has been normalized to the geometric surface area (*GSA*), while the kinetic current density (j_K^*) has been normalized to an electrochemical surface area (*ESA*) corresponding to a roughness factor (*RF*) of 10. Reprinted from Ref. [44] with permission by the American Chemical Society

Fig. 2 **a** SEM image of PtPd nanotubes. **b** TEM image and electron diffraction pattern (*inset*) of the PtPd nanotubes. Reprinted from Ref. [75] with permission by Wiley-VCH

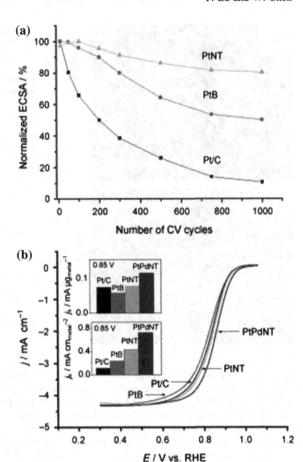

Fig. 3 **a** Loss of electrochemical surface area (ECSA) of Pt/C (E-TEK), platinum-black (PtB; E-TEK), and PtNT catalysts with number of CV cycles in Ar-purged 0.5 M H_2SO_4 solution at 60 °C (0–1.3 V vs. RHE, sweep rate 50 mV s^{-1}). **b** ORR curves (shown as current–voltage relations) of Pt/C, platinum black (PtB), PtNTs, and PdPtNTs in O_2-saturated 0.5 M H_2SO_4 solution at room temperature (1600 rpm, sweep rate 5 mV s^{-1}). *Inset* mass activity (*top*) and specific activity (*bottom*) for the four catalysts at 0.85 V. Reprinted from Ref. [75] with permission by Wiley-VCH

platinum-black and Pt/C catalysts lose about 51 and 90 % of their ECSA, respectively. However, the Pt nanotubes only lose about 20 % of its ECSA, indicating the enhanced electrochemical stability of the unsupported nanotubes. In addition, the durability tests indicated that the ECSA of PtPd NTs is 5.8 times higher than that of the Pt/C and 1.5 times higher than that of the Pt NTs. Such enhanced activity and durability of PtPd NTs toward ORR can be ascribed to the change of electronic structures induced by the addition of Pd into the platinum lattice.

The electrocatalytic results obtained from the Pt and PtPd nanotubes clearly indicated that the catalytic performance of 1D nanomaterials can be improved by decorating another metal. Recently, Alia et al. [79] synthesized Pt-coated Pd nanotubes (Pt/PdNTs) with a wall thickness of 6 nm, outer diameter of 60 nm, and length of 5–20 μm via the partial galvanic displacement of Pd nanotubes with Pt. ORR on Pt/PdNTs, Pt nanotubes (PtNTs), Pd nanotubes, and supported Pt nano-particle was studied to evaluate their electrocatalytic activities as PEMFCs cathodes. It was found that the Pt/PdNTs with Pt loading of 9 wt% (PtPd 9) produced an ORR mass activity 95 % of PtNTs. Taking account of the reduction of Pt loading in PtPd 9, the Pt mass activity of Pt/PdNTs is actually significantly higher than PtNTs. Compared to the dollar activity target (9.7 A $\$^{-1}$) of the United States Department of Energy (DOE), the calculated dollar activity of PtPd 9 (10.4) exceeds that of DOE by 7 %. On the other hand, the area activity of Pt/PdNTs outperformed the DOE target by greater than 40 %.

In a recent paper, Koenigsmann et al. [97] reported the synthesis, characterization, and electrochemical performance of novel 1D ultrathin Pt monolayer shell-Pd nanowires core catalyst. They found that the UV ozone-treated nanowires exhibited outstanding area and mass specific activities of 0.77 mA cm^{-2} and 1.83 A mg$_{Pt}^{-1}$ toward ORR, respectively, which were significantly enhanced as compared with conventional commercial Pt nanoparticles, core-shell nanoparticles, and acid-treated nanowires. This study also indicated that the methods to remove the organic residue on the surface of nanomaterials can affect their subsequent catalytic activity. In another report by Koenigsmann et al. [66] they successfully synthesized a series of bimetallic $Pd_{1-x}Au_x$ and $Pd_{1-x}Pt_x$ nanowires with control over composition and size through an ambient, template-based technique. In their report, the as-synthesized 1D alloy nanowires (ANWs) maintain significantly enhanced activity toward ORR as compared with commercial Pt nanoparticles and other 1D nanostructures. Specifically, the Pd_9Au and Pd_4Pt nanowires possess ORR activities of 0.49 and 0.79 mA cm^{-2}, which are larger than the analogous value from commercial Pt nanoparticles (0.21 mA cm^{-2}).

Recent studies have demonstrated that Pd catalysts can produce high ORR activity when combined with appropriate transition metals, such as Co, Fe, Mo, etc [59, 61, 98]. By an organic phase reaction of $[Pd(acac)_2]$ and thermal decomposition of $[Fe(CO)_5]$ in a mixture of oleyamine and octadecene at 160 °C, Li et al. [73] synthesized PdFe nanorods (PdFe-NRs) with tunable length. The morphology of the PdFe products can be tuned by altering the volumetric ratio of oleyamine (OAm) and octadecene (ODE) surfactants during the synthesis. As shown in Fig. 4a, under the OAm/ODE ratio of 1/3, there is only PdFe nanoparticles formation with an average particle size of 2–4 nm. With the ratio increasing to 1/1, PdFe nanorods with a diameter of 3 nm and length of 10 nm were formed (Fig. 4b). If only OAm was used in the synthesis, long PdFe nanorods with a length of 10 nm can be produced. Interestingly, around 20–40 PdFe nanorods were self-assembled to "flower"-like bundles (Fig. 4c, d), and the distance between two nanorods is around 2–4 nm. It was proposed that this bundle structure is likely to form a thin and dense catalyst layer in a membrane-electrode assembly (MEA), which can facilitate the mass transport of reactants. In the following electrocatalytic investigations, the as-synthesized PdFe-NRs demonstrated a better PEMFC performance than commercial Pt/C in the practical working voltage region (0.80–0.65 V), which can be attributed to their unique 1D morphology, high intrinsic activity toward ORR, reduced cell inner resistance, and improved mass transport.

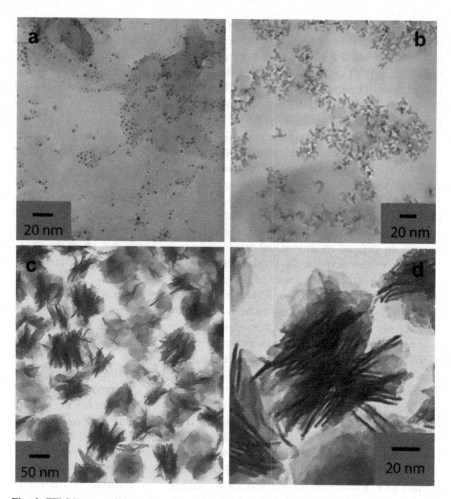

Fig. 4 TEM images of the PdFe nanoparticles (**a**) and nanorods with a length of 10 nm (**b**), 50 nm (**c**), and 50 nm in one bundle (**d**). Reprinted from Ref. [73] with permission by Elsevier

In another report, Li and co-workers [74] reported the preparation and characterization of PdFe nanoleaves (NLs) for ORR. The novel nanoleaf-structured materials were synthesized through a wet chemical reduction method with the presence of oleylamine, and the composition of the final products can be tuned by changing the precursors of Pd(acac)$_2$ and Fe(CO)$_5$. It was found that the Fe concentration largely affects the morphology of the as-synthesized nanostructures. From the TEM images shown in Fig. 5, the nanoleaves are actually composed of Fe-based nanosheets and Pd-rich nanowires embedded in the sheets. Lower Fe content results in the formation of 1D Pd nanowires with shorter lengths and larger diameters. The high resolution TEM (HRTEM) images in Fig. 5g shows that the side surfaces of these Pd-rich nanowires are composed of Pd(111) planes, while

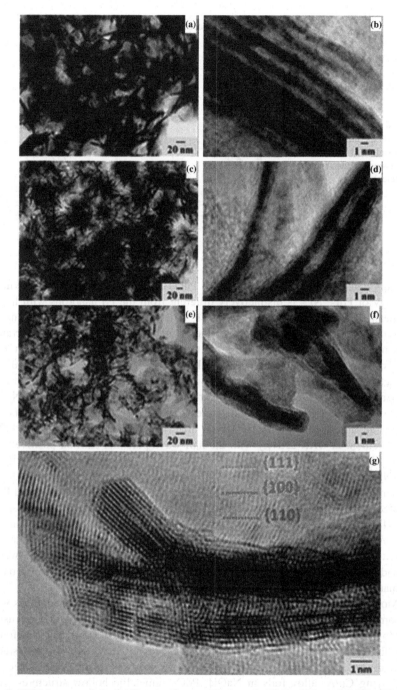

Fig. 5 TEM images of the Pd_xFe_y nanoleaves (NLs): **a, b** Pd_1Fe_1-NL, **c, d** Pd_2Fe_1-NL, **e, f** Pd_5Fe_1-NL, and **g** HRTEM image of Pd_1Fe_1-NL. Reprinted from Ref. [74] with permission by the American Chemical Society

Fig. 6 ORR polarization curves of commercial Pt/C (*black curve*), Pd/C (self-prepared by EG method, *red curve*), Pd$_1$-NL/C (*green curve*), and Pd$_2$-NL/C (*blue curve*) in oxygen saturated 0.1 M NaOH (conditions: 10 mV s^{-1}, 2500 rpm, and room temperature). Reprinted from Ref. [74] with permission by the American Chemical Society

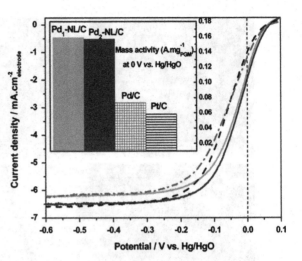

the tips and ends are predominated by Pd(110) and Pd(100) facets. The morphology of PdFe-NLs is distinctively different from the Pd nanorods with larger diameter (5–10 nm) prepared using PVP as a stabilizer. The varied Pd nanostructures suggest that the different synthesis conditions, i.e., different surfactants, presence/absence of Fe, etc., can affect the growth mechanisms of 1D nanomaterials. By etching away the enveloping Fe-rich sheets using an organic acid, the Pd-rich NWs are exposed on the surfaces of the nanoleaves, and they demonstrated high reactivity toward electrocatalytic reduction of oxygen in a 0.1 M NaOH electrolyte. The ORR polarization curves obtained from different catalysts are shown in Fig. 6. It can be observed that the Pd-NLs show a remarkable improvement in ORR activity with the half-wave potential shifting positively by ∼38 mV, compared to commercial Pt/C catalyst. Moreover, the mass activity of Pd$_1$-NL and Pd$_2$-NL are 0.159 and 0.157 A mg$_{Pd}^{-1}$, respectively, which are twice higher than that of Pd/C (0.0735 A mg$_{Pd}^{-1}$) and ∼2.7 times higher than that of Pt/C (0.0585 A mg$_{Pt}^{-1}$) at 0 V versus Hg/HgO. For the specific activity, the Pd$_1$-NLs and Pd$_2$-NLs at 0 V are 312 and 305 μA cm$_{Pd}^{-2}$, respectively, which are higher than that of Pd/C (207 μA cm$_{Pd}^{-2}$) and Pt/C (103 μA cm$_{Pt}^{-2}$). The enhanced electrocatalytic activity of Pd NLs may be attributed to the unique nanoleave structure, which provides more Pd(111) facets, a large surface area, and more resistance to oxide formation.

More recently, Xu et al. [77] prepared a novel PdCu bimetallic nanotubes with hierarchically hollow structures via a galvanic displacement reaction by using dealloyed nanoporous copper as both a template and reducing agent and found that the PdCu nanocomposites exhibit enhanced ORR activity. In this study, three-dimensional nanoporous copper (Fig. 7a, b) was firstly synthesized by dealloying Cu/Al alloy foils in NaOH. PdCu bimetallic hollow structures were then produced through a galvanic replacement reaction between nanoporous copper and K$_2$PdCl$_4$. The SEM and TEM measurements (Fig. 7c–f) show that the

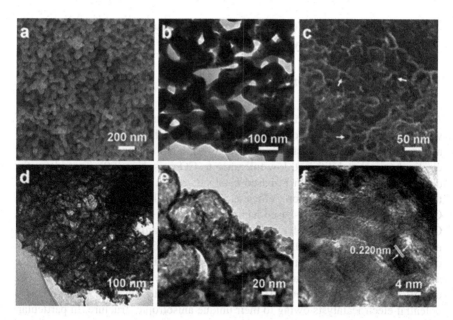

Fig. 7 SEM and TEM (HRTEM) images of nanoporous copper (**a, b**) and nanotubular mesoporous PdCu bimetallic nanostructure (**c–f**), respectively. Reprinted from Ref. [77] with permission by the American Chemical Society

formed PdCu bimetallic nanocomposites display nanotubular mesoporous nanostructures. The electrochemical polarization curves of the nanotubular mesoporous PdCu (NM-PdCu) revealed that the half-wave potential of the NM-PdCu for ORR was 0.840 V, which was 60 mV more positive than that of the commercial Pd/C catalysts. Moreover, the NM-PdCu catalyst is also superior to the commercial Pt/C (0.825 V) catalyst with more positive half-wave potential. The calculated specific activity of NM-PdCu at 0.8 and 0.85 V is 1.5 and 1.4 times that of the Pt/C catalyst at the respective potentials. From the Koutecky-Levich curves in rotating disk voltammetry measurements, a nearly complete reduction of O_2 to H_2O on the NM-PdCu surface via an efficient four-electron reaction process was obtained. More importantly, based on the experimental results, NM-PdCu catalyst showed enhanced methanol tolerance as compared with the commercial Pt/C and Pd/C catalysts. The enhanced ORR activity, stability, and methanol tolerance were ascribed to the presence of sublayer Cu atoms, which provide an electronic modification for the topmost Pd layer by a surface strain effect or an alloying effect. This effect could provide unique surface sites for the adsorption of O_2 molecules and be beneficial for their subsequent electro-reduction. Furthermore, the unique three-dimensional bicontinuous spongy structure and various hollow channels provide good transport channels for medium molecules and electrons, which may greatly facilitate the reaction kinetics of ORR on the catalytic surfaces.

3 One-Dimensional Pd-Based Nanomaterials as Effective Anode Electrocatalysts

3.1 Electrocatalysts for Alcohol Oxidation

Although Pt-based alloy catalysts, especially nanostructured PtRu, have been widely used as anode electrocatalysts in DMFC, the effectiveness and wide application of these catalysts are still largely limited due to their relatively low electrochemical stability and low tolerance to CO poisoning [99]. To address the disadvantages suffered from Pt-based catalysts, recent efforts have been devoted to the development of more efficient catalysts with high stability and high CO-tolerance. Recently, Pd-based nanomaterials have been found to exhibit high CO tolerance and superior electrocatalytic activity toward CO oxidation [38, 100], methanol oxidation [101, 102], and ethanol oxidation [69]. However, there are still obstacles in utilizing Pd nanoparticles because they often experience irreversible aggregation during electrocatalytic cycles, leading to a significant loss of catalytic activity and durability. Thus, 1D Pd-based nanomaterials have emerged as a potential electrocatalysts owing to their unique anisotropic structure. In particular, the ordered surface structure of one-dimensional nanostructured catalysts could affect the electrochemical and electrocatalytic properties. Among various synthetic approaches for one-dimensional nanostructure arrays, an anodic aluminum oxide (AAO) membrane-based method has received increasing attention because its uniform and reproducible porous structure as an ideal template can produce highly ordered nanotube or nanowire-type arrays.

Cheng et al. [71] prepared highly ordered Pd nanowire arrays (NWAs) using a porous AAO template by pulse electrodeposition method. It can be clearly observed from Fig. 8b–d that the Pd nanowires templated by AAO are highly ordered with uniform diameters of about 50 nm and length of 850 nm. Meanwhile, the as-prepared Pd NWAs retain the size and near cylinder shape of the pores of the AAO template as shown in Fig. 8a. More interestingly, the as-synthesized Pd NWAs exhibit good electrocatalytic activity toward isopropanol and methanol oxidation at room temperature. By using a similar AAO template-electrodepositon method and a subsequently magnetron sputtering techniques, Cheng's group also prepared highly ordered Pd/Pt core-shell nanowire arrays (Pd/Pt NWAs) [72]. In the method, Pd NWAs were first synthesized in AAO template. After removing the template in NaOH solution, Pt film was then coated on the surface of Pd NWAs through magnetron sputtering. Figure 9 shows the SEM images of the as-synthesized Pd and Pd/Pt NWAs. By using AAO as template, highly ordered Pd NWAs with smooth surface and uniform diameter and length were produced. However, after magnetron sputtering of Pt, the surface of Pd nanowires was covered by a thin layer (~ 1.7 nm) of cotton-like aggregated Pt nanoparticles. Nevertheless, the microstructure and morphology of Pt coated Pd NWAs is more or less the same as original Pd NWA core. Figure 10a shows the CVs of different electrodes in a nitrogen- saturated 0.5 M H_2SO_4 solution without methanol. It was found that the electrochemical active area

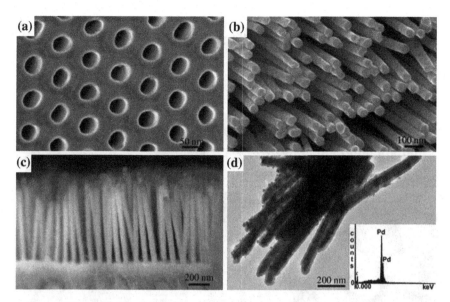

Fig. 8 Typical SEM images of **a** AAO template, **b** surface of Pd nanowire arrays (NWAs) with the diameter of 50 nm, **c** cross-section of Pd NWAs and **d** TEM image and EDX (*inset*) of Pd NWAs. Reprinted from Ref. [71] with permission by Elsevier

(EAA) of Pd/Pt core-shell NWAs electrode is 51 and 20 times larger than that of Pt thin film and PtRu/C electrode, respectively. This extremely high EAA from Pd/Pt core-shell NWAs shows that Pd/Pt core-shell NWAs nanostructure can substantially increase the effective electrochemical active sites most likely due to the very high surface-to-volume ratio of the highly ordered NWA core nanostructure and the high roughness of the Pt shell. By comparing the CVs in Fig. 10b, the Pd/Pt core-shell NWAs exhibit superior performance for methanol oxidation. It can be seen that the onset oxidation potential on Pd/Pt core-shell NWAs (0.19 V vs. SCE) is more negative than those on PtRu/C (0.21 V) and Pt thin film (0.39 V). The electrocatalytic activity, as measured by the peak current density in the forward scan is 22.7 mA cm^{-2} for the Pd/Pt core-shell NWAs, which is nearly 4.2 and 8.7 times higher than that of the E-TEK PtRu/C and Pt thin film electrodes. On the other hand, the peak potential for methanol oxidation on the Pd/Pt core-shell NWAs electrode is 170 mV more negative than that on the conventional PtRu/C electrode, indicating the enhanced electrode kinetics. Moreover, the Pt mass specific current peak density is 756.7 mA mg$_{Pt}^{-1}$ for the Pd/Pt core-shell NWAs electrode, which is four times higher than the E-TEK PtRu/C electrode (180.0 mA mg$_{Pt}^{-1}$). Since Pd is electrocatalytically inactive in acidic media, the high EAA and mass specific current for methanol oxidation can be mainly attributed to the Pt shell sputtered on the surface of Pd NWAs core.

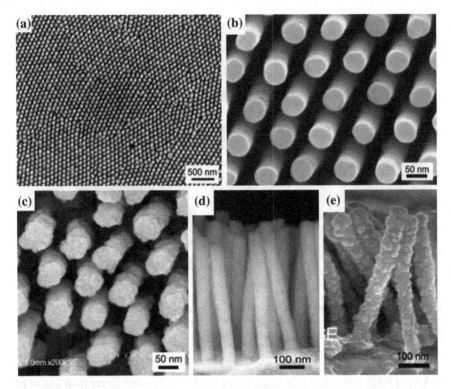

Fig. 9 SEM micrographs of **a** and **b** surface of Pd NWAs, **c** surface of Pd/Pt core-shell NWAs, **d** cross-section of Pd NWAs, and **e** cross-section of Pd/Pt core-shell NWAs. Reprinted from Ref. [72] with permission by Elsevier

Earlier studies have shown that 1D NWs have a strong interaction with carbon supports and are less vulnerable than conventional nanoparticles to dissolution, Ostwald ripening, and aggregation in strong acidic electrocatalytic conditions [71, 103]. Different from the bimetallic Pd-based alloys, Guo et al. [104] recently synthesized ultrathin (2.5 nm) trimetallic FePtPd alloy nanowires (NWs) with tunable compositions and controlled length (less than 100 nm). These FePtPd NWs exhibited composition-dependent catalytic activity and stability for methanol oxidation reaction. As shown in Fig. 11a, the as-prepared $Fe_{28}Pt_{38}Pd_{34}$ NWs exhibit the highest catalytic activity for methanol oxidation with the mass current density of 488.7 mA mg_{Pt}^{-1} and peak potential decreased from 0.665 V (vs. Ag/AgCl) obtained on Pt nanoparticle catalysts to 0.614 V. More interestingly, the $Fe_{28}Pt_{38}Pd_{34}$ displayed enhanced electrochemical stability with the mass current density (98.1 mA mg_{Pt}^{-1}) after i-t test for 2 h at 0.4 V (Fig. 11b). Figure 11c, d indicates that there was almost no noticeable morphology change before and after i-t tests, whereas the Pt nanoparticles experienced substantial aggregation. The authors attributed the enhanced stability of NWs versus NPs to the stronger NW interactions with carbon support and/or by better NW structure stability, which

Fig. 10 Cyclic voltammograms of Pt film electrode, E-TEK PtRu (33 wt%)/C electrode and Pd/Pt core-shell NWAs electrode in **a** a 0.5 M H_2SO_4 solution and **b** in a 1.0 M methanol +0.5 M H_2SO_4 solution at 298 K under a scan rate of 50 mV s^{-1}. Pt loading of the electrodes was 0.03 mg cm^{-2}. Reprinted from Ref. [72] with permission by Elsevier

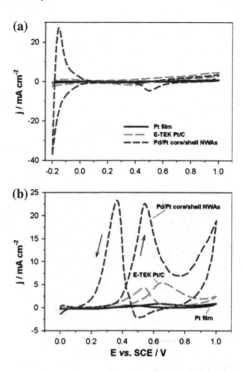

makes the NWs less subject to dissolution, Ostwald ripening, and aggregation in acidic solution.

Compared to methanol, ethanol is less toxic and can be produced in large quantities from agricultural products. Moreover, ethanol is the major renewable biofuel from the fermentation of biomass. Recent studies have shown that Pd-based nanomaterials are electrocatalytically active for ethanol oxidation in alkaline medium, representing an important alternative to Pt-based catalysts for direct ethanol AFCs [69, 80, 105]. For instance, Jiang and coworkers [69] successfully synthesized highly ordered Pd nanowires arrays (Pd NWA) through the anodized aluminum oxide (AAO) template-electrodeposition method. The SEM characterizations in Fig. 12 show that the as-synthesized Pd NWAs after AAO template fully dissolved are highly ordered with uniform diameter *ca.* 80 nm and length *ca.* 800 nm. The uniform porous structure of the nanowires can effectively improve their ECSA and thus enhance the active sites for the electrocatalytic reaction. Electrochemical experiments demonstrated that the as-synthesized Pd NWA exhibited excellent electrocatalytic activity and stability for ethanol oxidation in alkaline media compared to the conventional Pd film electrodes and the well-established commercial E-TEK PtRu/C electrocatalysts. As can be seen from Fig. 13, the onset potential for the ethanol oxidation on the Pd NWA electrode is −0.62 V, which is 150 mV more negative than the −0.45 V observed on the Pd film electrode and 40 mV more negative than the −0.58 V obtained from the conventional E-TEK PtRu/C

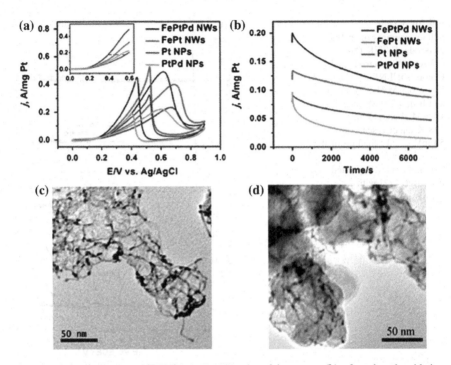

Fig. 11 CVs (**a**), linear sweep voltammetry (*inset*) and i-t *curves* (**b**) of methanol oxidation reaction catalyzed by $Fe_{28}Pt_{38}Pd_{34}$ NWs, FePt NWs, Pt NPs and PtPd NPs in 0.1 M $HClO_4$ + 0.2 M methanol solution. The CVs were obtained at a scan rate of 50 mV s^{-1} and the i-t *curves* were collected at a constant potential of 0.4 V. TEM images of the $Fe_{28}Pt_{38}Pd_{34}$ NWs/C before (**c**) and after (**d**) 2 h i-t test. Reprinted from Ref. [104] with permission by the American Chemical Society

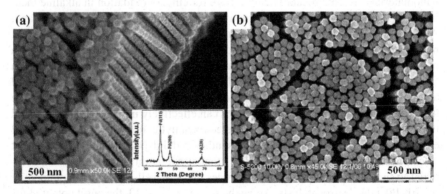

Fig. 12 SEM images of **a** cross section and **b** surface of Pd NWAs. *Inset* in (**a**): XRD pattern of Pd NWAs. Reprinted from Ref. [69] with permission by Wiley-VCH

Fig. 13 CVs measured on a Pd film electrode (*curves a*, Pd loading: 1.1 mg cm^{-2}), E-TEK PtRu (2:1 by weight)/C electrode (*curves b*, Pt loading: 0.24 mg cm^{-2}), and Pd NWA electrode (*curves c*, Pd loading: 0.24 mg cm^{-2}) in 1.0 M KOH + 1.0 M C$_2$H$_5$OH (*upper panel*) and 1.0 M KOH (*lower panel*) solution at a scan rate of 50 mV s^{-1}. Reprinted from Ref. [69] with permission by Wiley-VCH

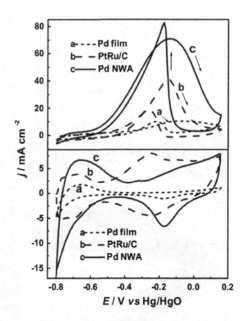

electrocatalyst. The negative shift of onset potential indicates the significant enhancement in the kinetics of the ethanol oxidation reaction on Pd NWAs. Similarly to methanol oxidation on Pt-based electrocatalysts in acid media, ethanol oxidation on Pd NWA electrode is also characterized by well-separated anodic peaks in the forward and reverse scans. The anodic peak current density on the Pd NWA electrode is seven times of that on the Pd film electrode and almost twice of that on the PtRu/C electrode. The obtained electrochemical results indicate that the 1D Pd NWA nanomaterials could be an efficient anode electrocatalysts for direct ethanol AFCs.

Interface formed in a bimetallic system plays an important role in the catalytic promotional effect and studies on bimetallic interface are also critical to investigate the fundamental mechanism of enhanced electrocatalytic activity for fuels oxidation. Recently, Yu et al. [80] designed unsupported Pd–Au bimetallic tubular nanostructures through one-step nonaqueous solvent electrodeposition method and studied their catalytic activity for electrooxidation of ethanol. With the electrodeposition method, tubular, dispersed, and unsupported Pd–Au bimetallic heterostructure tubes (BHTs) were formed with controlled atom percentage of Pd and with several micrometer-sized lengths as measured by SEM (Fig. 14a, b). The TEM image in Fig. 14c indicates the uniform wall thickness of the nanotubes, composed of many flocky-like spheres. Moreover, the unique porous structure of the BHTs could promote the mass transfer and effectively expose their inner and outer surfaces. From the HRTEM images shown in Fig. 14d–f, the single Au nanoparticle and the interfaces of Au–Pd and Au–Au can be observed clearly. The resultant tubular materials demonstrated high surface area, high stability, and

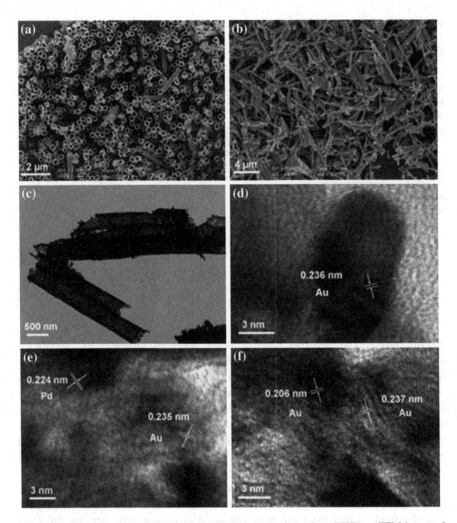

Fig. 14 a, b SEM images of Pd–Au bimetallic heterostructure tubes (BHTs). **c** TEM image of unsupported Pd–Au BHTs. HRTEM images of **d** a single particle **e** overlapped Pd–Au particles near the edge of the pore area, and **f** Au–Au particle interface, respectively. Reprinted from Ref. [80] with permission by the American Chemical Society

durability during the electrocatalytic studies. Figure 15 shows the CVs of various catalysts supported on glassy carbon electrode (GCE) in N_2-purged 1.0 M KOH aqueous solution with ethanol at a sweep rate of 50 mV s^{-1}. By comparison with Au nanotubes, Pd nanotubes, and commercial Pt/C catalysts, the synthesized Pd–Au BHTs exhibited larger ECSA (Fig. 15a), higher pseudo-area and mass activities (Fig. 15b–e), and higher electrochemical stability (Fig. 15f), suggesting that they can act as efficient Pt-free catalysts for fuel cells or other applications. The enhanced electrocatalytic performance was attributed to the electronic structure,

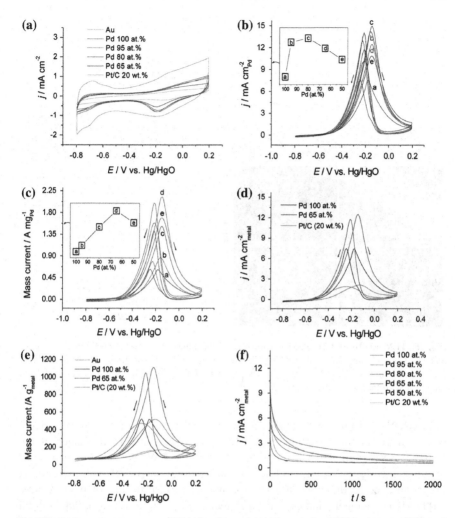

Fig. 15 Cyclic voltammograms (CVs) for determining the relative ECSA change of Pd and electrocatalytic activity. **a** CVs recorded in a N_2-purged 1.0 M KOH solution at room temperature. Pseudo-area activity (**b**) and mass activity (**c**) of unsupported Pd–Au BHTs with different Pd atom percentage for ethanol oxidation in 1.0 M KOH + 1.0 M C_2H_5OH. The *insets* show the peak current variation by increasing the Au atom percentage. **d** Pseudo-area activity and **e** mass activity of Pd NP tubes, unsupported Pd–Au BHTs, and Pt/C catalyst. **f** Chronoamperometric curves for ethanol electrooxidation at −0.3 V versus Hg/HgO on Pd NP tubes, unsupported Pd–Au BHTs, and Pt/C catalyst. Scan rate: 50 mV s^{-1}. Reprinted from Ref. [80] with permission by the American Chemical Society

local reactivity, and the significant coupling of d orbitals at the Pd–Au particle interfaces [80, 106].

More recently, Zhu et al. [64] developed a facile and general approach to synthesize high aspect ratio 1D Pd-based ANWs and found that the as-prepared

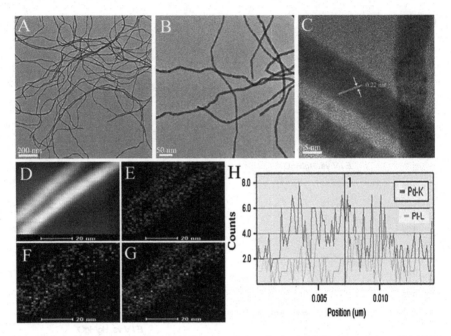

Fig. 16 TEM (**a, b**) and HRTEM (**c**) images of the obtained Pd$_{80}$Pt$_{20}$ nanowires. HAADF-STEM-EDS mapping images (**d–g**) of Pd$_{80}$Pt$_{20}$ nanowires. The cross-sectional compositional line profiles of individual Pd$_{80}$Pt$_{20}$ nanowire (**h**). **e** Pd-K; **f** Pt-L; **g** Pd-K + Pt-L. Reprinted from Ref. [64] with permission by Wiley-VCH

PdPt ANWs exhibit significantly enhanced activity and stability toward ethanol oxidation in alkaline medium. Figure 16a and b show the TEM images of the Pd$_{80}$Pt$_{20}$ ANWs templated from pre-synthesized Te NWs under different magnifications, clearly indicating the formation of uniform nanowires with high aspect ratios, an average diameter of 10.8 nm, and lengths up to tens of micrometers. HRTEM and elemental mapping measurements (Fig. 16c–h) revealed that both Pd and Pt elements are homogeneously distributed without significant segregation of each component. Electrochemical experiments are then performed to evaluate the catalytic activity of the PdPt ANWs toward alcohol oxidation. As shown in Fig. 17a, for the PdPt ANWs with different compositions, the Pd$_{45}$Pt$_{55}$ ANWs displays the highest activity toward ethanol oxidation in terms of onset potential and peak current. Based on the CV comparison in Fig. 17b, the mass activities on the Pd$_{45}$Pt$_{55}$ ANWs modified electrode is about 1.2 and 1.8 times of those on the Pd NWs and Pt nanotubes electrodes for ethanol oxidation. Moreover, the Pd$_{45}$Pt$_{55}$ ANWs also show a higher activity than commercial Pd/C electrocatalyst for ethanol oxidation. From the current–time curves recorded at −0.2 V shown in Fig. 17c, the Pd$_{45}$Pt$_{55}$ ANWs exhibit the highest catalytic stability among the studied electrocatalysts. On the other hand, the as-prepared 1D Pd$_{45}$Pt$_{55}$ catalyst also shows higher electrocatalytic activity and stability compared to the

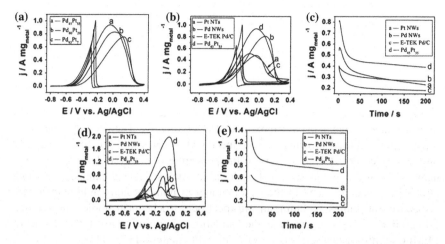

Fig. 17 a CVs of the Pd_xPt_y electrodes in a 0.5 M NaOH + 1 M ethanol solution; **b** CVs of the $Pd_{45}Pt_{55}$, Pd NWs, Pt NTs and E-TEK Pd/C electrodes in a 0.5 M NaOH + 1 M ethanol solution; **c** Current density–time curves of the $Pd_{45}Pt_{55}$, Pd NWs, Pt NTs and E-TEK Pd/C electrodes in a 0.5 M NaOH + 1 M ethanol solution at −0.2 V; **d** CVs of the $Pd_{45}Pt_{55}$, Pd NWs, Pt NTs and E-TEK Pd/C electrodes in a 1 M KOH + 1 M methanol solution; **e** Current density–time curves of the $Pd_{45}Pt_{55}$, Pd NWs, Pt NTs and E-TEK Pd/C electrodes in a 1 M KOH + 1 M methanol solution at −0.2 V. The loading amount of noble metal is 71.4 µg cm^{-2} for each catalyst and all the potential scan rates are 50 mV s^{-1}. Reprinted from Ref. [64] with permission by Wiley-VCH

monometallic catalysts Pt NTs and Pd NWs and the commercial E-TEK Pd/C catalyst toward methanol oxidation in alkaline conditions (Fig. 17d, e). The excellent catalytic performance of the PdPt nanowires could be attributed to three aspects: (1) the high aspect ratio; (2) the electronic effect; and (3) the synergistic effect [64].

3.2 Electrocatalysts for Formic Acid Oxidation

Except for alcohol, formic acid is another potential fuel for liquid fuel cells. Formic acid is a liquid at room temperature and dilute formic acid is on the US Food and Drug Administration list of food additives. Although the energy density of formic acid is lower than that of methanol, formic acid can be oxidized at less positive potential and with faster kinetics than methanol at room temperature. Moreover, formic acid can be easily handled and stored, and the low crossover through the polymer membranes allows fuel cells to work at relatively high concentrations of fuel and thin membranes. Therefore direct formic acid fuel cells (DFAFCs) have attracted increasing attention in recent years. Three possible reaction paths of formic acid oxidation have been widely accepted [107–110].

$$HCOOH_{ad} \rightarrow COOH_{ad} + H^+ + e^- \rightarrow CO_2 + 2H^+ + 2e^- \qquad (9)$$

$$HCOOH_{ad} \rightarrow HCOO_{ad} + H^+ + e^- \rightarrow CO_2 + 2H^+ + 2e^- \qquad (10)$$

$$HCOOH_{ad} \rightarrow CO_{ad} + H_2O \rightarrow CO_2 + 2H^+ + 2e^- \qquad (11)$$

In the first and second paths, through dehydrogenation with forming different intermediates formic acid can be oxidized to CO_2 directly. The Eq. (11) represents the indirect pathway, in which the produced CO intermediate can adsorb strongly on the surface of catalyst, leading to the poisoning of catalyst. To overcome the heavy CO-poisoning and high-cost of Pt-based catalysts, recent extensive research efforts have been devoted to the development of non-platinum anode electrocatalysts. Recent studies, including in situ spectroelectrochemical studies showed that Pd-based catalysts can catalyze the oxidation of formic acid at the anode of PEMFCs with greater resistance to CO than Pt catalysts [42, 111].

Recently, we successfully synthesized nanoneedle-covered 1D palladium-silver nanotubes through a galvanic displacement reaction with Ag nanorods at 100 °C (PdAg-100) and room temperature (PdAg-25) [76]. TEM and SEM measurements displayed that the synthesized PdAg nanotubes exhibit a hollow structure with a nanoneedle-covered surface, which provide the perfect large surface area for catalytic applications. From the HRTEM images of PdAg-25 shown in Fig. 18a, b, the surface of the PdAg-25 nanotubes is decorated with crystalline Pd nanoparticles with Pd(111) planes, and meanwhile, Ag and AgCl particles are dispersed in the inner space of the nanotubes. From the elemental mapping and cross-sectional line profiles shown in Fig. 18d–g, silver is dispersed in the core and Pd mainly distributes on the shell of the nanotubes. From the CVs in 0.1 M $HClO_4$ solution in Fig. 19a, more charge for hydrogen desorption was obtained with PdAg-25 nanotubes compared with that of PdAg-100 with the same loading on electrode surface, indicating rougher surface and thus larger ECSA of PdAg-25 nanotubes. By comparing the CV curves of the electrodes in 0.1 M $HClO_4$ + 0.5 M HCOOH solution (Fig. 19b), one can see that the PdAg-100 nanotubes exhibit more negative anodic peak potential (+0.285 vs. +0.316 V) and much larger anodic peak current density (3.82 vs. 1.97 mA cm^{-2}) of formic acid oxidation than those of PdAg-25 under same conditions. The higher electrocatalytic activity of PdAg-100 nanotubes possibly as the consequences of the higher ratio of Pd to Ag (36:64) in PdAg-100 compared with that in PdAg-25 (25:75) and the annealing process of the nanotube surface structures at 100 °C. Chronoamperometric analyses (Fig. 19c–e) were carried out to evaluate the activity and stability of the PdAg nanotubes for formic acid electrooxidation. It can be seen that at the three studied potentials, the maximum initial and steady-state oxidation current densities obtained from both of the PdAg nanotubes are much larger than that from the bulk Pd electrode over the entire time period. On the other hand, the initial current density on the bulk Pd electrode decays much more rapidly than those of the as-synthesized PdAg nanotubes. These results indicate that the synthesized PdAg nanotubes exhibit

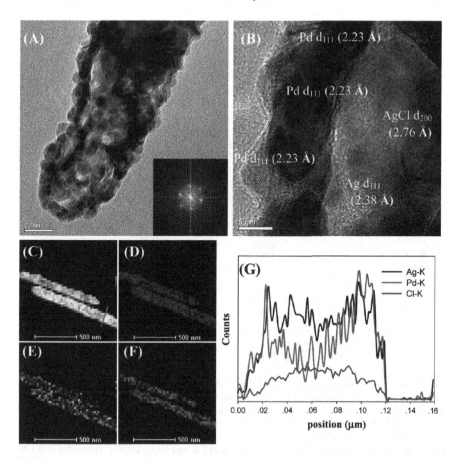

Fig. 18 a, b HRTEM images of PdAg-25 nanotubes. The *inset* in (**a**) shows the corresponding two-dimensional fast Fourier transform (FFT) pattern. **c** The high-angle annular dark-field (HAADF) STEM image of PdAg-25 nanotubes; the corresponding elemental mapping of Ag (**d**), Pd (**e**), and Cl (**f**) in the PdAg-25 nanotubes. **g** Cross-sectional compositional line profiles of a PdAg-25 nanotube. Reprinted from Ref. [76] with permission by the American Chemical Society

excellent catalytic activity and stability for formic acid oxidation due to their unique 1D nanostructures. In the electrochemical impedance spectroscopy studies, it was also found that the charge-transfer resistance (R_{CT}) at the PdAg-100 nanotubes is much smaller than that at the PdAg-25 nanotubes, indicating the electron-transfer kinetics for formic acid oxidation at the PdAg-100 nanotubes is much better facilitated.

In another study, we synthesized bimetallic PdAg ANWs based on a facile one-step wet chemical strategy [70]. Uniform PdAg nanowires were produced by heating the silver nitrate and $Pd(NO_3)_2 \cdot 2H_2O$ solution in ethylene glycol at 170 °C with the presence of poly(vinyl pyrrolidone). The HRTEM images in Fig. 20a, b

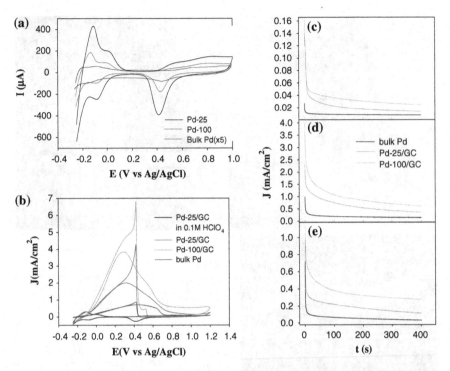

Fig. 19 a Cyclic voltammograms of the PdAg-25/GC (*black curve*), PdAg-100/GC (*red curve*), and bulk Pd electrode (*blue curve*) in 0.1 M HClO$_4$ solution. For comparison, the CV from the Pd bulk electrode is magnified 5×. **b** CVs of the PdAg-25/GC (*red curve*), PdAg-100/GC (*green curve*), and bulk Pd electrode (*blue curve*) electrodes in 0.1 M HClO$_4$ and 0.5 M HCOOH solution and the CV of PdAg-25/GC (*black curve*) in 0.1 M HClO$_4$. Potential scan rate 0.1 V s^{-1}. Chronoamperometric curves of the bulk Pd, PdAg-25/GC and PdAg-100/GC electrodes in 0.1 M HClO$_4$ + 0.5 M HCOOH at different electrode potentials: **c** −0.15, **d** +0.3, and **e** +0.68 V. Reprinted from Ref. [76] with permission by the American Chemical Society

show that worm-like nanowires were formed and the surface of the nanowires is predominated by Pd(111) planes. As mentioned previously, the (111) planes of Pd are less susceptible to oxidation and have a lower peak potential for formic acid oxidation, making the nanowires promising electrocatalysts for DFAFCs. The element maps revealed that the elements of Pd and Ag are homogeneously dispersed in the PdAg nanowires, indicating the formation of alloy structure. The electrochemical studies showed that although the onset potential of formic acid oxidation on PdAg nanowires is only a little more negative than that on the commercial Pd/C catalyst, the current density of formic acid oxidation in both forward and reverse potential sweeps is much larger (about 3.4 times) than that obtained at the Pd/C. Moreover, from the CV and chronoamperometric measurements, our synthesized PdAg nanowires exhibit high CO tolerance and

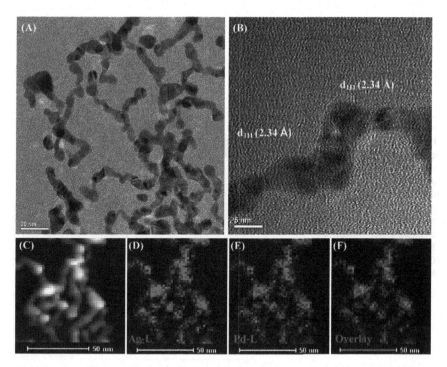

Fig. 20 a–b High-resolution TEM micrographs (HRTEM) of PdAg nanowires at different magnifications. The *scale bars* are **a** 20 nm and **b** 5 nm. **c** The high angle annular dark field (HAADF) STEM image of PdAg nanowires and the corresponding elemental mapping of Ag (**d**), Pd (**e**). **f** Overlay map of the elements in the PdAg nanowires. Reprinted from Ref. [70] with permission by the American Chemical Society

long-term electrochemical stability. On the other hand, the electrochemical impedance spectroscopy measurements showed that with the electrode potential increasing the impedance spectra of PdAg nanowires show arcs in the first quadrant firstly and then negative impedance was observed in the second quadrant (Fig. 21a). However, all the impedance spectra of Pd/C are located in the first quadrant within the entire potential range (Fig. 21b). The different impedance results obtained from the two catalysts suggest that PdAg nanowires have higher CO tolerance and fast formic acid oxidation kinetics. Also from Fig. 21c, it can be seen that the R_{CT} derived from PdAg ANWs is remarkably smaller than that from the Pd/C catalysts within the studied potential window. The smaller R_{CT} indicates that the electron-transfer kinetics for formic acid oxidation at the PdAg nanowires is much better facilitated than that at the Pd/C catalysts. All the results demonstrate that the as-synthesized PdAg nanowires have much better catalytic performance than that of commercial Pd/C catalysts.

Fig. 21 Nyquist impedance plots of formic acid oxidation on PdAg-NW/GC (**a**) and Pd/GC (**b**) electrodes in 0.1 M HClO₄ +0.5 M HCOOH at various electrode potentials. The *solid lines* show some representative fits to the experimental data based on the equivalent circuits. **c** Charge-transfer resistance (R_{CT}) of formic acid electrooxidation at different electrode potentials on PdAg-NW/GC (*black curve*) and Pd/GC (*red curve*) electrodes. Reprinted from Ref. [70] with permission by the American Chemical Society

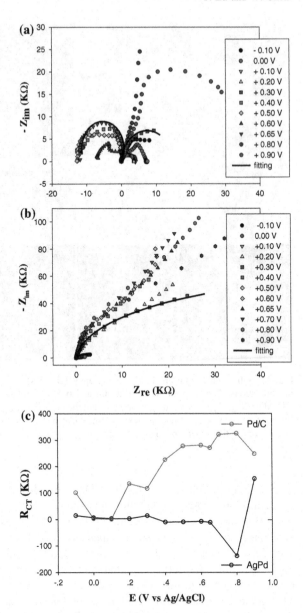

Recently, Song et al. [78] reported the fabrication of Pd nanotube electrodes by means of electrodeposition method as a function of applied current density. From the TEM and SEM images (Fig. 22) of the Pd array electrodes electrodeposited using different current densities, it can be seen that all the products exhibit tubular arrays having an average diameter of ∼220 nm and different lengths. Note that the surface of the as-synthesized Pd-NT becomes rougher with the increase in current

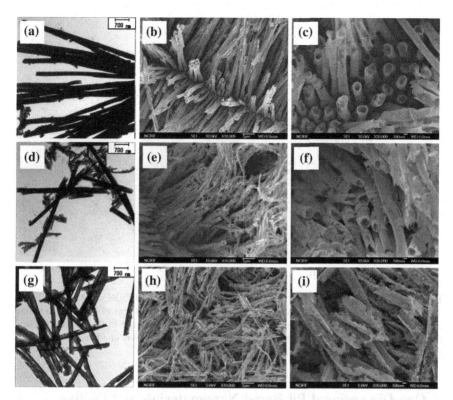

Fig. 22 TEM and SEM images of the Pd nanotube arrays fabricated at different current densities for 12 h: -1 mA cm^{-2} (Pd-NT-1) (**a, b, c**); -3 mA cm^{-2} (Pd-NT-3) (**d, e, f**) and -5 mA cm^{-2} (Pd-NT-5) (**g, h, i**). Reprinted from Ref. [78] with permission by Elsevier

densities and the Pd-NT-5 tube arrays electrodeposited under the highest applied current density display the roughest surface. This suggests that the fast tube array growth under higher current density may result in rougher surface structure. In the electrocatalytic studies for formic acid oxidation, it was found that the Pd-NT-5 exhibited the largest electrochemical active surface area and the highest electrocatalytic activity among the three samples, but the catalytic activities of all the Pd nanotubes are lower than that of the commercial Pd catalyst (20 wt%). The authors compared the electrochemical stability of the Pd-NT-5 and commercial Pd/C catalyst. As shown in Fig. 23a, the Pd-NT-5 exhibits nearly maintained electrocatalytic activity after the stability test. Simultaneously, the size and morphology of the Pd-NT-5 also remained after the stability test. In contrast, the electrocatalytic activity of the Pd/C deteriorated significantly after the stability test (Fig. 23b). Furthermore, the Pd/C showed increased particle size and nonuniform size distribution. The results strongly suggest that the Pd nanotubes have enhanced electrocatalytic stability, which may be ascribed to the less aggregation of 1D Pd nanotubes during the catalytic reaction in comparison with the Pd/C.

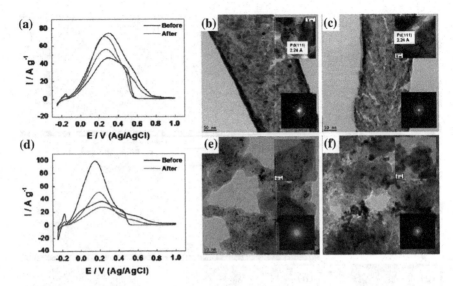

Fig. 23 a CVs of Pd-NT-5 before and after stability test in 0.5 M HCOOH and 0.1 M HClO₄ at 25 °C. TEM images and transmission electron diffraction (TED) patterns of Pd-NT-5, **b** before and **c** after stability test. **d** CVs of Pd/C before and after stability test in 0.5 M HCOOH and 0.1 M HClO₄ at 25 °C. TEM images and TED patterns of Pd/C, **e** before and **f** after stability test. Reprinted from Ref. [78] with permission by Elsevier

4 One-Dimensional Pd-Based Nanomaterials as Effective Hydrogen Storage Materials

Except for the physical methods, such as storage in tanks as compressed hydrogen and physical adsorption on large surface area adsorbents [112–114], chemical materials including metal-borohydrides [115], Mg-based alloys [116], carbon materials [117], and ammonia-boranes [118] etc. have also been widely used as potential hydrogen storage media. However, due to the strong binding between hydrogen and most of the materials, poor dehydrogenation kinetics and high temperatures needed for hydrogen release limit greatly the practical application of these materials. Recently, palladium-based alloys have been found to exhibit higher solubility and permeability of hydrogen than in pure Pd [119–121]. Especially, the nanostructured materials exhibit improved capacity of hydrogen storage compared to the bulk counterparts due to the large surface area and short hydrogen diffusion paths. Similar to the catalytic activity, the hydrogen storage capacity of the Pd-based nanomaterials is strongly related to their structure factors, such as size, composition, and morphology. For example, the previous study showed that for PdAg alloys, the highest permeability for hydrogen was obtained from the alloy with ~23 wt% of Ag when measured at a pressure of 1 atm and temperature above 473 K [122, 123]. Further simulation study suggested that faster hydrogen

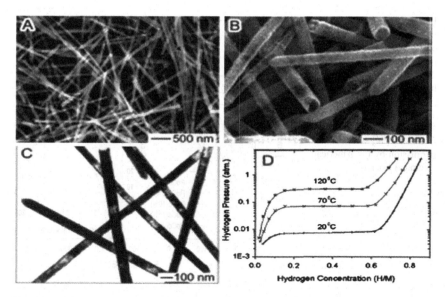

Fig. 24 **a, b** SEM images and **c** TEM image of silver nanowires with surface coated with Pd/Ag alloy sheaths. **d** PC isotherms for hydrogen desorption from the hydrides of the as-synthesized Ag@PdAg nanocables at 20, 70, and 120 °C. Here H/M is the hydrogen-to-metal ratio. Reprinted from Ref. [119] with permission by the American Chemical Society

diffusion can be realized from the PdAg alloys with Ag concentration higher than 63 % [124].

Sun et al. [119] synthesized silver nanowires with surface coated by thin sheaths of Pd/Ag alloys through the galvanic replacement reaction. From the SEM and TEM images shown in Fig. 24a–c, each Ag nanowire of ~ 60 nm in diameter is coated with Pd/Ag sheaths and is straight and uniform in diameter along the entire long axis. It was found that different from the Pd nanotubes formed from short Ag nanowires, only some segments of the long Ag wire have been converted to hollow structure after galvanic replacement reaction. The total percentage of Ag in the final product was determined to be as high as 92.2 wt%. Figure 24d shows the pressure-composition (PC) isotherms of the as-synthesized Ag/Pd nanostructure (Ag@PdAg) at different temperatures (20, 70, and 120 °C). It can be seen that the distinct plateaus were formed similar to those of polycrystalline Pd powders, indicating the existence of a broad metal-H miscibility gap for the Ag@PdAg wires. Moreover, the absorption and desorption of hydrogen were reversible and the reactions at room temperature were very fast for the Ag@PdAg system. By comparison, the hydrogen solubility of Ag@PdAg wires is larger than that of Pd powers, which can be ascribed to the unique structure of PdAg alloy sheath formed on the Ag nanowires.

Recently, Chen and coworkers [94] reported that the Pd–Cd nanomaterials with 10–15 % Cd exhibit the highest hydrogen storage ability, over 15 times greater than the pure Pd nanoporous materials. In their study, the hydrogen storage ability

Fig. 25 Cyclic voltammograms of the Pd–Cd electrodes in 0.1 M HClO$_4$ performed with a scan rate of 20 mV s^{-1}. The overall hydrogen desorption charge Q_H versus the normalized atomic composition of Cd is shown in the inset. Reprinted from Ref. [94] with permission by the American Chemical Society

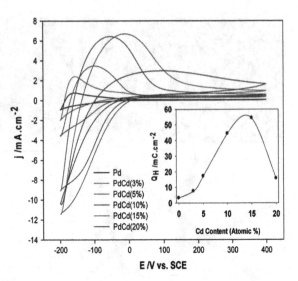

of Pd–Cd nanomaterials with different compositions was evaluated by using electrochemical cyclic voltammetry technique. Under the acid condition, it is easy to clearly separate the region of hydrogen sorption/desorption from the potential of the palladium oxide formation. However, it is not so easy to decouple adsorption from absorption of hydrogen. Thus, they used the total charge, Q_H, obtained by integrating the area under the anodic peaks in the CVs to determine the hydrogen storage ability. As shown in Fig. 25, the charges for hydrogen desorption on the Pd–Cd nanostructures displays a volcano shape with 10–15 % Cd possessing the highest capacity. The authors found that with the amount of Cd increasing from 0 to 15 %, the crystallite size decreased from 25.81 to 9.16 nm and the surface area was increased from 11.58 to 46.64 m^2 g^{-1}. However, when the amount of Cd was further increased to 20 %, larger PdCd nanopartilces were formed and decreased surface area was obtained, thus resulting in the lower hydrogen storage capacity. Therefore, the hydrogen storage ability of the PdCd nanostructures depends on the surface structure and crystallite size. The enhanced hydrogen storage capacity upon the addition of Cd can be ascribed to the formation of small dendritic structures, dilation of the lattice constant, and decrease of the crystalline size.

By using electrochemical cyclic voltammetry, our group studied the hydrogen storage properties of the PdAg nanotubes obtained by galvanic displacement between Ag nanorods and Pd(NO$_3$)$_2$ at different reaction times [125]. From the SEM images (Fig. 26a–d) of the PdAg-(10), PdAg-(90), PdAg-(150) and PdAg-(180) nanotubes, which were collected at reaction times of 10, 90, 150, and 180 min, respectively, the hollow structure could be seen gradually with the reaction time increasing. Based on the inductive coupling high frequency plasma-mass spectrophotometry (ICP-MS) measurements, the ratios of Pd to Ag with 5:95, 10:90, 15:85, and 18:82 were determined for the samples. That is, the Pd content in the PdAg nanotubes increases with the increase ingalvanic reaction time. In the

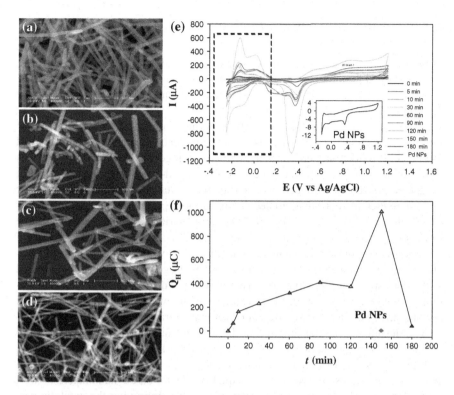

Fig. 26 Scanning electron microscopy (SEM) images of the PdAg nanotubes synthesized with different galvanic reaction times: 10 min (**a**); 90 min (**b**); 150 min (**c**) and 180 min (**d**). **e** Cyclic voltammograms of the PdAg nanotubes obtained at different galvanic reaction times and Pd nanoparticles in 0.1 M HClO$_4$ electrolyte solution. Potential scan rate 0.1 V s^{-1}. For clarity, the CV from Pd nanoparticles is shown in the *inset*. The *dashed frame* shows the hydrogen adsorption-desorption region obtained on the PdAg nanotubes. **f** The total charge (Q_H) of hydrogen adsorption and absorption on the PdAg nanotubes dependent on galvanic reaction time. Reprinted from Ref. [125] with permission by the Royal Society of Chemistry

CVs (Fig. 26e) of the synthesized PdAg nanotubes in a N$_2$ saturated 0.1 M HClO$_4$ aqueous solution, there is obvious hydrogen adsorption/desorption in the potential range of −0.25 to +0.1 V. From the dependence of Q_H on reaction time shown in Fig. 26f, it can be seen clearly that the hydrogen storage ability depends strongly on the composition of the PdAg nanostructured materials. The PdAg nanotubes with 15 % Pd possess the highest capacity for hydrogen absorption, which is over 200 times higher than that of pure Pd nanoparticles. Such significantly enhanced ability for hydrogen storage can be ascribed to the special tubular structures and alloying of Pd and Ag around the walls of the nanostructures. This work suggests that 1D Pd-based alloy nanotubes with low Pd content might represent a unique class of low-cost materials for efficient hydrogen storage.

5 Conclusions and Future Outlook

In this chapter, we discussed the technological challenges of the modern electrocatalysts and summarized recent research progress of 1D Pd-based nanomaterials as efficient electrocatalysts on both anode and cathode sides of fuel cells and the applications in hydrogen storage. Specifically, we presented the effect of structural parameters of Pd-based 1D nanomaterials on their electrocatalytic activities for cathode ORR, and anode small molecules (methanol, ethanol, and formic acid) oxidation, and on their hydrogen storage capacity. From the studies shown in this chapter, it can be concluded that 1D Pd-based nanomaterials exhibit enhanced electrocatalytic activity and improved electrochemical stability for oxygen reduction and small molecule oxidation. Except for the large electrochemically active surface area of 1D nanomaterials, the unique anisotropic structure of nanowire, nanorod, and nanotube can also facilitate the mass and electron-transfer during the catalytic reactions. Meanwhile, it has been found that 1D Pd-based nanomaterials possess excellent hydrogen storage ability compared to the corresponding 0D structures. Therefore, 1D Pd-based nanomaterials represent not only a class of non-Pt electrocatalysts with low cost and excellent catalytic performance, but also a class of novel hydrogen storage materials for fuel cells.

Yet, despite substantial progress in the electrocatalysts based on 1D Pd-based nanomaterials, some challenges remain in future work in this field. The emphases of future investigations should mainly include: (1) further developing different synthetic techniques to produce high quality 1D Pd-based nanomaterials with controlled size, shape, and composition; (2) further improving the catalytic performance and reducing Pd loading of 1D nanomaterials to achieve cost-effective fuel cell electrocatalysts; (3) further investigating theoretically the correlation between structures of 1D Pd-based materials and their catalytic and hydrogen storage properties to provide fundamental direction for nanomaterial design; (4) further enhancing the study of 1D Pd-based nanocatalysts in real fuel cells to test their catalytic performance in practical applications.

Acknowledgments This work was supported by the National Natural Science Foundation of China (Nos. 21275136, 21043013), the Natural Science Foundation of Jilin province, China (No. 201215090), and Scientific Research Foundation for Returned Scholars, Ministry of Education of China.

References

1. Dresselhaus MS, Thomas IL (2001) Nature 414(6861):332–337
2. Dillon R, Srinivasan S, Arico AS, Antonucci V (2004) J Power Sources 127(1–2):112–126
3. Lamy C, Lima A, LeRhun V, Delime F, Coutanceau C, Leger JM (2002) J Power Sources 105(2):283–296
4. Okada O, Yokoyama K (2001) Fuel Cells 1(1):72–77
5. Winter M, Brodd RJ (2004) Chem Rev 104(10):4245–4269

6. Vielstich W, Lamm A, Gasteiger HA (2003) Handbook of fuel cells: fundamentals, technology, and applications, vol 4. Wiley, Chichester
7. Chen AC, Holt-Hindle P (2010) Chem Rev 110(6):3767–3804
8. Chen W, Kim JM, Sun SH, Chen SW (2007) Langmuir 23(22):11303–11310
9. Zhang J, Yang HZ, Yang KK, Fang J, Zou SZ, Luo ZP, Wang H, Bae IT, Jung DY (2010) Adv Funct Mater 20(21):3727–3733
10. Yang HZ, Zhang J, Sun K, Zou SZ, Fang JY (2010) Angew Chem Int Ed 49(38):6848–6851
11. Chen W, Kim JM, Xu LP, Sun SH, Chen SW (2007) J Phys Chem C 111(36):13452–13459
12. Stamenkovic VR, Fowler B, Mun BS, Wang GF, Ross PN, Lucas CA, Markovic NM (2007) Science 315(5811):493–497
13. Xia BY, Wu HB, Wang X, Lou XW (2012) J Am Chem Soc 134(34):13934–13937
14. Chen W, Chen SW (2011) J Mater Chem 21(25):9169–9178
15. Chen W, Kim JM, Sun SH, Chen SW (2008) J Phys Chem C 112(10):3891–3898
16. Chen W, Kim J, Sun SH, Chen SW (2006) Phys Chem Chem Phys 8(23):2779–2786
17. Kang YJ, Murray CB (2010) J Am Chem Soc 132(22):7568–7569
18. Gasteiger HA, Markovic N, Ross PN, Cairns EJ (1994) J Electrochem Soc 141(7):1795–1803
19. Dinh HN, Ren XM, Garzon FH, Zelenay P, Gottesfeld S (2000) J Electroanal Chem 491(1–2):222–233
20. Oetjen HF, Schmidt VM, Stimming U, Trila F (1996) J Electrochem Soc 143(12):3838–3842
21. Frelink T, Visscher W, vanVeen JAR (1996) Langmuir 12(15):3702–3708
22. Goodenough JB, Hamnett A, Kennedy BJ, Manoharan R, Weeks SA (1988) J Electroanal Chem 240(1–2):133–145
23. Chen W, Xu LP, Chen SW (2009) J Electroanal Chem 631(1–2):36–42
24. Lu YZ, Chen W (2011) Chem Commun 47(9):2541–2543
25. Lim B, Jiang MJ, Camargo PHC, Cho EC, Tao J, Lu XM, Zhu YM, Xia YN (2009) Science 324(5932):1302–1305
26. Chen W, Chen SW (2009) Angew Chem Int Edit 48(24):4386–4389
27. Chen W, Ny D, Chen SW (2010) J Power Sources 195(2):412–418
28. Liu HS, Song CJ, Tang YH, Zhang JL, Zhang HJ (2007) Electrochim Acta 52(13):4532–4538
29. Zhang L, Zhang JJ, Wilkinson DP, Wang HJ (2006) J Power Sources 156(2):171–182
30. Morozan A, Jousselme B, Palacin S (2011) Energy Environ Sci 4(4):1238–1254
31. Chen ZW, Higgins D, Yu AP, Zhang L, Zhang JJ (2011) Energy Environ Sci 4(9):3167–3192
32. Wei WT, Lu YZ, Chen W, Chen SW (2011) J Am Chem Soc 133(7):2060–2063
33. Lu YZ, Wang YC, Chen W (2011) J Power Sources 196(6):3033–3038
34. Wu HB, Chen W (2011) J Am Chem Soc 133(39):15236–15239
35. Serov A, Kwak C (2009) Appl Catal B Environ 91(1–2):1–10
36. Serov A, Kwak C (2009) Appl Catal B Environ 90(3–4):313–320
37. Lu YZ, Chen W (2012) Chem Soc Rev 41(9):3594–3623
38. Antolini E (2009) Energy Environ Sci 2(9):915–931
39. Bianchini C, Shen PK (2009) Chem Rev 109(9):4183–4206
40. Cheng TT, Gyenge EL (2009) J Appl Electrochem 39(10):1925–1938
41. Zhou WP, Lewera A, Larsen R, Masel RI, Bagus PS, Wieckowski A (2006) J Phys Chem B 110(27):13393–13398
42. Larsen R, Ha S, Zakzeski J, Masel RI (2006) J Power Sources 157(1):78–84
43. Mazumder V, Sun SH (2009) J Am Chem Soc 131(13):4588–4589
44. Xiao L, Zhuang L, Liu Y, Lu JT, Abruna HD (2009) J Am Chem Soc 131(2):602–608
45. Chen XM, Lin ZJ, Jia TT, Cai ZM, Huang XL, Jiang YQ, Chen X, Chen GN (2009) Anal Chim Acta 650(1):54–58
46. Fu Y, Wei ZD, Chen SG, Li L, Feng YC, Wang YQ, Ma XL, Liao MJ, Shen PK, Jiang SP (2009) J Power Sources 189(2):982–987

47. Hu FP, Chen CL, Wang ZY, Wei GY, Shen PK (2006) Electrochim Acta 52(3):1087–1091
48. Wei WT, Chen W (2012) J Power Sources 204:85–88
49. Jiang YY, Lu YZ, Li FH, Wu TS, Niu L, Chen W (2012) Electrochem Commun 19:21–24
50. Chen XM, Wu GH, Chen JM, Chen X, Xie ZX, Wang XR (2011) J Am Chem Soc 133(11):3693–3695
51. Zhang J, Fang JY (2009) J Am Chem Soc 131(51):18543–18547
52. Bergamaski K, Pinheiro ALN, Teixeira-Neto E, Nart FC (2006) J Phys Chem B 110(39):19271–19279
53. Mayrhofer KJJ, Blizanac BB, Arenz M, Stamenkovic VR, Ross PN, Markovic NM (2005) J Phys Chem B 109(30):14433–14440
54. Tian N, Zhou ZY, Sun SG, Ding Y, Wang ZL (2007) Science 316(5825):732–735
55. Fernandez JL, Walsh DA, Bard AJ (2005) J Am Chem Soc 127(1):357–365
56. Suo YG, Zhuang L, Lu JT (2007) Angew Chem Int Edit 46(16):2862–2864
57. Schmidt TJ, Jusys Z, Gasteiger HA, Behm RJ, Endruschat U, Boennemann H (2001) J Electroanal Chem 501(1–2):132–140
58. Jiang L, Hsu A, Chu D, Chen R (2010) Electrochim Acta 55(15):4506–4511
59. Shao MH, Sasaki K, Adzic RR (2006) J Am Chem Soc 128(11):3526–3527
60. Jung CH, Sanchez-Sanchez CM, Lin CL, Rodriguez-Lopez J, Bard AJ (2009) Anal Chem 81(16):7003–7008
61. Fernandez JL, Raghuveer V, Manthiram A, Bard AJ (2005) J Am Chem Soc 127(38):13100–13101
62. Maiyalagan T, Scott K (2010) J Power Sources 195(16):5246–5251
63. He QG, Chen W, Mukerjee S, Chen SW, Laufek F (2009) J Power Sources 187(2):298–304
64. Zhu CZ, Guo SJ, Dong SJ (2012) Adv Mater 24(17):2326–2331
65. Zhu CZ, Guo SJ, Dong SJ (2012) J Mater Chem 22(30):14851–14855
66. Koenigsmann C, Sutter E, Chiesa TA, Adzic RR, Wong SS (2012) Nano Lett 12(4):2013–2020
67. Guo SJ, Dong SJ, Wang EK (2010) Chem Commun 46(11):1869–1871
68. Ksar F, Surendran G, Ramos L, Keita B, Nadjo L, Prouzet E, Beaunier P, Hagege A, Audonnet F, Remita H (2009) Chem Mater 21(8):1612–1617
69. Xu CW, Wang H, Shen PK, Jiang SP (2007) Adv Mater 19(23):4256–4259
70. Lu YZ, Chen W (2012) ACS Catal 2(1):84–90
71. Cheng FL, Wang H, Sun ZH, Ning MX, Cai ZQ, Zhang M (2008) Electrochem Commun 10(5):798–801
72. Wang H, Xu CW, Cheng FL, Zhang M, Wang SY, Jiang SP (2008) Electrochem Commun 10(10):1575–1578
73. Li WZ, Haldar P (2009) Electrochem Commun 11(6):1195–1198
74. Zhang ZY, More KL, Sun K, Wu ZL, Li WZ (2011) Chem Mater 23(6):1570–1577
75. Chen ZW, Waje M, Li WZ, Yan YS (2007) Angew Chem Int Edit 46(22):4060–4063
76. Lu YZ, Chen W (2010) J Phys Chem C 114(49):21190–21200
77. Xu CX, Zhang Y, Wang LQ, Xu LQ, Bian XF, Ma HY, Ding Y (2009) Chem Mater 21(14):3110–3116
78. Song YJ, Lee YW, Han SB, Park KW (2012) Mater Chem Phys 134(2–3):567–570
79. Alia SM, Jensen KO, Pivovar BS, Yan YS (2012) ACS Catal 2(5):858–863
80. Cui CH, Yu JW, Li HH, Gao MR, Liang HW, Yu SH (2011) ACS Nano 5(5):4211–4218
81. Koenigsmann C, Wong SS (2011) Energy Environ Sci 4(4):1161–1176
82. Huang XQ, Zheng NF (2009) J Am Chem Soc 131(13):4602–4603
83. Hoshi N, Kida K, Nakamura M, Nakada M, Osada K (2006) J Phys Chem B 110(25):12480–12484
84. Baldauf M, Kolb DM (1996) J Phys Chem 100(27):11375–11381
85. Smith PA, Nordquist CD, Jackson TN, Mayer TS, Martin BR, Mbindyo J, Mallouk TE (2000) Appl Phys Lett 77(9):1399–1401
86. Xia YN, Yang PD, Sun YG, Wu YY, Mayers B, Gates B, Yin YD, Kim F, Yan YQ (2003) Adv Mater 15(5):353–389

87. Garbarino S, Ponrouch A, Pronovost S, Gaudet J, Guay D (2009) Electrochem Commun 11(10):1924–1927
88. Reece SY, Hamel JA, Sung K, Jarvi TD, Esswein AJ, Pijpers JJH, Nocera DG (2011) Science 334(6056):645–648
89. Kudo A, Miseki Y (2009) Chem Soc Rev 38(1):253–278
90. Youngblood WJ, Lee SHA, Maeda K, Mallouk TE (2009) Acc Chem Res 42(12):1966–1973
91. Li Y, Zhang JZ (2010) Laser Photonics Rev 4(4):517–528
92. Maeda K, Domen K (2010) J Phys Chem Lett 1(18):2655–2661
93. Kruk M, Jaroniec M (2001) Chem Mater 13(10):3169–3183
94. Adams BD, Wu GS, Nigrio S, Chen AC (2009) J Am Chem Soc 131(20):6930–6931
95. Yeager E (1984) Electrochim Acta 29(11):1527–1537
96. Stamenkovic V, Mun BS, Mayrhofer KJJ, Ross PN, Markovic NM, Rossmeisl J, Greeley J, Norskov JK (2006) Angew Chem Int Edit 45(18):2897–2901
97. Koenigsmann C, Santulli AC, Gong KP, Vukmirovic MB, Zhou WP, Sutter E, Wong SS, Adzic RR (2011) J Am Chem Soc 133(25):9783–9795
98. Sarkar A, Murugan AV, Manthiram A (2008) J Phys Chem C 112(31):12037–12043
99. Liu HS, Song CJ, Zhang L, Zhang JJ, Wang HJ, Wilkinson DP (2006) J Power Sources 155(2):95–110
100. Jin MS, Liu HY, Zhang H, Xie ZX, Liu JY, Xia YN (2011) Nano Res 4(1):83–91
101. Lee YW, Ko AR, Han SB, Kim HS, Kim DY, Kim SJ, Park KW (2010) Chem Commun 46(48):9241–9243
102. Wu HX, Li HJ, Zhai YJ, Xu XL, Jin YD (2012) Adv Mater 24(12):1594–1597
103. Sun SH, Zhang GX, Geng DS, Chen YG, Li RY, Cai M, Sun XL (2011) Angew Chem Int Edit 50(2):422–426
104. Guo SJ, Zhang S, Sun XL, Sun SH (2011) J Am Chem Soc 133(39):15354–15357
105. Xu CW, Cheng LQ, Shen PK, Liu YL (2007) Electrochem Commun 9(5):997–1001
106. Roudgar A, Gross A (2004) Surf Sci 559(2–3):L180–L186
107. Capon A, Parsons R (1973) J Electroanal Chem 45(2):205–231
108. Neurock M, Janik M, Wieckowski A (2008) Faraday Discuss 140:363–378
109. Samjeske G, Miki A, Ye S, Osawa M (2006) J Phys Chem B 110(33):16559–16566
110. Kang YJ, Qi L, Li M, Diaz RE, Su D, Adzic RR, Stach E, Li J, Murray CB (2012) ACS Nano 6(3):2818–2825
111. Miyake H, Okada T, Samjeske G, Osawa M (2008) Phys Chem Chem Phys 10(25):3662–3669
112. McKeown NB, Budd PM (2006) Chem Soc Rev 35(8):675–683
113. Rosi NL, Eckert J, Eddaoudi M, Vodak DT, Kim J, O'Keeffe M, Yaghi OM (2003) Science 300(5622):1127–1129
114. Zhao XB, Xiao B, Fletcher AJ, Thomas KM, Bradshaw D, Rosseinsky MJ (2004) Science 306(5698):1012–1015
115. Nakamori Y, Li HW, Matsuo M, Miwa K, Towata S, Orimo S (2008) J Phys Chem Solids 69(9):2292–2296
116. Bardhan R, Ruminski AM, Brand A, Urban JJ (2011) Energy Environ Sci 4(12):4882–4895
117. Pumera M (2011) Energy Environ Sci 4(3):668–674
118. Stephens FH, Baker RT, Matus MH, Grant DJ, Dixon DA (2007) Angew Chem Int Edit 46(5):746–749
119. Sun YG, Tao ZL, Chen J, Herricks T, Xia YN (2004) J Am Chem Soc 126(19):5940–5941
120. Kobayashi H, Yamauchi M, Kitagawa H, Kubota Y, Kato K, Takata M (2008) J Am Chem Soc 130(6):1818–1819
121. Kobayashi H, Yamauchi M, Kitagawa H, Kubota Y, Kato K, Takata M (2010) J Am Chem Soc 132(16):5576–5577
122. Weiss A, Ramaprabhu S, Rajalakshmi N (1997) Z Phys Chem 199:165–212
123. Uemiya S, Matsuda T, Kikuchi E (1991) J Membr Sci 56(3):315–325
124. Barlag H, Opara L, Zuchner H (2002) J Alloys Compd 330:434–437
125. Lu YZ, Jin RT, Chen W (2011) Nanoscale 3(6):2476–2480

Low-Cost Nanomaterials for High-Performance Polymer Electrolyte Fuel Cells (PEMFCs)

S. M. Senthil Kumar and Vijayamohanan K. Pillai

1 Introduction

Production, storage and deployment of affordable and clean energy is one of the biggest challenges facing humanity. Although a majority of current energy requirements is obtained from fossil fuels, the supply is finite and could last only for a specified period depending on the nature of resources [1–4]. At the current consumption rate, coal will perhaps remain only for approximately 150 years, while oil and natural gas are expected to get exhausted in 50 years [5–8] assuming no new resources. The other forms of energy are nuclear, wind, hydrothermal, biomass, solar and geothermal where each has its own advantages and disadvantages based on the geographical locations of the country and its habitants. Among the different energy sources, solar energy presents the truly widespread and clean case where, despite being in its developmental stage, only 0.02 % of available resource is sufficient to replace fossil fuels and nuclear power together from the energy sector [9, 10]. However, due to the diffuse nature of solar energy, it demands a larger area to attain considerable energy density, and is thus not suitable for use as such in mobile and portable applications. Further, the huge requirement for landscape necessarily puts this technology into rural places from where the electricity has to be transported in grids that results in further energy loss. On the other hand, in today's scenario, the use of fossil fuels for meeting our energy needs cannot be ruled out due to the lack of other energy resources. However, it poses severe environmental concern due to the relentless emission of green house gases (CO_2). This fact is exacerbated by the low conversion efficiency (25–40 %) of current internal combustion engines (ICEs) based on the Carnot cycle. While Carnot cycle efficiency can reach much higher values in an ideal world, only

S. M. Senthil Kumar (✉) · V. K. Pillai (✉)
Central Electrochemical Research Institute, Karaikudi 630006, Tamilnadu, India
e-mail: senthilkumarsm@cecri.res.in; smsk_2k@yahoo.com

V. K. Pillai
e-mail: vijay@cecri.res.in

Z. Lin and J. Wang (eds.), *Low-cost Nanomaterials*, Green Energy and Technology, 359
DOI: 10.1007/978-1-4471-6473-9_13, © Springer-Verlag London 2014

25–40 % is achievable under practical conditions [11]. The significance of this problem can easily be understood from the sharp rise of CO_2 levels in the atmosphere from the 280 ppm of pre-industrial era, to 393 ppm in 2012 suggesting a 37 % increase from 1750s [12]. More importantly, while it took 215 years for initial 50 % of this increase in CO_2 content, it only took 33 years for the next 50 % and this rate of increase is only expected to grow further as observed from the Keeling curves [13, 14]. Thus, a clean energy conversion and/or storage device with almost no environmental impact and higher practical efficiency (ca. 70 % at least) is an urgent need for sustainable development. In this respect, electrochemical devices due to their intrinsic ability to convert chemical energy directly into electrical energy offer a unique opportunity to tackle many of these challenges in terms of innovative materials, processes and devices.

2 What Are Low-Cost Nanomaterials

The objective of this discussion is to identify relevant low-cost nanomaterials with high potential for next generation energy storage and conversion applications. The classification of such nanomaterials should be as simple as possible. For this reason the following material categories were defined based on their abundance, recyclability, cost-effective production methods as well as their intensive applications in various energy-related technologies. The material categories are (a) carbon-based nanomaterials; (b) nanocomposites; (c) metals and alloys; (d) nanopolymers and some of their hybrids. Carbon-based nanomaterials are one of the most widely employed candidates in various energy technology applications and have predicted to play a vital role in hydrogen storage and electrical energy storage.

3 Applications in Electrochemical Power Sources

Electrochemical power sources are devices that convert chemical energy stored in materials (or fuels) directly into electricity. Major electrochemical energy conversion and storage devices that are considered for the future energy needs are batteries, fuel cells and supercapacitors. While all these three technologies have different, material dependant, reactions for storage and conversion, the basic energy providing steps take place at the electrode–electrolyte interface as the case with any electrochemical system. Scheme 1 shows one major classification of electrochemical power sources with the exception of fuel cells which are postponed for a later discussion. Apart from the above categories, there are also overlapping systems such as metal–air and redox flow batteries where a battery electrode is combined with a fuel cell electrode (i.e. half-cell reaction) to realize the benefits of both the systems as illustrated by metal–air rechargeable batteries using Zn (or Fe) as the anode and air as the cathode [15, 16].

Scheme 1 Three types of electrochemical energy conversion and storage devices and their basic classification

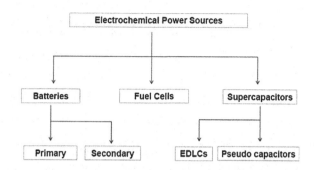

Fuel cells are electrochemical devices that convert the chemical energy stored in a fuel directly to electrical energy by sustaining separate redox reactions at the cathode and anode respectively separated by an electrolyte. The electrode reactions do take place at the three-phase interface (more specifically on the surface of electrocatalyst), leaving ions to pass through the electrolyte and electrons through the external circuit. Unlike batteries, fuel cells can operate continuously as long as the necessary reactant, fuel and oxidant, flows are maintained. By stacking hundreds of such single cells, known as membrane-electrode assemblies (MEAs) in a modular form, fuel cell power plants can be erected to provide electricity for a number of applications such as electric vehicles, large grid connected utility power plants for stationary and portable power applications [17].

Despite the electrochemical similarities between fuel cells and batteries, batteries have limited space for reactant storage and hence need to get recharged quite often for practical applications. For example, in fuel cells the reactants are continuously supplied externally which make them an ideal choice for applications that require the sustained delivery of power unlike that of batteries. Further, due to the external supply of fuel cells, filling up (recharging or more correctly replacing the feed) can be much faster than that of batteries. Also, fuel cells and batteries are not limited by the Carnot efficiency and hence have higher practical efficiency than the ICEs. However, many types of batteries have disposal and recyclability issues and hence special efforts are being made in recent times to make greener batteries. Fuel cells normally do not have any moving parts which reduce the wear and tear of their components. The only by-product when pure hydrogen is used as the fuel is water, thus making them environmental friendly. Due to those above-mentioned advantages, different types of fuel cells using innovative materials are under intense research and development for both stationary and transport applications.

4 Polymer Electrolyte Membrane Fuel Cells (PEMFCs)

PEMFCs utilize a solid polymer electrolyte membrane (hence PEM, also meaning proton exchange membrane) to transport protons from anode to cathode and restrict electrons from directly going to cathode from anode. PEMFCs operate

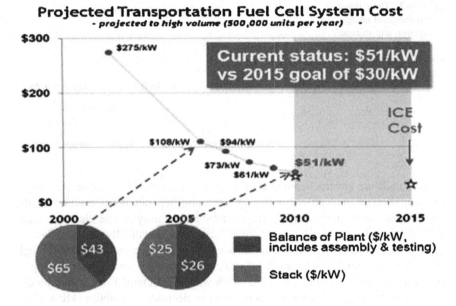

Fig. 1 DOE projected transportation fuel cell system cost (adapted from Ref. [26])

normally in the temperature range of 30–100 °C under humidified environment although PEMs that can operate up to 200 °C are also available. The easy start-up and flexible design has attracted interests in both stationary and portable applications [4, 18, 19]. In fact, it represents one potential case for reducing the green house gas emission from a renewable energy point of view especially if hydrogen is produced by solar technologies without any carbon footprint [20]. PEMFCs also have higher practical efficiencies over commercial engines especially if combined heat-power generation (cogent) is targeted [21].

A state-of-the-art PEMFC accordingly has five critical components: (i) Pt electrocatalyst, (ii) catalyst support carbon, (iii) gas diffusion layer (GDL) or backing layer, (iv) bipolar plates and (v) polymer electrolyte membranes. For the successful operation of PEMFCs, the reactant H_2 and O_2 must reach the catalyst site where the electrochemical reactions take place and the product water has to be expelled from the catalyst site to prevent water clogging and better access of reactant gases to the electrocatalyst. Similarly, the generated protons at the electocatalyst sites must reach the cathode through PEM while the electrons need to reach the cathode through an external circuit, where it does the useful work, to react with oxygen and protons to form water. Hence, the effective formation of triple-phase boundary (TPB) (reactant gases, electrocatalyst and electrolyte membrane) is an important criterion for successful PEMFC operation.

Despite many advantages, PEMFCs still have many challenges owing to the availability of critical materials and processes to effectively address the lack of commercialization aspects [22, 23]. For example, the use of Pt as electrocatalyst, graphite bipolar plate and perfluorosulphonic acid (PFSA) membranes makes the system very expensive, preventing its commercial applications. For example, the state-of-the-art PEMFC stacks use a Pt loading of 0.3 mg cm^{-2} which is too expensive for commercialization while US Department of Energy (DOE) target for 2015 lie at 0.03 mg cm^{-2} [24, 25]. Further, the use of PFSA membranes limits the operating temperature below 100 °C that necessitates the use of ultrapure hydrogen as the fuel which further increases the system cost. The net cost of a transportation fuel cell system (2010 technology) for high volume manufacturing (500,000 units per year) is \$51/kW (Fig. 1). This is a reduction of more than 80 % since 2002 and approaches the target of \$30/kW established for 2015. Research and development efforts appear to be on track to achieve cost-competitiveness with ICEs within the next few years.

Further, the PFSA-based PEMs have many drawbacks such as dependence on humidity for conductivity, high reactant permeability, tendency to disintegrate in the presence of hydroxyl radicals (an intermediate in the cathode reaction) and moderate mechanical, chemical stability. The use of carbon as the catalyst support and Pt as the electrocatalyst induce durability-related issues since Vulcan XC-72 carbon materials are reported to corrode after 150 h of continuous operation [27]. Similarly, other components that are used in fuel cells such as bipolar plates also have many restrictions related to fragility and affordability and therefore to realize a commercially feasible fuel cell many of these barriers should be surmounted using inexpensive options.

5 Nanoscale PEMFC Materials and Their Significant Properties

We will now describe recent developments in both fundamental and technological aspects of PEMFCs with a special emphasis on individual nanoscale materials as well as their assembling process into a single cell and subsequently stack fabrication. This is mainly due to the fact that recent synthesis of a gamut of nanomaterials with their unique properties have made a tremendous impact on many thrust areas like energy, electronics, biology and health care [28, 29]. For instance, few non-noble nanomaterials have been claimed to replace expensive electrocatalysts like Pt and Ru with similar electrocatalytic activity and nanocomposite polymer electrolytes offer several intrinsic advantages like increased thermal and chemical stability, enhanced mechanical strength and improved ion transport [30–32]. The section on individual nanoscale components particularly focuses on novel carbon supports and functionalized surfaces, electrocatalysts of Pt and non-Pt metals, composite polymer electrolyte membranes with tailor-made properties, bio-inspired electrocatalysts and composite bipolar plates. This section also deals with the development of both components and engineering process as well.

5.1 Novel Carbon Supports and Their Functionalized Surfaces

One of the major modes of failures of a PEMFC is the breakdown of membrane and the loss in active surface area of the Pt electrocatalyst with time in addition to the corrosion of the support. The lifetime and cost as a function of performance is intimately linked with design, materials and operation strategies and various targets are often used by fuel cell researchers and funding agencies for comparison and also for measuring progress. The cost targets of the US DOE for PEM fuel cell stack is an unrealistic \$30/kW by 2015 which is a way down from the current value of \$110/kW [33]. Most of this reduction has to be from the Pt catalyst, bipolar plate and PEM although total elimination of Pt has been recently indicated as a tangible possibility [34]. Development of different types of inexpensive carbon plays a critical role in accomplishing some of these objectives as it is well known that carbon with varying properties could be prepared to meet these technology-specific requirements. For example, the surface area of carbon can be varied from few metres to few thousand metres per gram although other useful properties such as pore size distribution, mechanical strength and electrical conductivity vary dramatically some times in an adverse manner. Consequently, carbon has been extensively engineered using diverse methods by a huge number of groups as a support material in PEMFC [35–40].

Earlier attempts to overcome these challenges were restricted mainly by selecting three different forms of carbon, i.e. activated carbon, carbon black and graphite or graphitized materials, as the primary choice of support for catalyst materials in different types of fuel cells [35]. The preferred form of carbon in fuel cell electrode including that in GDL is Vulcan XC-72, a kind of activated carbon, with moderate surface area (250 m^2/g) and good electrical conductivity, which is in stark contrast to the preference of activated carbon with a surface are of more than 3,000 m^2/g for certain other applications, perhaps due to poor electrical conductivity and different pore size distribution. However, the mesopores in Vulcan XC-72 result in part, of the Pt nanoparticles getting buried deeply inside the pores (especially if they are few nm) and hence becoming inaccessible for the TPB formation, which is essential for sustaining the electrode reactions in fuel cells. Further, Vulcan XC-72 undergoes corrosion (more important under peroxide intermediate formation conditions of fuel cell cathodes) resulting in the aggregation as well as dissolution of Pt nanoparticles [36–42].

Attempts to improve the carbon support by different strategies have generated mixed results. Coin like hollow carbon (Fig. 2) prepared by a simple solvothermal method has been used to support Pd electrocatalyst in methanol oxidation with an improved mass activity of 2,930 A g^{-1} against 870 A g^{-1} of Pd supported on Vulcan XC-72 carbon [37]. Similarly, carbon nanofibres and even scrolls are also attempted as a support due to its ease of fabrication [38–41]. Although all these improvements on carbon alone cannot solve most of the above challenges

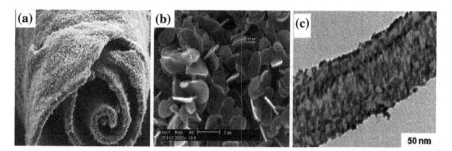

Fig. 2 **a** TEM image of carbon scrolls, **b** SEM image of coin like hollow carbon and **c** TEM images of Pt deposited carbon nanofibers (adapted from Refs. [37, 40] with permission from American Chemical Society)

associated with fuel cells mainly in terms of durability and performance, carbon nanotubes (CNTs) with its unique properties can actually do alleviate some critical problems.

5.2 Carbon Nanotubes as Electrode Material for PEMFCs

Carbon nanotubes comes under the carbonaceous material with distinct characteristics, like inertness under various chemical environments, highest Young's modulus, electrical conductivity, high surface area, lightweight and easy interfacing capability with many inorganic and organic compounds [43–47]. CNTs can be broadly classified into two types, single-walled and multiwalled CNTs with reports available on even double-walled nanotubes [48–52]. CNTs are considered as analogous to fullerenes due to the similarities in the electronic structure. Moreover, many recent reports have clearly illustrated that the unique electronic structure of CNTs helps in enhancing the catalytic activity of the supported metal in addition to providing mechanical integrity [53–56]. For example, nitrogen- or boron-doped CNTs can replace Pt as an electrocatalyst and there is a lot of excitement on developing these types of new nanostructured electrocatalysts. In this section we discuss current efforts on the use of CNTs in polymer electrolyte fuel cells in both electrodes and electrolytes illustrating their multifunctional role as catalyst layer, support and sometimes as a reinforcing component in polymer composite membranes. The impact of functionalized CNTs on the performance and durability of the MEAs in increasing the longevity and improved performance will be discussed in such manner to unravel their potential in reducing the cost of the stack per kW. Besides the discussion on materials and general procedures for functionalizing CNTs for PEMFC, the use of nanocomposite polymer electrolytes using surface-engineered CNTs in particular is also illustrated with their advantages and limitations using both single-walled and multiwalled CNTs.

Effective utilization of Pt nanoparticle is a key parameter in decreasing the cost of the fuel cell stacks as very low Pt loading (few hundred microgram/cm^2)

without any change in performance is essential for rapid progress in this area. This can be accomplished by the effective distribution of Pt on the supporting material with a high surface area as well as higher electrical conductivity [42]. For this purpose, a given support material for PEMFC electrodes should fulfil the following requirements, namely (i) high surface area, (ii) chemical stability under oxidative/reductive conditions, (iii) mechanical robustness under both open and closed circuit conditions and (iv) good electrical and thermal conductivity. Many of these conditions are met exceptionally well with CNTs although CNT is not cheap at present for large-scale applications. However, there are sufficient indications that CNT cost is likely to come down with increased production [57, 58].

One of the main problems associated with even commercial electrode formulations is the isolation of carbon particles by the use of Nafion as a binder in the catalyst layer. The insulating nature of this binder can indeed block Pt particles associated with carbon from accessing the external circuit due to the lack of electrical network resulting in the decrease of further Pt utilization. Pt nanoparticles deposited on CNTs, however, are almost certain to have electrical contact with external circuit which eliminates these types of problems associated with the presence of a thin insulating layer of Nafion covering the carbon particles [59]. Hence, a judicious use of CNTs could in fact overcome many of such issues that other forms of carbon-based electrodes struggle to overcome in power source related applications.

5.3 Carbon Nanotubes as Cathode Support Material

Sluggish kinetics of oxygen reduction reaction (ORR) sustained at the cathode of MEA typically necessitates the use of higher Pt loading in comparison with that in the anode. However, in many cases the use of CNTs has been shown to be profitable in terms of providing a better exchange current density towards ORR without causing any detrimental mechanical behaviour as a support material for the electrocatalyst. For example, Yan et al. have carried out pioneering work on the use of CNT in PEMFC electrodes especially for the cathode to improve Pt utilization [59–62]. Their initial study of depositing 4 nm Pt particles on CNTs has shown improved current and power density in the regions (i.e. activation, ohmic and mass transport domains), presumably due to the intrinsic properties of CNTs to increase the oxygen reduction kinetics. This is in accordance with the findings of Britto et al., where CNT/metal electrodes show higher exchange current density than that on other metal/C electrodes [63]. Enhanced mass transport is also anticipated to be beneficial in the case of CNT-based electrodes coupled with reduced ohmic loss, which is acceptable while comparing the electrical conductivity of CNTs, graphitic powders and other forms of conducting carbon. This has resulted in an enhanced Pt utilization of 58 % against 34 % of carbon-based electrodes under favourable conditions. However, water clogging remains as a critical problem in the cathode which severely restricts the performance of PEMFC under high humidity conditions [59].

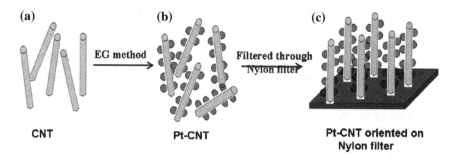

Fig. 3 **a** Scheme for the preparation of oriented Pt/CNT film-based MEAs. **b** TEM image of Pt deposited CNT. **c** SEM image of the oriented CNTs on Nafion membrane (adapted from Ref. [60] with permission from American Chemical Society). [In the above figure EG refers to ethylene glycol which acts as both reducing agent and solvent in the reaction]

Carbon nanotubes are generally hydrophobic in nature which helps in controlling gas diffusion properties. Oriented CNTs are shown to have increased hydrophobicity than that of disordered CNTs. Further, the electronic conductivity is higher along the tubes than across the tube along with increased gas permeability which would help in better mass transport conditions [63–65]. Considering all these benefits associated with oriented CNTs Yan et al. have developed a unique method to orient the CNTs by a filtration method followed by transfer to the membrane (Fig. 3) to prepare the GDL. In this method, the surface of CNTs is endowed with functional groups such as –COOH by refluxing with Con.HNO_3/ Con.H_2SO_4 mixture followed by in situ chemical reduction of Pt precursor solutions on them for the proper anchoring of Pt nanoparticles on the CNT surface. Subsequent to the chemical reactions, the functionalized CNTs are filtered through a hydrophobic nylon membrane with precisely controlled pore size/distribution in CNTs standing up with the preferred orientation and length.

The use of this type of oriented CNTs results in better fuel cell performance than that of randomly aligned CNT-based electrodes and Pt/C-based electrodes. Table 1 illustrates this use of CNTs to enhance the fuel cell performance compared to that of commercial Pt/C catalysts. Durability of a fuel stack is mainly restricted by the corrosion of carbon support under the operating conditions of the cathode especially due to the production of hydrogen peroxide (H_2O_2) as an intermediate. Several studies have established beyond doubt that Pt nanoparticles are expected to double in size with an operation time of around 200 h [62]. In this regard, the use of CNTs which is known for their chemical inertness and remarkable mechanical strength can increase the endurance of a fuel cell MEA. More specifically, an attempt to prove the durability of CNTs carried out by potentiostatic treatment for 168 h for Pt supported on CNTs and Vulcan XC-72 reveals that CNTs have lesser surface oxides than that of Vulcan XC-72, concomitantly demonstrating 30 % lower corrosion rates [62].

A more recent work from Lin et al. has attempted to reduce the cost of the fuel cell by decreasing the particle size as well as increasing the Pt distribution by a wet

Table 1 Variation of proton conductivity, domain size and yield strength with s-MWCNT composition and water content

Membrane composite	Proton conductivity (S cm^{-1})	Hydrophilic domain size (nm)	Water content (%)	Yield strength (MPa)
Nafion 115	0.028	48	29.2	2.68
Recast Nafion	0.020	51	42.3	2.40
Naf-s-MWCNT 0.01 %	0.029	54	39.7	2.50
Naf-s-MWCNT 0.05 %	0.036	72	33.4	2.62
Naf-s-MWCNT 0.1 %	0.032	70	30.4	2.67
Naf-s-MWCNT 0.5 %	0.031	67	28.1	3.10
Naf-s-MWCNT 1 %	0.030	35	27.2	4.30

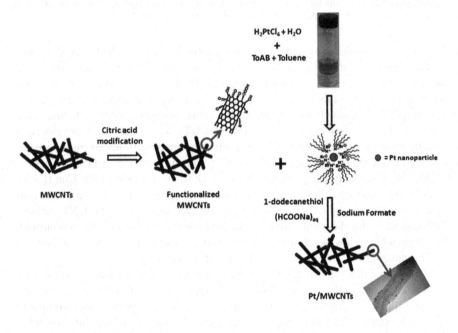

Fig. 4 Scheme for the surface modification of MWCNTs and Pt nanoparticles deposition (adapted from Ref. [66] with permission from Elsevier publishing company)

chemical modification route [66]. By this method Pt nanoparticles of 1–3 nm are stabilized at the same time maintaining a uniform Pt distribution due to the anchoring groups present on the CNT surface (Fig. 4). Interestingly, fuel cell polarization plot with this catalyst shows a power density of 1,110 mW cm^{-2} against 800 mW cm^{-2} (5 cm^2 area) observed for a commercial catalyst. Further, the activation loss observed at 50 mA cm^{-2} is only 50 mV from the OCV for this surface-modified process compared to that of 150 mV for unmodified CNTs. However, with the carbon black it is much higher, which signifies the more efficient use of CNTs in the catalyst layer of PEMFCs.

A large number of reports are available on preparing CNT-based electrodes especially to increase the Pt utilization by means of using different preparation conditions such as chemical reduction in a formaldehyde bath and electrochemical deposition of Pt on CNTs. Growing CNTs directly on the carbon paper support in order to reduce the ohmic resistance, and the modification of the reduction method to prepare smaller nanoparticles with narrow distribution on size have also been reported as an interesting alternative for Vulcan XC-72-based electrodes. Even though such type of CNT-based electrodes demonstrating performance better than that of Vulcan XC-72-based electrodes but their durability and chemical stability aspects need to be rigorously evaluated [66–75].

5.4 Nitrogen-Doped CNTs as ORR Catalysts to Replace Pt

Even though Pt is used currently as the benchmark catalyst for ORR, it has to be ultimately eliminated from the catalyst layer considering the very low abundance of Pt on the earth crust (3.7×10^{-6} %) and its fluctuating cost [76]. Interestingly, N_2-doped CNTs tend to give an option here although at present, it is only a partial solution waiting for confirmation from results of durability studies [77–82]. For example, vertically aligned CNTs containing nitrogen have been reported to show better ORR catalytic activity than that of Pt as proved by the cyclic voltammetry and RRDE experiments. Despite the fact that these results are in alkaline medium, the possibility of N_2-doped CNTs as a better support materials for Pt electrocatalysts on the cathode and anode of PEMFCs is really tempting where some of the degradation issues can be prevented due to robust mechanical properties of CNTs [79]. More promising results on N_2- and B-doped CNTs are expected to revolutionize this area in the near future.

5.5 Use of CNTs as Anode Support

Compared to the vast number of reports available on cathode support materials, only a few reports are available for anode supports mainly due to the highly facile nature of the hydrogen oxidation reaction and less critical materials requirements. However, the actual challenge in the anode is to obtain sustained performance using reformed H_2 having considerable CO content. In this regard, catalyst systems (e.g. Pt/Ru, Rh) that show good activity towards methanol oxidation reaction are expected to have a better tolerance for CO and substantial efforts are rendered towards improving the support metal interaction. Since stronger metal support interactions (SMSI) would help the electron transfer between the metal and support during electrochemical reactions, Pt nanoparticles deposited on CNTs tend to exhibit increased catalytic activity than that of unsupported metal due to the unique electronic structure. Interesting improvements have been observed for methanol oxidation on Pt surfaces with and without the presence of CNTs like significant enhancement in the oxidation current of 50–60 mA/cm^2 for CNT/Pt electrode, while unsupported Pt gives only 6 mA/cm^2. While this enhancement can have contribution from the increased surface area, the kinetic aspects demonstrate an unambiguous improvement in the catalytic activity of Pt that is supported on CNTs. Interestingly, the onset potential of methanol oxidation is also shifted in case of CNT/Pt electrode associated with enhancement of the anodic current. A similar shift observed with Pt/Ru alloy system is attributed to the reduced work function of Ru ($\Phi_{Ru} = 4.52$ eV) in comparison with that of Pt ($\Phi_{Pt} = 5.36$ eV) suggesting the possibility of a similar reasoning for the shift observed in CNT/Pt to the reduced work function of CNT ($\Phi_{CNT} = 5$ eV) [79–82]. In another report, Wu et al. have shown a remarkable enhancement in CO tolerance of Pt when supported on SWCNTs and MWCNTs over E-Tek Pt/C catalyst, a commercial sample often used by fuel cell companies for benchmarking. The peak potential for CO striping are observed at 0.75, 0.78 and 0.82 V respectively for Pt/SWCNT, Pt/MWCNT and E-Tek Pt/C catalysts suggesting a more easier removal of the adsorbed CO at a much lower onset potential [75]. This could help in achieving better performance even with increased CO level in the hydrogen stream especially using thermally stable polymer electrolyte.

6 Application of CNTs in Composite Electrolytes of PEMFCs

Solid polymer electrolyte membrane is one of the key materials that restrict the performance as well as the cost of the PEMFCs as electrolyte is a critical component of MEAs. A good PEM should have high protonic conductivity, yet electrically insulating in order to avoid short circuiting, and should have very low permeability towards fuels such as hydrogen, methanol and ethanol, in addition to having very high chemical stability to withstand high acidic conditions of the operating environment and sufficiently mechanical stability to withstand the stresses of stack fabrication. PEMs that are either used or being developed could be classified into two categories; PEM operating at temperatures less than 100 °C often with PFSA electrolyte and that operating above 100 °C with a variety of new thermally stable polymeric electrolytes. Consequently, these are far from commercialization although many prototype stacks are undergoing field trials in various parts of the world. Considering Nafion as a typical ionomer, we now discuss some promising aspects of CNT-based polymeric composite electrolytes for PEMFC applications especially for temperature less than 100 °C [82].

Nafion-based membranes are well known to show proton conductivities in the range of 0.1 S cm^{-1}. However, their conductivity relies mainly on the water content which restricts their operating temperatures to less than 100 °C. Further, the swelling and contraction of these membranes (dimensional change) with change in water content is of severe concern as it affects the integrity and durability of MEAs. Typical thickness values of membranes range 50–120 microns as indicated in the names (Nafion 112 and 115 etc.). A possible method to reduce the cost of PEM is reducing the thickness, which might also help in decreasing the membrane resistance thereby improving the performance of PEMFC. However, reduction in thickness could lead to increased hydrogen and methanol permeability coupled with reduced mechanical stability.

Efforts directed on improving the proton conductivity of these membranes include various approaches to mainly increase the water content by incorporating hygroscopic inorganic and organic additives such as SiO_2, ZrO_2, TiO_2, zirconium phosphate and zeolites which can keep the membrane humid at high temperatures during the operation [83–87]. However, most of these composite membranes show higher conductivity than Nafion only at higher temperatures, their base value being less than that of Nafion. More significantly, long-term operation of these membranes in MEAs faces severe limitations due to the agglomeration of these particles during operation resulting in extensive degradation in performance with time. Also as these membranes do not have any cross-linking, these dispersed particles often reduce the mechanical stability of the composite membranes [85]. Solution cast membranes made up of commercial Nafion solutions (also reinforced by fillers) and porous PTFE matrix help to some extent, in increasing the mechanical strength and structural integrity, finally enabling the use of reduced thickness of the membrane without any change in performance. However, the proton conductivity of the resulting membranes is very poor to limit their power density [87].

Even though CNTs have been used earlier to fabricate polymer composites with increased mechanical stability, applications as an additive for composite membrane electrolyte have not been tried mainly due to the fear of electrical short circuiting [88–90]. The addition of CNTs can be expected to give increased mechanical robustness and integrity as it is well known for its highest Young's modulus. Liu et al. studied the impact of CNTs on the electrolytic behaviour of Nafion membranes upon reinforcement to observe several improved features [88–90]. Fabrication of composite membranes of CNTs and Nafion at 1:99 wt% after ball-milling and solution casting show similar performance to pure Nafion membrane in terms of proton conductivity, but with significantly less dimensional change for the case of CNT-reinforced composite membranes. Similarly, Thomassin et al. have used melt extrusion to incorporate the CNTs on to Nafion membrane to observe reduced methanol permeability to about 60 % along with an unusual increase in the Young's modulus up to 140–160 % in comparison with that of commercial Nafion membranes [90].

In all the above applications, one has to naturally consider the effect of electronic conductivity of CNTs. Certain type of functionalization indeed enhances the electronic conductivity and it is always important to consider the risk of electrical short-circuiting despite the use of very low amounts of CNTs (<0.1 %). Proper dispersion is essential for ensuring uniform behaviour and the normally reported value of percolation threshold for CNTs in Nafion is around 5–11 % depending upon the dimensions of CNTs used and their mode of preparation. Lie et al. in yet another report have prepared a three-phase composite membrane by sandwiching a Pt-CNT-Nafion membrane between two pure Nafion membrane phases [88]. This has improved the performance of Nafion membranes significantly with respect to both mechanical stability and water management perspectives. Thus undoubtedly, the addition of CNTs results in both improved mechanical stability and chemical inertness while the Pt particles supported on it helps to produce water by reacting with H_2 and O_2 that penetrates the membrane and thereby keeping the membrane wet under dry conditions. However, critical lifetime data about Pt dissolution and carbon corrosion are necessary to comment on the benefits of these composite materials with respect to robustness and durability.

7 Role of Functionalization of Carbon Nanotubes

All the above reports are primarily intended to use CNTs as mechanical stability boosters or reinforcing phases albeit in small amounts, for Nafion in order to withstand the high processing conditions of stack fabrication and to prolong membrane life. As a result, the CNT-Nafion composite membranes have shown conductivity values similar or less than that of pure Nafion membranes, especially if pristine single or multiwalled CNTs are used for composite fabrication. However, CNTs can be functionalized on the sidewalls with desired groups through careful chemical process, which can be used in a constructive manner to enhance the proton conductivity of Nafion membranes. The main advantages of using functionalized CNTs for composite polymer electrolytes are (i) reduction in electronic conductivity: since all types of CNTs are known to have electronic conductivity and functionlization can effectively reduce the electronic conductivity. (ii) Tuning the interfacial structure: it is always desirable to attach molecules structurally analogous to that of the polymer backbone to ensure uniform properties in the composite. (iii) Better dispersion and adhesion: agglomeration and phase segregation of CNTs at isolated regions could be avoided using appropriate functionalization. (iv) Easy processability: since CNTs can be made soluble either in aqueous or organic solvents using different types of chemical functionalization approaches, these composites can be easily processed in the form of films of uniform thickness and properties. However, the choice of functionalization is critical since some functional methods could on the other hand enhance carrier density to finally yield better electronic conductivity while some others can damage the morphology [91, 92].

The choice of functionalization of CNTs for a composite electrolyte with Nafion is obvious. Since sulphonic acid groups on the side chains are responsible for the conductivity of PFSA membranes like Nafion via domain formation, CNTs are to be functionalized with sulphonic acid groups as any increase in their concentration is expected to improve the proton transport. We have utilized this concept and functionalized both single-walled and multiwalled CNTs with

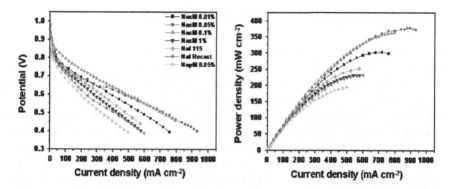

Fig. 5 Polarization plots of showing the potential and power density variations with increasing current densities

sulphonic acid moieties through microwave treatment in HNO_3–H_2SO_4 mixture [91, 92]. Interestingly, this sulphonic acid functionalized CNTs (s-CNTs) are more soluble in water and dimethylacetamide, offering additional advantages of flexibility in terms of membrane processability.

We have prepared several composite membranes based on Nafion-s-CNTs, with a systematic variation in CNT weight percentage from 0.01 to 1 %, beyond which there is saturation in proton conductivity. The results show that for both types of s-CNT-Nafion composite membranes, there is an increase in proton conductivity coupled with increased mechanical stability of the membrane. The cause for this enhancement is revealed by the small angle X-ray scattering experiments (SAXS) where the hydrophilic ionic domain size change shows a strong correlation with the increase of CNT content and proton conductivity. The domain size measured from SAXS measurements reveal approximately 50 Å clusters from commercial Nafion 115 membranes while composites exhibit a saturation limit of about 70 Å both for 0.05 and 0.1 % of s-CNT content (Table 1) in the composite.

These results are well supported by the fuel cell polarization experiments where 0.05 and 0.1 % composites have shown increased current and power densities than that of commercial and recast Nafion membranes of similar nature (Fig. 5). This result shed some light on the possibility of s-CNTs to play a vital role in enhancing the conductivity of composite electrolyte (reduced activation energy for proton transport) and also on the improved robustness of the membrane, presumably due to the amount of functionalized CNTs. Our further attempts to vary the –SO_3H content by changing the microwave treatment time and related parameters are compared in Table 1 with available data from other reports. For example, Peng et al. have prepared sulphonated CNTs by heat treatment with H_2SO_4 at 250 °C under N_2 atmosphere to attain a sulphonic acid content of 15 wt% which is much higher than the 2.5 wt% sulphonated CNTs that we used form composite membrane fabrication. However, there is no fuel cell polarization data and the presence of too much of sulphonic acid groups might create adverse effects in terms of corrosion and membrane degradation [44, 45].

8 Limitations of Carbon Nanotubes in PEMFC

Even though the use of CNTs have shown promising results both as an electrode support and as novel composite polymer electrolytes, it still has few limitations to overcome before being considered for widespread use. For example, the preparation of SWCNTs yields a mixture of metallic and semiconducting SWCNTs which may have significant variation in properties, Further, both SWCNTs and MWCNTs contain minimal amount of impurities which may have deteriorating effect on PEMFC performance for extended usage. It has been demonstrated that even trace level of Fe impurities can degrade Nafion due to the presence of sulphonic acid moieties in the backbone [49]. The use of highly purified CNTs is more expensive which will ultimately increase the cost of MEA and PEMFC stacks which is not desirable since cost is the single overriding concern preventing widespread application of PEMFCs.

Another important limitation of using CNTs is lack of enough data related to durability. Although the use of CNTs as electrode support has shown better corrosion resistance over Vulcan XC-72 in short time scale, its sustained utility for extended period of operation (such as 5,000 h or more) is yet to be evaluated. Moreover, CNTs are normally surface oxidized for increased binding with Pt nanoparticles. The stability of these surface groups under severe potential cycling has not been investigated. The effect of potential induced morphology changes is another major concern which may have severe impact on the use of CNT as a supporting material. The details regarding such attempt will be discussed in the following section with suitable examples. Similarly different processing conditions of catalyst preparation and MEA fabrication are expected to create defects on the CNT surface. For example, the use of ultrasonic homogenizer for prolonged time is known to chop CNTs into smaller dimensions to help dispersion although for certain applications this is harmful [51, 52]. Further, the non-toxicity of CNTs is yet to be completely established which necessitates cautions handling of them in bulk [53]. Despite the favourable improvements in performance by the use of CNTs both as electrode support and as a composite electrolyte, these limitations require further extensive research to fulfil the enormous application potential of CNTs and related carbonaceous materials in PEMFC.

9 Use of Graphene as Electrode Materials for PEMFC

Graphene is a one-atom-thick planar sheet of sp_2-bonded carbon atoms, densely packed in a honeycomb lattice, which has attracted tremendous attention for both fundamental researches and also possible applications in fuel cells, nanoelectronics, supercapacitors, solar cells and hydrogen storage [93, 94]. Graphene exhibits many exciting properties, like quantum confinement resulting in finite band gap and Coulomb blockade effects [95–98] which could be useful for making many novel electronic devices.

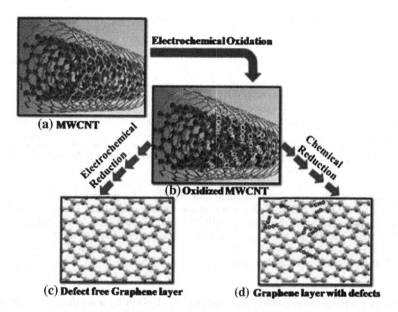

Scheme 2 Diagrammatic representation of the electrochemical transformation of GNRs from MWCNTs: **a** Pristine MWCNT; **b** MWCNT deposited on glassy carbon electrode after oxidation to generate functional groups on edges under controlled potential so that it gets broken; **c** electrochemical and **d** chemical reduction to graphene layers (adapted from Ref. [99] with permission from American Chemical Society)

However, in order to completely realize these properties and applications, a consistent, reliable and inexpensive method for preparing high-quality graphene layers is crucial, as the existence of residual defects will heavily impact their electronic properties, despite their expected insensitivity to impurity scattering. Unfortunately, many of the existing methods of graphene preparation have several major limitations. We recently reported a remarkable transformation of CNTs to nanoribbons (Scheme 2) composed of a few layers of graphene by a two-step electrochemical approach using the oxidation of CNTs at controlled potential (Fig. 6), followed by reduction to form graphene nanoribbons (GNRs) having smooth edges and fewer defects, as evidenced by multiple characterization techniques, including Raman spectroscopy, atomic force microscopy and transmission electron microscopy [99]. This type of 'electrochemical unzipping' of CNTs (single-walled, multiwalled) provides unique advantages with respect to the orientation of CNTs, facilitating possible the production of GNRs with controlled widths and fewer defects for energy storage applications.

In principle, it should be possible to make use of the high degree of graphitization, electrical conductivity and corrosion resistance of CNTs to impart high stability to ORR electrocatalysts. However, the ORR activities of carbon-nanotube-based catalysts have been found to be low in acids. Nitrogen-doped multiwalled CNTs or aligned carbon nanotube arrays have been made by feeding in nitrogen

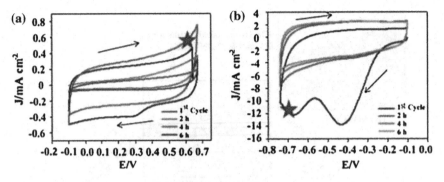

Fig. 6 a Cyclic voltammograms (oxidation) of MWCNTs in the potential window from 0.1 to 0.7 V versus MMS in 0.5 M H_2SO_4 using glassy carbon electrode at 100 mV/s scan rate. **b** Cyclic voltammograms (reduction) of MWCNTs in the potential window from -0.1 to -0.75 V versus MMS in 0.5 M H_2SO_4 at 100 mV/s scan rate. Regions marked with a *star* indicate the potentials at which the CNTs have been selectively oxidized or reduced (adapted from Ref. [99] with permission from American Chemical Society)

precursors during the growth of nanotubes [100–104] or by annealing pre-oxidized nanotubes in NH_3 at elevated temperatures [105]. The resulting catalysts exhibit superior ORR activity in alkaline electrolytes [103, 105, 106] but very low activity in acid electrolytes. Till date, majority of the nanotube-based ORR catalysts have exhibited inferior activities compared with those formed with carbon black and platinum/carbon in acidic solutions, due to availability of the relatively few catalytic sites formed on the CNTs. One of the possible approaches to enhancing ORR activity is to enrich defects and functional groups onto the CNTs by increasing the number of catalytic sites. However, severe oxidation conditions could lead to the loss of the structural integrity as well as their electrical conductivity which are desirable for faster charge transport during electrocatalysis. Taken into account of the above facts it is essential to identify a suitable protocol to afford abundant catalytic sites on CNTs while retaining the structure and electrical conductivity for producing advanced ORR electrocatalysts.

In connection with the above, Li et al. [107] developed a new type of ORR electrocatalyst based on few-walled (two to three walls) carbon nanotube–graphene (NT–G) complexes. They identified a unique oxidation condition to produce abundant defects on the outer walls of the CNTs through partial unzipping of the outer walls and the formation of large amounts of nanoscale graphene sheets, attached to the intact inner walls of the nanotubes. The edge- and defect-rich graphene sheets facilitate the formation of catalytic sites for ORR on annealing in NH_3. Iron impurities and nitrogen doping are found to be responsible for the high ORR activity of the resulting NT–G complex catalyst. Indeed, in acidic solutions the catalyst exhibits high ORR activity and superior stability, and in alkaline solutions its ORR activity closely approaches that of platinum. They have also employed annular dark-field (ADF) imaging and electron energy loss (EELS) spectrum imaging in aberration-corrected scanning transmission electron microscopy

(STEM) to investigate the chemical nature of the ORR catalytic sites on the atomic scale. Iron atoms are often found along the edges of the defective graphene sheets attached to the intact inner walls of few-walled nanotubes, and they often appear next to nitrogen atoms. This provides the first indication of the atomically resolved structure of the ORR catalyst.

Functionalized graphene sheets have been used as the cathode support for Pt electrocatalysts to show higher electrochemical surface area and oxygen reduction activity with improved stability as compared with that of the commercial catalysts. Similarly Pt-decorated graphene has been shown to have better methanol oxidation activity than its commercial counterpart [44–50]. It is worth to mention that in a recent review by Wu et al., the extensive use of graphene–transition metal oxide composites occupies an important role area and it may not be too far that the future devices could be made from any one of these composites [108].

10 Recent Developments in Pt-Based Electrocatalysts for PEMFC Applications

Platinum (Pt), especially in the form of small particles (<5 nm) on a support, plays an outstanding role as a multifunctional catalyst for many industrial reactions. Similarly, it is the only efficient fuel cell electrocatalyst (which is very hard to replace) for many reactions, irrespective of the nature of the reaction, i.e. oxidation or reduction. However, the high cost of platinum (currently about $1,500 per ounce) [109] remains a challenge that demands its full or partial replacement without affecting the performance for commercial applications [110]. The design of inexpensive and robust electrocatalysts for fuel cells for this replacement requires a thorough understanding of the behaviour of Pt in anodic and cathodic environments of PEMFCs. In this context, manipulation of the size and shape of platinum at the nanoscale can thus contribute to the lowering of Pt usage enabling the much-needed cost reduction. Interestingly, it has been established that the catalytic reactivity of platinum nanostructures depends highly on their morphology, and therefore, the design and synthesis of well-controlled shapes and sizes of platinum nanostructures is crucial for their applications, especially in the field of catalysis and electrocatalysis [111]. In our recent review [112], we have comprehended up to date endeavours in the area of shape selective synthesis of Pt nanostructures through several routes, concomitantly discussing some of the core issues related to stabilizing cubes, hexagons, multipods, discs, rods, etc. Importantly, recent accomplishments in the area of shape-dependent electrocatalysis of these nanostructures have been summarized, by giving special emphasis to electrocatalytic reactions relevant for microfuel cells.

Recently, our group has demonstrated an entirely different route for the synthesis of Pt multipods at room temperature by adopting a template-assisted electrodeposition using a porous alumina membrane (PAM) [113]. For example, Fig. 7b shows a distribution of multipods, where the number of arms of each

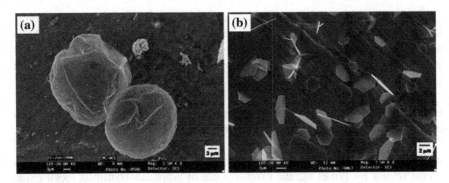

Fig. 7 SEM images of platinum **a** discs and **b** hexagons prepared through a template-assisted electrodeposition route (adapted from Ref. [113] with permission from American Chemical Society)

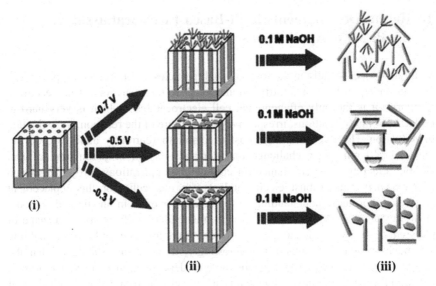

Scheme 3 Electric field-dependent morphological evolution of platinum structures using PAM. (*i*) PAM with one side evaporated with Au film for electrical contact. (*ii*) Formation of hexagonal, disc and multipod structures of Pt over alumina membrane at potentials −0.3, −0.5 and −0.7 V respectively. (*iii*) Dissolution of the membrane in 0.1 M NaOH and release of different morphology platinum structures along with the nanorods (adapted from Ref. [113] with permission from American Chemical Society)

multipod is more than 6, having a common origin with significantly longer length (ca. 500 nm). This way of preparing Pt multipods using an electric field-assisted evolution of morphology is depicted in Scheme 3. In addition to these multipods, nanorods of platinum are also formed inside the porous structures, as usually expected from template-assisted synthesis, which could be separated easily. More significantly, here the shape control was accomplished solely by the modulation of

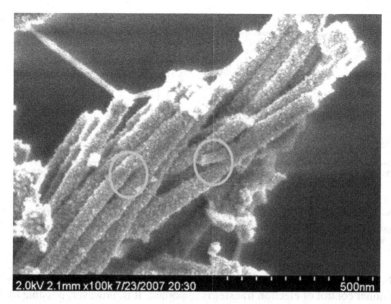

Fig. 8 FESEM image of the Pt Y-junction prepared using hierarchically designed alumina templates through electrodeposition (adapted from Ref. [114] with permission from American Chemical Society)

the electric field which provides important advantages in terms of purity of the systems (less surface contamination). This field induced growth facilitates flexibility to achieve several anisotropic structures of platinum which are thermodynamically not favourable. It is instructive to compare this approach with the previously demonstrated route for the formation of multipods, where shape control was achieved using surfactant/capping molecules. Even though most of the arms have uniform width (ca. 100 nm), some of the branches are formed by the assembly of nm-sized rods in contrast to the previous case.

In addition to the above-mentioned shapes, we have also demonstrated the synthesis of discs and hexagons through a template-assisted electrodeposition using PAM by merely tuning the deposition conditions [113]. Accordingly, the SEM images in Fig. 7a, b reveal the formation of discs with a diameter of ca. 2 mm, whereas hexagons are formed with an edge length of 0.5–1 mm. The morphological evolution of these structures as a function of the applied field is depicted in Scheme 3 and, as discussed in the case of multipod formation, these structures are also obtained without the assistance of any capping molecule/surfactants.

More intriguing shapes like the platinum Y-junction nanostructure has also been synthesized using hierarchically designed alumina templates through an electrodeposition route [114]. The FESEM image of such a Pt Y-junction after dissolving the membrane is shown in Fig. 8 to clearly reveal uniform Y-junctions with well-defined branches and stems having diameters ca. 100 and 50 nm, respectively. More interestingly, the angle between the branches is ca. 12° which

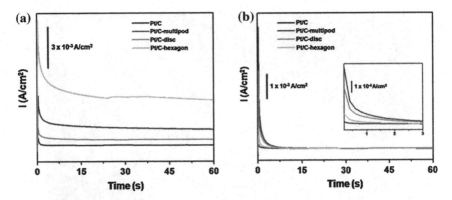

Fig. 9 Comparison of the transient current density curves of Pt/C-multipod, Pt/C-disc, Pt/C-hexagon and Pt/C towards **a** formic acid oxidation and **b** ethanol oxidation (adapted from Ref. [113] with permission from American Chemical Society)

might be helpful for some special applications like molecular interconnects, where directional control of electron transfer is important. It is, however, possible to tune these parameters by controlling the template design strategies. Similarly, the fabrication of other potential junction structures such as X and T is possible by a slight modification of the above template-design strategy.

In order to correlate the structure with activity, electrochemical oxidation reactions of formic acid and ethanol have been studied at different shaped Pt structures such as Pt multipods, Pt discs and Pt hexagons synthesized through the template-assisted route (vide supra) [113]. Accordingly, Fig. 9a reveals the comparison of the transient current density responses of these structures and commercial platinized carbon (Pt/C) towards formic acid oxidation at a particular potential, where the oxidation current density of Pt hexagons is significantly higher than those of Pt multipods, Pt discs and Pt/C. In contrast, for ethanol oxidation (Fig. 9b) the observed order of oxidation current density is as follows: Pt/C-multipods > Pt/C-discs > Pt/C-hexagons > Pt/C. Interestingly, the R value for these structures with respect to that of Pt/C is calculated from the steady-state current density at different potentials. The value of R for formic acid oxidation ranges up to 2,000 % for hexagons, whereas for multipods and disc, it is about 700 and 300 % respectively. Similarly, for ethanol oxidation, the calculated value of R varies up to 600 % for multipods, while for discs and hexagons this is 500 and 200 % respectively.

Hence, it is clear that for formic acid oxidation, Pt hexagons show better catalytic activity compared to other shapes. The origin of this shape-dependent electrocatalytic activity arises mainly due to the higher density ratio of (111)/(100) crystallographic planes present in Pt hexagons compared to that in other structures and commercial platinized carbon (calculated from XRD results). This could be correlated to the structural effect of Pt single crystal electrodes on formic acid oxidation, which reveal that formic acid oxidation to CO_2 proceeds favourably on

(111) planes with significantly less CO poisoning compared to that of other planes such as (100) and (110) respectively [115]. The geometrical arrangement of four sites on a square unit lattice of the (100) plane and on a rectangular unit lattice of the (110) plane are favourable for CO intermediate formation, whereas that on a hexagonal unit lattice of the (111) plane is not so favourable [116]. Thus, Pt/C-hexagons show higher activity towards formic acid oxidation compared to Pt multipods, Pt discs and commercial platinized carbon. In contrast, for ethanol oxidation, it is clear that Pt multipods show better electrocatalytic activity compared to that on Pt discs, Pt hexagons and commercial platinized carbon. The present observation could also be explained in terms of the proportion of crystallographic planes exposed on the surface. It is found that compared to the (111) and (110) planes (100) shows higher activity, while (111) shows least activity towards the C–C bond cleavage involved during ethanol oxidation [117] In the present case, comparison of the density ratio of (100)/(110) planes (calculated from XRD results) reveals a higher value for multipods than for discs, hexagons and commercial Pt/C. As a result, Pt multipods show enhanced activity for ethanol oxidation compared to other shapes.

In addition to these various anisotropic shapes of platinum, we have also compared the electrocatalytic capability of high aspect ratio nanostructures such as Pt Y-junction and Pt Nanowires [114]. Accordingly, Fig. 10a, c shows a comparison of transient current density response of Pt–Y/C, Pt–NW/C and Pt/C towards formic acid and ethanol oxidation reactions at a particular potential, where for both cases the oxidation current density on Pt–Y/C is significantly higher compared to those on both Pt–NW/C and Pt/C. Furthermore, these Pt–Y junction nanostructures show a significantly higher R, which varies up to a maximum of 270 % with respect to Pt/C and up to 200 % with respect to Pt–NW/C. Similarly for ethanol oxidation, the factor R increases up to a maximum of 180 % for Pt–Y with respect to Pt/C and 130 % for Pt–Y/C with respect to Pt–NW/C. This clearly suggests the importance of junction structures in controlling the kinetics of these oxidation reactions (shape-dependent reactivity). In addition to kinetic feasibility, these reactions are also thermodynamically more feasible on Y-junction nanostructures compared to that on nanowires and commercial Pt/C, as shown by the dotted lines in Fig. 10b, d. It is clear that at a given current density, the corresponding potential on Pt–Y/C is shifted negatively by ca. 90 mV with respect to that of Pt/C, whereas the shift is ca. 40 mV as compared with Pt–NW/C. Similarly, for ethanol oxidation, Pt–Y/C is shifted negatively by 70 mV as compared to that of Pt/C, while there is a 20 mV shift with respect to Pt–NW/C, thus indicating the order of thermodynamic stability.

Hence, Pt–Y nanostructures exhibit much enhanced catalytic activity per unit surface area for the oxidation of formic acid and ethanol. This could perhaps be due to the higher density of active sites on the surface of Y-junction Pt (large surface area is expected for these high aspect ratio nanostructures), and in addition it is presumed that the branched regions also enhance the activity due to a large field gradient. This is obvious on comparison of the performance of both Y-junctions and linear structures (nanowires) of Pt as electrocatalysts for the same reaction.

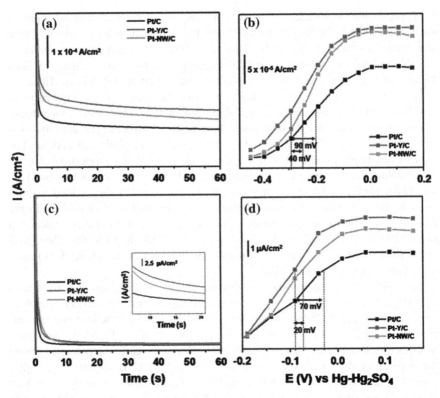

Fig. 10 Comparison of the electrocatalytic activity of Pt–Y/C, Pt–NW/C and Pt/C. **a** and **c** Transient current density *curves* of formic acid and ethanol oxidation. **b** and **d** Potential-dependent steady-state current density of formic acid and ethanol oxidation (adapted from Ref. [114] with permission from American Chemical Society)

11 Bio-inspired Catalysts Development for PEMFCs

Increased attention is being given nowadays to bio-inspired strategies, exploiting the similarities in the functional aspects of biological systems with some of the energy storage systems including fuel cell components [118]. For instance, a number of metalloenzymes catalysing the ORR [119] have been successfully investigated as cathode catalysts in H_2/O_2 PEFCs. However, their performance is suitable only for small power requirements like that of pacemakers and biosensors. As an alternative approach, the addition of metalloenzymes or their active sites to the state-of-the-art fuel cell cathodes is emerging rapidly. For example, alternate assemblies of metalloporphyrins with platinum nanoparticles have been identified as tunable electrocatalysts for the ORR [120]. On the other hand, cobalt phthalocyanine is found to increase the solubility of O_2 in the ionomer layer encapsulating the Pt particles in the cathode [121]. Similarly, the effect of adsorbing uracil, a nitrogenous base found in ribonucleic acids, on the ORR activity of Pt electrodes

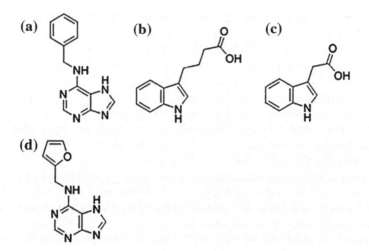

Fig. 11 Molecular structures of plant hormones used in the present study. **a** 6-benzylamino purine (BAP); **b** indole-3-butyric acid (IBA); **c** indole-3-acetic acid (IAA); **d** kinetin among which (**a**) and (**d**) are cytokinins whereas (**b**) and (**c**) are auxins

has been investigated [122]. However, almost all of the bio-inspired strategies discussed above, despite their fundamental interest, result in performance lower than that of the conventional PEFCs. Nevertheless, the performance of bio-inspired electrodes could be improved by choosing a different class of molecules other than porphyrins, nucleobases and metalloenzymes, which take part in physiological pathways more relevant to H_2/O_2 fuel cells.

In this context, plant hormones constitute one of the least explored classes of biomolecules for fuel cell applications probably due to the diversity in their molecular structures and comparatively unknown biochemical pathways compared to those in animal systems. Plant hormones are broadly classified as auxins, cytokinins and gibberellins based on their physiological functions. Some of these plant hormones could enhance proton transport in polymer membranes (Nafion for fuel cell applications), due to their involvement in proton transport-related processes in biological systems (Fig. 11). Accordingly, we have chosen two cytokinins, viz. kinetin and 6-benzylaminopurine, and two auxins, viz. indole-3-acetic acid (IAA) and indole-3-butyric acid (IBA), to test the above-said hypothesis [123].

A single cell with a geometric area of 5 cm^2 with Nafion 115 as the polymer electrolyte membrane was tested for performance. The GDL was prepared by brushing aslurry of Vulcan XC-72, PTFE, water and cyclohexane on a carbon cloth until a carbon loading of 4 mg cm^{-2} was achieved. The GDL was heat treated at 350 °C for 30 min. Then the catalyst ink was prepared by mixing 20 wt% Pt/C, Nafion, water and isopropyl alcohol in a homogenizer for 2 min at intervals of 20 s. The catalyst ink was then applied on the GDL by brushing so that Pt and Nafion loading were 0.5 and 0.6 mg cm^{-2} respectively for both cathode as well as anode. After applying a thin layer of Nafion over the catalyst layer for achieving optimum Pt catalyst utilization and mass transport of reactants to the catalyst layer [124], the MEA has been obtained by pressing the two electrodes uniaxially with the Nafion 115 membrane in between at 110 °C at a pressure of 1 ton for 4 min. Hormone-modified MEAs were fabricated by introducing different amounts of IAA in the catalyst layer by dissolving them in the catalyst ink (both in the cathode and in the anode). Proportions (by weight) of IAA with respect to the weight of Pt in the catalyst layer include $W_{Pt}:W_{IAA}$: 1:0, 3:1, 2:1, 1:1, 2:3 and 1:2. The MEAs were used to form single fuel cells by passing humidified H_2 (80 % RH) on one electrode and humidified O_2 on the other at 0.2 slpm through serpentine flow fields. The fuel cells were conditioned at 0.2 V for 30 min and polarization measurements were carried out at 60 °C.

Figure 12a shows the superimposed steady-state polarization plots of H_2/O_2 fuel cells (at 60 °C) containing Nafion 115 membrane as the polymer electrolyte with different hormones in the catalyst layer. While the maximum performance reported for the system built with state-of-the-art technology is 600 mW cm^{-2}, the performance obtained with the fuel cell design and testing procedures employed in the present study is 250 mW cm^{-2}, which is used as the benchmark (denoted as the 'reference system'). An interesting enhancement in the fuel cell performance (by 100 mW cm^{-2}) is observed for the MEA containing IAA in the catalyst layer compared to the reference system. On the other hand, the rest of the hormones exhibit only poor performance compared to the reference system with the cytokinins (BAP and kinetin) showing inferior performance compared to that with the auxin, IBA. While the pH of the composite dispersions and the electrochemical stability window of the corresponding membranes are almost similar, only IAA is capable of enhancing the performance of the PEFC. Although specific reasons for this discrimination could not be clearly identified, a probable reason could be the difference in chemical stability of the hormones in the presence of the reactive intermediates produced during the fuel cell reactions. Also, in this context, special mention is to be made of the auxin, IAA, a powerful plant hormone capable of stimulating a number of functions at in vivo concentrations as low as 10^{-8} M [125]. There is an astonishing correspondence of its physiological activity with the critical functional aspects of the catalyst layer [126]. More specifically, IAA is known to trigger proton pumps across plasma membranes resulting in the acidification of protoplasts to effect cell elongation [127], reduce molecular oxygen to superoxide radical ion and disintegrate H_2O_2 to a hydroxyl radical to create oxidative stress [128] and impart better permeability of ions through the cell membranes [129]. It should be noted that the disintegration of the H_2O_2 intermediate is the rate determining factor in the two-step O_2 reduction mechanism at fuel cell cathodes. Nevertheless, the indole derivative, IAA, is quite different in its reactivity compared to N-heterocycles such as imidazole, pyrazole and benzimidazole, deployed frequently as electrolytes in PEMs due to the presence of both proton donor and proton acceptor nitrogen atoms [129].

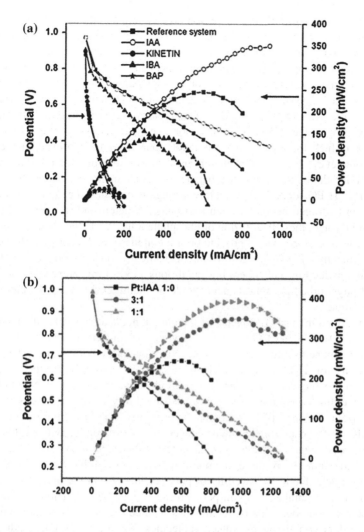

Fig. 12 Single cell polarization plots of MEAs measured at 60 °C using Nafion 115 as the polymer electrolyte membrane. **a** With different hormones and Nafion (reference system) binder in the catalyst layer. **b** Single cell polarization plots of MEAs with varying amounts of IAA in the catalyst (adapted from Ref. [123] with permission from The Royal Society of Chemistry)

12 Recent Advancements in High Temperature Polymer Electrolyte Membranes

Polybenzimidazole (PBI) was originally developed for flame retardation due to its very high thermal stability and oxidation resistance. Researchers at Case Western reported for the first time, the development of sulphonic acid- and phosphoric acid-doped PBI as a proton-conducting membrane. Moreover, they found that phosphoric acid in PBI plays a dual function, where it acts as a proton conductor and more significantly as a proton-conducting medium [130–134]. Phosphoric acid-doped PBI membranes represent one of the emerging proton-conducting membranes for the electrolyte applications in fuel cells that would operate in the range of 120–200 °C. Working at higher temperatures is extremely critical to prevent CO poisoning of Pt electrocatalyst (either present inadvertently in the hydrogen feed from a reformer or generated as an intermediate during oxidation of fuels like methanol and ethanol) apart from the improved kinetics and higher efficiency. Further, PBI membranes do not rely on water for its proton conductivity thus simplifying the balance of plant components greatly. Proton transport in H_3PO_4-doped PBI is mainly through two modes; the rapid exchange of proton via hydrogen bonds between phosphate, N-heterocycles of PBI (Grotthus type) and through the self-diffusion of phosphate ions. While the chemisorbed 2 % phosphoric acid alone cannot provide proton conductivity, higher phosphoric acid content due to physorption results in mechanically poor membranes and lack durability under fuel cell operating conditions [135, 136]. Hence, an optimum level of physisorbed phosphoric acid should be maintained to keep the membrane integrity intact.

Even though the H_3PO_4-doped PBI membrane holds advantages, like increased reaction kinetics, higher CO tolerance, easy water management and lesser balance of plant components but still there are several challenges. For example, the proton conductivity of H_3PO_4-doped PBI membranes is normally in the range of 0.01 S cm^{-1} which is one order of magnitude lesser than that of Nafion-based PEMs at room temperature. Moreover, the durability is also a serious issue in phosphoric acid-doped PBI membranes mainly due to the leaching out of physically adsorbed phosphoric acid that reduces the proton conductivity of the membrane thereby reducing the performance of the fuel cell stack. In addition, the mechanical strength of these membranes gets reduced by the adsorption of phosphoric acid due to the swelling of polymer matrix. On the other hand, lower phosphoric acid uptake results in poor proton conductivity. Attempts to increase the proton conductivity of PBI membranes include the addition of inorganic and organic fillers containing phosphate molecules resulted in proton conductivity improvement but hampered the mechanical stability [137–141]. On the other hand, the cross linkers added to improve the mechanical stability resulted in a sacrifice in proton conductivity [142–144].

Apart from the proton-conducting polymer electrolyte membranes, there are also other ions, such as Li$^+$-, Na$^+$- and hydroxyl-conducting PEMs that are being developed due to many advantages associated with solid PEMs over their liquid counterparts such as easy handling, removal of leakage problems, improved safety and flexibility. For example, crystalline complexes formed between alkali metal salts and poly(ethylene oxide) are capable of demonstrating significant Li$^+$ conductivity which highlight possible applications in LIB electrolytes. Various Li ion-conducting PEMs including LiClO$_4$-polyacrylonitrile and poly(vinylidene fluoride) have been reported [145–148]. Solid-state polymer-silicate nanocomposite electrolytes based on an amorphous polymer poly[(oxyethylene)$_8$ methacrylate], POM and lithium montmorillonite clay have also been tested as 'salt-free' electrolytes in lithium polymer batteries [148]. In a recent review appeared from Keith Scott's research team, it has consolidated the various research aspects and activities related to the development of solid acids as an electrolyte material for PEM water electrolysers [149].

Hydroxide-conducting polymer membranes—also termed as anion-exchange membranes (AEMs)—are another interesting variety of ion-conducting PEMs that recently started gaining more interest towards alkaline fuel cells and electrolyser. Several types of polymers, such as poly(2,6-dimethyl-1,4-phenylene oxide) (PPO), copolymer of chloromethylstyrene and divinylbenzene, PVDF-vinyl-benzyl chloride, and poly(vinyl alcohol) (PVA) poly(1,3-diethyl-1, 1-vinyl imidazolium bromide) 1,4-diazabicyclo-[2.2.2]-octane polysulphone, quaternary ammonium grafted poly vinyl benzyl chloride, have been used for the preparation of AEMs [150–154]. The preparation procedure for AEMs based on PVA and copolymer of poly(acrylonitrile (PAN)-dimethylamio ethylmethacrylate) (DMAEMA) with strongly basic quaternary ammonium in aqueous media has also been reported [155]. These studies reveal that different ion-conducting PEMs including proton are being developed for future energy needs and a breakthrough in any of these materials might open up the electrochemical power systems with improved efficiency.

13 Progress in Bipolar Plate Developments

The current graphite-based state-of-art bipolar plates in PEMFCs face severe concerns mainly due to their higher weight, fragility, brittleness and volume. Despite their inherent qualities, like desirable electrical conductivity, appreciable thermal and chemical properties but due to their inadequate mechanical properties are restricts their usage in PEMFC stacks. Such drawback necessitates the search for alternative options such as metallic alloys and conducting polymer-based composites. Conducting polymer-based composites provide a promising option, but needs higher level of carbon filler (>50 vol%) to achieve the required electrical and thermal conductivity. This also results in several manufacturing and processing related issues. However, CNTs offer significant advantages due to their high electrical properties and ability to form composites with conduction polymer [156–159]. For example, carbon nanotube composite with polyethylene terephthalte (PET)/polyvinylidene fluoride (PVDF) blend result in continuous conductive path provided by the CNTs while the PVDF phase offers crack bridging and PET/PVDF interface provides crack deflection for the composite. Due to this combination, the CNT-PET/PVDF composite has better electrical conductivity, strength and elongation and is considered as one of the promising materials for bipolar plate applications [159]. Recently, graphite-phenol formaldehyde resin show improved bend strength that is necessary for bipolar plate applications. Similarly, CNT-reinforced vinyl ester nanocomposite is shown to have bulk (in-plane)

Table 2 Comparison of electrical and mechanical properties of different types of bipolar plates [160]

Type of plate	Bulk density (g/cm)	Conductivity (S/cm)		Mechanical strength (MPa)		Weight loss at 200 °C
		In-plane	Through-plane	Tensile	Flexural	
CNC machined graphite plate (Shunk)	1.9	110	20	50	40	1.5 %
Compression moulded plates based on PET polymer (Virgina Tech)	<1.8	230	18–25	36.5	53	NA
Injection moulded plates (liquid crystalline polymer–graphite composite)	<1.8	100	NA	NA	NA	NA

electrical conductivity greater than 100 S cm^{-1}, the benchmark for good bipolar plates [156]. Composite of low crystalline polypropylene with CNTs shows electrical conductivity more than 100 S cm^{-1} along with improved mechanical stability and lower thermal expansion [159]. Thus, CNTs have helped in improving the desired features of conducting polymer composites towards bipolar plate applications.

For an industrial scale of production, the composite material granules which are obtained from the extruder can be employed to manufacture bipolar plates complete with flow field structures by conventional production methods, such as compression moulding or injection moulding. For small and intermediate series of bipolar plates, compression moulding still seems to be the most widely adopted production technology. Injection moulding, on the contrary, is a true mass production technique for large series (>50,000) of bipolar plates. Injection moulding of highly filled graphite composite materials are much more demanding than that of conventional plastics due to their high viscosity and high flow resistance. Injection pressure, injection velocity and nozzle temperature have to be optimized to the specific composite material. Injection moulding of bipolar plates, mostly with thermoplastic resins, has been demonstrated by a few companies and research institutes worldwide [160] (Table 2).

Injection moulding of thermosets and thermoplastics are quite different. For thermoplastics, the composites are heated above the melting point and the plasticized mixture is then injected into the mould which is kept at a lower temperature. In the case of thermosets, the compounded composite mixture is injection moulded at temperatures significantly below the melting point of the thermoset resin. Thermoset-bonded graphite composites are generally more complicated to process, as they have to be post-cured after injection moulding in order to achieve sufficient stability. Following the injection moulding of thermosets and thermoplastics, a thin polymer-rich skin layer has to be removed from the surface of the bipolar plates by processes like abrasive blasting in order to reduce the contact resistance to the GDL. Cycle times below 20 s for the production of a single plate have been reported for thermoplastic-bonded graphite composites. High production volumes of a specific type of bipolar plate would lead to its low production cost [160].

While bipolar plates based on graphitic material remains the main focus for PEMFC applications, further cost reduction and increase of power density is beneficial and bipolar plates based on metals offer a high potential to reduce costs and enhance power density (much thinner, yet stronger bipolar plates). However, it is well know that corrosion-resistant metals such as stainless steel form passive surface layers with intrinsically higher ohmic resistance under PEM fuel cell operating conditions. The direct use of these materials leads to a voltage drop in the fuel cell. In order to reduce the contact resistance of the metallic bipolar plates, various types of coatings and surface treatments have been applied to those metallic plates [161]. Among the different materials, stainless steel 316L is the material of choice for the bipolar plates application and recently Ti, Cr nitride [(Ti,Cr)N$_x$)] powders have been tested for coating applications [162, 163].

14 Development of the Bipolar Plate Less PEMFC by Microlithography

Even though the bipolar plate is one of the critical components in fuel cells which plays a multifunctional role, it is one of the most expensive and heavy components to fabricate which unfortunately adds substantial cost per kWh in addition to machining difficulties. Fabrication of three-dimensional structured MEAs with micropatterned electrodes can obviate the use of bipolar plates as the patterned channels can, in principle, provide well-defined pathways to enable homogeneous reactant distribution as well as product removal along the electrode surface with relatively quick heat dissipation. This can potentially reduce PEMFC cost by simplifying gas distribution components and improving performance through better mass transport. Microlithography approach enables the formation of three-dimensional, patterned microelectrodes with high fidelity micron-scale features. Within the extensive literature on PEMFCs, studies on micropatterning of electrode or electrolyte membranes are very few [164, 165]. Most of the micromachining work is concentrated on miniaturization of fuel cells as power sources for portable instruments [166, 167]. Mostly, photolithography is the technique used for fabrication of patterned microelectrodes in various kinds of microfuel cells [168–170]. Soft lithography has been shown to be a rapid and inexpensive way of forming and transferring patterns and structures (≥30 nm) onto or into other materials [171]. In the area of fuel cells, soft lithography has so far been used only for limited applications such as to fabricate microfuel cells or electrodes for microfuel cells [172, 173]. The benefits of designing MEAs possessing microchannelled electrodes by adopting the new microfabrication techniques can be better realized if applied for higher energy density power sources because the issue related to mass transfer of reactants and products is expected to be a matter of serious concern under rigorous reaction conditions in the electrodes which originate under increased power demanding situations.

In this direction in our recent work [174], we have demonstrated the advantages of micropatterning of an electrode to address the water management problem in conventional PEM fuel cell. The MEA with the micropatterned electrode has shown enhanced power density at a higher temperature as well as at a higher relative humidity when compared to a flat electrode. Although the maximum power density obtained at 60 °C with a flat electrode is a little higher than that using micropatterned electrode, better consistency in cell performance is observed in the case of micropatterned electrodes. More interestingly, this approach offers a strategy to eliminate the current use of bipolar plates to stack the MEAs and, thereby, making PEMFC systems much simpler to design and cheaper to fabricate.

15 Conclusions and Future Perspectives

From the above critical review, it is clear that several low-cost materials need to be developed further for making polymer fuel cells affordable, irrespective of the nature of the application. Of course, the requirements of these materials might vary for transportation or stationary applications but the emerging importance of low-cost nanotechnology cannot be neglected. It is worth to mention here that the current research attempts are intensively focused on some of the following cutting edge areas, For example

- Non-Pt-based (Fe, Co and C, N enriched) electrocatalysts for cathodes
- Pt monolayer-coated transition metal alloy catalysts for cathodes
- Metal-free conducting polymers and nitrogen-doped graphene catalysts
- Low-cost polymer composite membranes for low and high temperatures
- Carbon-coated lightweight polymer foam-based bipolar plates
- Micropatternized Al foil-based bipolar plates

and any further improvements will definitely pave a way to eliminate the existing bottlenecks or may provide economically viable, environmentally benign alternate technologies to realize fuel cell-based applications in our day-to-day life.

Acknowledgments The authors appreciate the financial support from Government of India, through CSIR NMITLI, and NWP0022 and from DST as research fellowships. The authors also thank all NMITLI team members from CSIR-NCL, CSIR-NPL and CSIR-CECRI (Chennai Unit) for their valuable contributions. We thank Dr. Sivaram for his constant support and useful discussions.

References

1. Karunadasa HI, Chang CJ, Long JR (2010) Nature 464:1329
2. Tollefson J (2010) Nature 464:1262
3. Lewis NS, Nocera DG (2006) Proc Natl Acad Sci USA 103:15729
4. Dresselhaus MS, Thomas IL (2001) Nature 414:332
5. http://www.eia.doe.gov/pub/international/iealf/table14.xls. Accessed 20 Sept 2010
6. http://www.eia.doe.gov/emeu/international/RecentPetroleumConsumptionBarrelsperDay. xls. Accessed 20 Sept 2010
7. http://www.eia.doe.gov/pub/international/iealf/table13.xls. Accessed 20 Sept 2010
8. http://en.wikipedia.org/wiki/Fossil_fuel#cite_note-14. Accessed 20 Sept 2010
9. Pagliaro M, Palmisano G, Cirminna R (2008) Flexible solar cells. Wiley VCH, Weinheim
10. Das BK (1985) Photovoltaic materials and devices. Wiley Eastern Ltd, New Delhi
11. Zhao TS, Kreuer KD, Nguyen TV (2007) Advances in fuel cells, vol 1. Elsevier, Oxford (Chapter 1)
12. http://www.epa.gov/climaechange/endangerment.html. Accessed 20 Sept 2010
13. http://cdiac.ornl.gov/trends/co2/graphics/mlo145e_thrudc04.pdf. Accessed 20 Sept 2010
14. http://en.wikipedia.org/wiki/Keeling_Curve#cite_ref-bbc_0-0. Accessed 20 Sept 2010
15. Smedley SI, Zhang XG (2007) J Power Sources 165:897
16. Sapkota P, Kim H (2010) J Ind Eng Chem 16:39
17. Viswanathan B, Scibioh MA (2006) Fuel cells: principles and applications. University Press, New Delhi
18. Steele BCH, Heinzel A (2001) Nature 414:345
19. Mallouk TE (1990) Nature 343:515
20. Schlapbach L (2009) Nature 460:809
21. Kunze J, Stimming U (2009) Angew Chem Int Ed 48:9230
22. Metha V, Cooper JS (2003) J Power Sources 114:32
23. Gsmburzev S, Appleby AJ (2002) J Power Sources 107:5
24. Mock P, Schmid SA (2009) J Power Sources 190:133
25. Kjelstrup S, Coppens MO, Pharoah JG, Pfeifer P (2010) Energy Fuels 24:5097
26. http://www1.eere.energy.gov/hydrogenandfuelcells/pdfs/2010_market_report.pdf. Accessed 20 Nov 2012
27. Wang X, Li W, Chen Z, Waje M, Yan Y (2006) J Power Sources 158:154
28. Rotello VM (2004) Nanoparticles: building blocks for nanotechnology. Springer Science and Business Media, Inc., New York, 2004
29. Heiz U, Landman U (2007) Nanocatalysis. Springer Science and Business Media, Inc., New York
30. Jensen BW, Jensen OW, Forsyth M, MacFarlance DR (2008) Science 321:671
31. Wu G, More KL, Johnston CM, Zelenay P (2011) Science 332:443
32. Wu ZS, Yang S, Sun Y, Parvez K, Feng X, Mullen K (2012) J Am Chem Soc 134:9082
33. http://www.eere.energy.gov/hydrogenandfuelcells/mypp/
34. Jenses BW, Jensen OW, Forsyth M, McFarlane DB (2008) Science 321:671
35. Litster S, McLean G (2004) J Power Sources 130:61
36. Auer E, Freund A, Pietsch J, Tacke T (1998) Appl Catal A 173:259
37. Yuan D, Xu C, Liu Y, Tan S, Wang X, Wei Z, Shen PK (2007) Electrochem Comm 9:2473
38. Lin JH, Ko TH, Yen MY (2009) Energy Fuels 23:4042
39. Pan CJ, Su WN, Senthil Kumar SM, Al Andra CC, Yang SJ, Chen HY, Hwang BJ (2012) J Chin Chem Soc 59:1303
40. Beena KB, Sreekittan MU, Sreekumar K (2009) J Phys Chem C 113:17572
41. Kakade BA, Allouche H, Mahima S, Sathe BR, Pillai VK (2008) Carbon 46:567
42. Sun X, Li R, Villers D, Dodelet JP, Desiles S (2003) Chem Phys Lett 379:99
43. Wang Y, Iqbal Z, Mitra S (2006) J Am Chem Soc 128:95
44. Du CY, Zhao TS, Liang ZX (2008) J Power Sources 176:9

45. Peng F, Zheng I, Wang H, Lv P, Yu H (2005) Carbon 43:2397
46. Yumura T, Kimira K, Kobayashi H, Tanaka R, Okumura N, Yamabe T (2009) Phys Chem Chem Phys 11:8275
47. Okamoto Y (2005) Chem Phys Lett 407:354
48. Seger B, Kamat PV (2009) J Phys Chem C 113:7990
49. Kou R, Shao Y, Wang D, Englehard MH, Kwak JH, Wang J, Viswanathan V, Liu J (2009) Electrochem Comm 11:954
50. Li Y, Tang L, Li J (2009) Electrochem Comm 11:846
51. Gu Z, Peng H, Hauge RH, Smalley R, Margrave JL (2002) Nano Lett 2:1009
52. Wang S, Liang R, Wang B, Zhang C (2009) Carbon 47:53
53. Jia G, Wang H, Yan L, Wang X, Pei R, Yan T, Zhao Y, Guo X (2005) Environ Sci Technol 39:1378
54. Li X, Hsing M (2006) Electrochim Acta 51:5250
55. Wu P, Li B, Du H, Gan L, Kang F, Zeng Y (2008) J Power Sources 184:381
56. Choi HC, Shim M, Bangsaruntip S, Dai H (2002) J Am Chem Soc 124:9058
57. http://www.electronics.ca/presscenter/articles/743/1/Carbon-Nanotube-Production-Dramatic-Price-Decrease-Down-to-150kg-for-Semi-Industrial-Applications/Page1.html
58. http://ipp.gsfc.nasa.gov/SS-ISM.html
59. Wang X, Waje M, Yan Y (2005) Electrochem Solid State Lett 8:A42
60. Li W, Wang X, Chen Z, Waje M, Yan Y (2005) Langmuir 21:9386
61. Wang C, Waje M, Wang X, Tang JM, Haddon RC, Yan Y (2004) Nano Lett 4:345
62. Wang X, Li W, Chen Z, Waje M, Yan Y (2006) J Power Sources 158:154
63. Britto PJ, Santhanam KSV, Rubio A, Allonso JA, Ajayan PM (1999) Adv Matt 11:154
64. Frank S, Poncharal P, Wang ZL, De Heer WA (1998) Science 280:1744
65. Liang W, Bockrath M, Bozovic D, Hafner JH, Tinkham M, Park H (2001) Nature 411:665
66. Lin JF, Kamavaram V, Kannan AM (2010) J Power Sources 195:466
67. Kakade BA, Pillai VK (2008) J Phys Chem C 112:3183
68. Kannan AM, Kanagala P, Veedu V (2009) J Power Sources 192:297
69. Huang JE, Guo DJ, Yao YG, Li HL (2005) J Electroanal Chem 577:93
70. Li W, Liang C, Zhou W, Qiu J, Zhou Z, Sun G, Xin Q (2003) J Phys Chem B 107:6292
71. Girishkumar G, Vinodgopal K, Kamat PV (2004) J Phys Chem B 108:19960
72. Leela Mohana Reddy A, Ramaprabhu S (2007) J Phys Chem C 111:16138
73. Wang J, Yin G, Liu H, Li R, Flemming RL, Sun X (2009) J Power Sources 194:668
74. Rajalashmi N, Ryu H, Shaijumon MM, Ramaprabhu S (2005) J Power Sources 140:250
75. Villers D, Sun SH, Serventi AM, Dodelet JP, Desilets S (2006) J Phys Chem B 110:25916
76. Retrieved from the web http://www.platinum.matthey.com/pgm.prices/price-charts/
77. Gong K, Du F, Xia Z, Durstock M, Dai L (2009) Science 323:760
78. Tang Y, Allen BL, Kauffman DR, Star A (2009) J Am Chem Soc 131:13200
79. Du HY, Wang CH, Hsu HC, Chang ST, Chen US, Yen SC, Chen LC, Shih HC, Chen KH (2008) Diam Relat Mater 17:535
80. Wu G, Xu BQ (2007) J Power Sources 174:148
81. Girishkumar G, Rettker M, Underhille R, Binz D, Vinodgopal K, Mcginn P, Kamat PV (2005) Langmuir 21:8487
82. Ma YL, Wainright JS, Litt MH, Savinell RF (2004) J Electrochem Soc 151:A8
83. Rhee CH, Kimm HK, Chang H, Lee JS (2005) Chem Mater 17:1691
84. Sahu AK, Selvarani G, Pitchumani S, Sridhar P, Shukla AK (2007) J Electrochem Soc 154:B123
85. Lee JH, Paik U, Choi JY, Kim KK, Yoon SM, Lee J, Kim BK, Kim JM, Park MH, Yang CW, An KH, Lee YH (2007) J Phys Chem C 111:2477
86. Adjemian KT, Dominey R, Krishnan L, Ota H, Majszrik P, Zhang T, Mann J, Kirby B, Gatto L, Simpson MV, Leahy K, Srinivaasan S, Benziger JB, Bocarsly AB (2006) Chem Mater 18:2238
87. Yu TL, Lin HL, Shen KS, Huang LN, Chang YC, Jung GB, Huang JC (2004) J. Polymer Res 11:217

88. Liu YH, Yi B, Shao ZG, Xing D, Zhang H (2006) Electrochem Solid State Lett 9:A356
89. Liu YH, Yi B, Shao ZG, Wang L, Xing D, Zhang H (2007) J Power Sources 163:807
90. Thomassin JM, Kollar J, Caldarea G, Germain A, Jerome R, Detrembleur C, Memb J (2007) Science 303:252
91. Kannan R, Kakade BA, Pillai VK (2008) Angew Chem 120:2693
92. Kannan R, Meera P, Maaraveedu SU, Kurungott S, Pillai VK (2009) Langmuir 25:8305
93. Novoselov KS, Jiang Z, Zhang Y, Morozov SV, Stormer HL, Zeitler U, Mann JC, Boebinger GS, Kim P, Geim AK (2007) Science 315:1377
94. Berger C, Song Z, Li X, Wu X, Brown N, Naud C, Mayou D, Li T, Hass J, Marchenkov AN, Conard EH, First PN, De Heer WA (2006) Science 312:1191
95. Novoselov KS, Jiang Z, Zhang Y, Morozov SV, Stormer HL, Zeitler U, Mann JC, Boebinger GS, Kim P, Geim AK (2007) Science 315:1379
96. Rao CNR, Sood AK, Voggu R, Subrahmanyam KSJ (2010) Phys Chem Lett 1:572
97. Novoselov KS, Geim AK, Morozov SV, Jiang D, Katsnelson MI, Grigorieva IV, Dubonos SV, Firsov AA (2005) Nature 438:197
98. Ozulmaz B, Jarillo Herrero P, Efetov D, Kim P (2007) Appl Phys Lett 91:192107
99. Shinde DH, Debgupta J, Kushwaha A, Aslam M, Pillai VK (2011) J Am Chem Soc 133:4168
100. Yang J, Liu DJ, Kariuki NN, Chen LX (2008) Chem Comm 21:329
101. Xiong W, Du F, Lin Y, Perez A, Supp M, Ramakrishnan TS, Dai L, Jiang L (2010) J Am Chem Soc 132:15839
102. Kundu S, Nagaiah TC, Xia W, Wang Y, Dommele SV, Bitter JH, Santa M, Grundmeier G, Bron M, Schuhmann W, Muhler M (2009) J Phys Chem C 113:14302
103. Geng D, Liu H, Chen Y, Li R, Sun X, Ye S, Knights S (2011) J Power Sources 196:1795
104. Wiggins-Camacho JD, Stevenson KJ (2011) J Phys Chem C 115:20002
105. Nagaiah TC, Kundu S, Bron M, Muhler M, Schuhmann W (2010) Electrochem Comm 12:338
106. Gong K, Du F, Xia Z, Durstock M, Dai L (2009) Science 323:760
107. Li Y, Zhou W, Wang H, Xie L, Liang Y, Wei F, Idrobo JC, Pennycook SJ, Dai H (2012) Nat Nanotech 7:394
108. Wu ZS, Zhou G, Yin LC, Ren W, Li F, Cheng HM (2012) Nano Energy 1:107
109. Alvarez GF, Mamlouk M, Senthil Kumar SM, Scott K (2011) J Appl Electrochem 41:925
110. Winther-Jensen B, Winther-Jensen O, Forsyth M, MacFarlane DR (2008) Science 321:671
111. Wieckowski A, Savinova ER, Vayenas CG (2003) Catalysis and electrocatalysis at nanoparticle surfaces. Marcel Dekker Inc, New York
112. Subhramannia M, Pillai VK (2008) J Mater Chem 18:5858
113. Subhramannia M, Ramaiyan K, Pillai VK (2008) Langmuir 24:3576
114. Subhramannia M, Ramaiyan K, Komath I, Aslam M, Pillai VK (2008) Chem Mater 20:601
115. Adzic RR, Tripkovic AV, O'Grady WE (1982) Nature 296:137
116. Motoo S, Furuya N (1985) J Electroanal Chem 184:303
117. Xia XH, Liess HD, Iwasita T (1997) J Electroanal Chem 437:233
118. Yamada M, Honma I (2006) Fuel Cells Bull 5:11
119. Malmstrom BG (1981) Annu Rev Biochem 51:21
120. Huang M, Shao Y, Sun X, Chen H, Liu B, Dong S (2005) Langmuir 21:323
121. Shoji M, Oyaizu K, Nishide H (2008) Polymer 49:5659
122. Saffarian HM, Srinivasan R, Chu D, Gilman S (2001) J Electroanal Chem 504:217
123. Meera P, Kannan R, Sreekumar K, Pillai VK (2010) J Mater Chem 20:9651
124. Ticianelli TA, Derouin CR, Srinivasan SJ (1988) J Electroanal Chem 251:275
125. Gazaryan IG, Lagrimini LM, Ashby GA, Thorneley RNF (1996) Biochem J 313:841
126. Sun W, Peppley BA, Karan K (2005) Electrochim Acta 50:3359
127. Tanimoto E (2005) Crit Rev Plant Sci 24:249
128. Kawano T (2003) Plant Cell Rep 21:829
129. Kreuer KD, Fuchs A, Ise M, Spaeth M, Maier J (1998) Electrochim Acta 43:1281
130. Wainright JS, Wang JT, Weng D, Savinell RF, Litt JM (1995) J Electochem Soc 142:L121

131. Kerres JA (2001) J Membr Sci 185:3
132. Kreuer KD, Paddison SJ, Spohr E, Schuster M (2004) Chem Rev 104:4637
133. Mustarelli P, Quartarone E, Grandi S, Carollo A, Magistris A (2008) Adv Mater 20:1339
134. Li Q, He R, Jensen JO, Bjerrum NJ (2003) Chem Mater 15:4896
135. Pu H, Meyer WH, Wegner G (2002) J Polym Sci, Part B: Polym Phys 40:663
136. Weng D (1996) Ph.D. thesis, Case Western Reserve University, Cleveland
137. Quartarone E, Mustarelli P, Carollo A, Grandi S, Magistris A, Gelbaldi C (2009) Fuel Cells 9:231
138. Hasiotis C, Qingfeng L, Deimede V, Kallitsis JK, Kontoyannis CG, Bjerrum NJ (2001) J Electrochem Soc 148:A513
139. He R, Li Q, Xiao G, Bjerrum NJ (2003) J Membr Sci 226:169
140. Jang MY, Yamazaki Y (2005) J Power Sources 139:2
141. Chuang SW, Hsu SLC, Hsu CL (2007) J Power Sources 168:172
142. Li Q, Pan C, Jensen JO, Noye P, Bjerrum NJ (2007) Chem Mater 19:350
143. Lu Y, Chen J, Cui H, Zhou H (2008) Compos Sci Technol 68:3278
144. Xu H, Chen K, Guo X, Fang J, Yin J (2007) J Membr Sci 288:255
145. Song JY, Wang YY, Wan CC (1999) J Power Sources 77:183
146. Jankova K, Jannasch P, Hvilsted S (2004) J Mat Chem 14:2902
147. Meyer WH (1998) Adv Mater 10:439
148. Kurian M, Galvin ME, Trapa PE, Sadoway DR, Mayes AM (2005) Electrochim Acta 50:2125
149. Urtiaga AG, Presvytes D, Scott K (2012) Int J Hyd Energy 37:3358
150. Xu T, Liu Z, Li Y, Yang W (2008) J Membr Sci 320:232
151. Tang B, Wu P, Siesler HW (2008) J Phys Chem B 112:2880
152. Tzanetakis N, Varcoe J, Slade RS, Scott K (2003) Electrochem Comm 5:115
153. Wang X, Li M, Golding BT, Sadeghi M, Cao Y, Yu EH, Scott K (2011) Int J Hyd Energy 36:10022
154. Cao Y, Wu X, Scott K (2012) Int J Hyd Energy 37:9524
155. Kumar M, Singh S, Shahi VK (2010) J Phys Chem B 114:198
156. Liao SH, Hung C, Zhu H, Ma CC, Yen CY, Lin YF, Weng CC (2008) J Power Sources 176:175
157. Liao SH, Yen CY, Weng CC, Lin YF, Ma CC, Yang CH, Tsai MC, Yen MY, Hsiao MC, Lee SJ, Xie XF, Hsiao YH (2008) J Power Sources 185:1225
158. Yin Q, Sun KN, Li AJ, Shao L, Liu SM, Sun C (2008) J Power Sources 175:861
159. Wu M, Shaw LL (2004) J Power Sources 136:37
160. Cunningham B, Baird DG (2006) J Mater Chem 16:4385
161. Hermann A, Chaudhuri T, Spagnol P (2005) Int J Hyd Energy 30:1297
162. Wind J, Spah R, Kaiser W, Bohm G (2002) J Power Sources 105:256
163. Choi HS, Han DH, Hong WH, Lee JJ (2009) J Power Sources 189:966
164. Zhou Z, Dominey RN, Rolland JP, Maynor BW, Pandya AP, DeSimone JM (2006) J Am Chem Soc 128:12963
165. Shah K, Shin WC, Besser RS (2004) Sens Actuators B97:157
166. Nguyen NT, Chan SH (2006) J Micromech Microeng 16:R1
167. Kjeang E, Djilali N, Sinton D (2009) J Power Sources 186:353
168. Kim HT, Reshentenko TV, Kweon HJ (2007) J Electrochem Soc 154:B1034
169. Bieberle A, Gauckler LJ (2000) Solid State Ionics 135:337
170. Yang J, Liu DJ (2007) Carbon 45:2845
171. Zhao XM, Xia Y, Whitesides GM (1997) J Mater Chem 7:1069
172. Shah K, Shin WC, Besser RS (2003) J. Power Source 123:172
173. Jiang X, Bent SF (2009) J Phys Chem C 113:17613
174. Deshmukh AB, Kale VS, Dhavale VM, Sreekumar K, Pillai VK, Shelke MV (2010) Electrochem Commun 12:1638

Cathode and Anode Materials for Na-Ion Battery

Lifen Xao, Yuliang Cao and Jun Liu

Abstract Energy storage is more important today than at any time in the human history. The battery systems that are pursued for clean renewable energy-based grid or the electrification of transportation need to meet the requirements of low cost and high efficiency. Li-ion battery is the most advanced battery system, but it is expensive and insufficient as a resource for widespread application. Na-ion battery is seen as a promising alternative due to the abundance of Na resource. However, the realization of the Na-ion intercalation/deintercalation mechanism is also challenging because Na ions are 40 % larger in radius than Li ions. This makes the finding of suitable host materials with high storage capacity, rapid ion uptaking rate, and long cycling life not easy. In the recent 3 years, several electrode materials were found to have energy density close to those used in Li-ion batteries. These scientific advances have greatly rekindled worldwide passion for Na-ion battery system. In this chapter, the development of the electrode materials for Na-ion batteries is briefly reviewed, with the aim of providing a wide view of the problems and future research orientations of this system.

L. Xao
College of Chemistry, Central China Normal University, Wuhan 430079, People's Republic of China

Y. Cao
College of Chemistry and Molecular Science, Wuhan University, Wuhan 430072, People's Republic of China

J. Liu (✉)
Pacific Northwest National Laboratory, Richland, WA 99352, USA
e-mail: Jun.Liu@pnnl.gov

Z. Lin and J. Wang (eds.), *Low-cost Nanomaterials*, Green Energy and Technology,   395
DOI: 10.1007/978-1-4471-6473-9_14, © Springer-Verlag London 2014

1 Introduction

Energy is not only the material foundation that sustains the advance of human civilization, but also the indispensable requirement for the development of modern society. With the rapid growth of the global economy, our society has become increasingly interdependent on energy. As data show 70 % of the world's annual energy consumption comes from fossil fuels, there can be no doubt that our reserves of fossil fuels are finite, and the resulting environmental pollution is worsening. Thus, we need to promote the harnessing of clean, renewable energy sources such as solar, wind, geothermal, and tidal energy. However, these renewable resources are commonly intermittent in time and diffusive in space; the electricity produced must be stored and made available on demand. So large-scale electric energy storage (EES) systems will be an essential part of the future renewable energy-based grid. Meanwhile, another pressing demand is to find appropriate EES for the electrification of transportation that does not emit greenhouse gases.

Electrochemistry can contribute to EES in several diverse ways. Electrochemical energy storage technologies include flow redox batteries, super capacitors, and rechargeable batteries (Pb–acid, Ni–Cd, Na–S, and Li-ion batteries, etc.), showing great advantages of high efficiency, low cost, and flexibility. Hereinto, Li-ion batteries have currently been considered as one of the most promising technologies due to their long lifetime and high energy density. However, for widespread EES applications, there is increasing concern about the cost and limitation of lithium terrestrial reserves. As a result, great efforts have been made to explore new low-cost and reliable electrochemical energy storage technologies.

Due to the global lithium resource constraint, scientists have turned their attention to rechargeable Na-based batteries. As Na is positioned in the same period with Li, its electrochemical properties are similar to Li. Na has very negative redox potential (−2.71 V, vs. SHE) and a small electrochemical equivalent (0.86 gAh^{-1}), which makes it the most advantageous element for battery applications after lithium. Na element instead has very abundant and widespread supply as the crustal abundance of Na is 2.64 %, compared to 0.006 % of Li. Moreover, there is 3.5 % of NaCl in the ocean, showing inexhaustible Na resource, thus Na is an ideal energy storage material.

In the past decades, high-temperature Na battery system such as Na/S and Na/$NiCl_2$ (ZEBRA battery) systems have been commercially developed for electric vehicles and MWh scale electric storage due to their high energy density and low cost. Na/S batteries produced by NGK Ltd. in Japan have entered in the market since 2000. The biggest units with a power output of 8 MW have been built and more than a hundred Na/S energy storage stations are operating over the globe. The biggest market of ZEBRA batteries is thought to be transportation. Benz Corp in German has carried out longtime tests since the 1990s, where the index of the ZEBRA batteries can fulfill the medium-term target of USABC. The ZEBRA battery powered vehicles have passed over 3,200 thousand miles' test.

A major obstacle hindering the broad market penetration of these two types of Na batteries is the long-term stability and endurance of the battery components at high temperatures of $\geq 300°C$.

2 Na-Ion Batteries and Problems

If a room temperature Na-ion rocking chair battery (Na-ion battery) (Fig. 1) can be achieved, it would bring about great improvement in safety and operational simplicity with respect to the conventional high-temperature Na batteries and also a remarkable decrease in cost with regard to Li-ion batteries, thus ensuring sustainable applications for large-scale electric energy storage. Since Na-ion batteries have the similar working principle as Li-ion batteries, the development of this system would benefit greatly from the knowledge as well as configurations gained within Li-ion batteries. However, it should be noted that: 1. the gravimetric capacity of Na is lower (1,165 mAh g^{-1}) than lithium (3,829 mAh g^{-1}) due to its larger molecular mass [1]. Thus, the Na-ion battery system will achieve lower energy density than Li-ion batteries. 2. Na ions are about 40 % larger than Li ions in radius, which make them more stable in rigid lattices [1]. Thus, it will be harder to find proper materials that are able to realize reversible Na storage and release. Due to its large dimensions, the insertion and extraction of Na ions can arouse more serious stress changes of the host materials, causing rapid collapse of lattice structure and therefore poor cycling stability. The diffusion of Na-ions in lattice is also slow, resulting in poor electrochemical utilization and rate capability. 3. Na metal is more chemically active than Li, which needs very strict restriction of the moisture and oxygen content on the experimental environment. In this chapter, recent research progress in electrode materials is reviewed for Na-ion batteries. The practical specific capacity and operating voltage of some of the electrode materials for Na-ion battery are summarized in Fig. 2.

3 Cathode Materials for Na-Ion Batteries

The electrochemical performance of Na-ion batteries mainly depends on the structures and properties of electrode materials and electrolytes. As one of the key components of Na-ion batteries, suitable cathode materials should meet the following requirements:

1. High redox potential, which is less influenced by the concentration of Na ions in the lattices, to maintain high and even charge/discharge voltage.
2. Small electrochemical equivalent, i.e., light molecular weight and more electrons transferred per unit of the active material, to provide high specific capacity.

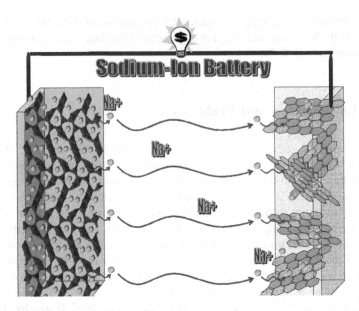

Fig. 1 Schematic diagram of Na-ion battery

Fig. 2 Pratical specific capacity and operating voltage in current Na-ion battery technologies

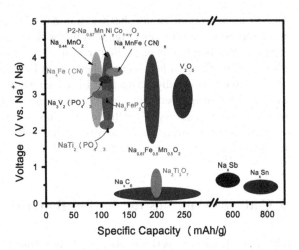

3. High structural and chemical stability to ensure good cyclability.
4. High electronic and Na-ion conductivity to realize rapid charge/discharge.
5. Mass production convenient, resource abundant, and environment benign.

A wide variety of compounds have been investigated for cathodes used in Na-ion batteries. In this section, we mainly focus on metal oxides, polyanion compounds, and metal hexacyanides.

3.1 Metal Oxides

Li metal oxide materials ($LiCoO_2$, $LiNi_xCo_yMn_{1-x-y}O_2$, $LiMn_2O_4$, etc.) have been widely investigated as Li-ion intercalation hosts. These materials exhibit good capacity and structural stability during repeated Li-ion insertion/extraction cycles. Analogously, Na-based metal oxides also have similar formula and crystal structure with Li-based oxides. Indeed, these metal oxides have been first considered as Na-ion hosts. Some typical crystal structures of Na-based metal oxides are illustrated in Fig. 3. Three types of layered structures (Fig. 3a–c) can be denoted as On or Pn (n = 1, 2, 3,...), depending on the insertion environment of alkali atoms and the repeat period (n) of the transition metal oxide (MO) sheets forming MO_6 edge-sharing octahedra [3]. Na atoms lie in the octahedral sites in O3 structure (Fig. 3a) while occupy the trigonal prismatic sites in P2 and P3 structures (Fig. 3b, c). Therefore, phase transitions between O3 and P3 will occur due to the movement of Na atoms from the octahedral sites to the trigonal prismatic sites along with the gliding of MO sheets. The phase transitions cause the multi-voltage plateaus of the layered Na_xMO_2 material during charge/discharge. However, the phase transitions between O3 and P2 cannot occur at room temperature because of the high cleavage energy of the M–O bonds [2]. In general, O3-type metal oxides can extract more Na atoms (hence discharge capacity) from the crystal structure; P2-type metal oxides have higher structural stability during repeated charge/discharge. Some other metal oxides have a three-dimensional tunnel structure such as orthorhombic $Na_{0.44}MnO_2$ shown in Fig. 3d. $Na_{0.44}MnO_2$ is made of MnO_5 square pyramids and MnO_6 octahedra, which are arranged to form two types of tunnels: large S-shaped tunnels and smaller pentagon tunnels (Fig. 3d). Na atoms can intercalate into these tunnels. 3D tunnels usually can not only offer stable structure to allow free movement of Na ions inside the crystal structure, but also provide fast channels for Na ion moving in/out of the crystal structure [4–7].

3.1.1 Na_xCoO_2

In the early 1980s, Na_xCoO_2 has been studied as cathode materials for Na batteries [8]. Braconnier et al. reported that Na_xCoO_2 has four different phases, O3 (0.9 < x < 1), O′3 (x = 0.75), P3 (0.55 < x < 0.6), and P2 (0.64 < x < 0.74) depending on the concentration of Na ions. Among them, P2–$Na_{0.7}CoO_2$ (Fig. 1b) shows the largest energy density (260 Wh kg^{-1}). Recently, Delmas et al. used Na batteries to convey a thorough investigation of the P2–Na_xCoO_2 phase diagram for x ≥ 0.50 (Fig. 4). It showed a succession of single phases or two-phase domains on Na ion intercalation [9]. Harharan et al. investigated the effect of synthesis methods on the electrochemical performance of P2–$Na_{0.7}CoO_2$ [10]. They found the sample prepared by sol–gel method has the highest discharge capacity (10 and 50 % higher than those prepared by solid state reaction and high-energy ball milling methods), due to the smaller particular size and larger surface area.

Fig. 3 Crystal structures of various Na_xMO_y: **a** P2–Na_xCoO_2 **b** O3–Na_xCoO_2 **c** P3–Na_xCoO_2, and **d** $Na_{0.44}MnO_2$ (Na: *yellow*, Co/Mn/V: *blue*, O: *red*) [2]

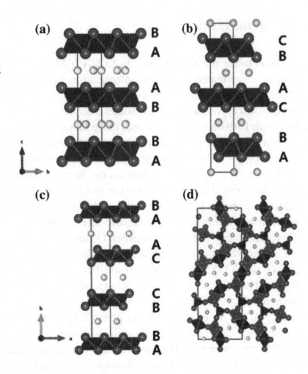

Though $LiCoO_2$ is the hottest commercial cathode material for Li-ion battery, Na_xCoO_2 unfortunately cannot match a good cathode either in energy density or in cyclability for Na ion batteries. Besides, Co as a precious metal does not meet the low-cost demand of Na-ion battery.

3.1.2 Na_xMnO_2

In the early 1970s, Mn–O–Na ternary system has been widely investigated and a wide structural Na_xMnO_2 (depending strongly on the concentration of Na and the preparation condition) has been reported [11]. The most typical structures are α–$NaMnO_2$ with lamellar structure of O′3 type (Fig. 3a), $Na_{0.7}MnO_2$ with lamellar structure of P2 type (Fig. 3b), and $Na_{0.44}MnO_2$ with three-dimensional channel structure (Fig. 3d). Delmas et al. studied the Na intercalation behavior in the above three types of Na_xMnO_2. They found that within a certain range of x ($0.45 \leq x \leq 0.85$ for α–$NaMnO_2$ and $Na_{0.70}MnO_2$, $0.30 \leq x \leq 0.58$ for $Na_{0.44}MnO_2$), these materials can maintain very well their pristine structures, showing potential to be used as cathode materials for Na-ion batteries. Caballero et al. [12] prepared P2–$Na_{0.6}MnO_2$ by using sol–gel method, which showed high purity and a well-defined layered structure (Fig. 3b). The material delivered a high specific capacity of *ca.* 140 mAh g^{-1} (corresponding to an intercalation amount of ~0.52) during the initial several cycles, and then the capacity declined on

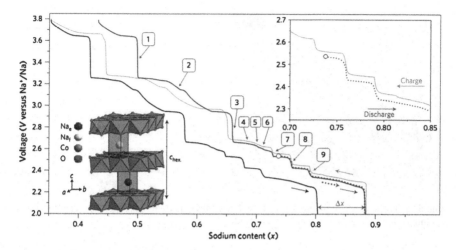

Fig. 4 Galvanostatic cycling curve of an Na/P2–Na$_x$CoO$_2$ battery giving an overview of the phase diagram [9]

successive cycles. It was considered that the intercalation/deintercalation of Na ions can cause repeated gliding of the MO sheets, leading to collapse of the layered structure into amorphous after the first eight cycles [12]. So, the structural stability is still an issue on the layered Na$_x$MnO$_2$ material.

Compared to layered Na$_x$MnO$_2$, orthorhombic Na$_{0.44}$MnO$_2$ shows higher stability in structure, the MnO$_6$ octahedra, and MnO$_5$ square pyramids form a framework possessing two types of Na-ion diffusion channels (Fig. 3d): large S-shaped tunnels and smaller pentagon tunnels. Na ions occupy three sites in this structure: the Na(1) site in the small tunnels is fully occupied while the Na(2) and Na(3) sites in the large S-shaped tunnels are half occupied. Na ions in the large S-shaped tunnels can be extracted, while Na ions in the small tunnels are immovable due to the small tunnel size and strong interaction with the surrounding oxygen atoms. Sauvage et al. synthesized well-crystallized Na$_{0.44}$MnO$_2$ with a rod shape (500 nm in width and 5–10 μm in length) as shown in Fig. 5a [5]. A specific capacity of about 140 mAh g^{-1} was obtained at a slow rate (C/200) within the potential range of 2–3.8 V (vs. Na/Na$^+$), indicating that the reversible insertion/extraction of Na in Na$_x$MnO$_2$ occurred with x varying between 0.18 and 0.64. Notably, the evident multi-potential plateaus in charge/discharge curves and multi-peaks in the incremental capacity curves suggest multiple phase transitions occurring during the Na intercalation/deintercalation. In situ X-ray diffraction measurements were carried out to detect the evolution of the multiphases. Six biphasic transition processes were observed during charge and discharge, which were consistent with the numbers of the potential plateaus on the charge/discharge curves. Unfortunately, when the current was increased to 1/10 C, only 80 mAh g^{-1} of capacity was obtained and faded to 50 % over 50 cycles (Fig. 5b). This notable hysteresis in Na intercalation/deintercalation dynamics probably

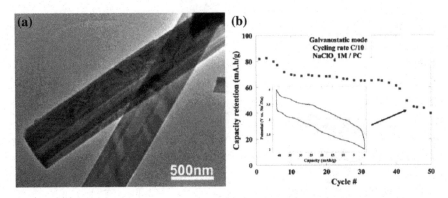

Fig. 5 **a** TEM image showing the rodlike $Na_{0.44}MnO_2$ particles. **b** Capacity retention upon cycling of a $Na_{0.44}MnO_2/C$ composite electrode at a C/10 rate [5]

resulted from its large diameter. Hence, a solution should be downsizing the particle of materials to shorten the diffusion path of Na ions.

Recently, Cao et al. reported the electrochemical properties of $Na_{0.44}MnO_2$ nanowires prepared by a polymer–pyrolysis method [4]. The $Na_{0.44}MnO_2$ sample treated at 750 °C was highly crystallized and showed quite uniform nanowire morphology (~50 nm in diameter) (Fig. 6a). The material delivered a reversible capacity of 128 mAh g^{-1} at 1/10 C and exhibited excellent cyclability (77 % capacity retention over 1,000 cycles at 0.5 C), showing great enhancement in the electrochemical performance than previous works [5]. The improvement was considered to be attributed to the high crystallinity and uniform nanowire structure, which can provide a mechanically stable structure as well as a short diffusion path for Na-ion insertion/extraction. An Na-ion full cell was tried by using $Na_{0.44}MnO_2$ as cathode and pyrolyzed carbon as anode. The full cell showed steep charge and discharge curves with an average discharge voltage of 2.7 V and 73 % capacity retention after 100 cycles (Fig. 7).

3.1.3 $Na_xV_yO_z$

Vanadium oxides have variable valences and open formwork structure, which are more suitably used as ion insertion hosts. In the late 1980s, the electrochemical Na-ion interaction behaviors of three types of vanadium oxides were investigated: β–$Na_xV_2O_5$ with channel structure, $Na_{1+x}V_3O_8$ and α–V_2O_5 with layered structure [13]. The reversible Na accommodation levels were ~1.7 mole of Na per mole of α–V_2O_5, ~1.2 mole of Na per mole of β–$Na_{0.33}V_2O_5$ and ~1.6 mole of Na per mole of $Na_{1+x}V_3O_8$ in the potential window of 1.0–3.6 V. Later, Bach et al. studied the electrochemical Na intercalation behaviors of β–$Na_{0.33}V_2O_5$ prepared by two procedures [14, 15]. It showed that β–$Na_{0.33}V_2O_5$ prepared by the sol–gel process can deliver reversibly ~0.7 mole of Na per mole of $Na_{0.33}V_2O_5$ (corresponding to 100 mAh g^{-1}) in the potential window of 2.0–3.4 V, while that prepared by the

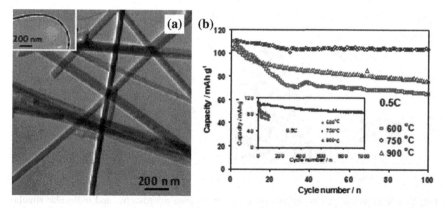

Fig. 6 a TEM images of $Na_4Mn_9O_{18}$ nanowires calcined at 750 °C. **b** Cycle performance of $Na_4Mn_9O_{18}$ samples calcined at 600, 750, and 900 °C at a current density of 60 mA g^{-1} (0.5C) [4]

Fig. 7 The charge/discharge profile (**a**) and cycle performance (**b**) of the $Na_4Mn_9O_{18}$/pyrolyzed carbon Na–ion battery in the potential range of 1.5–4.1 V from the second cycle at a constant current of 50 mA g^{-1} (~0.5 C) [4]

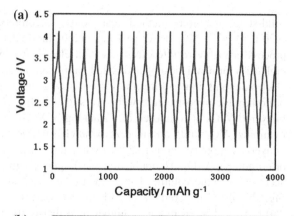

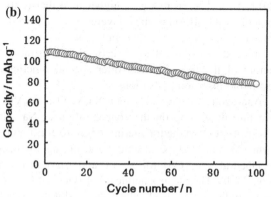

(a)

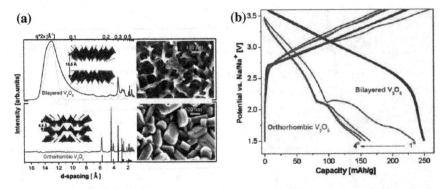

Fig. 8 a Synchrotron X-ray diffraction, scanning electron microscopy, and molecular simulations of electrodeposited vanadium oxide: **a** bilayered V_2O_5 annealed in vacuum at 120 °C (*blue*); **b** orthorhombic V_2O_5 annealed in oxygen at 500 °C (*green*). **b** First four charge/discharge cycles of bilayered V_2O_5 and orthorhombic V_2O_5 electrodes. Both cells were cycled at 20 mA g^{-1}, within the potential window of 3.8–1.5 V (vs. Na/Na$^+$) from 1 M NaClO$_4$ in PC solution [16]

solid-state reaction method can only accommodate 0.17 mole of Na (~20 mAh g^{-1}). The better performance of the material prepared by the sol–gel process probably resulted from its high electrical conductivity along the b-axis, high structural anisotropy and small particular size, which is favorable to fast electronic and ionic diffusion [15]. V_2O_5 with a bilayered structure and an orthorhombic nanostructure were investigated [16]. XRD results showed that the bilayered V_2O_5 is composed of a stacking of V_2O_5 bilayers based on the square–pyramidal VO$_5$ units arranged in parallel (Fig. 8a). The spacing distance of the bilayers is approximately 13.5 Å, significantly larger than that (only 4.4 Å) of the orthorhombic V_2O_5. Undoubtedly, the larger spacing would be more flexible to accommodate the volume change due to the intercalation of Na ions. As shown in Fig. 8b, the bilayered V_2O_5 electrode delivered a reversible capacity of 250 mAh g^{-1} at 20 mA g^{-1}, higher than that (only 150 mAh g^{-1}) of the orthorhombic V_2O_5. Besides, the bilayered V_2O_5 electrode demonstrated higher average discharge potential and reversible capacity on repeated cycling than its orthorhombic counterpart. This investigation illustrated that tailoring the nanoarchitecture of the materials can offer special functional properties to facilitate the reversible insertion of Na ions.

Analogous to Na$_x$CoO$_2$ and Na$_x$MnO$_2$, Na$_x$VO$_2$ also has different lamellar structure depending on the concentration of Na. The lamellar structure is composed of VO$_6$ octahedra sharing edges to form VO$_2$ layers, which are stacked to form O3 and P2-type structures according to the occupation sites of Na ions (octahedral site (O3) for x = 1 and trigonal prismatic site (P2) for x = 0.7) (Fig. 3a, b). The electrochemical behaviors of O3–NaVO$_2$ and P2–Na$_{0.7}$VO$_2$ have been studied [17]. Both electrodes exhibited similar charge/discharge profiles (Fig. 9). The intercalation concentration of Na ion for both materials is up to at least 0.5 mol of Na$^+$ per mole of NaVO$_2$, corresponding to a reversible capacity of

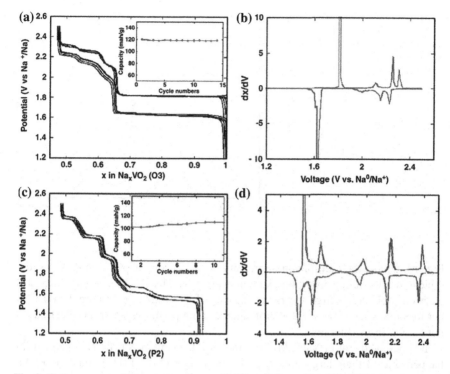

Fig. 9 Voltage–composition curves for **a** NaVO$_2$ (O3) **c** Na$_{0.7}$VO$_2$(P2) and associated respectively **b** and **d** derivative curves. Insets indicate the capacity evolution versus number of cycles. The cells using 5–6 mg of positive electrode composites were cycled at C/20. The small polarization between charge and discharge for the P2 phase suggests a high rate capability. This was confirmed by power rate measurements (e.g. signature curves) as this electrode can deliver 90 % of its capacity at 1 C rate [17]

about 130 mAh g^{-1}. For both materials, multiple step phase transitions can be easily observed on the charge/discharge curves and the derivative curves (Fig. 9). A careful observation can find that at low potential range, the voltage variation is about 200 mV for O3–NaVO$_2$, compared to 50 mV for P2–Na$_{0.7}$VO$_2$. The large difference probably due to O3–NaVO$_2$ is electron insulative, while P2–Na$_{0.7}$VO$_2$ is semiconductive [18, 19].

3.1.4 Mixed Metal Oxide

Except for Na$_x$MO$_2$, mixed metal oxides have also been surveyed in consideration of the combination of the respective properties of different elements. Carlier et al. used Mn to partly replace Co in P2–Na$_{0.67}$CoO$_2$ to form P2–Na$_{0.67}$Co$_{0.67}$Mn$_{0.33}$O$_2$ based on the consideration of the stabilization of Mn^{4+} ions in the lattice [20]. The charge/discharge curves of the P2–Na$_{0.67}$Co$_{0.67}$Mn$_{0.33}$O$_2$ electrode are shown in

Fig. 10 Cycling curves of
Na/Na$_x$Co$_{2/3}$Mn$_{1/3}$O$_2$ cells
obtained with a C/100 current
rate, starting by a charge (*red
curve*) or a discharge (*black
curve*) [20]

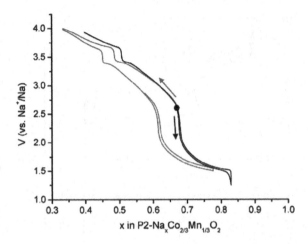

Fig. 10. Unlike the multiple phase transitions of P2–Na$_{0.67}$CoO$_2$, it only showed a typical solid-solution insertion/extraction process. The P2–Na$_{0.67}$Co$_{0.67}$Mn$_{0.33}$O$_2$ electrode delivered a high reversible capacity of ~130 mAh g^{-1} (more than 0.5 Na per formula unit). However, the prolonged cycling performance of this electrode is not shown in the literature.

It is believed that in NaNi$_x$M$_{1-x}$O$_2$, the displacement of Na$^+$ by Ni^{2+} can hardly happen due to their large difference in ion radius (Na$^+$: 1.02 Å, Ni^{2+}: 0.69 Å). Thus, for NaMnO$_2$, the part replacement of Mn by Ni could possibly result in an ordered structure. Komaba et al. prepared NaNi$_{0.5}$Mn$_{0.5}$O$_2$ via a co-precipitation method [21]. XRD results showed that only 0.4 % of site exchange of Na with Ni was detected, suggesting a stable layered structure and fast Na-ion diffusion pathway. The NaNi$_{0.5}$Mn$_{0.5}$O$_2$ electrode demonstrated an initial discharge capacity of 125 mAh g^{-1} in the potential range of 2.2–3.8 V and high cycling capacity of >100 mAh g^{-1} over 20 cycles (Fig. 11), showing better performance than NaMnO$_2$ [22]. However, the successive phase transition still occurred during charge/discharge, capacity loss with cycling. Thus, stoichiometric Li was introduced into the transition metal layer, to form a stabilizing charge ordering state between Ni^{2+} and Mn^{4+}, for example, Na$_{1.0}$Li$_{0.2}$Ni$_{0.25}$Mn$_{0.75}$O$_\delta$ [23]. Normalized Mn and Ni K-edge x-ray absorption near edge structure (XANES) spectra showed that the presence of Mn and Ni in the Na$_{1.0}$Li$_{0.2}$Ni$_{0.25}$Mn$_{0.75}$O$_\delta$ are predominantly Mn^{4+} and Ni^{2+}, respectively, resulting in no Jahn–Teller distortion in the structure. The Na$_{1.0}$Li$_{0.2}$Ni$_{0.25}$Mn$_{0.75}$O$_\delta$ electrode lost only 2 % of its initial capacity over 50 cycles (Fig. 12), exhibiting excellent cycling stability. The shapes of charge/discharge curves suggest a solid-solution insertion/extraction process, different from the multiple phase transitions of Na$_{0.67}$Ni$_{0.5}$Mn$_{0.5}$O$_2$ (Fig. 11). ICP–OES analysis found that after being charged to 4.2 V, less than 5 % of the total Li in Na$_{1.0}$Li$_{0.2}$Ni$_{0.25}$Mn$_{0.75}$O$_\delta$ was removed from the lattices. These results demonstrated that Li can stabilize the transition metal layer and restrict the phase transition during Na intercalation/deintercalation.

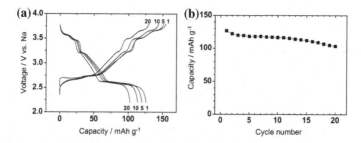

Fig. 11 a Galvanostatic charge and discharge curves and **b** variation in discharge capacity of $NaNi_{0.5}Mn_{0.5}O_2$ in Na cells at 4.8 mA g^{-1} [21]

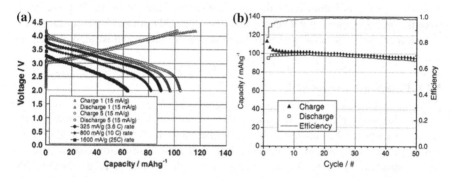

Fig. 12 a voltage profiles (first and fifth charge/discharge cycle; open circles and diamonds) for $Na/Na_{0.85}Li_{0.17}Ni_{0.21}Mn_{0.64}O_2$ cell between 4.2 and 2.0 V. Additional discharge voltage profiles for high-rate studies are also shown and labeled in the legend. The trickle charge data points have been removed for clarity, and **b** capacity versus cycle number for $Na/Na_{0.85}Li_{0.17}Ni_{0.21}Mn_{0.64}O_2$ cell between 4.2 and 2.0 V [23]

Recently, Komaba et al. used Fe to substitute Ni in $Na_{0.67}Ni_{0.5}Mn_{0.5}O_2$ to obtain P2–$Na_{0.67}Fe_{0.5}Mn_{0.5}O_2$ [24]. This material achieved a reversible capacity of 190 mAh g^{-1} with an average voltage of 2.75 V (Fig. 13). Hence, the energy density was estimated to be 520 Wh kg^{-1}, which is comparable to that of $LiFePO_4$ (about 530 Wh kg^{-1}) and slightly higher than that of $LiMn_2O_4$ (about 450 Wh kg^{-1}) [24]. If assembled with hard carbon (250–300 mAh g^{-1}) or alloy Sn/Sb (300–500 mAh g^{-1}) anode, the energy density of the system would be hopefully close to that of $LiFePO_4$/C battery, even higher than that of $LiMn_2O_4$/C battery. Besides, this material has the advantage of elemental abundance.

3.2 Polyanion-Based Cathode Materials

Over the last decade, polyanion compounds have attracted much attention as the cathode materials for Li-ion battery, due to their highly thermal and structural

Fig. 13 Galvanostatic charge/discharge (oxidation/ reduction) curves for Na/P2– $Na_{0.67}Fe_{0.5}Mn_{0.5}O_2$ cell at a rate of 12 mA/g in the voltage range of 1.5 and 4.3 V [24]

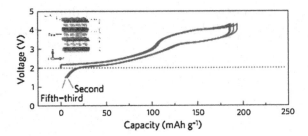

stabilities, such as $LiFePO_4$, $LiMnPO_4$, and so on [25, 26]. Analogously, there are a wide range of Na polyanion compounds with various structures based on the difference in the polyanion and the stoichiometry of the elements. In these compounds, the polyanion polyhedra constitute open 3D frameworks to form ion diffusion channels. Several typical polyanion compounds are illustrated in Fig. 14.

3.2.1 Phosphate Compounds

Among the various Na polyanion compounds, phosphate-based compounds were mostly studied for Na-ion intercalation/deintercalation. NASICON $Na_3V_2(PO_4)_3$, as one of the Na super ion conductors, has fast Na-ion diffusion channels in its crystal structure, [27, 32–34] and hence was widely studied. In the NASICON $Na_3V_2(PO_4)_3$ structure (Fig. 14a), the octahedral VO_6 links the tetrahedral PO_4 via corner to form $[V_2(PO_4)_3]$ unit, which is then interconnected via PO_4 to build up a three-dimensional framework. Na ions selectively occupy two Na sites (Na1 and Na2 in Fig. 14a). One Na ion occupies the Na1 sites, the other two Na ions occupy 2/3 of the Na2 sites. Because the valence of V in $Na_3V_2(PO_4)_3$ is +3, only two Na ions can freely move in/out, corresponding to the redox of the V^{4+}/V^{3+} couple, thus, the theoretical specific capacity of $Na_3V_2(PO_4)_3$ is 117.6 mAh g^{-1}. Uebou et al. first reported the Na intercalation behavior of $Na_3V_2(PO_4)_3$ [35]. In order to improve the electrochemical performance, carbon was used as coating by one-step solid-state reaction [33]. The initial discharge capacity of carbon coated $Na_3V_2(PO_4)_3$ reached 93 mAh g^{-1} with a voltage plateau at 3.4 V and the capacity maintained at 90.9 mAh g^{-1} after 10 cycles in the voltage range of 2.7– 3.8 V. Kim et al. synthesized highly crystallized, nano-scaled $Na_3V_2(PO_4)_3$ encapsulated by a conductive carbon-network (Fig. 15a) by using a polyol-assisted pyro-synthetic reaction [32]. The nanophase $Na_3V_2(PO_4)_3$ delivered a capacity of 117 mAh g^{-1} at 0.08 C (Fig. 15b). Balaya et al. reported a porous $Na_3V_2(PO_4)_3$/C composite obtained via a soft template approach [36]. TEM image (Fig. 16a) showed the $Na_3V_2(PO_4)_3$ nanoparticles were well dispersed in the carbon matrix (graphene clusters (SP2 type carbon)). This material not only exhibited a high discharge capacity (116 mAh g^{-1}), but also high rate capability (~65 mAh g^{-1} at 40 C) (Fig. 16b). Moreover, nearly 50 % of the initial capacity was retained after 30,000 cycles at 40 C (Fig. 16c).

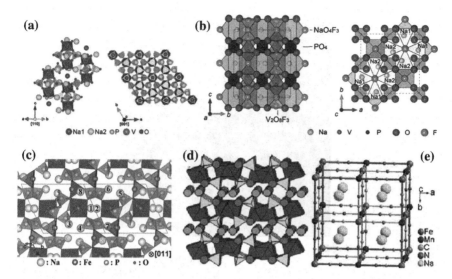

Fig. 14 Crystal structures of some typical polyanion compounds: **a** NASICON–type $Na_3V_2(PO_4)_3$ (R–3c), [27] **b** $Na_3V_2(PO_4)_2F_3$ (P42/mnm), [28] **c** $Na_2FeP_2O_7$ projected along the [011] direction (P1); [29], and **d** Na_2FePO_4F; [30] **e** cubic $NaMnFe(CN)_6$ [31]

Fluorophosphate materials commonly possess higher operating voltages than the corresponding phosphate counterparts [28, 37, 38] due to the strong inductive effect of the fluorine element. $Na_3V_2(PO_4)_2F_3$ (space group of P42/mnm) was first reported by Meins et al. [39]. In the structure, $[V_2O_8F_3]$ bi-octahedral units are linked by the fluorine atoms while the oxygen atoms are all interconnected through the $[PO_4]$ units, leading to the formation of the 3D formwork (Fig. 14b). Na ions have two types of sites in the tunnel: Na1 sites are fully occupied and Na2 sites are half occupied [28]. The charge/discharge profiles of $Na_3V_2(PO_4)_2F_3$ showed two voltage plateaus at about 3.7 and 4.2 V (Fig. 17a), with an average voltage of ~3.95 V [28]. The reversible capacity was ~110 mAh g^{-1} and showed almost no decline after 30 cycles (Fig. 17b). Thus, the energy density of $Na_3V_2(PO_4)_2F_3$ is comparable to that of $LiFePO_4$ and $LiMn_2O_4$ used in Li-ion batteries. Another high-voltage fluorophosphate is $NaVPO_4F$ [37]. Its structure is similar to that of $Na_3Al_2(PO_4)_3F_2$ (space group I4/mmm) [39]. $NaVPO_4F$ was found to have two voltage plateaus at 3.0 and 4.0 V, respectively, and a reversible capacity of ~80 mAh g^{-1} during charge/discharge [28]. This value is much lower than the theoretical capacity (142.5 mAh g^{-1}). Hence, there should still be a very large improvement margin by adopting different synthesis methods or structural tailoring.

A new family with the general formula $Na_3V_2O_{2x}(PO_4)_2F_{3-2x}$ was also studied for high voltage Na intercalation/deintercalation [38]. This type of compounds have similar crystal structure with $Na_3V_2(PO_4)_2F_3$ (tetragonal symmetry, P42/mnm space group) (Fig. 14b). $Na_3V_2O_{2x}(PO_4)_2F_{3-2x}$ showed two voltage plateaus

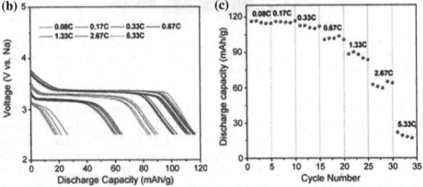

Fig. 15 a TEM images of the $Na_3V_2(PO_4)_3/C$ prepared by pyro–synthetic reaction. **b** The voltage profiles and **c** the discharge capacities obtained at various C–rates in the voltage range 3.8–2.5 V [32]

at 3.6 and 4.1 V, high rate capability (100 mAh g^{-1} at 1 C) and good cyclability (98 % capacity retention for 30 cycles).

Olivine LiFePO$_4$ has been widely investigated as cathode for Li-ion batteries. Unfortunately, NaFePO$_4$ does not present an olivine phase through conventional high temperature preparation conditions. NaFePO$_4$ usually shows a maricite structure, [40, 41] which is electrochemically inactive [30]. Moreau et al. synthesized olivine FePO$_4$ through chemical oxidation in acetonitrile with NO$_2$BF$_4$ [42]. FePO$_4$ is electrochemically active and can allow reversible Na-ion intercalation/deintercalation. In order to obtain electrochemically active olivine-type phosphate compound, Nazar et al. prepared metastable Na[Mn$_{1-x}$Fe$_x$]PO$_4$ nanorods by a low-temperature solid-state method [41]. This material showed a sloping voltage profile, indicating a single-phase reaction, which was different from the two-phase reaction of olivine LiFePO$_4$ [25].

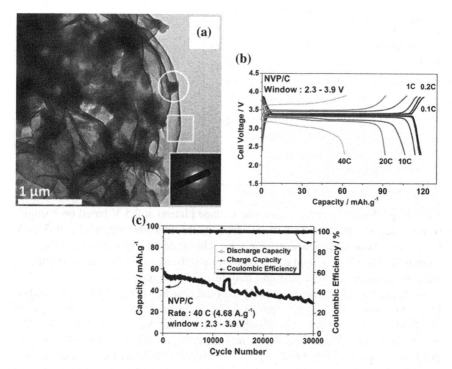

Fig. 16 a TEM images of the $Na_3V_2(PO_4)_3$ (NVP) particles anchored on the carbon matrix. **b** Galvanostatic cycling profiles of NVP/C at different current rates between 2.3 and 3.9 V. **c** Long term cycle life and coulombic efficiency for 30,000 cycles at 40 C [36]

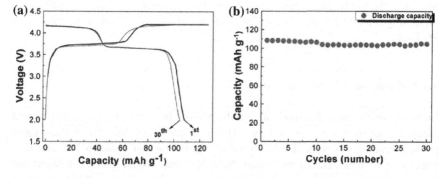

Fig. 17 Electrochemical performance of $Na_3V_2(PO_4)_2F_3$. **a** Charge/discharge profile at the first cycle and thirtieth cycle and **b** cycle performance under a C/10 rate [28]

It was found that $Na_2FeP_2O_7$ was electrochemically active [29]. $Na_2FeP_2O_7$ has a triclinic structure with the space group P1 (Fig. 14c). In the structure, metal polyhedra (FeO_6 and FeO_5) and pyrophosphate (P_2O_7) are interconnected with sharing corners that create various channel structures for Na-ion migration.

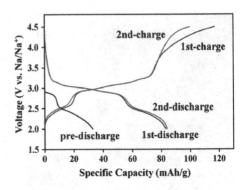

The charge/discharge curves show one voltage plateau at 2.5 V based on a single-phase reaction and another voltage slope in the potential range of 3.0–3.25 V based on a two-phase reaction (Fig. 18). The material can deliver a reversible capacity of 90 mAh g^{-1} and has good cycling stability, which make it a promising cathode candidate for low cost and long-term Na-ion batteries.

Fe-based mixed polyanion cathode material (Na$_4$Fe$_3$(PO$_4$)$_2$(P$_2$O$_7$)) was studied [43]. As shown in Fig. 19, this material exhibits higher reversible capacity (>100 mAh g^{-1}) and potential plateau (~3.0 V) than Na$_2$FeP$_2$O$_7$ [29]. (PO$_4$F)$^{4-}$ based Na–Fe compounds were also investigated [44–46]. Orthorhombic Na$_2$Fe-PO$_4$F is composed of [FePO$_4$F] layers, which are formed by the joining of bioctahedral Fe$_2$O$_7$F$_2$ units with PO$_4$ tetrahedra. The [FePO$_4$F] layers are stacked to form two-dimensional pathways providing Na-ion migration. The reversible Na-ion intercalation/deintercalation capacity reached 120 mAh g^{-1}, with an average voltage of 3.0 V and good cycling stability [44]. If Fe was partly substituted by Mn, the electrochemical performance of this material decreased rapidly with increasing Mn content, due to a strong tendency of structural transition from 2D to 3D [44].

3.2.2 Hexacyano-Type Compound

In the above-mentioned cathode materials, Na ions lie in the complex environment composed of oxides and polyanions. The immigration of Na ions in these close packed structures is difficult due to the strong interaction between Na$^+$ and O^{2-} ions. The replacement of O^{2-} ions by CN$^-$ ions would greatly weaken the interaction of the complexants with Na$^+$, leading to the reduction of the activation energy for Na$^+$ ion migration. Thus, Hexacyano-type compounds have attracted great attention [31, 47, 48].

Yang et al. proposed a new family of Na transition metal cyanides, such as hexacyanoferrate Na$_4$Fe(CN)$_6$ and Prussian blue Na$_x$M$_y$Fe(CN)$_6$ as Na-ion hosts. The as prepared Na$_4$Fe(CN)$_6$/C nanocomposite displayed a full utilization of its redox capacity of 87 mAh g^{-1} at a high potential of ~3.4 V, an excellent cycling

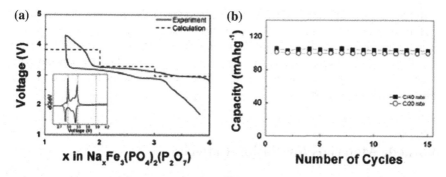

Fig. 19 **a** Galvanostatic charge/discharge profiles of $Na_4Fe_3(PO_4)_2(P_2O_7)$ under a C/40 rate and the calculated average voltage at each region. The inset shows the dQ/dV curve of the initial charge/discharge profile. **b** Cycle performance of a Na cell under C/40 and C/20 rates [43]

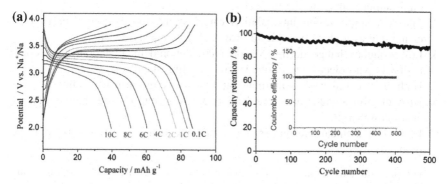

Fig. 20 Electrochemical characterizations of the $Na_4Fe(CN)_6$/C composite: **a** charge and discharge profiles at various C rates ($1C = 90$ mA g^{-1}); and **b** discharge capacity versus cycle numbers [48]

stability with 88 % capacity retention over 500 cycles and superior high rate capability with 45 % capacity delivery at a 10 C rate (Fig. 20) [48]. Prussian blue $Na_xM_yFe(CN)_6$ (M = Fe, Co, Ni, Mn) compounds showed very different electrochemical behaviors, when M was Fe, Co, and Mn, the specific capacities of $Na_xFe_yFe(CN)_6$, $Na_xCo_yFe(CN)_6$, and $Na_xMn_yFe(CN)_6$ reached 113, 120, and 113 mAh g^{-1}, respectively, indicating that both the $Fe(CN)_6^{4-}$ and M^{2+} ions in the Prussian blue lattices were electrochemically active. When M was Ni, Ni ions in the Prussian blue lattice were found to be electrochemically inactive and $Na_{x-}Ni_yFe(CN)_6$ could only deliver a specific capacity of 64 mAh g^{-1} but with quite stable cyclability [31, 47]. Goodenough et al. reported two types of Na manganese hexacyanoferrates: cubic $Na_{1.40}MnFe(CN)_6$ and rhombohedral $Na_{1.72}MnFe(CN)_6$ [31]. The rhombohedral $Na_{1.72}MnFe(CN)_6$ (Fig. 14e) showed higher discharge capacity of 130 mAh g^{-1} while the cubic $Na_{1.40}MnFe(CN)_6$ exhibited higher capacity retention of 96 % after 30 cycles. Overall, Hexacyano-type compounds with open framework have shown several advantages as Na-ion host materials: (1) high discharge voltage plateau; (2) excellent cycling stability due to their stable frameworks during Na intercalation/deintercalation; (3) high reversible capacity (>120 mAh g^{-1}); (4) abundant resources and low cost. So, Hexacyano-type compounds should be promising candidates as cathode hosts for low-cost and long-term Na-ion batteries.

4 Anode Materials for Na-Ion Batteries

Appropriate anode materials should meet the following requirements:

1. Na-ion intercalation/deintercalation potentials are close to Na/Na$^+$, and less influenced by the concentration of Na ions in the lattices, so as to guarantee high output voltage in the full cells.
2. The reversible Na-ion intercalation capacity and the discharge/charge efficiency are as high as possible to provide high capacity density.
3. Low volume change during Na-ion intercalation/deintercalation, good compatibility with electrolyte to obtain good cycling stability.
4. High chemical as well as thermal stability to ensure intrinsical safety.
5. High electron conductivity and Na-ion diffusion rate to offer high C rate at charge/discharge.
6. Low cost, abundant resources, environmental benignity, and easy to mass produce.

The possible anode materials for Na-ion batteries mainly include carbonaceous materials, alloy, and some metal oxides.

4.1 Carbonaceous Materials

Graphite and graphite-like materials were most often surveyed for alkali-ion intercalation anodes. The Na intercalation behavior in graphite was first investigated in the late 1970s [49, 50–52]. However, the intercalation amount of Na in graphite is low (~NaC$_{64}$) compared to that of LiC$_6$ and KC$_8$. Other carbon material with microcrystalline graphitic structure and microporous domains were found to have high Na storage capacity, such as carbon black (121 mAh g^{-1}), [53] petroleum coke, [51] and carbon fibers (209 mAh g^{-1}) [52, 54, 55].

A variety of precursors were used to obtain different structured hard carbon materials. Dahn et al. found that the hard carbon obtained by the glucose pyrolyzation showed reversible Na intercalation capacity of ~300 mAh g^{-1} [56]. Microspherical hard carbon particles with disordered nanostructure (Fig. 21a) synthesized by pyrolyzing resorcinol–formaldehyde compounds, [57] delivered an initial capacity of 285 mAh g^{-1}, and over 255 mAh g^{-1} after seven cycles in

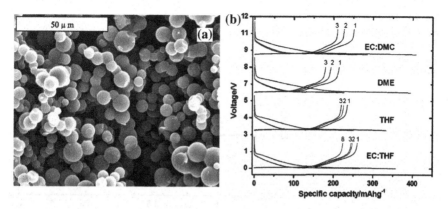

Fig. 21 SEM of carbon microspheres. **b** Voltage/capacity plots corresponding to the first discharge/charge cycles of carbon aerogel microspheres in Na cells, 1 M NaClO$_4$ dissolved in different solutions as electrolyte [57]

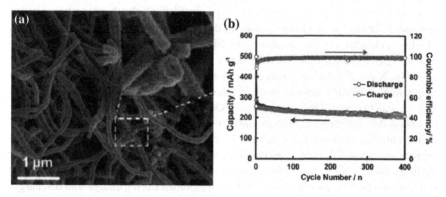

Fig. 22 **a** SEM images of the PANI–HNWs. **b** Cycle performance of the HCNW electrode at a current density of 50 mA/g (0.2 C) [58]

EC–THF mixture electrolyte (Fig. 21 b). Cao et al. reported hollow carbon nanowires (HCNWs) prepared by the pyrolyzation of a hollow polyaniline nanowire precursor (Fig. 22a) [58]. The HCNW electrode delivered high reversible capacity of 251 mAh g^{-1} and excellent cycling stability (82.2 % of capacity retention over 400 cycles) (Fig. 22b). Such excellent electrochemical performance was ascribed to the HCNWs that had a uniform hollow nanowire structure and an appropriate interlayer distance, leading to short diffusion distance for Na insertion, stable material structure, and feasible approach for Na-ion insertion into carbon layers. Besides, Komaba et al. reported the effect of electrolytes on Na insertion behavior in hard carbon [59]. It was found that in EC:DEC(1:1) solution containing 1 mol L^{-1} NaClO$_4$, the hard carbon electrode exhibited a high capacity of ca. 240 mAh g^{-1} with a stable capacity retention over 100 cycles. This was related

Fig. 23 Theoretical energy
cost for Na (*red curve*) and Li
(*blue curve*) ions insertion
into carbon as a function of
carbon interlayer distance
[58]

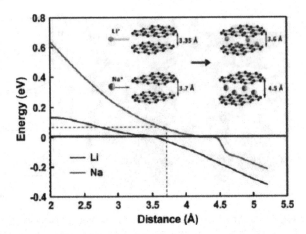

to the different chemical and electrochemical reaction of carbonate solution on the surface of the hard carbon material [59].

It was found that the charge/discharge profiles of the hard carbon anodes commonly comprise two parts: a slope at the high potential region and a plateau at the low potential region, as shown in Fig. 21b [56, 59]. Several spectroscopy technologies have been used to investigate the specific mechanism, such as ex-situ or in-situ X-ray diffraction, small-angle X-ray scattering (SAXS), [55, 60, 61] [23]Na magic angle spinning nuclear magnetic resonance (MAS NMR) spectra, [57] and Raman spectroscopy [59]. By in-situ SAXS investigation, Dahn et al. first demonstrated that the slope at high potential region corresponded to Na intercalation/deintercalation between the graphene layers and the plateau at the low potential region corresponded to Na adsorption/desorption in the nanopores of the carbon particles [60]. ^{23}Na MAS NMR spectra corresponding to the hard carbon microspheres also clearly showed that two signal responses during charge and discharge, which could be ascribed to Na insertion between misaligned graphene carbon framework and in the nanocavities or on a surface solid film, respectively [57]. In Raman spectroscopy, the shift of the G-band indicates the change of the state of negatively charged graphenes, the redshift observed during the voltage-sloping region corresponds to the Na insertion between the graphene layers. No shift of G-band in lower potential region can be ascribed to the formation of a nano-sized cluster of quasimetallic Na in the nanopores of the hard carbon [59].

It is obvious that the interplanar distances of the graphene layers in hard carbon will play a vital role on the Na intercalation behavior. Cao et al. carried out a theoretical simulation on the energy cost for the Na-ion insertion into carbon as a function of the carbon interlayer distance (Fig. 23) [58]. For comparison, the energy cost curve for Li-ion insertion into carbon was also calculated (Fig. 23). For graphite with an interlayer spacing of 0.335 nm, the energy cost for Li-ion and Na-ion insertion are 0.03 and 0.12 eV, respectively. In consideration that the energy of the thermal fluctuations at room temperature is 0.00257 eV, Li-ion intercalation into graphite layers is permissible while Na-ion is prohibitive.

Fig. 24 Na–M voltage curves calculated using DFT and known Na–M crystal structures. **a** M–Si **b** M–Ge **c** M–Sn, and **d** M–Pb [63]

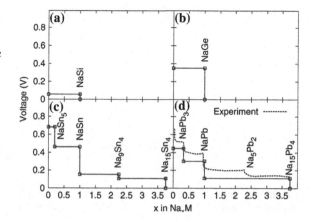

However, when the layer spacings increase to 0.37 nm, the energy barrier for Na-ion insertion drops markedly to 0.053 eV, indicating a feasibility of Na-ion insertion. Therefore, it is important to seek carbon materials with appropriate layer spacing to improve the Na insertion performance.

4.2 Alloy Materials

As shown above, the Na storage capacity of carbonaceous materials is commonly <300 mAh g^{-1}, it is difficult to further enhance their capacity due to the limited host sites in the carbon structure. Besides, the Na intercalation potentials in carbonaceous materials are close to the plating potential of Na ions, which would lead to a safety concern. Hence, it is necessary to seek alternative anode materials for high-capacity and high-safety Na-ion batteries. Analogous to Li alloy, Na can also alloy with some Group IVA and VA metal elements, such as Sn, Sb, Pb, and Ge [62]. These alloys are estimated to deliver high reversible capacity due to their high theoretical-specific capacities, such as 847 ($Na_{15}Sn_4$), 660 (Na_3Sb), 1108 (Na_3Ge), and 484 ($Na_{15}Pb_4$) mAh g^{-1}, respectively. Besides, the alloying reactions have higher thermodynamic potential than Na–C reactions, making them potentially safer. Recently, Ceder et al. reported the Na–M voltage curves calculated through density functional theory (DFT) for Si, Ge, Sn, and Pb (Fig. 24) [63]. The voltage curves for Si and Ge showed only one voltage plateau, indicating the formation of a single alloy phase. On the contrary, the voltage curves for Sn and Pb exhibited multi-voltage plateaus, corresponding to the formation of multiple alloy phases during the Na intercalation/deintercalation. These theoretical calculations are in good agreement with the experimental data in the literatures [62, 64–68] Jow et al. studied the electrochemical performance of $Na_{15}Pb_4$ alloy for Na intercalation [64]. The metallurgically formed $Na_{15}Pb_4$ gave out 86 % of the Na storage capacity at a low rate of 50 μA cm^{-3}.

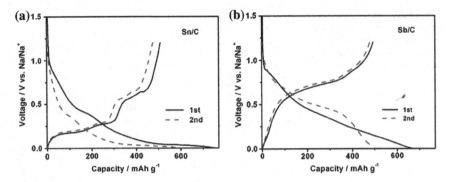

Fig. 25 The initial two discharge/charge profiles of the **a** Sn/C and **b** Sb/C nanocomposite electrodes between 0.0 and 1.2 V versus Na/Na$^+$ at a current rate of 100 mA g^{-1} [62]

Fig. 26 Cycling performance of the SnSb/C nanocomposite electrode at a cycling rate of 100 mA g^{-1} [62]

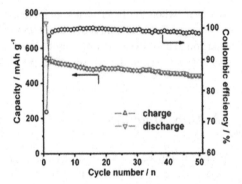

Xiao et al. first studied the Na storage properties in Sn and Sb (Fig. 25) [62]. As shown in Fig. 25a, the voltage profiles of Sn/C nanocomposites clearly showed two discharge and four charge plateaus, reflecting the stepwise Na–Sn alloy phase transition processes, similar to the calculated theoretical data (Fig. 24). The Sn/C electrode delivered a high reversible capacity of 509 mAh g^{-1} (Fig. 25a). After deducting the capacity associated with carbon black, Sn individually can offer a capacity of 653 mAh g^{-1}, which is about 77.1 % of the theoretical capacity of 847 mAh g^{-1} based on Na$_{15}$Sn$_4$. The Sb/C electrode also showed high initial capacity of 494 mAh g^{-1} (Fig. 25b). Sb individually can offer a capacity of 631 mAh g^{-1}, which is equivalent to 95.6 % of the theoretical capacity based on Na$_3$Sb (660 mAh g^{-1}). But both electrodes exhibited a rapid decrease in capacity with cycling [62]. This situation is commonly found in Li–alloy electrode, resulting from severe volume changes of the electrode during the alloying/dealloying processes which cause the pulverization of the metal particles, the loss of the electric conducting between the active materials. However, the SnSb/C composite exhibited high reversible capacity of 544 mAh g^{-1} and particularly excellent cycling capability with 80 % of capacity retention over 50 cycles (Fig. 26) [62]. It is proposed that when discharged to 0.4 V, Na ions insert into the SnSb alloy to form Na_3Sb and Sn phase, while the Sn phase is inactive at this time and acts as the conducting buffer to maintain structural integrity. When discharge continues to <0.4 V, alloying reaction of Na with Sn occurs and Na_3Sb is inactive and serves as a buffer matrix. So, such a self-supporting network can maintain the integrity and conductivity of the whole electrode material to provide good cycling stability of the electrode.

Yang et al. prepared an Sb/C composite by mechanical milling, which was composed of ~10 nm nanocrystallite Sb embedded in the carbon matrix [67]. The naocrystalline Sb/C electrode delivered an initial capacity as high as 610 mAh g^{-1} based on the pure Sb, very close to the theoretical capacity (Na_3Sb, 660 mAh g^{-1}). The Sb/C electrode in a 5 % FEC-containing electrolyte maintained an almost constant capacity of 575 mAh g^{-1} over 100 cycles [67]. The excellent electrochemical performances enable the Sb/C nanocomposite to be used as candidate anode for high capacity and high safety Na-ion batteries.

4.3 Metal Oxide Materials

In general, the Na storage capacities in materials with an intercalation reaction mechanism are limited by the number of host sites in their structures. One way to circumvent this intrinsic limitation and to achieve higher capacities would be the use of materials with a reversible conversion reaction mechanism. The Na storage capacity then depends on the change of the redox valence states of the material while not the material structure. Such materials are mainly oxides, [69–71] fluorides, [72] and sulfides [73]. Among them, oxides have the most appropriate Na storage potentials. Tirado et al. first reported the electrochemical characteristics of $NiCo_2O_4$ spinel for Na storage [70]. It showed that the $NiCo_2O_4$ electrode delivered a reversible capacity of ca. 200 mAh g^{-1} with an average potential of 1.5 V. Co_3O_4 was also investigated as Na storage anode [71]. This material delivered a reversible capacity of 444 mAh g^{-1}, but with poor cycling performance.

Johnson et al. synthesized amorphous titanium dioxide nanotube (TiO_2 NT) to investigate its utilization for Na-ion batteries [74]. It was found that only when the size of the nanotubes increased to >80 nm (wall thickness >15 nm), the amorphous TiO_2 could show relatively low specific capacity initially, which then self-improved as cycling proceeded. The electrode delivered a reversible capacity of 75 mAh g^{-1} in the first cycle and the capacity increased to 150 mAh g^{-1} after 15 cycles (Fig. 27a). This behavior can be explained that Na ions cannot adsorb on TiO_2 surface and screen the charge of electrons injected into the TiO_2 matrix in the relatively small-diameter tube, so that insufficient Na ions can not establish critical ion concentration that supports cycling. Thus, it is important to tailor the nanostructure of materials to enhance their Na insertion performance. Palacin et al. reported the use of $Na_2Ti_3O_7$ as anode for Na-ion batteries [75, 76]. $Na_2Ti_3O_7$

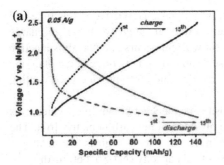

Fig. 27 **a** Charge/discharge galvanostatic curves of amorphous 80 nm I.D. TiO_2 NT in Na half cell (*red* for discharge and *black* for charge) cycled between 2.5 and 0.9 V versus Na/Na^+ at 0.05A g^{-1}. **b** SEM top–view images of TiO_2 NT electrodes [74]

consists of zigzag layers of Ti–O octahedra with Na ions in the interlayer spacing [77]. Additional Na ions (2 Na ions per formular unit) can insert into the interlayer space. The $Na_2Ti_3O_7$ electrode could reversibly deliver a capacity of 200 mAh g^{-1} at an average potential of 0.3 V and with good cycling stability.

5 Summary and Outlook

Although along with the development of Li intercalation materials, their Na analogs have also been investigated, the performance of Na intercalation materials is far inferior to the lithium counterparts, the energy density of the thus constructed Na-ion batteries then is far below that of Li-ion batteries. Moreover, some research on Na intercalation reactions focus only on the superficial exploratory study, deep consideration of materials and systems are bare. Time enters in the twenty-first century, and due to the shortage of fossil fuel and the worsening of environmental pollution, the development of renewable energy sources and electric vehicles has drawn more and more attention. Li-ion batteries that have occupied the portable electronic product market are considered as the ideal system for electric vehicle propulsion and renewable electric power storage. However, due to the concern of insufficient Li reserves, scientists again begin to show interest in Na-ion battery systems, which have no resource constraints. In the last 3 years, a wide range of new types of Na intercalation materials have been proposed and deeply investigated. The electrochemical performance of several materials was close to that of Li-ion batteries. For example, $P2–Na_{0.67}Fe_{0.5}Mn_{0.5}O_2$ cathode material can reach an energy density of 520 Wh kg^{-1}, which is comparable to that of $LiFePO_4$ (about 530 Wh kg^{-1}) and slightly higher than $LiMn_2O_4$ (about 450 Wh kg^{-1}). Rhombohedral $Na_{1.72}MnFe(CN)_6$ has a reversible capacity of 134 mAh g^{-1} with a high potential plateau of 3.4 V, very close to that of $LiFePO_4$ (140 mAh g^{-1} and 3.4 V potential plateau). For the anode, hard carbon with hollow nanowire structure can deliver a reversible capacity of 250 mAh g^{-1} with excellent cycling stability. The capacity of Sb/C nanocomposite can achieve 420 mAh g^{-1} based on the material including Sb and carbon. The electrochemical performance of these anode materials is close or superior to graphite commercially used in Li-ion batteries. If we use the above-mentioned materials to construct practical Na-ion batteries, the energy density of the batteries would be close to the commercial graphite/LiFePO$_4$ system. Although some achievements are obtained in Na intercalation materials, they are still far from realizing the real utility of Na-ion batteries. Some work should be further pursued, such as the industrial mass production of materials, balance and interaction of the cathode and anode, special design of the cell configuration, and safety protection of batteries. All in all, the development of Na-ion batteries is facing great chances as well as challenges, we believe that through the unremitting efforts of scientists, this newly emerging energy storage system will finally realize its application.

References

1. Slater MD, Kim D et al (2013) Sodium-ion batteries. Adv Funct Mater 23:947–958
2. Kim S-W, Seo D-H et al (2012) Electrode materials for rechargeable sodium-ion batteries: potential alternatives to current lithium-ion batteries. Adv Energy Mater 2:710–721
3. Delmas C, Fouassier C et al (1980) Structural classification and properties of the layered oxides. Physica B+C 99:81–85
4. Cao YL, Xiao LF et al (2011) Reversible sodium ion insertion in single crystalline manganese oxide nanowires with long cycle life. Adv Mater 23:3155–3160
5. Sauvage F, Laffont L et al (2007) Study of the insertion/deinsertion mechanism of sodium into Na$_{0.44}$MnO$_2$. Inorg Chem 46:3289–3294
6. Whitacre JF, Tevar A et al (2010) Na$_4$Mn$_9$O$_{18}$ as a positive electrode material for an aqueous electrolyte sodium-ion energy storage device. Electrochem Commun 12:463–466
7. Saint JA, Doeff MM et al (2008) Electrode materials with the Na$_{0.44}$MnO$_2$ structure: effect of titanium substitution on physical and electrochemical properties. Chem Mater 20:3404–3411
8. Braconnier J-J, Delmas C et al (1980) Comportement electrochimique des phases Na$_x$CoO$_2$. Mater Res Bull 15:1797–1804
9. Berthelot R, Carlier D et al (2011) Electrochemical investigation of the P2–Na$_x$CoO$_2$ phase diagram. Nat Mater 10:74–80
10. Bhide A, Hariharan K (2011) Physicochemical properties of Na$_x$CoO$_2$ as a cathode for solid state sodium battery. Solid State Ionics 192:360–363
11. Parant J-P, Olazcuaga R et al (1971) Sur quelques nouvelles phases de formule Na$_x$MnO$_2$ ($x \leq 1$). J Solid State Chem 3:1–11
12. Caballero A, Hernan L et al (2002) Synthesis and characterization of high-temperature hexagonal P2-Na$_{0.6}$MnO$_2$ and its electrochemical behaviour as cathode in sodium cells. J Mater Chem 12:1142–1147
13. West K, Zachau-Christiansen B et al (1988) Sodium insertion in vanadium oxides. Solid State Ionics 28–30(Part 2):1128–1131
14. Bach S, Baffier N et al (1989) Electrochemical sodium intercalation in Na$_{0.33}$V$_2$O$_5$ bronze synthesized by a sol-gel process. Solid State Ionics 37:41–49
15. Pereira-Ramos JP, Messina R et al (1990) Influence of the synthesis via a sol-gel process on the electrochemical lithium and sodium insertion in β-Na$_{0.33}$V$_2$O$_5$. Solid State Ionics 40–41(Part 2):970–973

16. Tepavcevic S, Xiong H et al (2012) Nanostructured bilayered vanadium oxide electrodes for rechargeable sodium-ion batteries. ACS Nano 6:530–538
17. Hamani D, Ati M et al (2011) Na_xVO_2 as possible electrode for Na-ion batteries. Electrochem Commun 13:938–941
18. Onoda M (2008) Geometrically frustrated triangular lattice system Na_xVO_2: superparamagnetism in x = 1 and trimerization in x approximate to 0.7. J Phys-Condens Matter 20:145205
19. McQueen TM, Stephens PW et al (2008) Successive Orbital Ordering Transitions in $NaVO_2$. Phys Rev Lett 101:166402
20. Carlier D, Cheng JH et al (2011) The $P2-Na_{2/3}Co_{2/3}Mn_{1/3}O_2$ phase: structure, physical properties and electrochemical behavior as positive electrode in sodium battery. Dalton Trans 40:9306–9312
21. Komaba S, Nakayama T et al (2009) Electrochemically reversible sodium intercalation of layered $NaNi_{0.5}Mn_{0.5}O_2$ and $NaCrO_2$. ECS Trans 16:43–55
22. Mendiboure A, Delmas C et al (1985) Electrochemical intercalation and deintercalation of $NaMnO_2$ bronzes. J Solid State Chem 57:323–331
23. Kim D, Kang SH et al (2011) Enabling sodium batteries using lithium-substituted sodium layered transition metal oxide cathodes. Adv Energy Mater 1:333–336
24. Yabuuchi N, Kajiyama M et al (2012) P2-type $Na_x[Fe_{1/2}Mn_{1/2}]O_2$ made from earth-abundant elements for rechargeable Na batteries. Nat Mater 11:512–517
25. Padhi AK, Nanjundaswamy KS et al (1997) Phospho-olivines as positive-electrode materials for rechargeable lithium batteries. J Electrochem Soc 144:1188–1194
26. Martha SK, Markovsky B et al (2009) $LiMnPO_4$ as an advanced cathode material for rechargeable lithium batteries. J Electrochem Soc 156:A541–A552
27. Lim SY, Kim H et al (2012) Electrochemical and thermal properties of NASICON structured $Na_3V_2(PO_4)_3$ as a sodium rechargeable battery cathode: a combined experimental and theoretical study. J Electrochem Soc 159:A1393–A1397
28. Shakoor RA, Seo DH et al (2012) A combined first principles and experimental study on $Na_3V_2(PO_4)_2F_3$ for rechargeable Na batteries. J Mater Chem 22:20535–20541
29. Kim H, Shakoor RA et al (2013) $Na_2FeP_2O_7$ as a promising iron-based pyrophosphate cathode for sodium rechargeable batteries: a combined experimental and theoretical study. Adv Funct Mater 23:1147–1155
30. Ellis BL, Makahnouk WRM et al (2007) A multifunctional 3.5 V iron-based phosphate cathode for rechargeable batteries. Nat Mater 6:749–753
31. Wang L, Lu Y et al (2013) A superior low-cost cathode for a Na-ion battery. Angew Chem Int Ed 52:1964–1967
32. Kang J, Baek S et al (2012) High rate performance of a $Na_3V_2(PO_4)_3$/C cathode prepared by pyro-synthesis for sodium-ion batteries. J Mater Chem 22:20857–20860
33. Jian Z, Zhao L et al (2012) Carbon coated $Na_3V_2(PO_4)_3$ as novel electrode material for sodium ion batteries. Electrochem Commun 14:86–89
34. Jian Z, Han W et al (2013) Superior electrochemical performance and storage mechanism of $Na_3V_2(PO_4)_3$ cathode for room-temperature sodium-ion batteries. Adv Energy Mater 3:156–160
35. Uebou Y, Kiyabu T et al (2002) The reports of institute of advanced material study (vol 16). Kyushu University, Fukuoka, p 1
36. Saravanan K, Mason CW et al (2013) The first report on excellent cycling stability and superior rate capability of $Na_3V_2(PO_4)_3$ for sodium ion batteries. Adv Energy Mater 3:444–450
37. Barker J, Saidi MY et al (2003) A sodium-ion cell based on the fluorophosphate compound $NaVPO_4F$. Electrochem Solid-State Lett 6:A1–A4
38. Serras P, Palomares V et al (2012) High voltage cathode materials for Na-ion batteries of general formula $Na_3V_2O_{2x}(PO_4)_2F_{3-2x}$. J Mater Chem 22:22301–22308

39. Le Meins JM, Crosnier-Lopez MP et al (1999) Phase Transitions in the $Na_3M_2(PO_4)_2F_3$ Family (M = Al^{3+}, V^{3+}, Cr^{3+}, Fe^{3+}, Ga^{3+}): Synthesis, Thermal, Structural, and Magnetic Studies. J Solid State Chem 148:260–277

40. Zaghib K, Trottier J et al (2011) Characterization of Na-based phosphate as electrode materials for electrochemical cells. J Power Sources 196:9612–9617

41. Lee KT, Ramesh TN et al (2011) Topochemical synthesis of sodium metal phosphate olivines for sodium-ion batteries. Chem Mater 23:3593–3600

42. Moreau P, Guyomard D et al (2010) Structure and stability of sodium intercalated phases in olivine $FePO_4$. Chem Mater 22:4126–4128

43. Kim H, Park I et al (2012) New iron-based mixed-polyanion cathodes for lithium and sodium rechargeable batteries: combined first principles calculations and experimental study. J Am Chem Soc 134:10369–10372

44. Recham N, Chotard JN et al (2009) Ionothermal synthesis of sodium-based fluorophosphate cathode materials. J Electrochem Soc 156:A993–A999

45. Ellis BL, Makahnouk WRM et al (2010) Crystal structure and electrochemical properties of A_2MPO_4F Fluorophosphates (A = Na, Li; M = Fe, Mn Co, Ni). Chem Mater 22:1059–1070

46. Kawabe Y, Yabuuchi N et al (2011) Synthesis and electrode performance of carbon coated Na_2FePO_4F for rechargeable Na batteries. Electrochem Commun 13:1225–1228

47. Wessells CD, McDowell MT et al (2012) Tunable reaction potentials in open framework nanoparticle battery electrodes for grid-scale energy storage. ACS Nano 6:1688–1694

48. Qian J, Zhou M et al (2012) Nanosized $Na_4Fe(CN)_6$/C composite as a low-cost and high-rate cathode material for sodium-ion batteries. Adv Energy Mater 2:410–414

49. Besenhard JO (1976) The electrochemical preparation and properties of ionic alkali metal- and NR4-graphite intercalation compounds in organic electrolytes. Carbon 14:111–115

50. Ge P, Fouletier M (1988) Electrochemical intercalation of sodium in graphite. Solid State Ionics 28–30(Part 2):1172–1175

51. Doeff MM, Ma YP et al (1993) Electrochemical insertion of sodium into carbon. J Electrochem Soc 140:L169–L170

52. Thomas P, Billaud D (2000) Effect of mechanical grinding of pitch-based carbon fibers and graphite on their electrochemical sodium insertion properties. Electrochim Acta 46:39–47

53. Alcantara R, Jimenez-Mateos JM et al (2001) Carbon black: a promising electrode material for sodium-ion batteries. Electrochem Commun 3:639–642

54. Thomas P, Ghanbaja J et al (1999) Electrochemical insertion of sodium in pitch-based carbon fibres in comparison with graphite in $NaClO_4$-ethylene carbonate electrolyte. Electrochim Acta 45:423–430

55. Thomas P, Billaud D (2001) Sodium electrochemical insertion mechanisms in various carbon fibres. Electrochim Acta 46:3359–3366

56. Stevens DA, Dahn JR (2000) High capacity anode materials for rechargeable sodium-ion batteries. J Electrochem Soc 147:1271–1273

57. Alcantara R, Lavela P et al (2005) Carbon microspheres obtained from resorcinol-formaldehyde as high-capacity electrodes for sodium-ion batteries. Electrochem Solid-State Lett 8:A222–A225

58. Cao Y, Xiao L et al (2012) Sodium ion insertion in hollow carbon nanowires for battery applications. Nano Lett 12:3783–3787

59. Komaba S, Murata W et al (2011) Electrochemical Na insertion and solid electrolyte interphase for hard-carbon electrodes and application to Na-ion batteries. Adv Funct Mater 21:3859–3867

60. Stevens DA, Dahn JR (2000) An in situ small-angle X-ray scattering study of sodium insertion into a nanoporous carbon anode material within an operating electrochemical cell. J Electrochem Soc 147:4428–4431

61. Stevens DA, Dahn JR (2001) The mechanisms of lithium and sodium insertion in carbon materials. J Electrochem Soc 148:A803–A811

62. Xiao L, Cao Y et al (2012) High capacity, reversible alloying reactions in SnSb/C nanocomposites for Na-ion battery applications. Chem Commun 48:3321–3323

63. Chevrier VL, Ceder G (2011) Challenges for Na-ion negative electrodes. J Electrochem Soc 158:A1011–A1014
64. Jow TR, Shacklette LW et al (1987) The role of conductive polymers in alkali-metal secondary electrodes. J Electrochem Soc 134:1730–1733
65. Xu Y, Zhu Y et al (2013) Electrochemical performance of porous carbon/tin composite anodes for sodium-ion and lithium-ion batteries. Adv Energy Mater 3:128–133
66. Komaba S, Matsuura Y et al (2012) Redox reaction of Sn-polyacrylate electrodes in aprotic Na cell. Electrochem Commun 21:65–68
67. Qian J, Chen Y et al (2012) High capacity Na-storage and superior cyclability of nanocomposite Sb/C anode for Na-ion batteries. Chem Commun 48:7070–7072
68. Wu L, Pei F et al (2013) SiC–Sb–C nanocomposites as high-capacity and cycling-stable anode for sodium-ion batteries. Electrochim Acta 87:41–45
69. Chadwick AV, Savin SLP et al (2007) Formation and oxidation of nanosized metal particles by electrochemical reaction of Li and Na with $NiCo_2O_4$: X-ray absorption spectroscopic study. J Phys Chem C 111:4636–4642
70. Alcantara R, Jaraba M et al (2002) $NiCo_2O_4$ spinel: First report on a transition metal oxide for the negative electrode of sodium-ion batteries. Chem Mater 14:2847–2848
71. Kuroda Y, Kobayashi E et al (2010) Electrochemical properties of spinel-type oxide anodes in sodium-ion battery. In: 218th ECS meeting abstract #389
72. Nishijima M, Gocheva ID et al (2009) Cathode properties of metal trifluorides in Li and Na secondary batteries. J Power Sources 190:558–562
73. Kim TB, Choi JW et al (2007) Electrochemical properties of sodium/pyrite battery at room temperature. J Power Sources 174:1275–1278
74. Xiong H, Slater MD et al (2011) Amorphous TiO_2 nanotube anode for rechargeable sodium ion batteries. J Phys Chem C 2:2560–2565
75. Senguttuvan P, Rousse G et al (2011) $Na_2Ti_3O_7$: lowest voltage ever reported oxide insertion electrode for sodium ion batteries. Chem Mater 23:4109–4111
76. Wang W, Yu CJ et al (2013) Single crystalline $Na_2Ti_3O_7$ rods as an anode material for sodium-ion batteries. RSC Adv 3:1041–1044
77. Andersson S, Wadsley AD (1961) The crystal structure of $Na_2Ti_3O_7$. Acta Crystallogr A 14:1245–1249

Chemical Routes to Graphene-Based Flexible Electrodes for Electrochemical Energy Storage

Fei Liu and Dongfeng Xue

Abstract Due to their many fascinating properties and low-cost preparation by chemical reduction method, particular attention has been paid to the graphene-based materials in the application of energy storage devices. In the present chapter, we focus on the latest work regarding the development of flexible electrodes for batteries and supercapacitors based on graphene as well as graphene-based composites. To begin with, graphene as the sole or dominant part of flexible electrode will be discussed, involving its structure, relationship between structure and performance, and strategies to improve their performances; The next major section deals with graphene as conductive matrix for flexible electrode, the role of graphene to offer efficient electrically conductive channels and flexible mechanical supports will be discussed. Another role of graphene in flexible electrode is as active additives to improve the performance of cellulose and carbon nanofiber papers, examples will be given and such strategy is promising for further reducing the cost of flexible electrodes. Finally, prospects and further developments in this exciting field of graphene-based flexible energy storage devices will be also suggested.

Keywords Graphene · Composite · Flexible · Supercapacitor · Li-ion battery

1 Introduction

There has been an increasing interest recently in soft or bendable portable electronic equipment, such as roll-up displays, wearable devices and implanted medical devices, to meet the various requirements of advanced applications [1–4].

F. Liu · D. Xue (✉)
State Key Laboratory of Rare Earth Resource Utilization, Changchun Institute of Applied Chemistry, Chinese Academy of Sciences, 5625 Renmin Street, Changchun 130022, China
e-mail: dongfeng@ciac.ac.cn

F. Liu · D. Xue
School of Chemical Engineering, Dalian University of Technology, Dalian 116024, China

Z. Lin and J. Wang (eds.), *Low-cost Nanomaterials*, Green Energy and Technology,   425
DOI: 10.1007/978-1-4471-6473-9_15, © Springer-Verlag London 2014

The fabrication of such devices requires the development of new type of batteries or supercapacitors that are flexible as their power sources. Typically, batteries and supercapacitors consist of several major parts, which are electrodes (including a positive electrode and a negative electrode), separator, and electrolyte, in which two electrodes are spaced by a separator, and all these parts are soaked in an electrolyte solution or gel [5]. A conventional electrode for battery or supercapacitor is fabricated on a metal substrate that is coated with a thin layer mixture of active material, electrical conductor, and binder, as illustrated in Fig. 1. This kind of electrode configuration is not suitable for flexible or bendable application conditions, because the heterogeneous interface between the active material layer and the metal substrate is not strong enough to endure the shape changing process [4]. The active material layer will be damaged or peeled off from the substrate thereby causing the rapid performance fading.

Recent reports show that thin film or paper-like materials could be adopted as key elements for application in flexible energy storage devices; several routes toward the development of flexible batteries are being explored [6–9]. A range of studies focus on batteries developed for disposable-card applications. However, primary batteries can only produce current by an irreversible chemical reaction and cannot be recharged, which therefore hindered their application in portable electronic equipment [1]. Secondary Li-ion batteries (LIBs) and supercapacitors are generally used to power portable electronic equipment [10–15]. Fabrication of LIBs and supercapacitors into thin films while maintaining their high energy and power density is crucial for their application in flexible devices. Making a flexible LIB requires the development of soft electrode containing electrochemical-active materials, such as metal oxide nanoparticles for cathodes and nanocarbons for anodes, while flexible supercapacitor requires electrodes with high surface area for electro double layer capacitance (EDLC) or high electrochemical activity for pseudocapacitance [16–19]. There has been a great number of research works on the development of electrode materials designed for LIBs and supercapacitors in regular use conditions, in which conventional electrodes fabricated on metal substrates are applied, and tremendous achievements have been accomplished [13, 20–27]. But the researches on the flexible electrode for the application in the soft or bendable LIB and supercapacitors have just been carried out in the very recent years.

For flexible devices, polymers can be used as substrate [4]. However, most polymers cannot offer stability for electrode fabrication, since they degrade easily at a wide range of environments. Also, poor adhesion between polymer and oxide materials would be problematic upon long-term battery cycling. The free-standing paper-like carbon-based materials, featured by light-weight, flexible, and high conductivity, appear to be promising electrodes for wearable or rolling-up devices. Flexible electrode materials made of carbon nanotubes (CNT) and their composites have been extensively studied [6, 28, 29]. Although these binder-free electrodes afford wonderful mechanical properties and desirable electrochemical properties, the still high production cost of CNTs and the difficulty in making

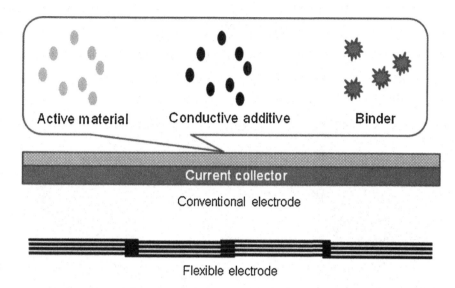

Fig. 1 Schematic illustration of conventional electrode and flexible electrode

stable CNT dispersions (which is essential to make CNT papers) have limited their practical application as flexible electrodes.

Following the discoveries of fullerene and CNT in the earlier decades, the successful fabrication of graphene has recently opened up an astonishing new field in the research of materials science and technology [30–38]. Graphene is a two-dimensional (2D) monolayer sheet of sp^2-bonded carbon, which possesses unique optical, electrical, mechanical, and electrochemical properties. The surface area of graphene is calculated to be 2,630 m^2 g^{-1}; graphene has high structural stability, high electric conductivity, and easy to be functionalized with other molecules, and all these characters are highly favorable for energy storage applications [39–41]. A family of graphene-related materials, can all be commonly called "graphene" by the research community, consists of structural or chemical derivatives of graphene. These include double- and few-layer graphene and chemical reduced graphene oxide (reduced GO, RGO) [42]. Inspired by their promising properties and enormous potential applications, great attentions have rapidly been paid to explore new significant scientific problems about graphene since its discovery. Actually, graphene-based materials have the enormous potential to rival or even surpass the performance of their CNT-based counterparts, given that cheap, large-scale production and processing methods for high-quality graphene have already been achieved [42–44].

Micromechanical cleavage from bulk graphite was firstly to be used to produce graphene by Geim et al. [30], but the yield of this method is extremely low and the process is uncontrollable. After that, various synthesis methods have been developed, including epitaxial growth, chemical vapor deposition (CVD), etc. [42, 45, 46]; however, these methods usually require high temperature, high-cost

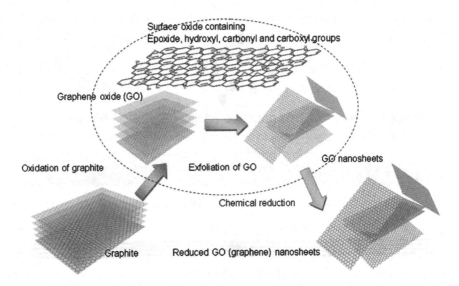

Fig. 2 Illustration of the production of graphene from the graphite oxidation route

apparatus, and accurate control over operating, which hindered the large-scale manufacturing of graphene. Instead of the high-cost methods for generating graphene, the production of graphene sheets by oxidative exfoliation of graphite and subsequent reduction (Fig. 2) can offer the high-volume production of RGO (graphene) [42]. Furthermore, functionalized graphene materials are highly dispersible in a wide range of solvents, make it easy processable and can be assembled into various desired macroscopic structures or incorporated with other nanomaterials into functional composites [32, 41, 44, 47]. Although graphene materials produced by the chemical route possess such advantages, they are still far inferior to products prepared by mechanical cleavage or CVD due to the high density of plane defect, which may be undesirable for some technological applications such as in electronics. However, the defects induced by the chemical process can provide chemical active sites on the graphene surface, which may enhance the performance of graphene in the application where high chemical activity is required, combined with its low cost and large yield, the chemical reduction generated graphene is especially suitable in the area of energy storage, catalyst, and chemical sensors, etc.

Recently, Ruoff and Wallace et al. demonstrated that vacuum filtration of the as-prepared GO or graphene dispersion can result in the formation of free-standing paper-like materials exhibiting smooth surfaces, which can be called graphene paper (GO paper) or film [44, 48]. A typical graphene paper possesses a layered structure through the entire cross section (SEM image in Fig. 3a) [49]. And in X-ray diffraction pattern, the graphene paper obtained by vacuum filtration displayed a weak, broad diffraction peak of at about $2\theta = 23°$ (Fig. 3b), which corresponds to a layer–layer distance (d-spacing) of 0.379 nm, such value is a little larger than

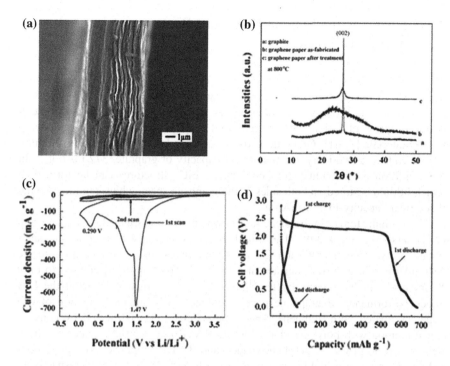

Fig. 3 a SEM image of a graphene paper obtained by vacuum filtration method. **b** XRD patterns of graphite, graphene paper and graphene paper after treatment at 800 °C. **c** Cyclic voltammogram (CV) curves of graphene paper in LiPF$_6$ with Li as counter and reference electrode (scan rate: 0.1 mV s^{-1}). **d** Charge/discharge profiles of graphene paper at a current density of 50 mA g^{-1}. Reprinted with permission from [49]. Copyright (2009)

the d-spacing of graphite (0.336 nm). The increased d-spacing of graphene paper can be ascribed to the presence of residual oxygen-containing functional groups and structural defects formed in the chemical process. The graphene papers obtained by vacuum filtration of graphene dispersion are mechanically strong and electrically conductive, with Young's modulus of 41.8 GPa, tensile strength of 293.3 MPa, and conductivity of 351 S cm^{-1} [49]. These flexible, robust graphene papers are expected to be promising candidates as electrodes in flexible energy storage devices.

In this review chapter, we will focus on recent developments on the graphene and graphene composite-based flexible electrodes for electrochemical energy storage, including LIB and supercapacitor applications. The following discussion will be categorized into three sections by the role of graphene in the flexible electrode, which are (a) graphene as the sole or dominant part of flexible electrode; (b) graphene as conductive matrix for flexible electrode; and (c) graphene as active additives for flexible electrode. The applications in LIBs and supercapacitors are mentioned to be discussed together due to their many commons from the material science viewpoint.

2 Graphene as the Sole or Dominant Part of Flexible Electrode

The charging/discharging process of batteries and supercapacitors is generally dominated by the electron and counter-ion transport at the surface of the electrodes [5, 50]. Since the invention of LIB in 1990s, there has been an intensive research on the insertion of Li-ions into the lattice of graphite and, recently, in graphene-based materials [2, 51]. Graphitic carbon can form LiC_6 structures with lithium, which leads to a relatively low theoretical capacity of graphite, 372 mAh g^{-1}. In the condition of individual graphene sheets, LiC_3 structures can be formed in which lithium is stored on both sides of the graphene sheet, which can increase the theoretical capacity to 744 mAh g^{-1}.

The lithium storage in graphene papers was firstly investigated by Wallace and coworkers (Fig. 3c, d) [49]. The initial discharge capacity of a graphene paper obtained by vacuum filtration of graphene dispersion was measured to be 680 mAh g^{-1} at a current density of 50 mA g^{-1}, but rapidly decreased to only 84 mAh g^{-1} in the second cycle. Nguyen et al. investigated in detail the LIB anode performance of graphene paper prepared via hydrazine reduction of GO paper compared with graphene powder-based electrodes fabricated by conventional method with polyvinylidene fluoride (PVDF) binder [52]. Under a current density of 50 mA g^{-1}, graphene paper and PVDF–graphene powder anodes exhibited initial reversible capacities of 84 and 288 mAh g^{-1}, respectively (Fig. 4a, b). The larger reversible capacity for the graphene powder anode may be attributed to the more disordered packing of the graphene sheets (Fig. 4d), which is conducive to anisotropic Li diffusion. Significant increases in the reversible capacity of graphene paper (from 84 to 214 mAh g^{-1}) were observed as the current density decreased from 50 to 10 mA g^{-1} (Fig. 4c), suggesting the well-ordered structure of the graphene nanosheets in the bulk paper form would create a kinetic barrier for the diffusion of Li-ions during electrochemical process.

Graphene with its maximal surface area of 2,630 m^2 g^{-1} is an ideal medium for supercapacitors as the EDLC is directly proportional to the surface area. Their application as supercapacitor electrodes was first explored by Ruoff and coworkers, who found that chemically derived graphene powder exhibits specific capacitances of 135 and 99 F g^{-1} in aqueous and organic electrolytes, respectively [53], and then different forms of graphene with improved performances have been reported [54–56]. The different values obtained with different forms of graphene mainly because the graphene nanosheets used tends to restack thus reducing the surface areas as only the surfaces can contact with the electrolyte contribute to the specific capacitance.

The supercapacitor application of graphene-based flexible electrodes was first studied by Chen et al. [57]. The ultrathin graphene films were also prepared using the vacuum filtration method. The thickness of graphene films can be tuned by choosing different volume of graphene solution and films of 25, 50, 75, and 100 nm thick can be obtained. The supercapacitor tests were performed using a

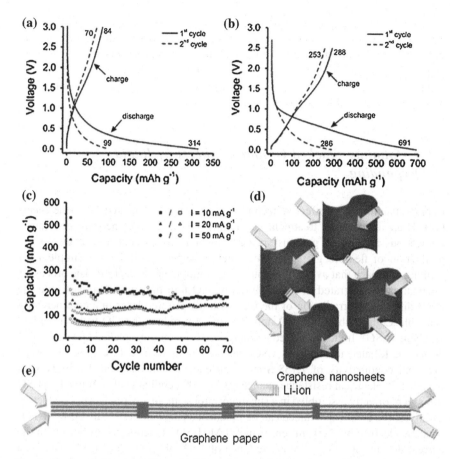

Fig. 4 Charge–discharge profiles of a Li/graphene half-cell at a current rate of 50 mA g^{-1}, where the graphene component is either **a** graphene paper or **b** graphene powder. **c** Cycle performance of a Li/graphene paper half-cell at three different current densities. **d, e** Schematic illustration of Li-ion storage capability of graphene powder and graphene paper. Reprinted with permission from [52]. Copyright (2010)

three-electrode half-cell system in aqueous 2 M KCl solution. At a scan rate of 10 mV s^{-1}, specific capacitances of 111, 105, 102, and 99 F g^{-1} were obtained for the 25, 50, 75, and 100 nm films, respectively. Such values are relatively low compared with many other forms of graphene.

Therefore, the application of graphene paper as flexible electrodes for energy storage faces a key scientific and technical challenge. The performance of bulk material strongly depends on the manner the individual sheets are arranged. During the process of graphene nanosheets assembled into bulk form, the intersheet van der Waals attractions inevitably cause aggregation or restacking of individual graphene sheets. Consequently, many of the unique properties that individual sheets possess, including high surface area and high ion conductivity are significantly compromised in a bulk form of graphene such as thin film or paper. Effective prevention of intersheet restacking or creating fast ion/electrolyte transportation ways are essential to allow the individual sheets in multilayered graphene structures to behave as graphene in electrochemical energy storage applications. Up-to-date, several strategies have been applied to enhance the performance of flexible graphene electrode.

2.1 Exfoliating Stacked Graphene Sheets by Rapid Expanding

Rapid expanding is an effective technique for exfoliating stacked graphene sheets [42]. High temperature treatment can reduce GO and spontaneously generate a tremendous amount of gaseous species. Kaner et al. present a strategy for the production of flexible graphene electrode for supercapacitor by a simple laser scribing approach that avoids the restacking of graphene sheets [58]. The process is schematically illustrated in Fig. 5. Initially, GO thin film was drop-cast onto the DVD disk. Then, irradiation of the film with the infrared laser in a commercially available LightScribe CD/DVD optical drive can reduce the GO to laser-scribed graphene (LSG). In this rapid reduction process, the closely stacked GO sheets can be well-exfoliated to a porous LSG sheets. The as-formed LSG film possessed excellent conductivity of 1,738 S m^{-1} and excellent mechanical flexibility with only ~1 % electrical resistance change after 1,000 bending cycles. When two LSG films were directly assembled in a supercapacitor without any binders or conductive additives, the areal capacitance can be obtained as 3.67 mF cm^{-2} in 1.0 M H$_3$PO$_4$ aqueous electrolyte (4.04 mF cm^{-2} in 1.0 M H$_2$SO$_4$ aqueous electrolyte) with the current density of 1 A g^{-1}, a capacitance of 1.84 mF cm^{-2} can still be achieved even when operated at an high charge/discharge rate of 1,000 A g^{-1}, indicating the high rate capability of such electrode. Additionally, the LSG electrode retained 96.5 % of its initial capacitance after 10,000 cycles. The liquid electrolyte can further be replaced with poly(vinyl alcohol) (PVA)–H$_3$PO$_4$ polymer gelled electrolyte, which also acts as the separator. In this condition, the performance of electrode was comparable with those obtained with an aqueous electrolyte. The high performance of LSG electrode can be attributed to the porous structure of the graphene film, which can provide an electrolyte reservoir to facilitate ion transport and minimize the diffusion distance to the interior surfaces.

2.2 Separating Graphene Sheets with "Nanospacers"

Another approach to prevent the restacking of graphene nanosheets is blending graphene sheets with other nanoparticles as "spacers" to form composites with high porosity and surface area. For example, Lian et al. demonstrated that flexible

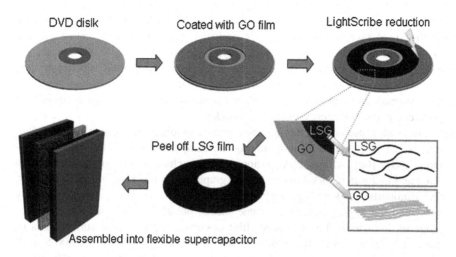

Fig. 5 Schematic illustration of the fabrication of laser-scribed graphene (LSG)-based electrochemical capacitors

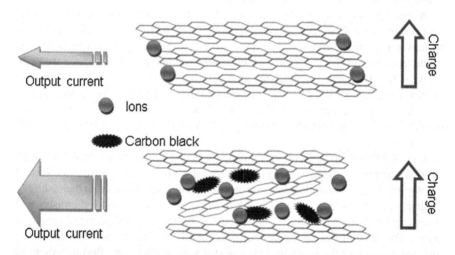

Fig. 6 Schematic demonstrating the concept of manipulating the geometry and performance of graphene paper by pillared with carbon black

graphene paper pillared by carbon black (CB) nanoparticles can be obtained by introducing inexpensive CB nanoparticles as spacer [59]. With the similar vacuum filtration method, restacking of graphene sheets can be greatly mitigated during the assembly process (Fig. 6). The pillared graphene paper exhibit excellent electrochemical performances and cyclic stabilities compared with graphene paper without the addition of CB nanoparticles when used as supercapacitor electrodes. The specific capacitances of the pillared GP electrode were calculated to be

138 F g^{-1} in 6 M KOH aqueous electrolyte and 83.2 F g^{-1} in 1 M $LiPF_6$ organic electrolyte at a scan rate of 10 mV s^{-1}, and 80 F g^{-1} can be reached at a high scan rate of 500 mV s^{-1} in aqueous electrolyte. Meanwhile, after 2,000 cycles, the degradations of its specific capacitance in aqueous and organic electrolytes are only 3.85 and 4.35 %, respectively. The ability to maintain high conductivity and a high surface area provides an effective route to improve the performance of graphene paper for energy-storage electrodes.

Shi et al. reported that flexible mesoporous graphene films with high conductivities can be prepared by graphitizing the composite films of GO and nanodiamond (ND) [60]. After graphitization, ND was changed into onion-like carbon (OC) and GO was reduced to conductive graphene. In this kind of graphene films, OC nanoparticles were sandwiched between graphene sheets, which not only prevented the aggregation of graphene sheets, but also formed mesopores in the range of 2–11 nm. The mesopores film possesses high specific surface area of about 420 m^2 g^{-1} and high conductivities in the range of 7,400–20,300 S m^{-1}. Such films are flexible and can be directly used as the electrodes of supercapacitors. In a three electrode configuration with 1.0 M H_2SO_4 aqueous electrolyte, the supercapacitor showed specific capacitances of 143 and 78 F g^{-1} at the current density of 0.2 and 10 A g^{-1}, respectively.

A three-component, flexible electrode has also been developed for supercapacitors utilizing CNT decorated with γ-MnO_2 nanoflowers (γ-MnO_2/CNT) as spacers for graphene nanosheets [61]. Such electrode can deliver a high specific capacitance of 308 F g^{-1} for two-electrode symmetric supercapacitors with 30 wt% KOH aqueous electrolyte, at a scan rate of 20 mV s^{-1}. A maximum energy density of 43 Wh kg^{-1} can be obtained for this kind of supercapacitors at a constant current density of 2.5 A g^{-1}. The fabricated supercapacitor device also exhibits excellent stability by retaining $\approx$90 % of the initial specific capacitance after 5,000 cycles.

Instead of the above-mentioned "hard" spacers, Li et al. developed a "soft" approach to prevent the restacking of graphene sheet in the paper form by delicately employing the solvent molecule as spacers [62]. They discovered that the assembly of graphene can be manipulated using the principles of colloidal chemistry and water can serve as an effective "soft spacer" to prevent the restacking of graphene sheets. This kind of hydrated graphene paper can be obtained when the filtration is just completed and the resultant paper is still wet. Without any drying, graphene paper is found to contain $\approx$92 wt% water. The hydrated graphene paper containing 0.045 mg cm^{-2} of graphene gives a sheet resistivity of 1,860 Ω $square^{-1}$. Hydrated graphene sheets can remain significantly separated as assembled in the bulk paper form. The high specific surface area of graphene sheets can be effectively reserved. When testing the supercapacitance properties of hydrated graphene paper in a two-electrode configuration, a specific capacitance of up to 215.0 F g^{-1} in 1.0 M H_2SO_4 aqueous electrolyte can be obtained at the current density of 1.08 A g^{-1}, and more importantly, a capacitance of 156.5 F g^{-1} can still be retained and even discharged at an ultrahigh current density of 1,080 A g^{-1}. And the hydrated graphene paper can retain over 97 % of its initial capacitance after cycling in a high operation current of

100 A g^{-1} for 10,000 times. Thus, this strategy has been proven to be very effective to make graphene nanosheets remain largely separated in a solvated state, providing a simple strategy for addressing the key challenge that has limited the large-scale application of graphene.

2.3 Creating Fast Ion/Electrolyte Transportation Ways in Graphene Paper

Besides preventing graphene nanosheets from restacking, generating fast ion/ electrolyte transporting ways is another approach to improve the electrode performance of graphene paper. As an illustration, Kung et al. reported an approach by introducing in-plane carbon vacancy defects (pores) into graphene sheets using a facile solution method to enhance the electrochemical energy storage performance of graphene paper electrodes [63]. In-plane defects was firstly introduced into the basal planes of GO by treating it in hot HNO$_3$ with the assistance of ultrasonic vibration. As illustrated in Fig. 7a. Under such conditions, sections of GO can be transformed into soluble polyaromatic hydrocarbons, left a holey GO (HGO) sheet. Thermal reduction of a free-standing HGO paper led to an electrically conducting graphene paper constructed by holey graphene sheets. The in-plane porosity can provide a high density of cross-plane ion diffusion channels that facilitate ion transport and storage at high rates. The electrochemical performance of holey graphene papers with a thickness of ~ 5 μm was examined as LIB anode; reversible capacities of 454 and 178 mAh g^{-1} can be obtained with optimized reduction conditions at the current densities of 50 and 2,000 mA g^{-1}, respectively.

Recently, Ruoff et al. reported that KOH activation can generate holes or defects on the surface of graphene [64]. Based on the similar method, activated graphene films can also be fabricated (Fig. 7b) [65]. First, an "ink paste" was prepared by adding KOH into GO colloidal suspension and then concentrated by heating. Films composed of stacking GO nanosheets decorated with KOH were obtained through vacuum filtration method. The activation step was carried out under flowing argon at 800 °C for 1 h. The final activated graphene film was obtained after washing and drying, which is flexible free-standing porous carbon film with high specific surface areas of up to 2,400 m^{2}g^{-1} and a very high in-plane electrical conductivity of 5,880 S m^{-1}. Such electrode showed a capacitance normalized to BET surface area of 14 μF cm^{-2} at the scan rates of 50 mV s^{-1} and can be retained at 11 μF cm^{-2} at 400 mV s^{-1}. The high rate performances of the above two graphene film electrodes is closely related to their highly interconnected 3D structure and short diffusion pathway, which favor fast ion transportation.

Up to now, the mentioned fabrications of graphene thin film or paper are all based on the vacuum filtration method, as an alternative strategy, Xue et al. developed a novel graphene aerogel pressing approach to fabricate graphene paper with folded structured graphene sheets, the process is illustrated in Fig. 8 [66]. The graphene aerogel was prefabricated by freeze-drying GO aqueous dispersion and

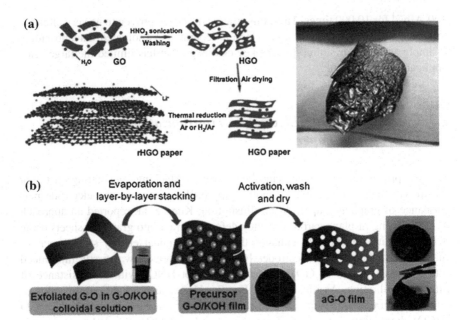

Fig. 7 Schematic illustration and digital images of graphene paper with plane defects (holes) fabricated by **a** acid oxidation and **b** KOH activation method. Reprinted with permission from [63 and 65]. Copyright (2011 and 2012)

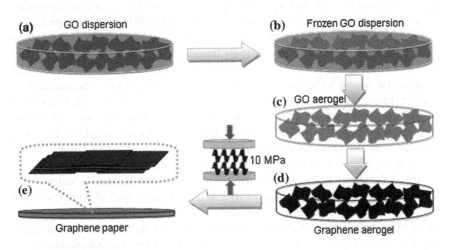

Fig. 8 Illustration of the formation process of graphene paper **a** GO aqueous dispersion. **b** GO dispersion frozen at −50 °C. **c** GO aerogel obtained by freeze drying (**b**) in vacuum. **d** Graphene aerogel obtained by treating (**c**) at 200 °C in air. **e** Mechanical press graphene aerogel to form graphene paper

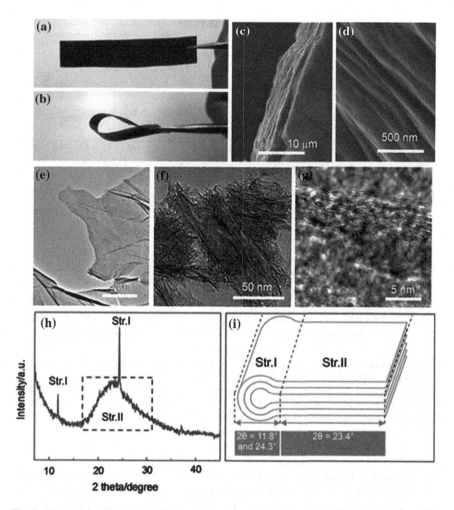

Fig. 9 Digital images (**a**, **b**), SEM images (**c**, **d**) and TEM images (**e–g**) of folded structured graphene paper. **h** XRD pattern and **i** structure illustration of folded structured graphene paper

subsequent thermal reduction. As shown in Fig. 9a and b, this new type of graphene paper is freestanding and highly flexible. SEM images of the graphene paper in Fig. 9c show flat surface, which is similar with graphene paper fabricated by vacuum filtration method, and assembled graphene nanosheets can be observed in the entire paper sample. The SEM image taken from the cross section in Fig. 9d revealed many close edges of graphene nanosheets instead of open edges in graphene paper obtained by vacuum filtration, and TEM images in Fig. 9e–g indicated that besides layer stacking, the folding of graphene layers is also a common structure feature in the graphene paper. The X-ray diffraction pattern in

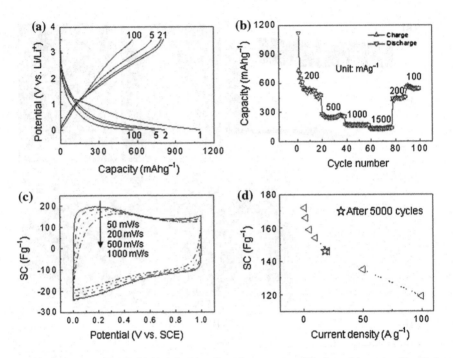

Fig. 10 Performance test of folded structured graphene paper. **a** Discharge/charge profiles (1st, 2nd, 5th and 100th cycle) at the current density of 100 mA g^{-1} as LIB anode. **b** Cycling stability and rate capacity of the graphene paper as LIB anode at different current densities. **c** CV curves of the graphene paper as supercapacitor electrode at different scan rates. **d** Specific capacitance of the graphene paper as a function of charge/discharge rate

Fig. 9h show a broad peak centered at $2\theta = 23.4°$, with the interlayer spacing calculated to be 0.38 nm, which is slightly higher than that of graphene paper obtained by vacuum filtration (0.37 nm). Besides the broad peak, this type graphene paper is distinguished with two sharp and strong XRD peaks at $2\theta = 11.8°$ and 24.3°. As shown in Fig. 9i, these two peaks correspond to the folding edge of multilayer graphene sheets. The peak at $2\theta = 24.3°$ is a result of more regular d-spacing of graphene layers at the curved edge, and the peak at $2\theta = 11.8°$ originates from the uniform spacing of the channel at the folding axis, which is similar with the intertube space of multi wall CNT (MWCNT).

When test the performance of folded structured graphene paper as anode in LIB, the first discharge and charge capacities of graphene sheets are 1,091 and 864 mAh g^{-1} at the current density of 100 mA g^{-1}, reached a Coulombic efficiency of 79.2 % (Fig. 10a), the discharge and charge capacities are as high as 815 and 806 mAh g^{-1} at the second cycle, and high Coulombic efficiencies that all over 98 % can be obtained in the following cycles. These results indicate that the reversible capacity of folded structured graphene paper is much higher than that of regular graphene paper obtained by vacuum filtration (84 mAh g^{-1}). Such graphene paper also possesses high rate capability and stability; at the current densities of 200, 500, 1000, and 1500 mA g^{-1}, the corresponding reversible specific capacities are 557, 268, 169 and 141 mAh g^{-1}, respectively. And 568 mAh g^{-1} can still be reached after 100 cycles at 100 mA g^{-1} (Fig. 10b).

The performance of folded structured graphene paper as supercapacitor electrode has also been tested in a two-electrode configuration with 1 M H$_2$SO$_4$ aqueous electrolyte. Figure 10c shows the CV curves at the scan rates of 50–1,000 mV s^{-1} over the voltage range from 0 to 1 V, which all display quasirectangular shape, indicating the excellent capacitance characteristics of the folded structured graphene paper. From the galvanostatic charge/discharge test, a specific capacitance up to 172 F g^{-1} can be obtained at the current density of 1 A g^{-1}, which can still be retained at 110 F g^{-1} even when the supercapacitor is operated at a fast rate of 100 A g^{-1} (Fig. 10d). Additionally, after 5,000 times cycling at a current density of 20 A g^{-1}, the graphene paper can retain over 99 % of its initial capacitance, indicating its excellent stability.

The high performance of such graphene paper is originated from its folded structured graphene sheets, and when used as LIB anode, such folds can provide slightly increased intersheet spacing and nucleation sites, which can facilitate Li-ion diffusion and SEI formation, leading to higher reversible capacity. And in the application as supercapacitor electrode, the folded structure is helpful for graphene paper to contact the electrolyte and remarkably improve the capacitance properties.

3 Graphene as Conductive Matrix for Flexible Electrode

To further increase the specific energy density of graphene paper as flexible LIB and supercapacitor electrodes, incorporation of materials such as metal oxides or conducting polymers into the graphene matrix has been proved to be an effective approach [67]. From another point of view, due to the large specific surface area and the conductive robust structure, graphene can be a promising matrix for dispersing functional materials in which the charge transfer, redox reaction, as well as the mechanical stability will be enhanced. Moreover, the aggregation of graphene sheets can be partly prevented by sandwiching with other nanomaterials [67]. Therefore, anchoring redox active materials on graphene will yield highly porous composites attractive for fabricating high performance LIBs and supercapacitors.

3.1 Fabricating Graphene Composite Papers by In Situ Reaction

Strategies for preparing graphene-based composites can be generally categorized into two types: in situ reaction and blending. Many graphene-based composites

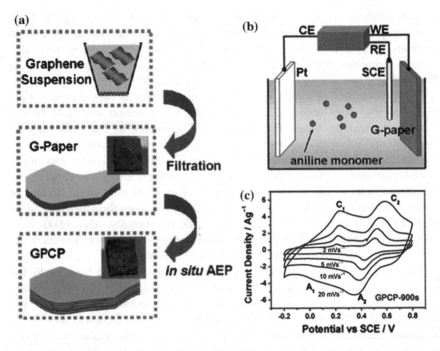

Fig. 11 a, b Illustration of the fabrication of graphene/polyaniline composite paper through in situ anodic electro-polymerization. **c** CV curves of graphene/polyaniline composite paper from 2 to 20 mV s^{-1} in 1 M H$_2$SO$_4$ aqueous electrolyte. Reprinted with permission from [68]. Copyright (2009)

with conducting polymers or inorganic nanoparticles were prepared via in situ chemical or electrochemical reactions. In chemical syntheses, positively charged metal ions (e.g. Sn^{2+}) or monomer molecules (e.g., aniline hydrochloride) can adsorb on the negatively charged graphene sheets via electrostatic or π–π stacking interactions. Therefore, crystallization or polymerization preferentially occurred on graphene sheets to form the corresponding composite materials [67]. Cheng et al. reported a freestanding and flexible graphene/polyaniline composite paper can be prepared by in situ anodic electro-polymerization of a polyaniline film on graphene paper (Fig. 11a and b) [68]. In the electro-polymerization process, the adsorbed aniline monomers on the surface of graphene sheets can be anodized to form polyaniline interspacing in the layers of assembled graphene nanosheets. The composite paper can maintain the flexibility of the original graphene paper, and the mechanical and electrochemical properties of composite paper can be tuned by simply adjusting the electro-polymerization time. This composite paper, which combines flexibility, conductivity, and electrochemical activity, exhibits a gravimetric capacitance of 233 F g^{-1} and a volumetric capacitance of 135 F cm^{-3}

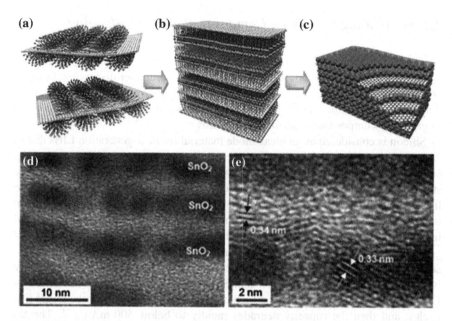

Fig. 12 Illustrations of the ternary self-assembly approach to ordered metal oxidegraphene nanocomposites. **a** Adsorption of surfactant hemimicelles on the surfaces of the graphene. **b** The self-assembly surfactant with Sn^{2+} and the transition into the lamella mesophase of SnO_2–graphene nanocomposites. **c** Nanocomposites composed of alternating layers of SnO_2 nanocrystals and graphene. **d**, **e** TEM and HRTEM images of SnO_2–graphene nanocomposites. Reprinted with permission from [69]. Copyright (2010)

when tested as supercapacitor electrode in a three-electrode cell with 1 M H_2SO_4 aqueous electrolyte (Fig. 11c).

Aksay et al. employed surfactants to modify the surface of graphene since they can not only assist the dispersion of the graphene nanosheets in aqueous media but also direct the self-assembly of metal oxides into nanostructures [69]. The surfactants assembled on the graphene nanosheets can bind to Sn^{2+} and SnO_2 was then crystallized between graphene nanosheets forming an ordered nanocomposite in which alternating layers of graphene and SnO_2 nanocrystals were assembled into layered nanostructures (Fig. 12). Free-standing flexible films can be obtained by vacuum filtration of the layered SnO_2/graphene nanocomposites. This kind of composite films featured both overall layer structure from 2D graphene and the locally ordered alternating layers of graphene nanosheets and SnO_2, which are favorable for conductivity and structural integrity. When tested as LIB anode, a steady specific capacity of 760 mAh g^{-1} for the nanocomposite electrode can be obtained at a current density of 0.008 A g^{-1}, close to the theoretical capacity of SnO_2 (780 mAh g^{-1}). And the capacity has no significant fading over 100 charge/discharge cycles, which is superior to both pure SnO_2 nanocrystals and SnO_2/graphene composites with disordered structures.

3.2 Fabricating Graphene Composite Papers by Blending

Blending is a convenient and effective route to graphene-based composites. GO nanosheets are rich in functional groups and highly soluble or dispersible in water or organic solvents [67]. Thus, they can easily composite with organic molecules, polymers, or inorganic nanostructures by one-step solution-blending and converted to graphene composites by chemical reduction.

Silicon is considered as an ideal anode material for next-generation LIBs due to its low discharge potential and the highest theoretical capacity (4,200 mAh g^{-1}). However, silicon suffers from a seriously irreversible capacity and a poor cycling stability, which results from the large volume change (about 300 %) during the lithium ion insertion/extraction process [70, 71]. Dispersing silicon into carbon matrix is believed to be an effective approach to buffer the volume changes and improve the electronic and ionic conductivities. Fan et al. reported that Si/graphene nanocomposite films can be prepared by vacuum filtration and a subsequent thermal reduction, in which silicon nanoparticles were dispersed directly into an aqueous suspension of GO nanosheets (Fig. 13a–c) [72]. When tested as LIB anodes, pure silicon exhibits a high discharge capacity only for the initial five cycles, and then the capacity degrades rapidly to below 500 mAh g^{-1}. The Si/graphene nanocomposite film delivers a reversible specific capacity of 1,040 mAh g^{-1} and a significantly improved cyclic stability (Fig. 13d). After 30 cycles, the capacity that can be obtained is 977 mAh g^{-1} (Fig. 13e) and 786 mAh g^{-1} can still be obtained after 300 cycles (at the current density of 50 mA g^{-1}). The improved electrochemical performance of such composite film can be attributed to the graphene matrix, which offers an efficient carrier channel and a robust mechanical support for strain release.

On the other hand, the large conjugated basal plane of graphene provides an ideal substrate for assembling aromatic organic molecules and conducting polymers through π–π stacking. And graphene nanosheets derived from chemical reduction can be easily functionalized with positively or negatively charged surfaces. Thus, they can form composites with other charged components via direct blending or layer-by-layer (LBL) assembly. Wang et al. reported that flexible graphene/polyaniline nanofibers composite paper can be prepared via a two-step route composed of electrostatic adsorption between negatively-charged poly(-sodium 4-styrenesulfonate) decorated graphene and positively-charged polyaniline nanofibers and then vacuum filtration of the as-prepared graphene/polyaniline nanofibers suspension [73]. The effective synergy can remarkably improve electrochemical properties of graphene by introducing polyaniline nanofibers, which can ascribe to high surface area of graphene and good contact between the components. The highest specific capacitance of the composites reaches 301 F g^{-1} when tested in a three-electrode configuration in 1 M H$_2$SO$_4$ aqueous electrolyte at the current density of 0.5 A g^{-1}.

Chen et al. employed a LBL filtration technique to assembly hybrid graphene–MnO$_2$ nanotube thin films and studied as anodes for LIBs (Fig. 14) [74].

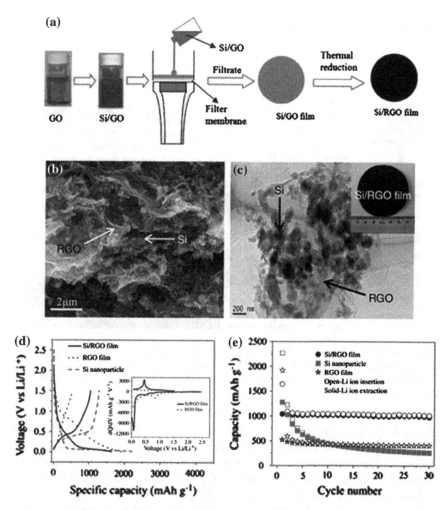

Fig. 13 a Schematic diagram of the fabrication of Si/graphene film. **b** SEM and **c** TEM images of the Si/graphene film. **d** The first charge/discharge curves of Si/graphene film, graphene film and Si nanoparticles at a current density of 50 mA g^{-1}. **e** Cycling stabilities of Si/graphene film, graphene film and Si nanoparticle electrodes at a current density of 50 mA g^{-1}. Reprinted with permission from [72]. Copyright (2011)

Composite films were assembled LBL by 20-layer films (10 layers graphene, 10 layers MnO$_2$ nanotube, and the weight ratio of graphene/MnO$_2$ nanotube is 1:1), with each layer prepared by filtration of graphene or MnO$_2$ nanotube dispersion. By this method, the number of layers and composition of each were properly controlled. Each thin layer of graphene provides not only conductive pathways to accelerating the electrochemical reaction, but also buffer layers for strain release during lithium insertion/extraction in MnO$_2$ (Fig. 14b). The graphene–MnO$_2$ nanotube films as anode present excellent cycle and rate capabilities with a

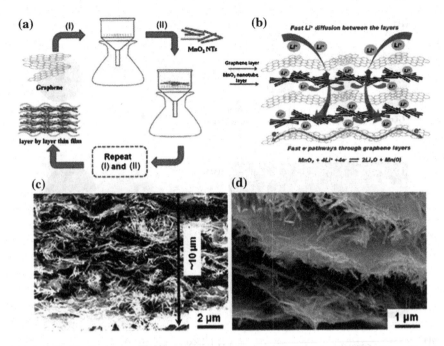

Fig. 14 Schematic illustration of **a** the process for layer-by-layer constructing of graphene–MnO₂ composite paper, and **b** the enhancement of electrode performance. **c, d** Cross-sectional SEM images of graphene–MnO₂ composite paper. Reprinted with permission from [74]. Copyright (2011)

reversible specific capacity based on electrode composite mass of 495 mAh g^{-1} at 100 mA g^{-1} after 40 cycles. On the contrary, graphene-free MnO₂ nanotube electrodes demonstrate only 140 mAh g^{-1} at 80 mA g^{-1} after 10 cycles. Furthermore, at a high current rate of 1,600 mA g^{-1}, the charge capacity of graphene–MnO₂ nanotube film can still reach 208 mAh g^{-1}.

Graphene paper-based composite electrodes fabricated by methods like LBL deposition need tedious procedure, and developing facile and versatile strategy to fabricate graphene composite paper electrodes is still a challenge. Xue et al. extended the method mentioned in Sect. 2.3 to the fabrication of graphene composite papers [75]. Graphene composite aerogels are prefabricated by freeze-drying GO/active material aqueous dispersions and subsequent thermal reduction. By pressing such aerogels at the pressure of 10 MPa can produce graphene composite papers, which are robust and flexible. Figure 15a and b show the digital and SEM images of graphene composite papers containing MnO₂ nanowires and LiFePO₄ particles fabricated by such method. SEM images show that introduced MnO₂ nanowires and LiFePO₄ particles are interspersed between layers of graphene nanosheets. The good contact between introduced active nanomaterials and graphene layers can significantly enhance the overall electrical conductivity and stability of electrodes, which is favorable for their applications in flexible LIBs.

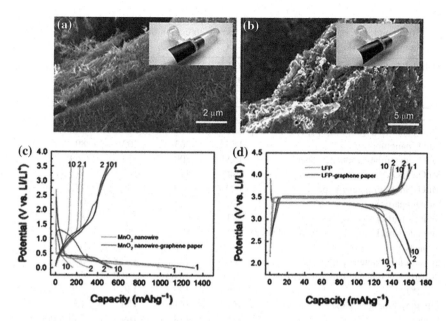

Fig. 15 Digital and SEM images of graphene composite papers containing **a** MnO₂ nanowires and **b** LiFePO₄ particles. **c** Comparison of galvanostatic charge and discharge curves at different cycles for the graphene–MnO₂ composite paper and MnO₂ nanowire electrode prepared by conventional method. **d** Comparison of galvanostatic charge and discharge curves at different cycles for the graphene–LiFePO₄ composite paper and LiFePO₄ nanoparticle electrode prepared by conventional method

Figure 15c shows the comparison of galvanostatic charge/discharge curves for the graphene–MnO₂ composite paper and MnO₂ nanowire electrode prepared by conventional method with CB and PVDF binder. The initial reversible capacities achieved for the free-standing graphene–MnO₂ composite paper was 531 mAh g^{-1} as LIB anode, the capacity is much higher than that of graphene-free MnO₂ nanowire electrode (320 mAh g^{-1}), indicating that the interspaced graphene layers can facilitate electron and Li-ion transport in the electrode. After 10 cycles, the capacity of graphene–MnO₂ reached 548 mAh g^{-1}, but the capacity of graphene-free MnO₂ nanowire electrode rapidly faded to 174 mAh g^{-1}. Imply the effective protection of the MnO₂ nanowires by graphene matrix. Similar to the results of the graphene–MnO₂ composite paper, graphene–LiFePO₄ composite paper also shows higher reversible capacity (160 mAh g^{-1}) compared with electrode prepared by traditional method (140 mAh g^{-1}), when tested as LIB cathode (Fig. 15d). The improved performance can also be attributed to the acceleration of electron transport by graphene matrix. Such method is quite versatile and a wide range of nanomaterials can be applied to composite with graphene to fabricate flexible electrode for LIBs. Table 1 compares of the performance of several

Table 1 Comparison of the performance of several nanomaterials in its composite in graphene paper form and in the conventional electrode form with carbon black and PVDF binder, current densities are all at 100 mA g^{-1}

Material form	Reversible capacitance at different cycles (mAh g^{-1})			
	1st	2nd	5th	50th
MnO$_2$ nanowire	1,108	320	206	91
MnO$_2$ nanowire–graphene paper	1,321	531	501	546
MnO$_2$ nanotube	1,221	303	202	89
MnO$_2$ nanotube–graphene paper	1,292	500	453	369
MnO$_2$/SnO$_2$	1,400	445	362	113
MnO$_2$/SnO$_2$–graphene paper	1,119	510	435	272
Sn nanoparticle	726	175	57	0
Sn nanoparticle–graphene paper	593	672	573	83
Sn nanoparticle–graphene paper (calcined)	1,107	823	715	437
Si particle	2,498	198	71	0
Si particle–graphene paper	2,540	2,080	1,768	886
Si particle–graphene paper (calcined)	3,009	2,663	2,100	44
LiFePO$_4$	136	137	136	149
LiFePO$_4$–graphene paper	161	161	159	159
V$_2$O$_5$	185	178	169	50 (200th)
V$_2$O$_5$–graphene paper (calcined)	25	25	30	198 (200th)

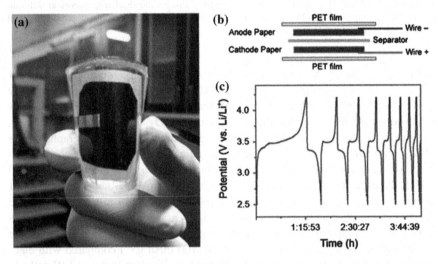

Fig. 16 **a** Digital photograph of the flexible battery. **b** Schematic illustration of the flexible battery configuration. **c** Galvanostatic charge/discharge curves of the flexible battery

nanomaterials in its composite in graphene paper form and in the conventional electrode form with CB and PVDF binder, from which we can see both reversible capacity and cycling stability can be improved by composite into graphene paper.

A flexible battery can further be built employing the obtained flexible cathode and anode (Fig. 16a). The graphene–LiFePO$_4$ paper cathode and the graphene–MnO$_2$ paper anode were separated by a polypropylene membrane (Celgard 2400) dipped in liquid electrolyte and sealed by two PET films. Such battery is able to be bended or rolled up (Fig. 16b). The charge–discharge cycles of the assembled battery shows the typical charge–discharge behavior of a LIB with charge/discharge occurring reversibly (Fig. 16c). The capacity fading observed in the full cell is due to various engineering factors in a lab scale fabrication.

According to the discussion above, assembled graphene nanosheets can be used as high performance matrix to disperse various functional components to form flexible electrodes. The incorporation of graphene into the composites can provide them with the unique properties of graphene and also possibly induce new properties and functions based on synergetic effects. The composites exhibited improved electrical conductivity, mechanical stability, and enhanced electrochemical reactivity, which are promising in the energy storage application.

4 Graphene as Active Additives for the Flexible Electrode

The introduction of an active phase into a flexible matrix such as cellulose paper can make it electrically active for applications such as electrochemical energy storage. Traditional cellulose-based electric papers are commonly mixed with CB, metal powders or carbon fibers as the electric active or conductive additives. Such papers suffer from weak adhesion with additives and performance degradation upon long services or repeatedly bending [4, 76]. Composite papers fabricated by dispersing CNTs into cellulose papers or imprinting them as a conductive layer have been extensively studied as high performance electrode materials in flexible supercapacitor and battery devices [6]. However, the high cost of CNTs and difficulty to be dispersed in solution hindered their wide application. Due to the easy preparation and low cost of chemical derived graphene, it can displace CNTs in many conditions. And the flexible 2D nanosheet geometry of graphene, in combination with its high mechanical and electric properties, makes it an attractive additive material for the development of novel electric or multifunctional flexible electrodes.

By directly dispersing chemically reduced graphene nanosheets into cellulose pulp and then subjected to a typical paper-making process by infiltration, Li et al. fabricated a conductive and electrochemically active composite paper of graphene nanosheets/cellulose [76]. The graphene nanosheets introduced were coated onto the surface of cellulose fibers and form a continuous conductive network through the interconnected cellulose fibers. The composite paper is flexible as cellulose paper and has a conductivity of 11.6 S m^{-1}. The application of the composite paper as a flexible double layer supercapacitor in LiPF$_6$ organic electrolyte displays a high capacity of 252 F g^{-1} at the current density of 1 A g^{-1} (respect to the mass of graphene). Moreover, the paper can be used as the anode in a LIB, a

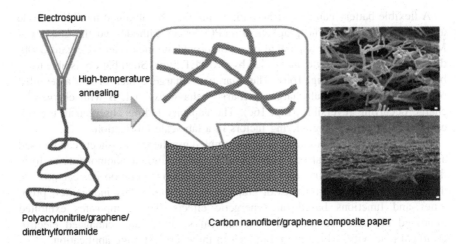

Electrospun

High-temperature annealing

Polyacrylonitrile/graphene/ dimethylformamide

Carbon nanofiber/graphene composite paper

Fig. 17 Schematic illustration of the formation process of carbon nanofiber/graphene composite paper. Reprinted with permission from [78]. Copyright (2012)

reversible capacity of 257 mAh g^{-1} can be obtained after 10 cycles, which overpassed the performance of a CNT/cellulose composite paper (140 mAh g^{-1}).

Cheng et al. also reported a different method to fabricate graphene/cellulose composite paper by simply filtering a graphene nanosheets suspension through a cellulose filter paper [77]. During filtration, the graphene suspension penetrated throughout the filter paper, and graphene nanosheets were anchored to the cellulose fibers by electrostatic interaction. As the process continued, the voids of the filter paper were gradually filled with graphene nanosheets, a composite paper with graphene nanosheets distributed through the macroporous texture of the filter paper was achieved. Benefiting from the structural characteristics, when tested as supercapacitor electrode with H$_2$SO$_4$–PVA gel electrolyte at 1 mV s^{-1}, the graphene/cellulose paper electrodes show good rate capability and long cyclic stability with a capacitance per geometric area of 81 mF cm^{-2}, which is equivalent to a gravimetric capacitance of 120 F g^{-1} of graphene, and retains >99 % capacitance over 5,000 cycles.

Different with cellulose, carbon nanofibers, which have high surface area, controllable electronic conductivity, and low cost have already been playing the leading role in supercapacitor electrodes. However, the performance of carbon nanofiber as supercapacitor electrodes is still limited by their low power and energy density. To improve the performance of carbon nanofiber-based supercapacitors, modifying carbon nanofiber with other electroactive materials is an effective route. Yan et al. prepared flexible and freestanding carbon nanofiber/graphene composite paper via a high throughput electrospinning method followed by high-temperature annealing, in which polyacrylonitrile/graphene/dimethylformamide mixture was used as electrospun precursor (Fig. 17) [78]. The structure characterizations show that graphene nanosheets homogeneously distributes in the carbon nanofiber,

Table 2 Summarization of the LIB performances of graphene based electrodes mentioned in this review chapter

Material form	Reversible capacity $(mAh\ g^{-1})$	Cycling stability $(mAh\ g^{-1}/$ Cycles$)$	Reference number
Graphene powder	288 (50 mAh g^{-1})	–	Abouimrane et al. [52]
Graphene paper	84 (50 mAh g^{-1}) 214 (10 mAh g^{-1})	–	Wang et al. [49], Abouimrane et al. 52
Holey graphene paper	454 (50 mAh g^{-1}) 178 (2,000)	–	Zhao et al. [63]
Folded structured graphene paper	864 (100 mAh g^{-1}) 169 (1,000 mAh g^{-1})	568/100	Liu et al. [66]
SnO$_2$/graphene composite paper	760 (8 mAh g^{-1})	No fading after 100 cycles	Wang et al. [69]
Si/graphene composite film	1,040 (50 mAh g^{-1})	977/30 786/300	Tao et al. [72]
graphene–MnO$_2$ nanotube thin films	581 (100 mAh g^{-1})	495/40	Yu et al. [74]
MnO$_2$ nanowire– graphene paper	531 (100 mAh g^{-1})	546/50	Liu et al. [75]
LiFePO$_4$–graphene paper	161 (100 mAh g^{-1})	159/50	Liu et al. [75]

forming a thin, light-weight, and flexible composite paper. Due to the reinforcing effects coming from carbon nanofiber and graphene nanosheets, the specific surface area, conductivity, and capacitance performance are significantly improved after compositing. When tested as supercapacitor electrode in 6 M KOH aqueous electrolyte at the scan rate of 100 mV s^{-1}, the composite paper exhibits a specific capacitance of 197 F g^{-1}, 24 % higher than that of pure carbon nanofiber paper.

Although graphene itself can form paper like bulk forms without the assistant of other substances, using commercialized flexible matrix such as cellulose and carbon nanofiber paper to load graphene nanosheets is an effective way to further reduce the cost of flexible electrodes. The composite papers can also reserve the macro-porous texture of the cellulose or carbon nanofiber paper, which helps to overcome the low porosity of graphene papers. Furthermore, the cellulose fibers in the composite electrode can significantly absorb electrolyte and act as electrolyte reservoirs to facilitate ion transport, which is of significant importance for the application as electrochemical energy storage electrodes.

5 Conclusions and Outlook

The LIB and supercapacitor performances of graphene-based flexible electrodes mentioned in this review chapter have been summarized in Tables 2 and 3. From which we can see specific capacities, specific capacitances, current densities, and

Table 3 Summarization of the supercapacitor performances of graphene based electrodes mentioned in this review chapter

Graphene form	Cell configuration electrolyte	Specific capacitance ($F\ g^{-1}$)	Reference number
Chemically modified graphene	Two-electrode (5.5 M KOH aq) Two-electrode (TEABF4/AN)	165 (10 mA g^{-1}) 99 (10 mA g^{-1})	Stoller et al. [53]
Graphene hydrogel	Two-electrode (5 M KOH aq)	220 (1 A g^{-1}) 165 (100 A g^{-1})	Zhang and Shi [54]
Functionalized graphene	Three-electrode (1 M H_2SO_4 aq)	276 (0.1 A g^{-1}) 205 (5 A g^{-1})	Lin et al. [55]
Graphene film (25 nm thick)	Three-electrode (2 M KCl aq)	111 (10 mV s^{-1})	Yu et al. [57]
Laser-scribed graphene	Two-electrode (1 M H_3PO_4 aq)	3.67 mF cm^{-2} (1 A g^{-1}) 1.84 mF cm^{-2} (1,000 A g^{-1})	El-Kady et al. [58]
	Two-electrode (1 M H_2SO_4 aq)	4.04 mF cm^{-2} (1 A g^{-1})	
Graphene paper pillared by carbon black	Two-electrode (6 M KOH aq)	138 (10 mV s^{-1}) 80 (500 mV s^{-1})	Lin et al. [59]
Mesoporous graphene films	Three-electrode (1 M H_2SO_4 aq)	143 (0.2 A g^{-1}) 78 (10 A g^{-1})	Sun et al. [60]
Graphene/γ-MnO_2/ CNT paper	Two-electrode (30 wt% KOH aq)	307 (20 mV s^{-1})	Rakhi et al. [61]
Hydrated graphene paper	Two-electrode (1 M H_2SO_4 aq)	215(1.08 A g^{-1}) 156 (1,080 A g^{-1})	Yang et al. [62]
Porous graphene paper	Two-electrode (1 M KOH aq)	14 μF cm^{-2} (50 mV s^{-1}) 11 μF cm^{-2} (400 mV s^{-1})	Zhang et al. [65]
Folded structured graphene paper	Two-electrode (1 M H_2SO_4 aq)	172 (1 A g^{-1}) 119 (100 A g^{-1})	Liu et al. [66]
Graphene-polyaniline composite paper	Three-electrode (1 M H_2SO_4 aq)	233 (20 mV s^{-1})	Wang et al. [68]
Graphene-polyaniline composite paper	Three-electrode (1 M H_2SO_4 aq)	301 (0.5 A g^{-1})	Liu et al. [73]
Graphene/cellulose composite paper	Two-electrode ($LiPF_6$ organic electrolyte)	252 (1 A g^{-1})	Kang et al. [76]
Graphene/cellulose composite paper	Two-electrode (H_2SO_4–PVA gel)	120 (1 mV s^{-1})	Weng et al. [77]
Graphene/carbon nanofiber composite paper	Two-electrode (1 M H_2SO_4 aq)	197 (100 mV s^{-1})	Tai et al. [78]

cycling numbers up to 1080 mAh g^{-1}, 307 F g^{-1}, 1080 A g^{-1}, and 10,000 cycles, respectively, have been reported for prototype devices based on different types of graphene papers, indicating that graphene-based flexible LIBs and supercapacitors can indeed compete with many other energy storage devices. At present, there is still an interesting ongoing development involving the design of flexible graphene-based energy storage devices, either with or without other active materials. These types of devices hold great promise for a number of new applications, which are incompatible with conventional contemporary battery and supercapacitor technologies.

Although graphene-based flexible electrodes have already become one of the most promising candidates for bendable or roll-up electrochemical energy storage devices, the following challenges still remain. First, high-quality graphene nanosheets are still required. New techniques have to be developed for producing graphene nanosheets with highly controlled chemical and physical properties. Second, the properties and functions of the electrodes depend strongly on their microstructures. Therefore, the assembly behaviors of graphene sheets and with other functional building blocks require more intensive investigation to achieve the precise control. Finally, the applications of graphene-based flexible electrodes are at their initial stages. They need to be studied systematically from both theoretical and experimental aspects, for example, although many research works tested the mechanical stability of different graphene papers, but the performance of graphene papers built in a flexible device especially with liquid electrolytes after repeat bending is still mistiness. With the multidisciplinary efforts from chemistry, physics, and materials science, we believe that many applications of these materials will become reality in the near future.

Acknowledgments Financial support from the National Natural Science Foundation of China (grant nos. 50872016, 20973033, and 51125009) and National Natural Science Foundation for Creative Research Group (grant no. 20921002) is acknowledged.

References

1. Nishide H, Oyaizu K (2008) Materials science—toward flexible batteries. Science 319(5864):737–738
2. Tarascon JM, Armand M (2001) Issues and challenges facing rechargeable lithium batteries. Nature 414(6861):359–367
3. Feng J, Sun X, Wu C, Peng L, Lin C, Hu S, Yang J, Xie Y (2011) Metallic few-layered VS_2 ultrathin nanosheets: high two-dimensional conductivity for in-plane supercapacitors. J Am Chem Soc 133(44):17832–17838
4. Nyholm L, Nystrom G, Mihranyan A, Stromme M (2011) Toward flexible polymer and paper-based energy storage devices. Adv Mater 23(33):3751–3769
5. Liu F, Song S, Xue D, Zhang H (2012) Selective crystallization with preferred lithium-ion storage capability of inorganic materials. Nanoscale Res Lett 7:149
6. Pushparaj VL, Shaijumon MM, Kumar A, Murugesan S, Ci L, Vajtai R, Linhardt RJ, Nalamasu O, Ajayan PM (2007) Flexible energy storage devices based on nanocomposite paper. Proc Natl Acad Sci 104(34):13574–13577

7. Sukjae J, Houk J, Youngbin L, Daewoo S, Seunghyun B, Byung Hee H, Jong-Hyun A (2010) Flexible, transparent single-walled carbon nanotube transistors with graphene electrodes. Nanotechnology 21(42):425201

8. Gwon H, Kim H-S, Lee KU, Seo D-H, Park YC, Lee Y-S, Ahn BT, Kang K (2011) Flexible energy storage devices based on graphene paper. Energy Environ Sci 4(4):1277–1283

9. Huang Z-D, Zhang B, Liang R, Zheng Q-B, Oh SW, Lin X-Y, Yousefi N, Kim J-K (2012) Effects of reduction process and carbon nanotube content on the supercapacitive performance of flexible graphene oxide papers. Carbon 50(11):4239–4251

10. Zhang Y, Xue D (2012) Mild synthesis route to nanostructured aplha-MnO_2 as electrode materials for electrochemical energy storage. Funct Mater Lett 5(3):1250030

11. Lu P, Liu F, Xue D, Yang H, Liu Y (2012) Phase selective route to $Ni(OH)_2$ with enhanced supercapacitance: performance dependent hydrolysis of $Ni(Ac)_2$ at hydrothermal conditions. Electrochim Acta 78:1–10

12. Liu J, Zhou Y, Liu F, Liu C, Wang J, Pan Y, Xue D (2012) One-pot synthesis of mesoporous interconnected carbon-encapsulated Fe_3O_4 nanospheres as superior anodes for Li-ion batteries. RSC Adv 2(6):2262–2265

13. Chen Y, Huang Q, Wang J, Wang Q, Xue J (2011) Synthesis of monodispersed $SnO_2@C$ composite hollow spheres for lithium ion battery anode applications. J Mater Chem 21:17448–17453

14. Liu J, Xue D (2010) Hollow nanostructured anode materials for Li-ion batteries. Nanoscale Res Lett 5:1525–1534

15. Liu J, Xia H, Xue D, Lu L (2009) Double-shelled nanocapsules of V_2O_5-based composites as high-performance anode and cathode materials for Li ion batteries. J Am Chem Soc 131(34):12086–12087

16. Cheng F, Liang J, Tao Z, Chen J (2011) Functional materials for rechargeable batteries. Adv Mater 23:1695–1715

17. Liu C, Li F, Ma L, Cheng H (2010) Advanced materials for energy storage. Adv Mater 22:E28–E62

18. Song H, Lee K, Kim M, Nazar L, Cho J (2010) Recent progress in nanostructured cathode materials for lithium secondary batteries. Adv Mater 20:3818–3834

19. Arico A, Bruce P, Scrosati B, Tarascon J, Schalkwijk W (2005) Nanostructured materials for advanced energy conversion and storage devices. Nat Mater 4:366–377

20. Liu J, Liu F, Gao K, Wu J, Xue D (2009) Recent developments in the chemical synthesis of inorganic porous capsules. J Mater Chem 19:6073–6084

21. Liu J, Xia H, Lu L, Xue D (2010) Anisotropic Co_3O_4 porous nanocapsules toward high-capacity Li-ion batteries. J Mater Chem 20:1506–1510

22. Liu J, Xue D (2010) Sn-based nanomaterials converted from SnS nanobelts: facile synthesis, characterizations, optical properties and energy storage performances. Electrochim Acta 56:243–250

23. Liu J, Zhou Y, Liu C, Wang J, Pan Y, Xue D (2012) Self-assembled porous hierarchical-like $CoO@C$ microsheets transformed from inorganic-organic precursors and their lithium-ion battery application. CrystEngComm 14(8):2669–2674

24. Bruce P, Scrosati B, Tarascon J (2008) Nanomaterials for rechargeable lithium batteries. Angew Chem Int Ed 47:2930–3946

25. Wang H, Yang Y, Liang Y, Cui L, Casalongue H, Li Y, Hong G, Cui Y, Dai H (2011) $LiMn_{1-x}Fe_xPO_4$ nanorods grown on graphene sheets for ultrahigh-rate-performance lithium ion batteries. Angew Chem Int Ed 50:7364–7368

26. Goodenough J, Kim Y (2010) Challenges for rechargeable Li batteries. Chem Mater 22:587–603

27. Ji L, Lin Z, Alcoutlabi M, Zhang X (2011) Recent developments in nanostructured anode materials for rechargeable lithium-ion batteries. Energy Environ Sci 4:2682–2699

28. Beguin F, Szostak K, Lota G, Frackowiak E (2005) A self-supporting electrode for supercapacitors prepared by one-step pyrolysis of carbon nanotube/polyacrylonitrile blends. Adv Mater 17(19):2380–2384
29. Chen J, Minett AI, Liu Y, Lynam C, Sherrell P, Wang C, Wallace GG (2008) Direct growth of flexible carbon nanotube electrodes. Adv Mater 20(3):566–570
30. Novoselov KS, Geim AK, Morozov SV, Jiang D, Zhang Y, Dubonos SV, Grigorieva IV, Firsov AA (2004) Electric field effect in atomically thin carbon films. Science 306(5696):666–669
31. Kim KS, Zhao Y, Jang H, Lee SY, Kim JM, Kim KS, Ahn J-H, Kim P, Choi J-Y, Hong BH (2009) Large-scale pattern growth of graphene films for stretchable transparent electrodes. Nature 457(7230):706–710
32. Stankovich S, Dikin DA, Dommett GHB, Kohlhaas KM, Zimney EJ, Stach EA, Piner RD, Nguyen ST, Ruoff RS (2006) Graphene-based composite materials. Nature 442(7100):282–286
33. Geim AK, Novoselov KS (2007) The rise of graphene. Nature Mater 6(3):183–191
34. Avouris P, Chen Z, Perebeinos V (2007) Carbon-based electronics. Nat Nanotechnol 2(10):605–615
35. Bae S, Kim H, Lee Y, Xu X, Park J-S, Zheng Y, Balakrishnan J, Lei T, Kim HR, Song YI, Kim Y-J, Kim KS, Ozyilmaz B, Ahn J-H, Hong BH, Iijima S (2010) Roll-to-roll production of 30-in. graphene films for transparent electrodes. Nat Nanotechnol 5(8):574–578
36. Geim AK (2009) Graphene: status and prospects. Science 324(5934):1530–1534
37. Lin YM, Dimitrakopoulos C, Jenkins KA, Farmer DB, Chiu HY, Grill A, Avouris P (2010) 100-GHz transistors from wafer-scale epitaxial graphene. Science 327(5966):662
38. Novoselov KS, Jiang Z, Zhang Y, Morozov SV, Stormer HL, Zeitler U, Maan JC, Boebinger GS, Kim P, Geim AK (2007) Room-temperature quantum hall effect in graphene. Science 315(5817):1379
39. Meyer JC, Geim AK, Katsnelson MI, Novoselov KS, Booth TJ, Roth S (2007) The structure of suspended graphene sheets. Nature 446(7131):60–63
40. Eda G, Fanchini G, Chhowalla M (2008) Large-area ultrathin films of reduced graphene oxide as a transparent and flexible electronic material. Nat Nanotechnol 3(5):270–274
41. Li X, Zhang G, Bai X, Sun X, Wang X, Wang E, Dai H (2008) Highly conducting graphene sheets and Langmuir-Blodgett films. Nat Nanotechnol 3(9):538–542
42. Park S, Ruoff RS (2009) Chemical methods for the production of graphenes. Nat Nanotechnol 4(4):217–224
43. Hernandez Y, Nicolosi V, Lotya M, Blighe FM, Sun Z, De S, McGovern IT, Holland B, Byrne M, Gun'ko YK, Boland JJ, Niraj P, Duesberg G, Krishnamurthy S, Goodhue R, Hutchison J, Scardaci V, Ferrari AC, Coleman JN (2008) High-yield production of graphene by liquid-phase exfoliation of graphite. Nat Nanotechnol 3(9):563–568
44. Li D, Mueller MB, Gilje S, Kaner RB, Wallace GG (2008) Processable aqueous dispersions of graphene nanosheets. Nat Nanotechnol 3(2):101–105
45. Emtsev KV, Bostwick A, Horn K, Jobst J, Kellogg GL, Ley L, McChesney JL, Ohta T, Reshanov SA, Roehrl J, Rotenberg E, Schmid AK, Waldmann D, Weber HB, Seyller T (2009) Towards wafer-size graphene layers by atmospheric pressure graphitization of silicon carbide. Nature Mater 8(3):203–207
46. Sutter PW, Flege J-I, Sutter EA (2008) Epitaxial graphene on ruthenium. Nature Mater 7(5):406–411
47. Ramanathan T, Abdala AA, Stankovich S, Dikin DA, Herrera-Alonso M, Piner RD, Adamson DH, Schniepp HC, Chen X, Ruoff RS, Nguyen ST, Aksay IA, Prud'homme RK, Brinson LC (2008) Functionalized graphene sheets for polymer nanocomposites. Nat Nanotechnol 3(6):327–331
48. Dikin DA, Stankovich S, Zimney EJ, Piner RD, Dommett GHB, Evmenenko G, Nguyen ST, Ruoff RS (2007) Preparation and characterization of graphene oxide paper. Nature 448(7152):457–460

49. Wang C, Li D, Too CO, Wallace GG (2009) Electrochemical properties of graphene paper electrodes used in lithium batteries. Chem Mater 21(13):2604–2606
50. Goodenough J, Kim Y (2011) Challenges for rechargeable batteries. J Power Sources 196:6688–6694
51. Tarascon J (2010) Key challenges in future Li-battery research. Phil Trans R Soc A 368:3227–3241
52. Abouimrane A, Compton OC, Amine K, Nguyen ST (2010) Non-annealed graphene paper as a binder-free anode for lithium-ion batteries. J Phys Chem C 114(29):12800–12804
53. Stoller MD, Park S, Zhu Y, An J, Ruoff RS (2008) Graphene-based ultracapacitors. Nano Lett 8(10):3498–3502
54. Zhang L, Shi G (2011) Preparation of highly conductive graphene hydrogels for fabricating supercapacitors with high rate capability. J Phys Chem C 115(34):17206–17212
55. Lin Z, Liu Y, Yao Y, Hildreth OJ, Li Z, Moon K, Wong C (2011) Superior capacitance of functionalized graphene. J Phys Chem C 115(14):7120–7125
56. Sun Y, Wu Q, Shi G (2011) Graphene based new energy materials. Energy Environ Sci 4(4):1113–1132
57. Yu A, Roes I, Davies A, Chen Z (2010) Ultrathin, transparent, and flexible graphene films for supercapacitor application. Appl Phys Lett 96(25):253103–253105
58. El-Kady MF, Strong V, Dubin S, Kaner RB (2012) Laser scribing of high-performance and flexible graphene-based electrochemical capacitors. Science 335(6074):1326–1330
59. Lin J, Zhong J, Bao D, Reiber-Kyle J, Wang W, Vullev V, Ozkan M, Ozkan CS (2012) Supercapacitors Based on Pillared Graphene Nanostructures. J Nanosci Nanotechnol 12(3): 1770–1775
60. Sun Y, Qiong W, Xu Y, Bai H, Li C, Shi G (2011) Highly conductive and flexible mesoporous graphitic films prepared by graphitizing the composites of graphene oxide and nanodiamond. J Mater Chem 21(20):7154–7160
61. Rakhi RB, Chen W, Cha D, Alshareef HN (2012) Nanostructured ternary electrodes for energy-storage applications. Adv Energy Mater 2(3):381–389
62. Yang X, Zhu J, Qiu L, Li D (2011) Bioinspired effective prevention of restacking in multilayered graphene films: towards the next generation of high-performance supercapacitors. Adv Mater 23(25):2833–2838
63. Zhao X, Hayner CM, Kung MC, Kung HH (2011) Flexible holey graphene paper electrodes with enhanced rate capability for energy storage applications. ACS Nano 5(11):8739–8749
64. Zhu Y, Murali S, Stoller MD, Ganesh KJ, Cai W, Ferreira PJ, Pirkle A, Wallace RM, Cychosz KA, Thommes M, Su D, Stach EA, Ruoff RS (2011) Carbon-based supercapacitors produced by activation of graphene. Science 332(6037):1537–1541
65. Zhang LL, Zhao X, Stoller MD, Zhu Y, Ji H, Murali S, Wu Y, Perales S, Clevenger B, Ruoff RS (2012) Highly conductive and porous activated reduced graphene oxide films for high-power supercapacitors. Nano Lett 12(4):1806–1812
66. Liu F, Song S, Xue D, Zhang H (2012) Folded structured graphene paper for high performance electrode materials. Adv Mater 24(8):1089–1094
67. Bai H, Li C, Shi G (2011) Functional composite materials based on chemically converted graphene. Adv Mater 23(9):1089–1115
68. Wang D-W, Li F, Zhao J, Ren W, Chen Z-G, Tan J, Wu Z-S, Gentle I, Lu GQ, Cheng H-M (2009) Fabrication of graphene/polyaniline composite paper via in situ anodic electropolymerization for high-performance flexible electrode. ACS Nano 3(7):1745–1752
69. Wang D, Kou R, Choi D, Yang Z, Nie Z, Li J, Saraf LV, Hu D, Zhang J, Graff GL, Liu J, Pope MA, Aksay IA (2010) Ternary self-assembly of ordered metal oxide-graphene nanocomposites for electrochemical energy storage. ACS Nano 4(3):1587–1595
70. Szczech J, Jin S (2011) Nanostructured silicon for high capacity lithium battery anodes. Energy Environ Sci 4:56–72
71. Teki R, Datta M, Krishnan P, Parker T, Lu T, Kumta P, Koratkar N (2009) Nanostructured silicon anodes for lithium ion rechargeable batteries. Small 5:2236–2242

72. Tao H-C, Fan L-Z, Mei Y, Qu X (2011) Self-supporting Si/Reduced Graphene Oxide nanocomposite films as anode for lithium ion batteries. Electrochem Commun 13(12): 1332–1335

73. Liu S, Liu X, Li Z, Yang S, Wang J (2011) Fabrication of free-standing graphene/polyaniline nanofibers composite paper via electrostatic adsorption for electrochemical supercapacitors. New J Chem 35(2):369–374

74. Yu A, Park HW, Davies A, Higgins DC, Chen Z, Xiao X (2011) Free-standing layer-by-layer hybrid thin film of graphene–MnO$_2$ nanotube as anode for lithium ion batteries. J Phys Chem Lett 2(15):1855–1860

75. Liu F, Xue D (2012) Flexible composite electrodes upon aerogel derived graphene paper towards lithium-ion batteries. Energ Environ Focus 1(2):93–98

76. Kang Y-R, Li Y-L, Hou F, Wen Y-Y, Su D (2012) Fabrication of electric papers of graphene nanosheet shelled cellulose fibres by dispersion and infiltration as flexible electrodes for energy storage. Nanoscale 4(10):3248–3253

77. Weng Z, Su Y, Wang D-W, Li F, Du J, Cheng H-M (2011) Graphene-cellulose paper flexible supercapacitors. Adv Energy Mater 1(5):917–922

78. Tai Z, Yan X, Lang J, Xue Q (2012) Enhancement of capacitance performance of flexible carbon nanofiber paper by adding graphene nanosheets. J Power Sources 199:373–378

Phase Change Material Particles and Their Application in Heat Transfer Fluids

J. J. Xu, F. Y. Cao and B. Yang

Abstract Phase change materials (PCMs) have received considerable attention for the application of thermal energy storage and transfer. This chapter discusses synthesis and characterization of several types of PCM particles, as well as the use of PCMs to enhance the performance of heat transfer fluids. Two different PCM microcapsules are introduced first: one comprises solid–liquid PCM paraffin encapsulated in polymer shell; the other involves solid–solid PCM neopentyl glycol (NPG) core and silica shell. Then the synthesis of low-melting metallic nanoparticles and NPG nanoparticles without shells are discussed. The last part of this chapter is dedicated to a new type of phase-changeable fluids, nanoemulsion fluids, in which the dispersed nanodroplets can be liquid–vapor PCM or liquid–solid PCM, depending on the PCM properties and the operating temperature. Material synthesis and property characterizations of these phase-changeable fluids are two main aspects of this chapter.

1 Introduction

Phase change materials (PCMs) have received considerable attention for the application of thermal energy storage and transfer, which offer the potential to reduce energy consumption and in turn lower the related environmental impact [1–7]. PCMs are capable of absorbing and releasing large amounts of thermal energy when they undergo phase transition. Latent heat storage materials provide much higher energy storage density with a smaller temperature difference between storing and releasing processes, than the sensible thermal storage materials [4, 8–10]. PCMs' latent heat storage can be normally achieved through solid–solid, solid–liquid, and liquid–gas phase change.

Xu and Cao contributed equally to this work.

J. J. Xu · F. Y. Cao · B. Yang (✉)
Department of Mechanical Engineering, University of Maryland,
College Park, MD 20742, USA
e-mail: baoyang@umd.edu

Z. Lin and J. Wang (eds.), *Low-cost Nanomaterials*, Green Energy and Technology,
DOI: 10.1007/978-1-4471-6473-9_16, © Springer-Verlag London 2014

The strategy of adding PCM particles to improve thermal performance of heat transfer fluids has been pursued [5, 7, 11–16]. The need for high-performance heat transfer fluids is driven by the increasing thermal management demand. The advances in semiconductor materials and more precise fabrication techniques have the unfortunate side effect of generating higher amounts of waste heat within a smaller volume. The need for higher capacity and more complex cooling systems is limiting the full potential of the advances in electronics and power electronics. The solution is the development of significantly improved heat transfer systems and their kernel components. One important component of these heat transfer systems is the cooling fluid used inside. The inherently poor heat transfer properties of some of the coolants, lubricants, oils, and other heat transfer fluids used in today's thermal systems limit the capacity and compactness of the heat exchangers that use these fluids. Therefore, there is an urgent need for innovative heat transfer fluids with improved thermal properties over those currently available.

This chapter discusses several types of PCM particles that can be potentially used in the application of heat transfer fluids. These PCM particles can be categorized into three groups: solid–liquid phase change particles, solid–solid phase change particles, and liquid–vapor phase change droplets. This chapter starts with the introduction of two types of PCM microcapsules: paraffin–polymer and neopentyl glycol (NPG)-silica microcapsules. Then the low-melting metallic nanoparticles and NPG nanoparticles, which both are synthesized by physical methods, are discussed [5, 7]. The last part of this chapter is dedicated to phase-changeable nanoemulsion fluids, in which the dispersed nanodroplets can be liquid–vapor PCM or liquid–solid PCM, depending on the PCM and the operating temperature [11–16]. Both material synthesis and property characterization of these PCM microcapsules and particles will be covered. This chapter is not intended to serve as a complete description of all phase-changeable particles available for heat transfer applications. The selection of the coverage was influenced by the research focus of the authors and reflects their assessment of the field.

2 Use of PCMs to Increase the Fluid Thermal Properties

Most research on thermal fluids to date has focused on how to increase thermal conductivity [17–19]. This chapter will introduce the concept of using PCM particles to increase the effective heat capacity as well as the effective thermal conductivity of thermal fluids. Both the heat capacity and thermal conductivity of the fluids strongly influence the heat transfer coefficient. For example, for a fully developed turbulent flow of a single-phase fluid, the convective heat transfer coefficient, h, can be described in terms of thermal conductivity, k_f, and specific heat, C_f, as $h \propto k_f^{0.6} C_f^{0.4}$ [20]. If the particles are made of PCMs, the effective specific heat of the PCM fluid will be increased by a factor of $1 + \frac{\alpha \cdot H_{PCM}}{\Delta T \cdot C_f}$, where α is the weight fraction of the PCM nanoparticles in the fluid, H_{PCM} is the latent heat of the PCM per

Table 1 PCMs with potential use in thermal fluids [21, 22]

	Compound name	Phase change temperature (°C)	Thermal conductivity (W/mK)	Latent heat (J/g)
Solid–liquid	Water	0	0.6 (liquid)	334
	Myristic acid + Capric	24	0.15 (liquid)	147
	$CaCl_2.6H_2O$	29	0.56 (liquid)	192
	$Na_2SO_4.10H_2O$	32	0.54 (liquid)	251
	Paraffin C17	22	0.15 (liquid)	215
	Paraffin C18	28	0.15 (liquid)	245
	Paraffin C19	32	0.15 (liquid)	222
	Capric Acid	32	0.15 (liquid)	152
	$Zn(NO_3)_2.6H_2O$	36	0.47 (liquid)	147
	Indium	157	82 (solid)	29
	FC-72	56	0.06 (liquid)	88
Liquid–vapor	Ethanol	78	0.2 (liquid)	855
	Water	100	0.6 (liquid)	2260
	Trihydroxy methyl-aminomethane	134.5	0.22	285
Solid–solid	Pentaglycerin	81	0.22	193.2
	Neopentyl glycol	43	0.22	131.5

unit mass, and ΔT is the temperature difference between the heat transfer surface and the bulk fluid. For example, a 5 % mass fraction of solid–liquid phase-changeable hexadecane (melting point, 291 K) may enhance the effective heat capacity of FC-72 by up to 100 % at $\Delta T = 10$ K. Therefore, the use of PCMs as the dispersed particles will be able to boost the thermal properties of thermal fluids much more significantly than non-PCM particles at the same volume fraction.

Current PCMs available for thermal fluid application can be divided into three main categories: solid–solid, solid–liquid, and liquid–vapor transitions. The thermophysical properties of the state-of-the-art PCMs with potential use in thermal fluids are listed in Table 1. Liquid–vapor PCMs, such as ethanol and water, have a much larger latent heat and therefore result in a higher heat transfer rate, but a condenser is needed for vapor condensation. Solid–liquid PCMs have small volume change during phase transition, and therefore the suspension of solid–liquid phase change particles can be used in the thermal management system designed for single-phase fluids. Solid–solid PCMs do not involve the liquid phase, so there is no concern about liquid leakage and thermal expansion during phase transition.

3 PCM Microcapsules

PCM microcapsules comprise a PCM core and a polymer or inorganic shell. The microcapsules can avoid the PCM leakage and the possible interaction between the PCM and the matrix. This section will discuss synthesis and characterization of two types of PCM microcapsules; one is paraffin–polymer microcapsules that are synthesized using the in situ polymerization method, and the other is NPG-silica microcapsule fabricated by the interfacial hydrolysis method.

3.1 Paraffin–Polymer Microcapsules (Solid–Liquid PCM)

Paraffins, the n-alkanes (C_nH_{2n+2}) with different numbers of carbon atoms in their chain, are common solid–liquid PCMs used as the core material of microcapsules because of their appropriate phase transition temperature, large latent heat of fusion, chemical stability, and capability of being microencapsulated [9, 23–27].

3.2 Synthesis of Paraffin–Polymer Microcapsules

In situ polymerization method has been widely used to synthesize microcapsules comprising PCM paraffin encapsulated in polymer shells [28]. This method generally involves two immiscible liquids, e.g., paraffin and water, in which polymerization reaction occurs at the interface between the PCM particles and the continuous phase water. Melamine–formaldehyde (M/F) and urea–formaldehyde are often used as the shell materials due to their good chemical stability and mechanical strength. Figure 1 shows the fabrication process of microcapsules comprising solid–liquid paraffin core and melamine–formaldehyde shells. In the process, melamine and formaldehyde were first added into alkaline water solution with a certain weight ratio [28]. The mixture was heated to make a transparent aqueous solution of melamine–formaldehyde prepolymer. In the meantime, an oil-in-water emulsion of paraffin oil-in-acidic water was made with a homogenizer and heated up to 60 °C. A certain amount of the hot melamine–formaldehyde prepolymer solution was added into the paraffin oil-in-water emulsion, and a polymerization was processed for hours to form shells of melamine–formaldehyde resin at the oil–water interface.

Two reaction steps are involved in the formation of the microcapsule resin shell. The first step is melamine–formaldehyde precondensation in basic environment.

(1)

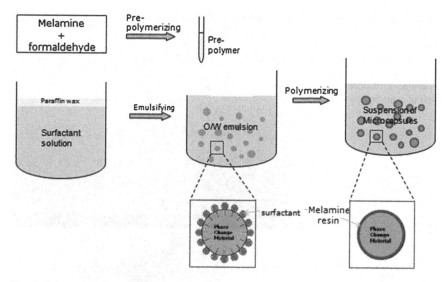

Fig. 1 Schematic illustrating the formation of the microcapsules comprising paraffin core and melamine–formaldehyde shell [28]

It is followed by in situ polymerization of the prepolymer in acidic environment, in which the melamine–formaldehyde resin shell is formed on the surface of the octadecane droplets,

$$(2)$$

The microstructure of the melamine–formaldehyde shell can be characterized by the degree of crosslink and bridge types between two triazine rings, which can be either methylene ether ones or methylene ones [29, 30]. The parameters of the encapsulation process, including surfactant concentration, F:M ratio, the pH values of prepolymer solution, and HAc concentration in emulsion, affect significantly the microstructure and phase changing properties of the microcapsules.

Fig. 2 **a** SEM image of the
as-produced paraffin
microcapsules with
melamine–formaldehyde
shell. **b** TEM image of the
melamine–formaldehyde
shell of a microcapsule [28]

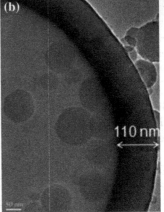

Figure 2a shows the SEM image of the as-produced microcapsule comprising paraffin core and melamine–formaldehyde shell. It can be found that the microcapsules are shaped spherically, with their sizes in about 5–15 microns. The surface of the microcapsules is relatively smooth, with small concaves on the surface of some particles, which are caused by contraction during the temperature drop from the experimental temperature to room temperature. Figure 2b shows the core–shell structure of a microcapsule particle, in which the melamine–formaldehyde shell is uniform in thickness of about 110 nm.

3.3 Phase Change Behavior of Paraffin–Polymer Microcapsules

Figure 3 shows the cyclic DSC heating and cooling curves for the paraffin–polymer microcapsules and the bulk paraffin. The latent heat of the microcapsules

Fig. 3 DSC curves of bulk
and microencapsulated
paraffin n-$C_{18}H_{38}$

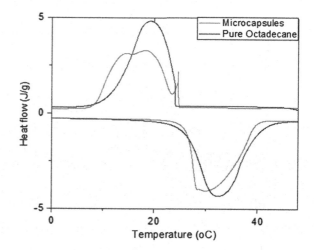

is found to be 209 J/g by measuring the integral area of the single endothermic peak on the bottom. The weight percentage of paraffin in the microcapsules is about 88 %, by comparing the latent heat of the microcapsules and the bulk paraffin.

Three peaks were observed on the cooling curve of the microcapsules, which is different from the bulk curve with a single endothermic peak and a single exothermic peak [28]. These DSC curves suggest that the microencapsulated n-$C_{18}H_{38}$ first changes to a metastable rotator phase from melt rather than directly into the solid triclinic phase, as for bulk n-$C_{18}H_{38}$. Later on, with temperature further dropping, the metastable rotator phase converts to the stable triclinic phase. Based on this analysis, the second exothermic peak with smaller subcooling is assigned to the bulk crystallization from melt to a metastable rotator phase, while the third one is the phase transition from the metastable rotator phase to the stable triclinic phase.

3.4 NPG–Silica Microcapsules (Solid–Solid PCM)

Certain molecular crystals, such as polyalcohols, undergo solid-state crystal transformations that absorb sufficient latent heat; they can be used for practical thermal energy storage and transfer application [9, 31]. Those polyalcohols, including neopentyl glycol (NPG), pentaerythritol (PE), etc., will transform from heterogeneous crystals to homogeneously face-centered cubic (FCC) crystals with high symmetry when the temperature rises across a certain point. A sufficient latent heat is associated with such a solid–solid phase transition, due mainly to the formation of hydrogen bonds among these molecules [9, 31]. Compared to conventional solid–liquid PCMs, the solid–solid PCMs do not involve the liquid phase, so there is no concern about liquid leakage and thermal expansion during

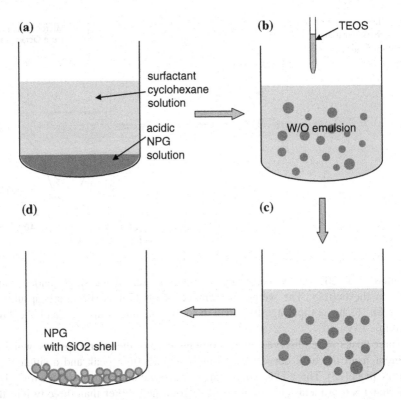

Fig. 4 Process of synthesizing NPG microcapsules. **a** mix water into cyclohexane with surfactant; **b** add TEOS; **c** hydrolysis of TEOS to form silica shell; and **d** collection of microcapsules [32]

phase transition. One interesting feature of these polyalcohol PCMs is that their phase transition temperature can be continuously tuned by mixing two or more polyalcohols. This flexibility is important in order to address many applications that have different operation temperatures.

In some applications, NPG microcapsules are needed to protect the NPG core from the reactive base fluids. The NPG microcapsules can be synthesized by the method of interfacial hydrolysis of TEOS in W/O emulsions [32]. The synthesis process is illustrated in Fig. 4. Highly concentrated NPG aqueous solution with HCl was used as the water phase of the W/O emulsion. The silicon oxide shell is formed by the hydrolysis of TEOS, as given in the equation below:

$$(C_2H_5O)_4Si + H_2O \rightarrow C_2H_5OH + SiO_2 \cdot xH_2O \qquad (3)$$

As the precursor TEOS is hydrophobic and the product silica is hydrophilic, the hydrolysis reaction of TEOS occurs at the interface between aqueous droplets and the bulk cyclohexane. During this reaction, water in the aqueous solution is

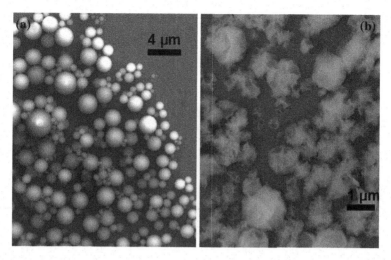

Fig. 5 SEM image of **a** as-synthesized microcapsules of NPG in silica shell, and **b** wrinkled silica shell after removal of NPG [32]

consumed by the hydrolysis of TEOS. These microcapsules are dried in a vacuum oven to remove the residual ethanol.

The SEM image of the NPG–silica microcapsules is shown in Fig. 5a. It can be seen that these microcapsules are spherical in shape with smooth surface. The diameter of these microcapsules is in the range of 0.2–4 μm. The NPG microcapsules have a relatively wider size distribution due to the large viscosity of the aqueous phase, NPG solution [33]. After removing NPG from the microcapsule, the silica shell collapses without rupture on the shell, as shown in Fig. 5b. The flexibility of the silica shell is due to its thin thickness, about 30 nm.

3.5 Phase Change Behavior of Heat Transfer Fluids with Microcapsules

The effective heat capacity of heat transfer fluids can be enhanced by dispersing appropriate PCM particles into them. Figure 6 shows the DSC curves of the PAO dispersion with 20 wt% of paraffin microcapsules. PAO has been widely used as dielectric heat transfer fluids and lubricants. It remains oily in a wide temperature range due to the flexible alkyl branching groups on the $C-C$ backbone chain, but it has relatively poor thermal properties. The overall latent heat of the fluid is measured as 23.5 J/g. The effective specific heat capacity is about 3.37 J/g k when the temperature difference between the heat transfer surface and the fluid is assumed to be 20 °C. This value indicates a more than 50 % increase in effective heat capacity in the microcapsule dispersions, compared to the pure PAO.

Fig. 6 DSC curves of
dispersions of paraffin
microcapsules in PAO

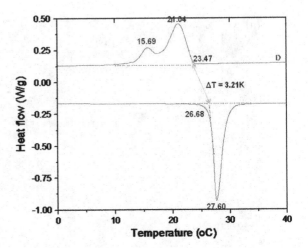

It is interesting to note that on the cooling side of the DSC curve, the liquid-rotator
and rotator-triclinic phase transition processes are well separated from each other,
peaking at 19.7 and 21.0 °C, representatively. The latent heat of the liquid-rotator
phase transition dominates in the freezing process, and the subcooling for the
liquid-rotator phase transition is about 3.2 K, as shown in the DSC curve in Fig. 6.

4 PCM Nanoparticles Without Shell

PCM microcapsules are often synthesized using the chemical process, as discussed
in the previous section. Physical methods, including direct emulsification and
spray drying, are capable of producing PCM particles smaller than 1 μm without
shell. The following sections will discuss two types of PCM nanoparticles: low-
melting-point metallic nanoparticles and NPG nanoparticles.

4.1 Low-Melting-Point Metallic Nanoparticles
(Solid–Liquid PCM)

Low melting metals, such as Indium and BiSn alloy, have thermal conductivity
much higher than conventional dielectric thermal fluids; for example, thermal
conductivities of Indium and PAO are 82 and 0.14 W/(mk), respectively. The use
of low-melting metallic PCM nanoparticles provides a way to simultaneously
improve the effective thermal conductivity and heat capacity of the base fluids [5,
7]. When the metallic PCM particles are small enough (e.g., below 50 nm in size)
and stabilized with surfactant molecules, the nanoparticle suspensions could
remain stable during freezing and melting of the PCM particles.

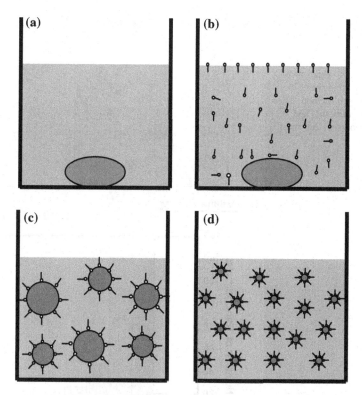

Fig. 7 Schematic illustrating the formation of metallic PCM nanoparticles. **a** PAO and molten metal are in the reaction vessel. These two liquids are immiscible and phase separate. **b** The polymer surfactant is soluble in PAO and preferentially adsorbs at the interface. **c** The mixture is stirred using a magnetic stirrer and the bulk molten metal breaks into microscale droplets. **d** The microscale emulsion is exposed to high-intensity ultrasonication until the PCM nanoparticles are formed [5]

4.1.1 Synthesis of Low-Melting-Point Metallic Nanoparticles

A one-step, nanoemulsification technique has been developed to prepare suspensions of metallic PCM nanoparticles [5]. The fabrication process of metallic PCM nanoparticles is illustrated in Fig. 7. Briefly, this technique exploits the extremely high shear rates generated by the ultrasonic agitation and the relatively large viscosity of the continuous phase–PAO, to rupture the molten metal down to diameter below 100 nm. As an example, the preparation of Indium (melting point: 156.6 °C) and Field's metal (melting point: 63.2 °C) metallic PCM fluids is discussed below.

A large number of factors can affect the metallic particle size in the emulsification process. These include selecting an appropriate composition, controlling the reaction temperature, choosing the order of addition of the components, and applying the shear in an effective manner. However, the fundamental relationship

Fig. 8 TEM images (*bright field*) and size distributions of **a** the Field's metal nanoparticles, and **b** Indium nanoparticles, which were fabricated using the nanoemulsion method. The scale bars in both TEM images represent 200 nm [5]

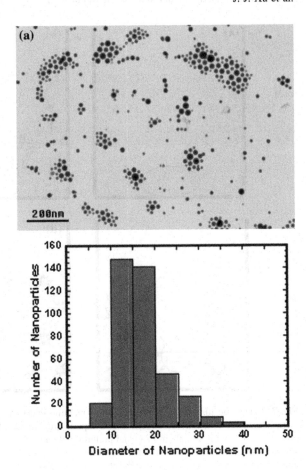

governing how the dispersed phase can be ruptured in another immiscible liquid under a shear stress is given simply by Taylor's formula [34]:

$$d \approx \frac{2 \cdot \sigma}{\eta_c \cdot \hat{\gamma}}, \tag{4}$$

where d is the droplet diameter, σ is the interfacial tension between the droplet and continuous phase, η_c is the viscosity of the continuous phase, and $\hat{\gamma}$ is the shear rate. Based on Taylor's formula, it is possible to estimate the shear rate required to form PCM nanoparticles. The nanoparticle size may be effectively regulated by changing the synthesis temperature in order to vary the viscosity of the continuous phase.

The size distribution of the Field's metal and Indium nanoparticles was examined using transmission electron microscopy (TEM, JEOL 2100F). TEM bright field (BF) images of Field's and Indium nanoparticles are shown in Fig. 8a and b, respectively. These nanoparticles are spherical because the liquid nanodroplets

Fig. 8 continued

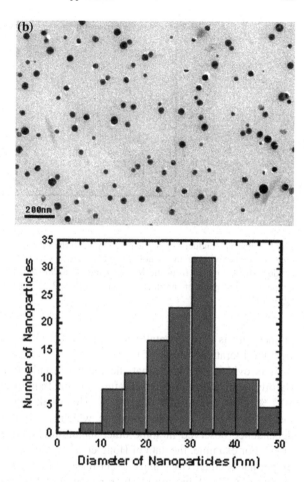

have a positive interfacial tension (i.e., surface energy) in emulsion. In addition, these nanoparticles are highly dispersed. The polymer surfactants appear to provide sufficient steric stabilization despite the strong cohesion forces among molten metal nanodroplets.

4.1.2 Thermal Conductivity of Metallic PCM Fluids

The thermal conductivity of the pure PAO and dispersions of Indium nanoparticles in PAO, and as well as the relative thermal conductivity, are plotted against temperature in Fig. 9. Results estimated from the Maxwell Model are also shown for comparison. The relative thermal conductivity is defined as k_{nf}/k_o, where k_o and k_{nf} are thermal conductivity of the base fluids and dispersions of Indium nanoparticles in PAO, respectively. The thermal conductivity of the PAO at room

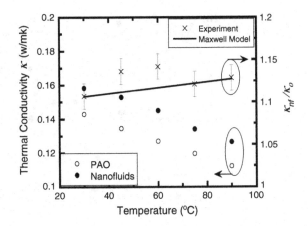

Fig. 9 Thermal conductivity of the pure PAO and the indium-in-PAO nanofluid (*left x-axis*) and relative conductivity of the nanofluid (*right x-axis*) versus temperature. The relative thermal conductivity estimated from the Maxwell model (*solid line*) is also shown for comparison. A temperature-independent interfacial resistivity $R_b = 3.22 \times 10^{-8}$ m²k/W is used in this calculation [5]

temperature is experimentally found to be 0.143 W/(mK), which compares well with the literature values [35].

It is evident in Fig. 9 that the thermal conductivity enhancement increases slightly with increasing temperature in the Indium-In-PAO PCM fluids about 10.7 % at 30 °C and 12.9 % at 90 °C. This weak dependence on temperature was measured with the 3ω-wire technique, which is very different from the very strong dependence observed in some other nanofluids obtained with the hot-wire technique. For example, the data of Hong et al. showed a factor of about three increase in thermal conductivity enhancement for a temperature increase of 50 °C [36]. As also seen in Fig. 9, the solid Indium nanoparticles increase thermal conductivity of the pure PAO by about 12 % for volume fraction of 8 %. The experimental data are well captured in the temperature range of 30−90 °C by the Maxwell Model,

$$k_{nf} = (1 + 3\Phi \frac{\alpha - 1}{\alpha + 2})k_o \qquad (5)$$

when a constant interfacial resistivity $R_b = 3.22 \times 10^{-8}$ m²k/W is used [37]. Here, ϕ is the volume fraction of the particles, $\alpha = {}^{r_p}/_{R_b k_o}$, r_p is the particle radius, and R_b is the thermal resistance per unit area of the particle/fluid interface. The interface resistance appears to play a significant role in thermal transport in the metallic PCM fluid. The interface resistance plays a significant role in thermal transport in the nanofluid, as predicted by the Maxwell Model. For example, for a negligible interface resistance, $R_b = 0$, the conductivity enhancement would increase to 24 % in the PAO-based metallic PCM fluids containing 8 vol% of nanoparticles. These experimental data suggest that the thermal conductivity enhancement observed in the Indium–PAO PCM fluids using the 3ω-wire technique can be explained using the Maxwell model, and no anomalous enhancement of thermal conductivity is observed in this study. However, many other studies in nanofluids have shown a thermal conductivity increase beyond the Maxwell predictions [38].

4.1.3 Phase Change Behavior of Metallic PCM Fluids

Knowledge of the phase change behavior of these low-melting metallic nanoparticles is critical for their use as thermal fluids [39–42]. The melting–freezing phase transition of the as-prepared Field's metal and Indium nanoparticles is measured using DSC measurements which were taken at an ordinary cyclic ramp mode, with a scan rate of 10 °C/min. Figure 10 shows the cyclic DSC heating and cooling curves for the nanoemulsions containing Field's metal and Indium nanoparticles and similar data for each material in the bulk state.

A relatively large melting–freezing hysteresis, about 45 °C for Field's metal nanoparticles and about 50 °C for Indium nanoparticles, can be seen in Fig. 10. Based on the classical nucleation theory, the melting and freezing of these nanoparticles are dependent on the interface energies between the solid metal nanoparticles and the oil matrix, the liquid metal and oil matrix, and the solid and liquid metals [43]. The observed T_m slightly below the bulk value implies that $\gamma_{SM} > \gamma_{SL} + \gamma_{LM}$ (or $\gamma_{LM} < \gamma_{SM} + \gamma_{SL}$), where γ is the interfacial energy and the subscripts S, L, and M represent the solid phase, the liquid phase, and the oil matrix. In this situation, the molten phase would not "presolidify" at the interfaces, instead, it would require critical nuclei inside these nanoparticles, i.e., homogeneous nucleation. Therefore, these liquid nanoparticles supercool to tens of degrees below the bulk melting temperature, until critical nuclei associated with solidification are formed. This characteristic may provide a mechanism to tailor the phase transition behavior of nanoparticles by varying their interfacial energy or size for different applications.

The phase change within the metallic PCM fluid has its greatest impact on the effective specific heats of these fluids. The effective specific heat can be defined as $C_{\text{eff}} = C_0 + \phi \cdot H_{\text{particle}}/\Delta T$, where ϕ is the volume fraction of the phase-changeable nanoparticles, H_{particle} is the latent heat of the phase-changeable nanoparticles per unit volume, and ΔT is the temperature difference between the heat transfer surface and the bulk fluid or the difference between the nanoparticle melting and freezing temperature. If assuming $\Delta T = 47$ °C, the effective volumetric-specific heat can be increased by about 20 % for the PCM fluids containing 8 vol% Indium nanoparticles. If ΔT could reduce to 10 °C, which is feasible by introducing external nucleating agents to suppress the melting–freezing hysteresis, the effective volumetric-specific heat of the metallic PCM fluids would be increased by up to 100 %.

Fig. 10 DSC heating and cooling curves for as-prepared **a** field's metal and **b** indium nanoparticles dispersed in PAO oil. For comparison, the experimental data for the corresponding bulk metals are also plotted. The melting−freezing phase transition points correspond to the endothermal and exothermal peaks. Subscripts are as follows: b = bulk metal; n = nanoparticles; m = melting; f = freezing [5]

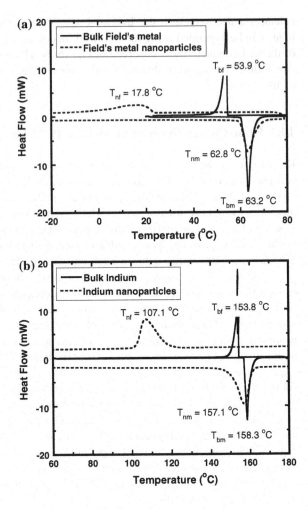

4.2 NPG Nanoparticles (Solid–Solid PCM)

NPG is a solid–solid PCM, which can eliminate potential issues with having a liquid dispersed phase. The section will discuss the synthesis of NPG nanoparticles.

4.2.1 Synthesis of NPG Nanoparticles

An aerosol system has been employed to synthesize NPG PCM nanoparticles. The schematic setup is shown in Fig. 11. In this system, PCM droplets are generated by the Atomizer, dried in the Diffusion Dryers, and then directly bubbled to the base fluids to form PCM nanofluids. The PCM nanoparticle size can be *real-time* monitored by passing them to the Differential Mobility Analyzer (DMA) and a

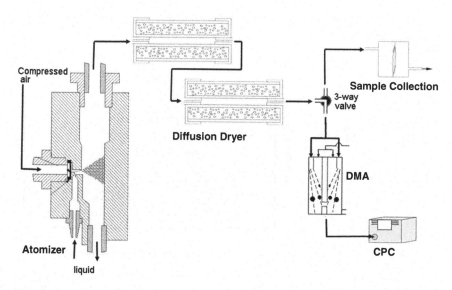

Fig. 11 An aerosol system for production of PCM nanoparticles

Condensation Particle Counter (CPC). Advantages of this method include its ability to control particle size and chemical composition, and also its high reliability.

4.2.2 Size Control of NPG Nanoparticles

The NPG PCM nanoparticle size can be effectively regulated by changing the air flow rate (or inlet air pressure) and the solution concentration in our aerosol system. For example, higher inlet air pressure would generate higher shear force in the air flow, and then rupture solution droplets to smaller size. The nanoparticle size distribution can be *real-time* monitored using the DMA. The as-prepared PCM nanoparticles are polydisperse in nature, as shown in Fig. 12. However, mono-disperse nanoparticles (2–1,000 nm) can be generated by using an electrostatic band-pass filter from the initial wider distribution. The filtration is accomplished using a DMA or a nano-DMA developed most recently.

5 Phase-Changeable Nanoemulsion Fluids

A new type of PCM fluids, nanoemulsion fluids, has been recently developed for thermal energy storage and transfer [44]. The synthesis and properties of these fluids are fundamentally different from traditional PCM fluids, such as microcapsule slurries and nanofluids, as discussed below. One fluid is dispersed into another immiscible fluid as nanosized structures such as droplets and tubes to create a

Fig. 12 Size distribution of the as-prepared PCM nanoparticles

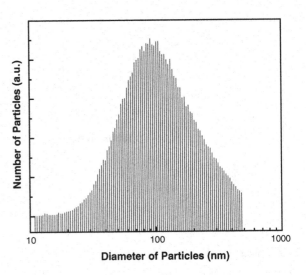

"nanoemulsion fluid." Those nanostructures are reverse micelles swollen with the dispersed phase and stabilized by the surfactant molecules. Nanoemulsion fluids are part of a broad class of multiphase colloidal dispersions [11–16]. Some nano-emulsion fluids can be spontaneously generated by self-assembly without the need of external shear and are suited for thermal management applications [12–15, 45–58]. Self-assembled nanoemulsion fluids are thermodynamically stable. Figure 13 shows a picture of Ethanol-in-polyalphaolefin (PAO) nanoemulsion heat transfer fluids: both PAO- and PAO-based nanoemulsion fluids are transparent but the nanoemulsion exhibits the Tyndall Effect [11–16, 59, 60]. Note that nanoemulsions are emulsions are very difference in terms of microstructures and materials prop-erties. Emulsions are dispersions of micron-sized droplets and are stable only for a relatively short time [56, 61–68]. Table 2 is the comparison between self-assembled nanoemulsion fluids and conventional emulsions. Self-assembled nanoemulsion fluids are thermodynamically stable system, and possess long-term stability.

Phase change nanoemulsion fluids can be divided into two groups, liquid–vapor nanoemulsions and liquid–solid nanoemulsions, depending on the phase transition behavior of the dispersed particles and the fluid operating temperature. These two types of nanoemulsion fluids are discussed in the following sections.

5.1 Liquid–Vapor Phase Change Nanoemulsion Fluids

Ethanol-in-Polyalphaolefin (PAO) is a liquid–vapor phase change nanoemulsion fluid, in which the ethanol nanodroplets could evaporate explosively and thus enhance the heat transfer rate of the base fluid PAO [12]. The microstructure and thermophysical properties of the Ethanol-in-PAO nanoemulsion fluids are dis-cussed below.

Fig. 13 Pictures of ethanol-in-PAO nanoemulsion fluids (*Bottle A*) and pure PAO (*Bottle B*). Liquids in both bottles are transparent. The Tyndall effect (i.e., a light beam can be seen when viewed from the side) can be observed only in *Bottle A* when a laser beam is passed through *Bottles A* and *B* [11, 44]

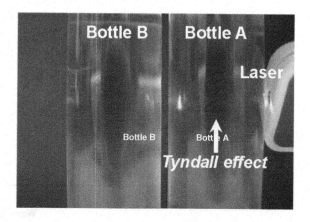

Table 2 Comparison of nanoemulsion fluids and emulsions [44]

Sample	Property	Nanoemulsion	Emulsion
1	Appearance	Transparent	Turbid
2	Interfacial tension	Ultra low (usually $\ll 1$ mN/m)	low
3	Droplet size	<50 nm	>500 nm
4	Stability	Thermodynamically stable, long shelf-life	Thermodynamically unstable
5	Preparation	Self-assembly	Need of external shear
6	Viscosity	Newtonian	Non-newtonian

5.1.1 Microstructure of Ethanol-in-PAO Nanoemulsion Fluids

The microstructure of nanoemulsion fluids is affected by many factors, including surfactant type and concentration, dispersed liquid type and concentration, molar ratio of dispersed liquid to surfactant, temperature, pH value, and salinity [56, 69–75]. The characterization of the microstructure of nanoemulsion fluids is a challenging task. Small angle neutron scattering (SANS) and Small angle X-ray scattering (SAXS) are often used because they can be applied to the "concentrated" colloidal suspensions (e.g., >1 vol%) [76–82]. Figure 14 shows the SANS data for ethanol-in-PAO nanoemulsion fluid. The wave vector is given by:

$$q = 4\pi \sin(\theta/2)/\lambda \qquad (6)$$

where λ is the wavelength of the incident neutrons and θ is the scattering angle. The analysis of the SANS data suggests that the inner cores of the swollen micelles, i.e., the ethanol droplets are spherical and have a radius of about 0.8 nm for 9 vol%.

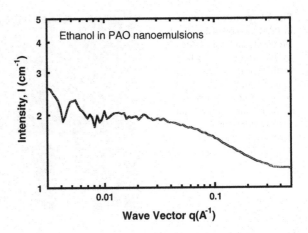

Fig. 14 Intensity *I* versus wave vector *q* measured in ethanol-in-PAO nanoemulsion fluids [11]

5.1.2 Thermal Conductivity of Ethanol-in-PAO Nanoemulsion Fluids

Figure 15 shows the relative thermal conductivity in Ethanol-in-PAO nano-emulsion fluids as a function of the ethanol loadings. The observed conductivity increase in the Ethanol-in-PAO nanoemulsion fluids is rather moderate. The prediction by the Maxwell model is also plotted in Fig. 15 for comparison. The relative thermal conductivity is defined as k_{eff}/k_o, where k_o and k_{eff} are thermal conductivities of the base fluid and nanoemulsion fluids, respectively. The effective medium theory reduces to Maxwell's equation for suspensions of well-dispersed, noninteracting spherical particles,

$$\frac{k_{eff}}{k_o} = \frac{k_p + 2k_o + 2\phi(k_p - k_o)}{k_p + 2k_o - \phi(k_p - k_o)} \tag{7}$$

where k_o if the thermal conductivity of the base fluid, k_p is the thermal conductivity of the particles, and ϕ is the particle volumetric fraction. This equation predicts that the thermal conductivity enhancement increases approximately linearly with the particle volumetric fraction for dilute nanofluids or nanoemulsion fluids (e.g., $\phi < 10\,\%$), if $k_p > k_o$ and no change in particle shape. It can be seen in this figure that the relative thermal conductivity of Ethanol-in-PAO nanoemulsion fluids appears to be linear with the loading of ethanol nanodroplets over the loading range from 0 to 9 vol%. However, the magnitude of the conductivity increase is rather moderate in the ethanol-in-PAO nanoemulsion fluids, e.g., 2.3 % increase for 9 vol% ($k_{PAO} = 0.143$ W/mK and $k_{alcohol} = 0.171$ W/mK [35, 83]). No strong effects of Brownian motion on thermal transport are found experimentally in those fluids although the nanodroplets are extremely small, around 0.8 nm.

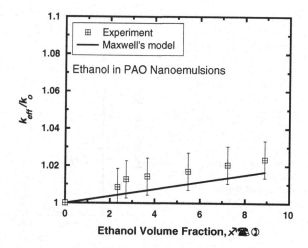

Fig. 15 Relative thermal conductivity increase with ethanol loading in the ethanol-in-PAO nanoemulsion fluids. The estimate from Maxwell's model is shown for comparison [11]

5.1.3 Phase Change Behavior in Ethanol-in-PAO Nanoemulsion Fluids

The pool boiling of these Ethanol-in-PAO nanoemulsion fluids have been investigated, in which the dispersed ethanol nanodroplets undergo liquid–vapor phase transition. In these Ethanol-in-PAO nanoemulsion fluids, the dispersed phase ethanol has a boiling point of 78 °C at 1 atm which is much lower than the boiling point of the base fluid PAO (277 °C) [35]. The pool boiling heat transfer curves are plotted in Fig. 16 for the pure PAO and PAO-based nanoemulsion fluids. When the wire temperature is less than 170 °C, the heat transfer coefficient values of the pure PAO and the PAO-based nanoemulsion fluids appear to be the same. This indicates that these ethanol nanodroplets have little effect on the fluid heat transfer efficiency if there is no phase transition in these nanodroplets. This is also consistent with the measured thermal conductivity shown in Fig. 15. When the heater temperature is further increased, an abrupt increase in convective heat transfer coefficient is observed in the PAO nanoemulsion fluids, compared to that of the pure PAO case. For example, the dissipated heat flux q is found to be 90 and 400 W/cm^2 at $T_{wire} = 200$ °C for the pure PAO and the PAO nanoemulsion fluid, respectively. What is more interesting is that the CHF of the PAO nanoemulsion fluids is significantly larger than that of their pure components ethanol and PAO.

The causes of the observed abrupt increase in the heat transfer coefficient can be first examined by evaluating the Morgan correlation that works for free convection over a long cylinder [84],

$$\overline{Nu_D} = CRa_D^n \tag{8}$$

where Ra is the Rayleigh number, and C and n are constants. The Rayleigh number is in the range $10^{-10} \sim 10^{-2}$ for the nanoemulsion experiment, so $n = 0.058$. A direct impact of the nanodroplet vaporization would be the enhanced effective heat

Fig. 16 Surface heat flux as a function of heater temperature for pure PAO, pure ethanol, and PAO nanoemulsion fluids. These data were measured in a pool boiling setup with an untextured heater surface (Pt wire, 25.4 μm in diameter), where the bulk liquid was at atmospheric pressure and room temperature (25 °C, not saturated state). The CHF is determined within an accuracy of 5 % [44]

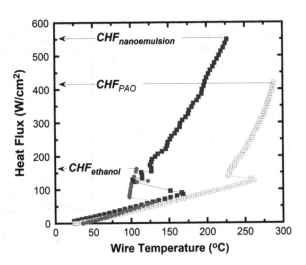

capacity due to the heat of vaporization. The effective heat capacity can be evaluated using the formula:

$$C_{eff} = C_0 + \phi \cdot H_{droplet}/\Delta T \tag{9}$$

where ϕ is the volume fraction of the phase-changeable nanodroplets, $H_{droplet}$ is the heat of vaporization of the nanodroplets per unit volume, and ΔT is the temperature difference between the heat transfer surface and the bulk fluid. In this experiment, if it is assumed that $\Delta T = 10$ °C, the effective volumetric-specific heat can be increased by up to 162 % for the 4 vol% nanoemulsion fluid when the ethanol nanodroplets undergo liquid–vapor phase transition. However, this would provide only 2.7 % enhancement of the heat transfer coefficient according to the Morgan correlation.

The nanodroplet vaporization can enhance heat transfer mainly through inducing drastic fluid motion within the thermal boundary layer around the heat transfer surface. The Ethanol-in-PAO interface constitutes a hypothetically ideal smooth surface, free of any solid motes or trapped gases, so the heterogeneous nucleation and ordinary boiling are suppressed. In this case, the ethanol nano-droplets can be heated to a temperature about 120 °C above their normal atmospheric boiling point (78 °C). Such a temperature is very close to the thermodynamic limit of superheat or the spinodal state of ethanol, and is only about 10 % below its critical point. The spinodal states, defined by states for which $\frac{\partial P}{\partial V}\big|_{T,n} = 0$, represent the deepest possible penetration of a liquid in the domain of metastable states [85–88]. When those ethanol nanodroplets vaporize after reaching their limit of superheat, the energy released could create a sound-shock wave, a so-called vapor explosion. This sound wave would lead to strong fluid mixing within the thermal boundary layer, therefore enhancing the fluid heat transfer. These fluids with enhanced heat transfer rate are expected to find applications in various energy conversion systems, heat exchange systems, air-conditioning, refrigeration and heat pump systems, and chemical thermal processes [84, 86, 89–93].

5.2 Liquid–Solid Phase Change Nanoemulsion Fluids

Water in FC-72 nanoemulsion fluids are discussed below as an example of liquid–solid phase change fluids, in which water could undergo liquid–solid transition with appropriate operation temperature range and thus increase heat transfer rate of the base fluid FC-72. FC-72 is one of a line of Fluorinert™ Electronic Liquids developed by 3 M™, which is used as the cooling fluids in liquid-cooled thermal management systems [94]. But it has poor thermal conductivity and heat capacity, compared to other fluids such as water.

5.2.1 Microstructure of Water in FC-72 Nanoemulsion Fluids

Water in FC-72 nanoemulsion fluids are generated by emulsifying deionized water into FC-72 with a small amount of perfluorinated amphiphiles. Figure 17 shows the picture of the prepared water in FC-72 nanoemulsion fluids and the pure FC-72 liquid and schematic diagram of a water nanodroplet dispersed in 3 M's FC-72 thermal fluid.

The Dynamic Light Scattering (DLS) technique is used to measure the size and Brownian diffusivity of the nanodroplets in the prepared water in FC-72 nano-emulsion fluid [95]. The autocorrelation function of the scattered light for the 12 vol% water in FC-72 nanoemulsion fluids is plotted in Fig. 18. The curve shows a typical exponential decay of the correlation function versus time [15, 16]. The Brownian Diffusivity and effective hydrodynamic radius of the nanodroplets are found to be 3.5×10^{-7} cm²/s and 9.8 nm at $T = 25$ °C, respectively.

5.2.2 Thermal Conductivity of Water in FC-72 Nanoemulsion Fluids

Thermal conductivity of the water in FC-72 nanoemulsion is measured for different water loadings, and the results are shown in Fig. 19. It can be seen in Fig. 19 that a very large increase in thermal conductivity is achieved in the prepared water-in-FC72 nanoemulsion fluids, with thermal conductivity enhancements of up to 52 % observed in the nanoemulsion fluid containing 12 vol% (or 7.1 wt%) of water nanodroplets. The observed enhancement in thermal conductivity is much larger than that predicted by the effective medium theory (EMT) with assumption of spherical droplets [96]. This suggests that the water droplets are column-like

Fig. 17 a Pictures of water in FC-72 nanoemulsion fluids (*Bottle A*) and pure FC-72 (*Bottle B*). Liquids in both *bottles* are transparent. The Tyndall effect can be observed only in *Bottle A* when a laser beam is passed through *Bottles A* and *B*. **b** Schematic diagram of a water nanodroplet dispersed in 3 M's FC-72 thermal fluid. A micelle of amphiphiles surrounds the nanodroplet, with the polar head interacting with water [15]

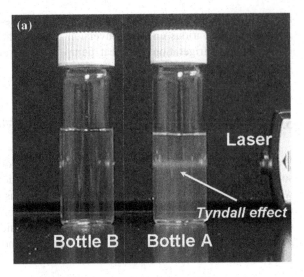

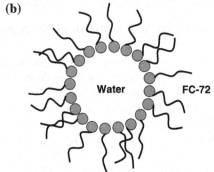

with high aspect ratio of length to radius, which leads to a higher thermal conductivity enhancement in nanoemulsion fluids than the spherical droplets.

5.2.3 Phase Change Behavior in Water in FC-72 Nanoemulsion Fluids

The heat capacity of nanoemulsion fluid can be enhanced through two different mechanisms: one is due to the high specific heat of the dispersed phase; the other is due to the latent heat of the dispersed PCMs. The latter one, i.e., use of PCMs, is much more efficient for the heat capacity enhancement. In water in FC-72 nanoemulsion fluids, the fluid's heat capacity can be increased by the high specific heat of water (i.e., $C_{water} = 4.2$ J/g C, $C_{FC\,72} = 1.1$ J/g C) or the latent heat of water ($\Delta H = 334$ J/g), depending on the operating temperature of the fluids [94].

Fig. 18 Correlation function of the scattered light for the 12 vol% water in FC-72 nanoemulsion fluids. Measurements taken by a Photocor-Complex DLS instrument [15]

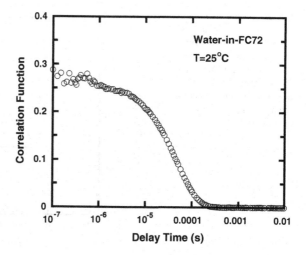

Fig. 19 Relative thermal conductivity of FC-72 nanoemulsion fluids versus water volume concentration. The effective thermal conductivity estimated from the effective medium theory for spherical particles is shown for comparison [15]

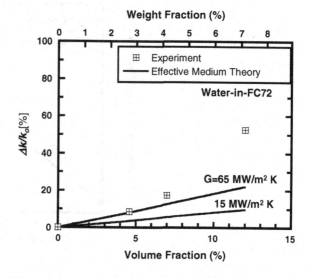

When the water nanodroplets do not undergo solid–liquid phase change, the specific heat of the water in FC-72 nanoemulsion fluids follows the simple mixture rule:

$$C_{\text{nanoemulsion}} = (1 - \Phi)c_{\text{oil}} + \Phi C_{\text{water}} \tag{10}$$

where Φ represents the concentration of the "water" phase. The specific heat of the pure FC-72 and water in FC-72 nanoemulsion fluids are measured using a Differential Scanning Calorimetry (DSC). The specific heat of FC-72 and water-in-FC-72 nanoemulsion fluids is measured using a TA-CC100 DSC. The measured and calculated heat capacities of the water in FC-72 nanoemulsion fluids are

Fig. 20 a DSC cyclic curves
of the pure FC-72 and FC-72
nanoemulsion fluids. b the
measured and calculated heat
capacity of the water in FC 72
nanoemulsion fluids. The
dispersed water droplets
remain in liquid phase during
operation [44]

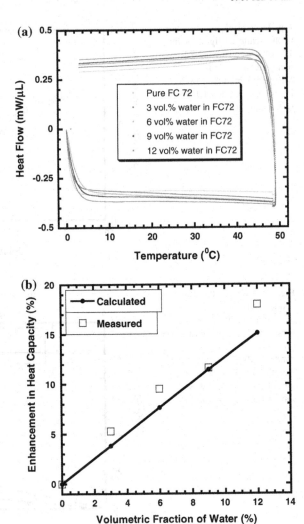

shown in Fig. 20. It can be seen that over 15 % increase in heat capacity can be achieved for a water volumetric fraction of 12 %.

In contrast with single-phase water droplets, the solid–liquid phase change of these droplets will significantly enhance the effective specific heat of the nano-emulsion fluids. The effective specific heat of the nanoemulsion fluid will be increased by a factor of $1 + \frac{\alpha \cdot H_{water}}{\Delta T \cdot C_{FC72}}$, where α is the water volume fraction, H_{water} is the latent heat of fusion of water per unit volume, and ΔT is the temperature difference between the heat transfer surface and the bulk fluid. The heat of fusion H_{water} of pure water is 334 J/ml. The measured $\alpha \cdot H_{water}$ values of the water in FC-72 nanoemulsions for water loadings of 3, 6, 9, and 12 vol% are 10.52, 15.44, 25.48, 39.78 J/ml, respectively, as shown in Fig. 21a and b. The effective heat

Fig. 21 a DSC cyclic curves
of the water in FC-72
nanoemulsion fluids with
different water loadings.
b The measured and
calculated heat capacity
enhancement in the water in
FC-72 nanoemulsion fluids.
The dispersed water droplets
undergo solid–liquid phase
transition [44]

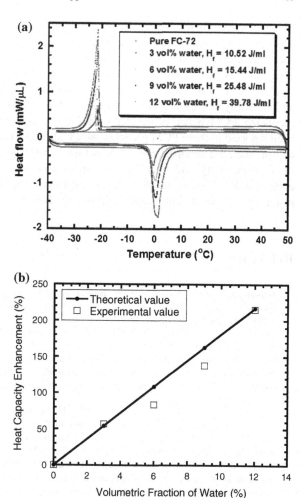

capacity of FC-72 can be enhanced by more than 200 % for a 12 % volume
fraction of water, according to these experimental data. However, a critical issue
for this type of nanoemulsion fluids is the large subcooling, more than 20 C
observed in water in FC-72 nanoemulsion fluids. More work needs to be conducted
in the future to reduce the subcooling of the dispersed PCM nanodroplets.

6 Conclusions

The use of PCMs in thermal fluids has emerged in recent years as a way to enhance
the performance of heat transfer fluids. This chapter discusses several types of
PCM microcapsules and nanoparticles that can be potentially used in application

of heat transfer fluids. Synthesis processes of PCM particles are usually categorized into two groups: chemical processes and physical processes. A major chemical process is the in situ polymerization, which has been widely used to fabricate polymer–paraffin microcapsules. The interfacial hydrolysis method was recently developed to synthesize silica-NPG microcapsules. Physical methods include spray drying and direct emulsification processes, which are capable of producing PCM particles smaller than 1 µm. Many interesting properties have been reported in these PCM particles and PCM fluids. For example, the use of low-melting metallic nanoparticles, such as Field's metal and Indium, can increase the fluid conductivity and heat capacity simultaneously in the base fluids. In the ethanol-in-PAO nanoemulsion fluids, explosive vaporization of the dispersed nanodroplets would significantly improve the heat transfer.

References

1. Buddhi D, Sawhney RL, Sehgal PN, Bansal NK (1987) A simplification of the differential thermal-analysis method to determine the latent-heat of fusion of phase-change materials. J Phys D-Appl Phys 20:1601–1605
2. Prakash J, Garg, HP, Datta G (1985) A solar water heater with a built-in latent-heat storage. Energy Convers Manage 25:51–56
3. Shaikh S, Lafdi K (2010) C/C composite, carbon nanotube and paraffin wax hybrid systems for the thermal control of pulsed power in electronics. Carbon 48:813–824
4. Mondal S (2008) Phase change materials for smart textiles—an overview. Appl Therm Eng 28:1536–1550
5. Han ZH, Yang B, Qi Y, Cumings J (2011) Synthesis of low-melting-point metallic nanoparticles with an ultrasonic nanoemulsion method. Ultrasonics 51:485–488
6. Han ZH, Cao FY, Yang B (2008) Synthesis and thermal characterization of phase-changeable indium/polyalphaolefin nanofluids. Appl Phys Lett 92:243104
7. Cao F, Kalinowski P, Lawler L, Lee H, Yang B (2014) Synthesis and heat transfer performance of phase change microcapsule enhanced thermal fluids. ASME J Heat Trans
8. Baetens R, Jelle BP, Gustavsen A (2010) Phase change materials for building applications: A state-of-the-art review. Energy Build 42:1361–1368
9. Farid MM, Khudhair AM, Razack SAK, Al-Hallaj S (2004) A review on phase change energy storage: materials and applications. Energy Convers Manage 45:1597–1615
10. Zhao CY, Zhang GH (2011) Review on microencapsulated phase change materials (MEPCMs): fabrication, characterization and applications. Renew Sustain Energy Rev 15:3813–3832
11. Xu JJ, Wu CW, Yang B (2010) Thermal- and phase-change characteristics of self-assembled ethanol/polyalphaolefin nanoemulsion fluids. J Thermophys Heat Transfer 24:208–211
12. Xu J, Yang B, Hammouda B (2011) Thermal conductivity and viscosity of self-assembled alcohol/polyalphaolefin nanoemulsion fluids. Nanoscale Res Lett 6:2011
13. Xu J, Hammouda B, Yang B (2012) Thermophysical properties and pool boiling characteristics of water in polyalphaolefin nanoemulsion fluids. In: Proceedings of ASME micro/nanoscale heat & mass transfer international conference 2012: ASME
14. Xu J, Yang B (2012) Novel heat transfer fluids: self-assembled nanoemulsion fluids. In: Govil DJN (ed) Nanotechnology. Volume: synthesis and characterization. Studium Press LLC, New Delhi

15. Yang B, Han ZH (2006) Thermal conductivity enhancement in water-in-FC72 nanoemulsion fluids. Appl Phys Lett 88:261914
16. Han ZH, Yang B (2008) Thermophysical characteristics of water-in-FC72 nanoemulsion fluids. Appl Phys Lett 92:093123
17. Choi SUS (1995) Enhancing thermal conductivity of fluids with nanoparticles. In: Siginer DA, Wang HP (eds) Developments and Applications of Non-Newtonian Flows. ASME, New York, pp 99–105
18. Eastman JA, Phillpot SR, Choi SUS, Keblinski P (2004) Thermal transport in nanofluids. Annu Rev Mater Res 34:219–246
19. Keblinski P, Eastman JA, Cahill DG (2005) Nanofluids for thermal transport. Mater Today 8:36–44
20. Mills AF (1999) Basic heat and mass transfer, 2nd edn. Prentice Hall Inc, New Jersey
21. Sharma S, Kitano H, Sagara K (2004) Phase change materials for low temperature solar thermal applications. Res Rep Fac Eng Mie Univ 29:31–64
22. Osterman E, Tyagi VV, Butala V, Rahim NA, Stritih U (2012) Review of PCM based cooling technologies for buildings. Energy Build 49:37–49
23. Mehling H, Cabeza LF (2008) Heat and cold storage with PCM: an up to date introduction into basics and applications. Springer, Berlin
24. Hasnain SM (1998) Review on sustainable thermal energy storage technologies, part I: heat storage materials and techniques. Energy Convers Manage 39:1127–1138
25. Chen Z-H, Yu F, Zeng X-R, Zhang Z-G (2012) Preparation, characterization and thermal properties of nanocapsules containing phase change material n-dodecanol by miniemulsion polymerization with polymerizable emulsifier. Appl Energy 91:7–12
26. Rao Z, Wang S, Peng F (2012) Self diffusion of the nano-encapsulated phase change materials: a molecular dynamics study. Appl Energy 100:303–308
27. Kalinowski P, Lawler J, Yang B, Cao F (2010) Heat transfer performance of a phase change microcapsule fluid. In: ASME 2012 3rd micro/nanoscale heat and mass transfer international conference, Atlanta, Georgia, USA, 2012, pp MNHMT2012-75190
28. Cao F, Yang B (2014) Supercooling suppression of microencapsulated phase change materials by optimizing shell composition and structure. Appl Energy 113:1512–1518
29. Kumar A, Katiyar V (1990) Modeling and experimental investigation of melamine formaldehyde polymerization. Macromolecules 23:3729–3736
30. Coullerez G, Leonard D, Lundmark S, Mathieu HJ (2000) XPS and ToF-SIMS study of freeze-dried and thermally cured melamine-formaldehyde resins of different molar ratios. Surf Interface Anal 29:431–443
31. Wang XW, Lu ER, Lin WX, Liu T, Shi ZS, Tang RS, Wang CZ (2000) Heat storage performance of the binary systems neopentyl glycol/pentaerythritol and neopentyl glycol/trihydroxy methyl-aminomethane as solid-solid phase change materials. Energy Convers Manage 41:129–134
32. Cao F, Ye J, Yang B Synthesis and characterization of solid state phase change material microcapsules for thermal management applications. ASME J Nano Eng Med
33. Wang J-X, Wang Z-H, Chen J-F, Yun J (2008) Direct encapsulation of water-soluble drug into silica microcapsules for sustained release applications. Mater Res Bull 43:3374–3381
34. Taylor GI (1934) The formation of emulsions in definable fields of flow.In: Proceedings of the royal society of London. Series A, containing papers of a mathematical and physical character 146:501–523, 1934/10/01 1934
35. Synfluid PAO Databook (2002) Synfluid PAO databook, Chveron Philips Chemical LP, Pasadena
36. Hong T, Yang H, Choi CJ (2005) Study of the enhanced thermal conductivity of Fe nanofluids. J Appl Phys 97:64311-1-4
37. Yang B (2008) Thermal conductivity equations based on Brownian motion in suspensions of nanoparticles (nanofluids). ASME J Heat Transfer 130:042408/1-4

38. Shima PD, Philip J, Raj B (2010) Synthesis of aqueous and nonaqueous iron oxide nanofluids and study of temperature dependence on thermal conductivity and viscosity. J Phys Chem C 114:18825–18833

39. Eastman JA, Choi SUS, Li S, Yu W, Thompson LJ (2001) Anomalously increased effective thermal conductivities of ethylene glycol-based nanofluids containing copper nanoparticles. Appl Phys Lett 78:718–720

40. Nagano K, Ogawa K, Mochida T, Hayashi K, Ogoshi H (2004) Thermal characteristics of magnesium nitrate hexahydrate and magnesium chloride hexahydrate mixture as a phase change material for effective utilization of urban waste heat. Appl Therm Eng 24:221–232

41. Han ZH, Yang B, Kim SH, Zachariah MR (2007) Application of hybrid sphere/carbon nanotube particles in nanofluids. Nanotechnology 18:105701–105704

42. Khodadadi JM, Hosseinizadeh SF (2007) Nanoparticle-enhanced phase change materials (NEPCM) with great potential for improved thermal energy storage. Int Commun Heat Mass Transfer 34:534–543

43. Vehkamaki H (2006) Classical nucleation theory in multicomponent systems, 1st edn. Springer, New York

44. Xu J, Yang B (2013) Nanostructured phase changeable heat transfer fluids. Nanotech Rev 2:289–306

45. Chen SJ, Evans DF, Ninham BW (1984) Properties and structure of 3-component ionic microemulsions. J Phys Chem 88:1631–1634

46. Ruckenstein E (1986) The surface of tension, the natural radius, and the interfacial-tension in the thermodynamics of microemulsions. J Colloid Interface Sci 114:173–179

47. Siano DB, Bock J, Myer P, Russel WB (1987) Thermodynamics and hydrodynamics of a nonionic microemulsion. Colloids Surf 26:171–190

48. Rosano HL, Cavallo JL, Chang DL, Whittam JH (1988) Microemulsions—a commentary on their preparation. J Soc Cosmet Chem 39:201–209

49. Chen ZQ, Chen LD, Hao C, Zhang CZ (1990) Thermodynamics of microemulsion.1. The effect of alkyl chain-length of alkyl aromatics. Acta Chim Sinica 48:528–533

50. Moulik SP, Das ML, Bhattacharya PK, Das AR (1992) Thermodynamics of microemulsion formation.1. Enthalpy of solution of water in binary (triton-x 100 + butanol) and ternary (heptane + triton-x 100 + butanol) mixtures and heat-capacity of the resulting systems. Langmuir 8:2135–2139

51. Moulik SP, Ray S (1994) Thermodynamics of clustering of droplets in water/aot/heptane microemulsion. Pure Appl Chem 66:521–525

52. Ray S, Bisal SR, Moulik SP (1994) Thermodynamics of microemulsion formation.2. enthalpy of solution of water in binary-mixtures of aerosol-ot and heptane and heat-capacity of the resulting systems. Langmuir 10:2507–2510

53. Strey R (1994) Microemulsion microstructure and interfacial curvature. Colloid Polym Sci 272:1005–1019

54. Bergenholtz J, Romagnoli AA, Wagner NJ (1995) Viscosity, microstructure, and interparticle potential of aot/h2o/n-decane inverse microemulsions. Langmuir 11:1559–1570

55. Mukherjee K, Mukherjee DC, Moulik SP (1997) Thermodynamics of microemulsion formation.3. Enthalpies of solution of water in chloroform as well as chloroform in water aided by cationic, anionic, and nonionic surfactants. J Colloid Interface Sci 187:327–333

56. Kumar P, Mittal KL (1999) Handbook of microemulsion science and technology. Marcel Dekker, New York

57. Talegaonkar S, Azeem A, Ahmad FJ, Khar RK, Pathan SA, Khan ZI (2008) Microemulsions: a novel approach to enhanced drug delivery. Recent pat drug delivery formulation 2:238–57

58. Wu C, Cho TJ, Xu J, Lee D, Yang B, Zachariah MR (2010) Effect of nanoparticle clustering on the effective thermal conductivity of concentrated silica colloids. Phys Rev E 81:011406

59. Tyndall J (1868) On the blue colour of the sky, the polarization of sky-light, and on the polarization of light by cloudy matter generally. Proc R soc lond 17:223

60. He GS, Qin H-Y, Zheng Q (2009) Rayleigh, Mie, and Tyndall scatterings of polystyrene microspheres in water: wavelength, size, and angle dependences. J Appl Phys 105:023110/ 1–10
61. Rosele ML Boiling of dilute emulsions. Ph.D.dissertation, University of Minnesota
62. Bulanov NV, Skripov VP, Khmylnin VA (1984) Heat transfer to emulsion with superheating of its disperse phase. J Eng Phys 46:1–3
63. Bulanov NV, Skripov VP, Khmylnin VA (1993) Heat transfer to emulsion with a low-boiling disperse phase. Heat Transfer Res 25:786–789
64. Bulanov NV (2001) An analysis of the heat flux density under conditions of boiling internal phase of emulsion. High Temp 39:462–469
65. Bulanov NV, Gasanov BM (2005) Experimental setup for studying the chain activation of low-temperature boiling sites in superheated liquid droplets. Colloid J 67:531–536
66. Bulanov NV, Gasanov BM, Turchaninova EA (2006) Results of experimental investigation of heat transfer with emulsions with low-boiling disperse phase. High Temp 44:267–282
67. Bulanov NV, Gasanov BM (2008) Peculiarities of boiling of emulsions with a low-boiling disperse phase. Int J Heat Mass Transf 51:1628–1632
68. Lunde DM (2011) Boiling dilute emulsions on a heated strip. MS thesis, University of Minnesota
69. De M, Bhattacharya SC, Panda AK, Moulik SP (2009) Interfacial behavior, structure, and thermodynamics of water in oil microemulsion formation in relation to the variation of surfactant head group and cosurfactant. J Dispersion Sci Technol 30:1262–1272
70. Moulik SP, Paul BK (1998) Structure, dynamics and transport properties of microemulsions. Adv Colloid Interface Sci 78:99–195
71. Chen SH (1986) Small-angle neutron-scattering studies of the structure and interaction in micellar and microemulsion systems. Ann Rev Phys Chem 37:351–399
72. Smith RD, Blitz JP, Fulton JL (1989) Structure of reverse micelle and microemulsion phases in near-critical and supercritical fluid as determined from dynamic light-scattering-studies. Acs Symp Ser 406:165–183
73. Nyden M, Soderman O (1995) Structures and emulsification failure in the microemulsion phase in the didodecyldimethylammonium sulfate/hydrocarbon/water system—a self-diffusion NMR-study. Langmuir 11:1537–1545
74. Hirai M, Takizawa T, Yabuki S, Kawaihirai R, Oya M, Nakamura K, Amemiya Y (1995) Structure and reactivity of aerosol-OT reversed micelles containing alpha-chymotrypsin,". J Chem Soc-Faraday Trans 91:1081–1089
75. Hirai M, Hirai RK, Iwase H, Arai S, Mitsuya S, Takeda T, Nagao M (1999) Dynamics of w/o AOT microemulsions studied by neutron spin echo. J Phys Chem Solids 60:1359–1361
76. Kotlarchyk M, Chen SH, Huang JS (1982) Temperature-dependence of size and polydispersity in a 3-component micro-emulsion by small-angle neutron-scattering. J Phys Chem 86:3273–3276
77. Kaler EW, Bennett KE, Davis HT, Scriven LE (1983) Toward understanding microemulsion microstructure—a small-angle X-ray-scattering study. J Chem Phys 79:5673–5684
78. Howe AM, Toprakcioglu C, Dore JC, Robinson BH (1986) Small-angle neutron-scattering studies of microemulsions stabilized by aerosol-OT .3. The effect of additives on phase-stability and droplet structure. J Chem Soc Faraday Trans I 82:2411–2422
79. Hammouda B, Krueger S, Glinka CJ (1993) Small-angle neutron-scattering at the national-institute-of-standards-and-technology. J Res Natl Inst Stand Technol 98:31–46
80. Hammouda B (2010) A new Guinier-Porod model. J Appl Crystallogr 43:716–719
81. Hammouda B (2010) SANS from polymers-review of the recent literature. Polym Rev 50:14–39
82. Regev O, Ezrahi S, Aserin A, Garti N, Wachtel E, Kaler EW, Talmon Y (1996) A study of the microstructure of a four-component nonionic microemulsion by cryo-TEM, NMR, SAXS, and SANS. Langmuir 12:668–674
83. Touloukian YS, Liley PE, Saxena SC (1970) Thermal conductivity for nonmetallic liquids & gases, vol 3. Thermophy Propert Matter. IFI/PLENUM, Washiongton

84. Incropera FP, DeWitt DP, Bergman TL, Lavine AS (2001) Fundamentals of heat and mass transfer, 5th edn. Wiley, New York, p 944
85. Bradfiel WS (1967) On effect of subcooling on wall superheat in pool boiling. J Heat Transfer 89:269
86. Shepherd JE, Sturtevant B (1982) Rapid evaporation at the superheat limit. J Fluid Mech 121:379–402
87. Park HC, Byun KT, Kwak HY (2005) Explosive boiling of liquid droplets at their superheat limits. Chem Eng Sci 60:1809–1821
88. Dong ZY, Huai XL, Liu DY (2005) Experimental study on the explosive boiling in saturated liquid nitrogen. Prog Nat Sci 15:61–65
89. Kandlikar SG, Shoji M, Dhir VK (1999) Handbook of phase change:boiling and condensation. Taylor&Francis, UK
90. Forster HK, Zuber N (1955) Dynamics of vapor bubbles and boiling heat transfer. Aiche J 1:531–535
91. Zuber N (1957) On the correlation of data in nucleate pool boiling from a horizontal surface. Aiche J 3:S9–S11
92. Zuber N, Fried E (1962) 2-Phase flow and boiling heat transfer to cryogenic liquids. Ars J 32:1332–1341
93. Mudawar I, Bowers MB (1999) Ultra-high critical heat flux (CHF) for subcooled water flow boiling—I: CHF data and parametric effects for small diameter tubes. Int J Heat Mass Transf 42:1405–1428
94. 3 M Fluorinert™ Electronic Liquid FC-72 Product Information. http://solutions.3m.com/wps/portal/3M/en_US/Electronics_NA/Electronics/Products/Product_Catalog/~/3M-Fluorinert-Electronic-Liquid-FC-72?N=4294412721+5153906/Nr=AND(hrcy_id%3A7JV9NFW9KRgs_0HBRWJKZQJ_N2RL3FHWVK_GPD0K8BC31gv)/rt=d
95. Chu B (1974) Laser light scattering. Academic Press, New York
96. Evans W, Fish J, Keblinski P (2006) Role of Brownian motion hydrodynamics on nanofluid thermal conductivity. Appl Phys Lett 88:093116

Printed in the United States
By Bookmasters